**Systems, Networks,
and Computation:
Basic Concepts**

SYSTEMS, NETWORKS, and COMPUTATION: BASIC CONCEPTS

Michael L. Dertouzos
Michael Athans
Richard N. Spann
Samuel J. Mason

Members of the Faculty
Department of Electrical Engineering
Massachusetts Institute of Technology

McGraw-Hill Book Company

New York St. Louis San Francisco Düsseldorf Johannesburg
Kuala Lumpur London Mexico Montreal New Delhi
Panama Rio de Janeiro Singapore Sydney Toronto

TO
HADWIG, ZENE,
ANNETTE, JEANIE

Systems, Networks, and Computation: Basic Concepts

Library of Congress Catalog Card Number 76-161665

07-016530-0

1234567890 KPKP 798765432

This book was set in Primer by Black Dot, Inc., and printed and bound by Kingsport Press, Inc. The designer was Richard Paul Kluga; the drawings were done by John Cordes, J & R Technical Services, Inc. The editors were Michael Elia and Madelaine Eichberg. Sally Ellyson supervised production.

Contents

Preface

This book is the result of an educational experiment which started five years ago in the department of electrical engineering, Massachusetts Institute of Technology. Our first class, made up of 20 volunteers, was exposed to what at that time seemed to be a radical departure from the traditional course in network analysis. They were introduced to digital systems, computational methods, nonlinear systems, difference equations—all in the same breath with the more classical approaches to network theory.

While our experiment was evolving from these early days of relatively high confusion and intense preoccupation with the selection, development, and organization of material, to the more organized and hopefully less confused class of today, a very fundamental question was being raised with increasing frequency: With the declining use of electric networks and the increasing use of computers for information processing, communication, and control, what should be the directions and goals for the electrical engineers of the future? *This question which transcends the locality of a particular university is still unresolved. It is important because it influences the nature of, in fact the need for, a first sophomore technological course. While we are not confident that we have isolated* the *set of necessary fundamentals for the electrical engineers of the future, our motivation for developing this course was based on this question. Further refined, our motivation, which is explained in more detail in Chap. 1, rests on our conviction that the electrical engineer of the future should be able to understand, choose, and utilize a variety of building blocks ranging over network elements, operational amplifiers, logic gates, and computer algorithms. He should be able to move freely among a variety of mathematical techniques and physical equipment, choosing what seems to be best suited for accomplishing purposeful tasks in communication, computation, and control. Moreover, he should be able to do this without fear or prejudice for a particular class of systems. The techniques that he will learn should also be sufficiently basic to apply to a broader class of problems that will confront him in the future.*

At MIT, our experiment culminated in two one-semester courses. In the first course, which is covered by this book, we have concentrated on what seem to be the basic concepts of continuous systems and networks, digital systems, and computational methods for system analysis. Since the emphasis is on basic ideas, the systems and approaches of this book are restricted to a small number of variables—typically one or two. In the second course, covered by the companion volume "Systems Networks and Computation: Multivariable Methods," emphasis is placed on the techniques which extend the basic ideas to large systems with many variables. Some of our students, especially those interested in control, take both courses. Other students, primarily interested in computers, take only the first course.

In this book, we have combined the traditionally diverse areas of continuous systems and networks, digital systems, and computational methods for several reasons: The alternative of offering three distinct courses, besides being uneconomical, places more emphasis per area than seems to be necessary for subsequent specialization. Teaching this material under one cover economizes on joint concepts such as the formulation of equations that describe system behavior, or the notion of system equivalence. In addition, certain tasks such as "addition" can be accomplished with resistive networks, operational amplifiers, digital adders, and computer programs; their joint study stimulates comparisons which are both educational and practically desirable. Finally, joint exposure imparts to the student a feeling of equal value for several approaches and techniques, thereby broadening his point of view. Naturally, the teaching of this material in this particular way results in a student whose outlook differs substantially from that of his traditional classmate. While he knows less about many topics which were more important in the past, he knows more about other topics and approaches—for example, what can be done with digital systems and computational methods. His mathematical and engineering attitude is different too, since emphasis is placed on certain basic notions such as function, systematic formulation of equations, existence, linearity; which apply to many different types of systems.

Although the book is addressed to the electrical engineering sophomore who is interested in communication, computation or control, it may be found useful by the established engineer who wishes to review modern fundamentals, and by the worker in other possibly nonengineering disciplines, such as economics, who is using the engineering methodology.

Some of the problems in the book can be best solved, in fact sometimes only solved, by a computer. At MIT, and only during the last two years, our students have been using conversational terminals with the APL language, and several "canned" utility packages which make possible the solution of systems of algebraic or differential equations, the evaluation of system functions, and the design and analysis of proposed iteration rules. The use of computers, although desirable and motivational, is not essential. Batch facilities and other programming languages could be used equally well, or in their absence, computer use could be altogether excluded—the concentration on basic concepts in terms of one and two variables makes this possible without substantial sacrifice.

The book is organized in 11 chapters. The first two form a technical preface and an introduction to the subject matter. Chapters 3 and 4 deal with static or memoryless, continuous (analog) systems and resistive networks; Chap. 5 with the computational analysis of such systems; Chap. 6 with memoryless discrete (combinational or digital) systems; Chaps. 7 and 8 with dynamic continuous networks and systems (from the state and input-output points of view, respectively); Chap. 9 with the computational analysis of these systems; Chap. 10 with dynamic digital systems; and Chap. 11 with the special case of linear systems with delays. The problems appearing at the end of each chapter were chosen to show applications, and to explore new ideas. Problem numbers are keyed to chapter sections and are generally ranked in order of increasing difficulty within each section.

At MIT, we have been teaching in one semester, Chaps. 3, 4, 5, 6, 7, 8, 10, and the appendix. For students who have already had exposure to traditional networks, a one-semester course on computational methods and digital systems can be based on Chaps. 3, 5, 6, 9, and 10. For students who have been, or plan to be, exposed elsewhere to digital systems, a one-semester course on continuous systems and networks and on computational methods can be based on Chaps. 3, 4, 5, 7, 8, 9, and 11.

Both this book, and the companion volume to be published have been the result of cooperative work among all four co-authors. Each chapter has been rewritten several times with different authors taking prime responsibility. Major contributions have been as follows: Athans to Chapters 4, 5, 7, and 8; Spann to 6, 8, 9, and 10; Mason to 3, 4, and 11; Dertouzos to 1, 2, 3, 4, 5, 7, and 9, as well as primary responsibility for the entire book.

We wish to extend our gratitude to all the people who made this book possible. First, we wish to acknowledge the valuable help of Les Kramer who, starting as a sophomore in our first class stayed with us, throughout the experiment as teaching assistant, instructor, and author of the Solutions Manual. His suggestions, criticisms, and enthusiasm have had a direct impact on the progress of the experiment and on the contents of the book. Our colleagues, L. D. Braida, W. B. Davenport, Jr., R. G. Gallagher, L. A. Gould, F. C. Hennie III, P. L. Penfield, Jr., W. M. Siebert, and D. C. White, all of the MIT faculty who taught the course, have given us valuable suggestions and ideas. Outside MIT, we wish to thank our colleagues R. A. Rohrer and L. A. Zadeh of the University of California at Berkeley, J. G. Truxal of Brooklyn Polytechnic Institute, and M. E. VanValkenburg of Princeton University, for helpful criticism, manuscript reviews, and encouragement. We are thankful to the Department of Electrical Engineering at MIT, and more specifically to P. Elias and L. D. Smullin for providing the environment and resources that made our experiment possible. We also wish to thank our secretaries who typed and retyped five different manuscript versions—they are Donna Guy, Susan

Issacs, Nancy Read, Maureen Stanton, Ann Wilkening and Lynn Yaple. Finally, we wish to express our gratitude to our students who, unaware that they should be afraid of radical approaches, not only survived but supplied valuable criticism and much appreciated enthusiasm.

Michael L. Dertouzos
Michael Athans
Richard N. Spann
Samuel J. Mason

chapter 1
Introduction

The purpose of this chapter is to help the reader understand why he is exposed to these particular subjects in this particular way. In a sense, this chapter may be thought of as an extension of the preface, since it reviews for the instructor, who already knows most of the material, the origin of ideas and the direction of approach. For the student, who is not generally aware of this material, the chapter provides some history and background and hopefully some motivation.

1.1 MOTIVATION AND HISTORY

There are many intangible factors that determine the quality of an academic subject. To the extent that they can be identified, there seem to be three basic components: *pedagogy, relevance,* and *permanence.* Pedagogy includes the methodology of teaching; the apportionment of subject matter into lectures, recitations, laboratory, homework, and tutorial sessions; and the simplicity and continuity of what is presented. In addition, and perhaps more importantly, through the instructor's interest and enthusiasm for the subject, good pedagogy gives the student courage to attack problems that he is not quite ready to solve. The relevance component indicates how closely the subject reflects the exciting research and development frontiers of the field, the meaningful application of fundamentals to the real world, and the preparation that a student receives for more intensive specialization and graduate study in elective courses. Finally, the permanence component reflects the immunity of the subject matter to what may be fashionable today but less important in the future.

The result of measuring the fundamental courses in present electrical engineering curricula against these three yardsticks was the prime motivation for developing the subject matter of this book. In particular, the relevance and permanence values of classical electrical engineering fundamentals seemed questionable in light of the technological changes brought about by digital computers and other advances in control and communications. These recent changes, and their consequences, are perhaps better understood through the historical perspective of electrical engineering curriculum developments during the last 70 years.

If we consider, for example, the EE undergraduate curricula appearing in the MIT annual catalogs since 1900, we note that they may be divided into three periods of more or less homogeneous requirements. The first period (1900–1945) was primarily concerned with the generation, distribution, and use of electric power. During this period, we find courses dealing with magnetic circuits, transformers, and ac circuits; and thesis titles such as "The Electrification of the Pennsylvania Railroad System." The second period (1945–1965) was concerned with the use of electronics for communication, information processing, and control. This is the time in which radar flourished, information theory was developed, and electronic networks and filters were used for the execution of almost every desired "task." The third period (1965–), the one that we have now entered, is concerned with digital computers, the digital processing of information, and automatic decision making.

While the technological or predominantly electrical content of the EE curricula seems to have progressed through the above three periods, the mathematical content has remained essentially unchanged. Specifically, two or three semesters of calculus, one semester of differential equations, and one semester of applied calculus for engineers appear in the curriculum of 1900 as well as in the curriculum of 1970.

There are two basic conclusions or lessons that emerge from this historical view. First, the mathematical fundamentals seem invariant to technological change and therefore possess a high relevance and permanence value; hence, we should think twice before subjecting them to major perturbations. Second, although through the years, the technological subjects seem to undergo big changes for preservation of relevance, they are not totally discarded when new subjects are introduced. Instead, the old subjects are abstracted, compressed, and retained in a more fundamental sense in new courses. For example, while today's student is not subjected to two semesters of electric motor design, he nevertheless knows enough electromagnetic theory and electromechanical energy conversion for rapid comprehension of a book on this special subject.

The primary motivation for putting together this book is one of relevance and permanence. In particular, it became apparent to us around 1965 that our students of electrical engineering could not reconcile their fundamental training in classical "continuous" electrical engineering fundamentals with a surrounding world which was progressively doing more and more system analysis and system design, either by digital computer or by the use of special digital systems. The alternative of taking enough different electives to acquire competence in the modern technology seemed costly and unfair; for example, it accented as fundamental one part of technology, and treated as secondary or not as fundamental another part, which both student and instructor knew to be significant. An even more important factor in our motivation, however, was the danger of a possible future obsolescence caused by not studying the emerging new fundamentals in their proper perspective. We shall discuss what these fundamentals entail and what the objectives of this book are in the following section.

1.2 OBJECTIVES

The main objective of this book is to introduce the reader to system-oriented fundamentals of electrical engineering. Modern electrical engineering encompasses many diverse areas such as control, materials, electromagnetics and dynamics, communication theory, circuits and signals, bioelectronics, and computers. We believe that a student, regardless of his ultimate specialization, should be exposed to a certain subject matter which is common and fundamental to such otherwise diverse areas.

The material in this book consists of certain technological fundamentals and their accompanying applied mathematics, taught through the vehicle of system theory. This vehicle and, in particular, the so-called *state model* view of systems are part of the methodology for achieving our objectives, and are discussed in more detail in Chap. 2. The contents of the book may be further characterized along three main dimensions which we have called *continuous-discrete*, *linear-nonlinear*, and *analytical-algorithmic*. These dimensions or components can be visualized as making up the cube of Fig. 1.2-1 which, in turn, contains all the relevant subject matter.

For example, traditional electric networks and analog building blocks are linear and continuous, while digital systems are nonlinear and discrete. Alternatively, the techniques for analyzing linear systems are primarily analytical, while the computational methods of iteration and numerical integration—useful in the analysis of continuous nonlinear systems—are

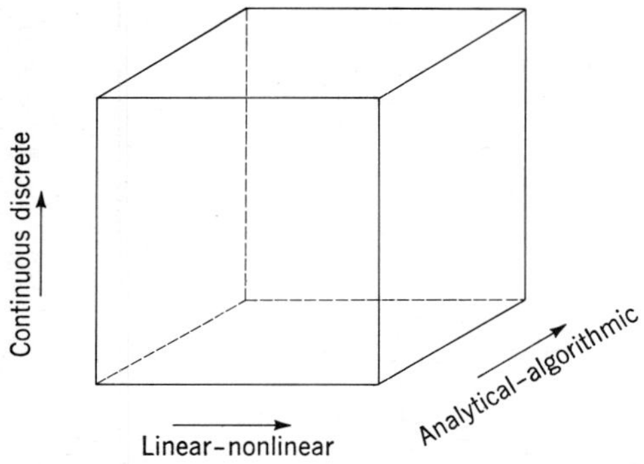

FIG. 1.2–1 Characterization of subject matter.

algorithmic. To the reader who has not encountered these terms before, the meaning of each of these three components is perhaps best explained through the following three examples.

The Analytical-Algorithmic Dimension

Consider the simple first-degree polynomial:

For all x,

$$f(x) = ax + b$$

with a and b satisfying the constraints

$$\left|\frac{b}{a}\right| < c \quad \text{and} \quad a > 0 \tag{1}$$

The statement

$f(x)$ has one and only one zero† s, given by $s = -b/a$, such that $-c < s < c$

is a theorem that can be easily proved, and forms the foundation for an *analytical* approach to finding the zero of $f(x)$. That is, given the two real numbers a and b, we can provide a closed-form answer for the zero s using the operations "division" and "multiplication by a constant." This is illustrated in Fig. 1.2-2 where the blocks denoting division and multiplication may correspond physically to manual division and multiplication or to a program that does the same operations or to electronic boxes that do these tasks (that is, impressing a voltage which represents a and a voltage which represents b on the division block causes the appearance of an output voltage which represents b/a).

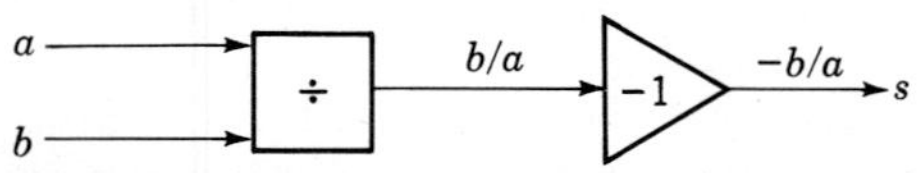

FIG. 1.2–2 Illustration of the analytical solution.

†A zero s of $f(x)$ is a value $x=s$ for which $f(s)=0$.

In a sense, this analytical solution has resulted in inversion of the implicit form $0 = ax + b$, to the explicit form $s = -b/a$.

Now consider the following *algorithm* or set of rules for finding the zero s.

ALGORITHM 1.2-1

1. Let $i = 0$ and $u_0 = 0$.
2. (*a*) If $f(u_i) > \epsilon$ ($\epsilon > 0$), then $u_{i+1} = u_i - c/2^{i+1}$; go to step 3.
   (*b*) If $f(u_i) < -\epsilon$, then $u_{i+1} = u_i + c/2^{i+1}$; go to step 3.
   (*c*) If $|f(u_i)| \le \epsilon$, then $s \cong u_i$. Stop.
3. Increase i by one, and go to step 2.

As shown in Fig. 1.2-3*a*, this set of rules generates a sequence 0, u_1, u_2, . . . , where at each iteration (each new value of i), u_i moves closer to the solution s by one-half the amount by which u_{i-1} moved toward the solution in the preceding iteration. It can be easily shown that this sequence approaches s to any desired accuracy, which is determined by the choice of a tolerable error ϵ.

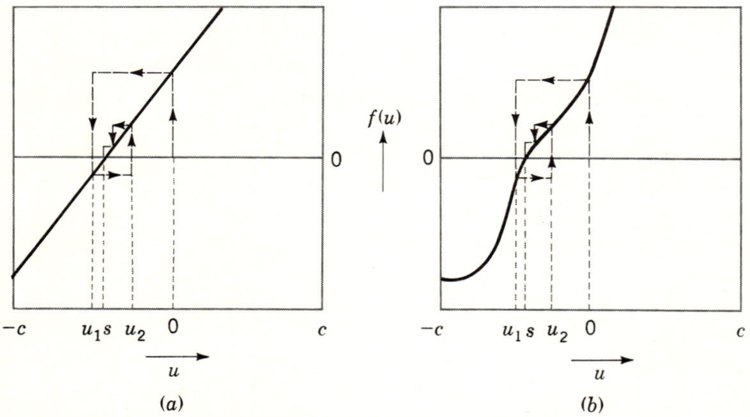

FIG. 1.2–3 Illustration of algorithm progress.

If the reader shows this, he will have developed a *theorem* concerning the *convergence* of this algorithm.

At this point it may appear highly artificial and computationally costly to find s through Algorithm 1.2-1 when it can be obtained "exactly" through the relationship

$$s = -\frac{b}{a} \tag{2}$$

But it is conceivable that a and b are not known, since we will be usually dealing with an unknown or inaccessible device, which when given x, responds with $f(x)$. Alternatively, $f(x)$ may not be a first-degree polynomial but may be known to obey the constraints

For all x,

$$0 < m_1 \leq f'(x) \leq m_2 < \infty \qquad |s| < c \tag{3}$$

That is, $f(x)$ increases with x, and its slope ranges between two positive real numbers. The reader may again wish to show that in this case Algorithm 1.2-1 will converge on the zero s, as illustrated in Fig. 1.2-3*b*. Observe that now Algorithm 1.2-1 becomes quite useful, since we generally do not know of any analytical means for finding the zero; indeed, if the polynomial is of degree higher than four, we know (Galois, 1811–1832) that there exists no analytical solution that can be computed, using square roots, cube roots, etc. Nevertheless, Algorithm 1.2-1 will approximate the zero to any desired accuracy, regardless of polynomial degree, provided that Eq. (3) holds.

The foregoing examples illustrate in terms of simple cases that *an algorithm is a set of directions for performing mathematical operations designed to lead to the solution of a given problem.* It is not difficult to imagine the existence of algorithms for solving networks or systems of arbitrary complexity where analytical solutions are known. The reader may rightfully wonder at this point or even later on why he should study the fundamentals of networks and systems in the first place when system analysis could be done for him by some computer through algorithmic means. Algorithms, however, are limited to solve certain classes of problems. [For example, Algorithm 1.2-1 cannot generally find the zero of $f(x)$ if constraints (3) are removed.] Furthermore, any constructive and creative work that might be undertaken, whether in algorithm or system development or in the process of result evaluation and decision making, is based by necessity on a thorough understanding of system and algorithm fundamentals.

Before we leave the algorithmic dimension of our cube, we would like to take a brief historical glimpse at the evolution of algorithmic approaches in mathematics. There was a time when most, if not all, of mathematics was motivated by and applied to real problems of everyday life. Euler (1707–1783) who was one of the leaders of this algorithmic age, kept foremost in his mind the actual *use* of his theorems and algorithms. In the nineteenth century, a progressively increasing emphasis was placed on abstraction and the logical trend of mathematics, while algorithmic issues assumed a secondary role. Mathematicians of that time were perfectly satisfied to prove existence of solutions, even if the related computations were downright impossible. In the late 1930s, however, a significant technological development, the digital computer, made its appearance. A quantitative significance of this development is shown in Fig. 1.2-4, where we see that computational speed, which had remained virtually constant for centuries, experienced an increase of six to seven orders of magnitude. Computations that would have taken years in the past could now be performed in a few seconds.

This dramatic speed change, accompanied by the ability of computers to make elementary decisions and to do, without protest, highly repetitive tasks, has had a great impact on the revival of algorithmic mathematics and, more generally, on technology and on our everyday life. It is certainly not possible in a book of this scope to discuss and pursue the widespread nature of all of these implications. We have therefore concentrated our objectives on

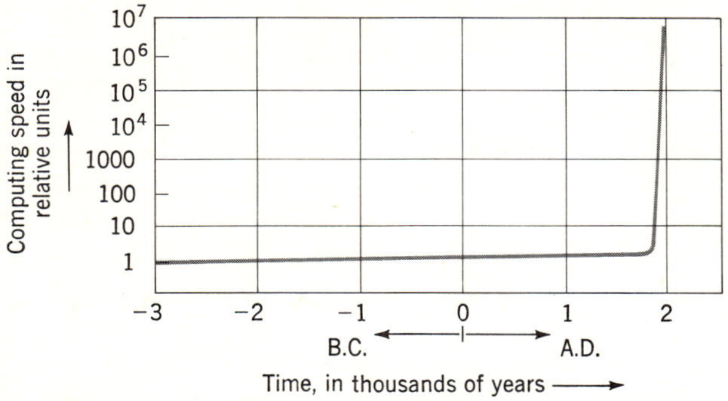

FIG. 1.2–4 Historical progress of computational speed.

treating fundamentals of algorithms which apply to the analysis of continuous systems on the technological side and to elementary analysis on the mathematical side. We proceed now with the next dimension of the cube.

The Continuous-Discrete Dimension

Let us suppose that we wish to add two integers each between 0 and 49 using physical devices. We can represent each of these numbers by the amount of water in each of two containers A and B. Addition is then performed by pouring the contents of container A into container B and measuring the resultant contents. The addition process, however, is not "perfect," since it is subject to inaccuracies caused by careless spilling of water, evaporation, measurement errors, and so forth. We shall refer to these undesirable disturbances collectively as *noise*. To find out what the sum is in the presence of such noise, we round off the measured contents to the nearest integer. Thus, if we measure 67.6 units of water, we say that the sum is 68, whereas if we measure 67.4, we say that the sum is 67.†

Next, suppose that we represent *each* of the above two integers by the contents of *two* containers, one for the units and one for the tens. We shall refer to these contents as A_{10}, A_1 and B_{10}, B_1. Thus, $A = 18$ corresponds to $A_{10} = 1$ and $A_1 = 8$. In order to add A and B we observe the rules of decimal addition. That is:

1. If $9.5 \leq A_1 + B_1 < 10.5$, then we throw away $A_1 + B_1$ units. If $A_1 + B_1 \geq 10.5$, then we throw away ten units from these contents and in either case, add one unit of water to the combined tens contents $A_{10} + B_{10}$.
2. If $A_1 + B_1 < 9.5$ we merely pour A_1 into B_1, and A_{10} into B_{10}.

We "measure" the sum again by rounding off to the nearest integers. That is, if the contents after addition are $A_{10} + B_{10} = 5.7$ and $A_1 + B_1 = 8.3$, we say that the sum is 68.

†67.5 is rounded off to 68.

In comparing the foregoing two approaches, we see that with the two-container method we can tolerate† noise that changes each container's content by less than ±5 percent of each container's full capacity. In the one-container method, the corresponding figure is ±0.5 percent of the container's full capacity. Hence, the two-container method can tolerate ten times as much noise as the single-container method. These observations can be carried out to two extremes as follows: At one extreme, we represent integers A and B by as many containers as necessary, keeping the number of levels per container to a minimum (i.e., two levels—full or empty) in order to achieve maximum noise immunity. This results in the binary system of 1s and 0s, which we shall study further in the sequel and which represents what we mean by a *discrete* system. At the other extreme, we could represent a *real* number by a level in one container; for example, the number π would be represented by π units of water. This represents what we call a *continuous* system. In general, replacing the continuous approach of representing, communicating, or processing information by a discrete approach increases the noise immunity of the process. Naturally, the discrete systems that we shall study will not entail water and containers, but corresponding electrical entities such as the electrical state of a flip-flop or the magnetization state of a magnetic core.

The Linear-Nonlinear Dimension

For our purposes, an *operator* (device, element, system) L with input x and output y is *linear* if and only if for every choice of real numbers a, b, x_1, and x_2, or real-valued functions x_1 and x_2, it is the case that

$$L(ax_1 + bx_2) = aL(x_1) + bL(x_2) \tag{4}$$

where $+$ is ordinary arithmetic addition. If an operator does not obey Eq. (4), it is called *nonlinear*. For example, the following operations,

1. $y = x$
2. $y = 4x$
3. For all t, $y(t) = \int_0^t x(u)\, du \qquad y(0) = 0$

are linear since they satisfy Eq. (4), whereas the following,

1. $y = |x|$
2. $y = \dfrac{1}{x}$
3. For all t, $y(t) = \int_0^t |x(u)|\, du$

do not satisfy Eq. (4) and therefore are nonlinear.

The simple constraint of Eq. (4) is remarkably powerful and applicable to many physical devices and systems. It gives rise to a structure that is

†That is, after roundoff of the contents of each container, we will have the correct sum.

both elegant and useful in the analysis and synthesis of engineering systems. Nevertheless, by virtue of the fact that Eq. (4) is a constraint, it limits the class of problems to which it can provide satisfactory solutions.

Consider, for example, the problem of finding a *linear* operator which converts *every* triangular "signal" $x(t)$, where t is time, into a sinusoidal signal $y(t)$ as shown in Fig. 1.2-5, *regardless of the period* T (that is, regardless of whether T is one year or a millionth of a second, $y(t)$ should be sinusoidal with period T).

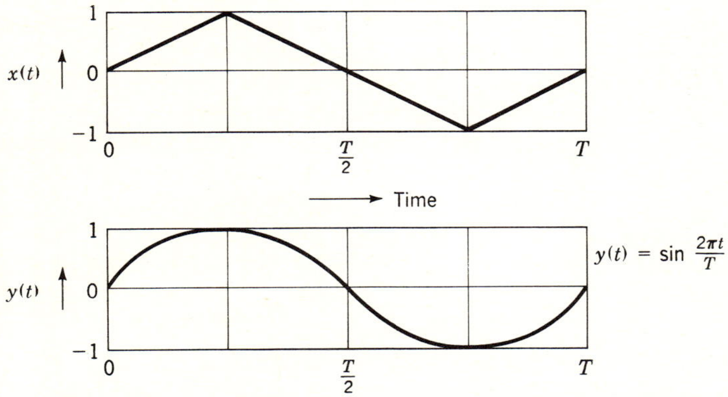

FIG. 1.2–5 Example illustrating the linear-nonlinear dimension.

The reader may enjoy searching for such an operator, but he may also appreciate knowing that such a *linear* operator does not exist. On the other hand, the simple nonlinear operator shown in Fig. 1.2-6 accomplishes the desired objective simply and exactly. That is, as x varies in the form of the given triangular wave, y varies as the desired sinusoidal signal.

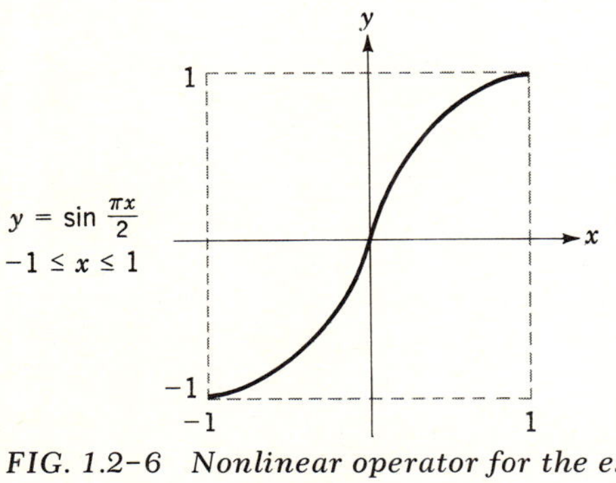

FIG. 1.2–6 Nonlinear operator for the example of Fig. 1.2–5.

A number of useful and interesting tasks can often, indeed sometimes only, be accomplished by nonlinear means. Moreover, the great majority of discrete systems that make up a computer or a digital processor are normally nonlinear. In the past, useful tasks involving the processing and communication of information were done almost exclusively through linear operators. With the advent of computers and algorithmic techniques, the analysis and design of nonlinear systems are no longer as formidable a task, although they still are more difficult than linear system analysis and design.

In summary, the objectives set forth by the linear-nonlinear dimension of our cube entail the study of both linear and nonlinear systems, regardless of whether they are continuous or discrete and whether they are analyzed analytically or algorithmically.

Summary of Objectives and Methodology

Using the three dimensions of the cube, we can now see that the traditional fundamentals of electrical engineering occupy that eighth of the cube denoted by the terms linear, continuous, analytical and shown shaded in Fig. 1.2-7.

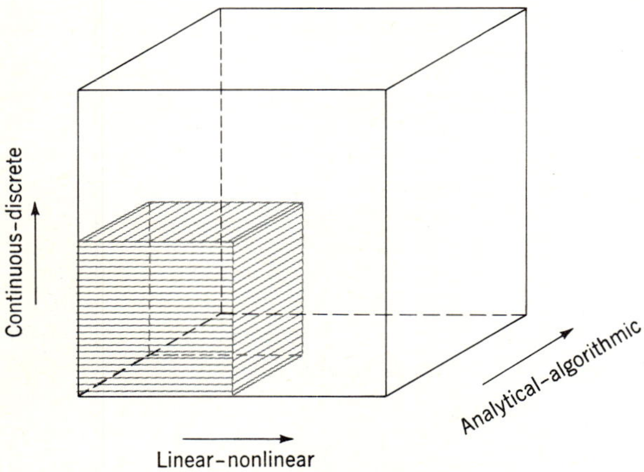

FIG. 1.2-7 Traditional fundamentals (shaded cube).

Our objectives in this book and in the companion volume are to further explore the cube, without necessarily spending equal time on all of its parts, but concentrating our attention with varying degrees of intensity on what we feel are fundamental areas. These fundamentals entail three classes of systems—electric networks, analog block diagram systems, and digital systems organized according to whether they do or do not have memory. The accompanying mathematics include Boolean algebra for discrete systems and for continuous systems, the solution of algebraic and differential equations by analytical means (when they are linear) and by iteration and numerical integration, otherwise. Basic concepts which are emphasized throughout include the formal definition of systems, the formulation and solution of equations that characterize a system, the existence and uniqueness of solutions to such equations, the convergence of algorithms, and the equivalence of systems.

chapter 2

Systems and the State Model

This chapter introduces and defines systems, discusses their analysis and synthesis, and classifies them according to whether they do or do not have memory. The state model is developed as a means for characterizing system behavior. In contrast to Chap. 1 which dealt with history, motivation, objectives, and methodology, this chapter forms a technical introduction for, and a brief summary of some basic ideas which are used throughout the book.

2.1 DEFINITION AND ANALYSIS OF SYSTEMS

Analysis and *synthesis* are words which are often encountered in the study of engineering systems. *Analysis* usually means a process of decomposition or simplification of a relatively complex structure into simpler aggregates with the objective of understanding what the structure is or does. *Synthesis* is the task of specifying how a structure should be constructed out of simpler structures in order to accomplish a certain objective.

Whether in analysis or synthesis, these structures, which are called *systems* are made up of basic building blocks—the *elements*—which are said to be *connected* or *interconnected.* However, elements and systems are not always distinguishable, since that which is considered a system from one point of view may be considered an element from another. For example, a *system* of resistors, transformers, and transistors, such as an audio amplifier, may be and often is called a component or *element* of an audio entertainment system. The notions of element and system are then to some extent interchangeable, with the appropriate hierarchy chosen according to necessity, convention, convenience, or ease of understanding.

The way in which elements are combined to form a system will generally vary from one class of systems to another. For example, some electrical elements such as resistors may be connected in any arbitrary way, and the resulting aggregate will be a resistor (possibly with many terminals). Logic gates, on the other hand, may not be arbitrarily interconnected; for example, in most digital systems the outputs of two logic gates may not be connected to each other. It is desirable then to specify in more precise terms what objects may be used and how they may be interconnected to form a particular type of system. This leads us to the following general inductive definition form which will be used to establish a particular class of systems.

DEFINITION 2.1-1 A system class S is defined as follows:

Basis Certain indivisible systems called *elements* are members of S. They form the element set E.

Induction Certain interconnection rules R are specified. These rules when applied to members of S yield a system which is also a member of S.

Closure Only those systems belonging to E or formed by a finite application of R are members of S.

The above definition says that certain elements which cannot be further subdivided (the basis) may be interconnected only according to a prescribed set of rules. The resulting systems may be interconnected to elements or to other systems such as themselves (induction), resulting in systems which may again be interconnected to build yet larger systems, and so forth. Finally, we are told that this process of generating and interconnecting systems must stop after a finite number of steps, and that the only systems which are in the defined class are the ones so constructed (closure). Such precision might seem unwarranted, and it is true that it is unnecessary for informal discussions about systems. It is frequently desirable, however, to ensure that a certain system analysis technique, typically in the form of a computer program, is guaranteed to analyze *any system* within a prescribed class. In this and in related cases the precise definitions will be essential. But enough about generalities; let us proceed to illustrate the above definition in terms of two well-known types of systems.

Example 2.1-1 The system class S of single-output AND-OR-NOT logic networks is defined as follows:

The element set E contains three elements or gates, called AND, OR, NOT, each possessing input(s) u_i and output y, which are shown in Fig. 2.1-1.

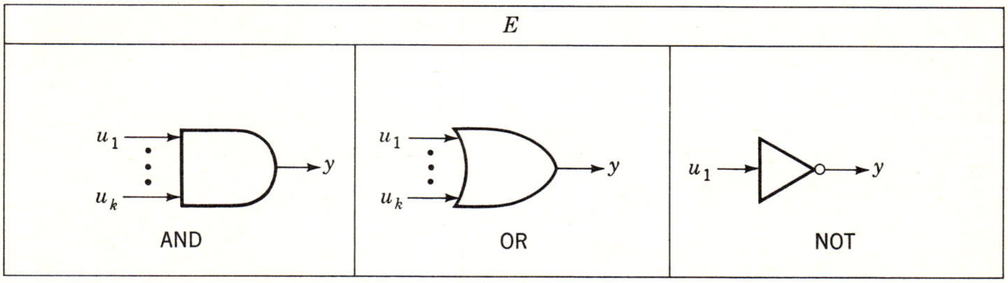

FIG. 2.1-1 Element set for Example 2.1-1.

The variables u_i and y can each take one of two values, 0 or 1. The output y is related to the input(s) u_i in the following way: For the AND gate, y is (takes the value) 1 if and only if all the u_i are 1; for the OR gate, y is 0 if and only if all the u_i are 0; and for the NOT gate, y is 1 only if u_1 is 0.

Looking a little bit ahead, we remark that all the systems, which according to this definition are in class S, will be similar in form to the above gates; that is they will have several (possibly only one) inputs and one output. Denoting any such system in class S, as well as the above AND-OR-NOT gates by s_i $(i = 1, 2, 3, \ldots)$, we are now ready to state the second part of the definition—the permissible rules for connecting the s_i with each other. For this class of systems, *connecting any output y_i to any input u_j means that the value of u_j is set equal to the value of y_i.* This is graphically depicted by joining the arrowed lines corresponding to u_j and y_i. Accordingly, a system N in class S can be constructed only through the rules given in Table 2.1-1.

TABLE 2.1-1 Interconnection Rules

R
1. Outputs of the s_i may not be connected to each other.
2. Loops are not allowed; that is, a connection is prohibited if it makes possible the tracing of a *closed* path (following the arrows and passing through the connected members).
3. Of all the s_i that make up N, exactly one output shall not be connected to any other s_j input, and will be called the output of N.
4. At least one s_i input shall not be connected to an s_j output. Any such inputs will be called the input(s) of N.

Figure 2.1-2*a* shows a system which is not in S because of interconnection of two element outputs—a violation of Rule 1. Figure 2.1-2*b* shows a system which is not in class *S* because of the existence of a "loop," which violates Rule 2. The systems of Fig. 2.1-2*c* and *d*, however, are both in *S*, as the reader may readily verify. In particular, the system of Fig. 2.1-2*c* is called an *exclusive-OR*, i.e., $y = 1$ if and only if exactly one of the two inputs has the value 1. It occurs so frequently that it is represented as a single "gate" with a new symbol. In Fig. 2.1-2*d* the system of Fig. 2.1-2*c* is used to construct a "parity checking" AND-OR-NOT network, where $y = 1$ if and only if exactly an odd number of the four inputs have the value 1. This system is shown in Fig. 2.1-2*d* using the exclusive-OR symbol, and in Fig. 2.1-2*e* using the three basic elements

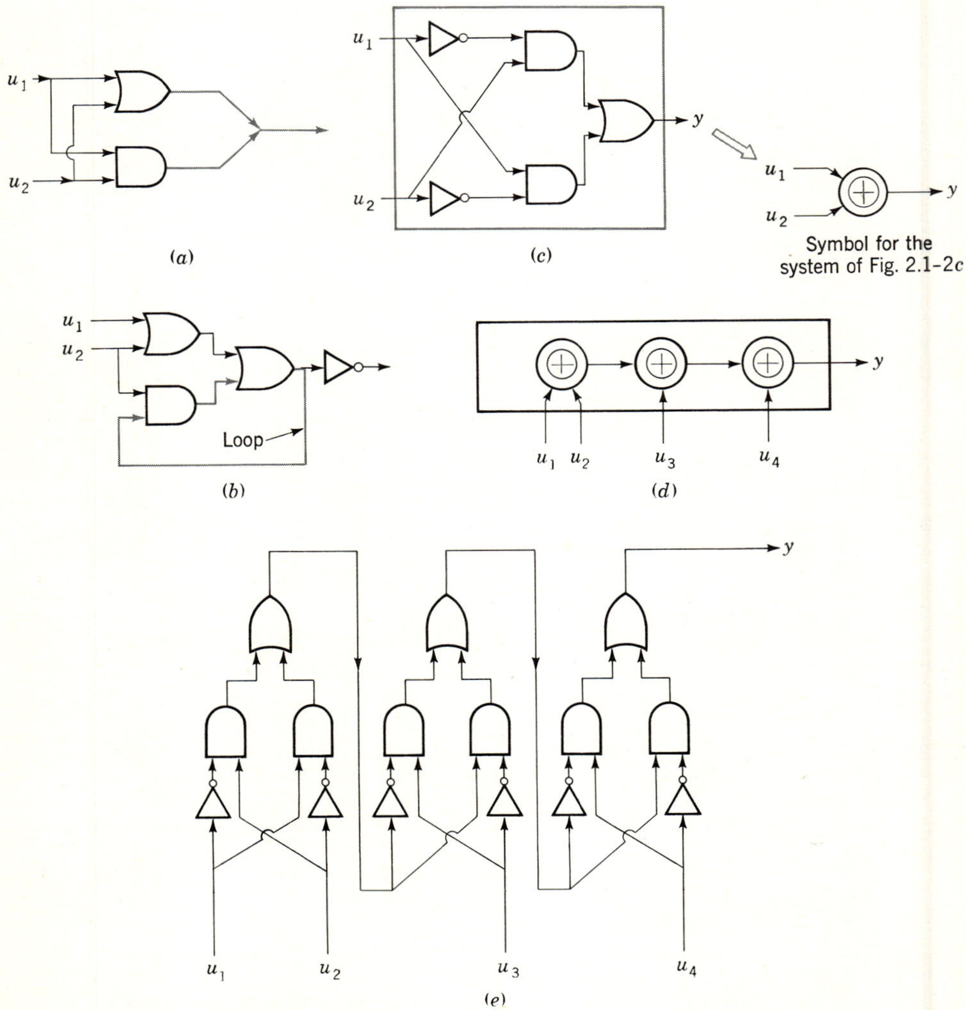

FIG. 2.1-2 *Examples of systems that are or are not in class S.*

of S. Note that this system is indeed in class S and may be used to construct more complicated systems through the same set of interconnection rules.

Let us now turn to an example which illustrates the analytical power made possible by the form of Definition 2.1-1.

Example 2.1-2 The system class Q of series-parallel nonlinear spring networks is defined as follows:

The element set E contains the basic elements which in this case are mechanical springs. Associated with each such element a are two variables, x_a and y_a, whose physical meaning is displacement and force, respectively. These variables take their values from the set of real numbers, subject to the spring constraint A; that is, $y_a = A(x_a)$, as shown in Fig. 2.1-3. Here A is a con-

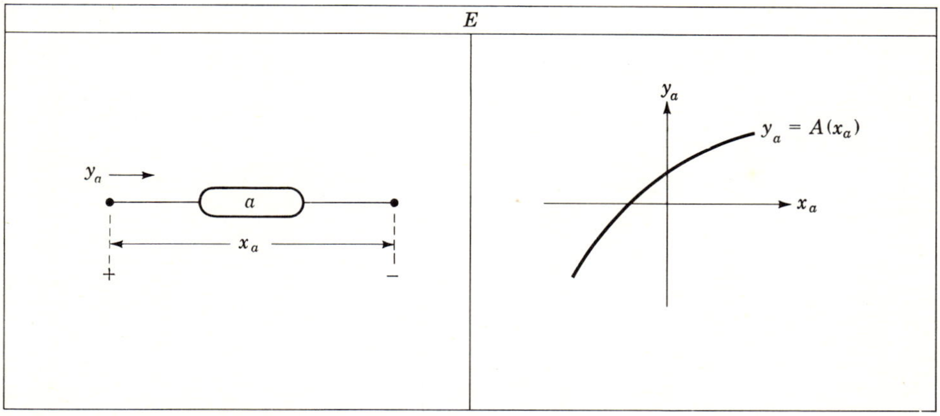

FIG. 2.1-3 Element set for Example 2.1-2.

tinuous strictly monotonic function with bounded "increase,"† which, given a value x_a for displacement, returns a value y_a for the force. Because of the conditions on A, it follows that the inverse function A^{-1} exists; this function, given a value y_a for the force, returns the value x_a given by $A^{-1}(y_a)$. Graphically, the inverse function can be portrayed by interchanging the x and y axes on the graph of A.

The interconnection rules permit the construction of a spring "system" c, in Q from exactly two springs a and b of Q, by one of two rules—series or parallel connection—as shown in Fig. 2.1-4.

The parallel-connection rule means that the two springs are fastened together in such a way that the length of spring c is identical to the lengths of each component spring, i.e.,

$$x_a = x_b = x_c > 0 \tag{1}$$

†There exist constants L and H, such that for any two values x_1, x_2

$$0 < L \leq \frac{A(x_1) - A(x_2)}{x_1 - x_2} \leq H < \infty$$

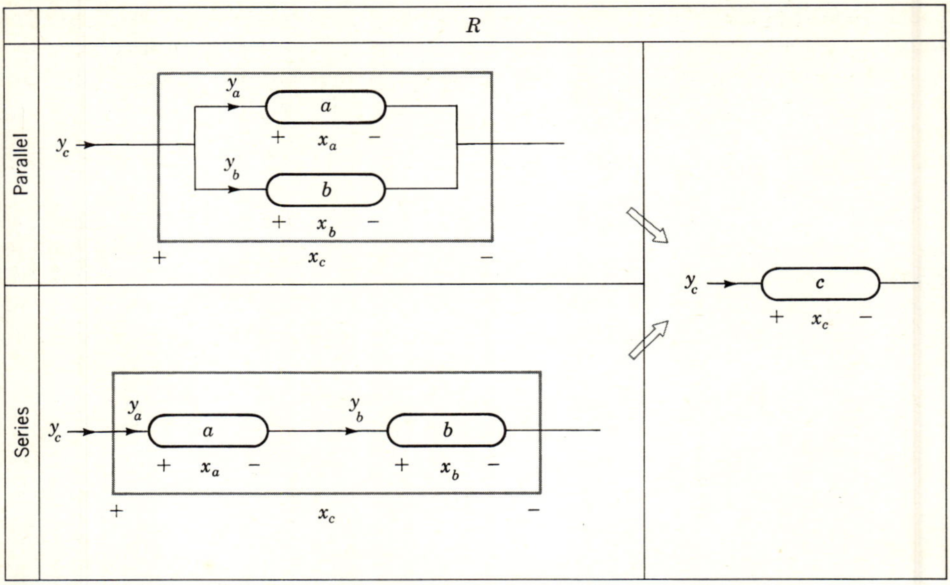

FIG. 2.1-4 Interconnection rules for Example 2.1-2.

From elementary mechanics it follows that the net force of spring c is the sum of the forces of the constituent springs, i.e.,

$$y_c = y_a + y_b \tag{2}$$

It is also known that a and b are characterized by functions A and B:

For all $x_a > 0$,

$$y_a = A(x_a)$$

(3)

For all $x_b > 0$,

$$y_b = B(x_b)$$

From Eqs. (1), (2), and (3) we conclude that

For all $x_c > 0$,

$$y_c = A(x_c) + B(x_c)$$

which we can rewrite as:

For all $x_c > 0$,

$$y_c = C(x_c) \tag{4a}$$

where function C is given by

$$C = A + B \tag{4b}$$

For example, if A and B are as shown in Fig. 2.1-5, then function C is graphically obtained by adding, for each value of $x_c = x_a = x_b > 0$, the corresponding y_a and y_b values. Thus, the *parallel interconnection* rule, when applied to two systems characterized by functions A and B, gives rise to a system characterized by $C = A + B$. The reader is invited to verify that the resultant function meets all the conditions of the constituent functions; hence the new "spring" c is also in class Q.

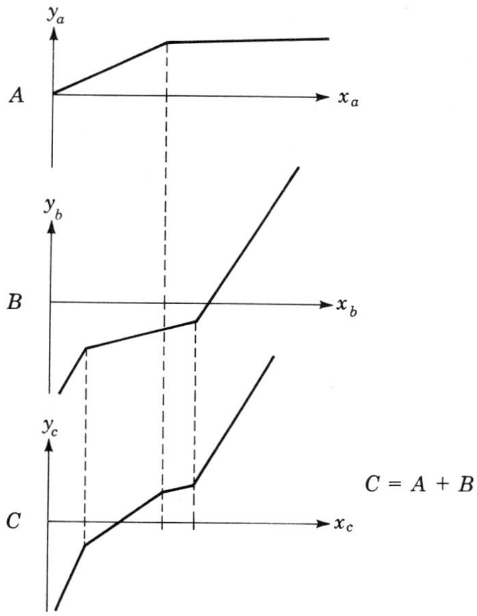

FIG. 2.1-5 Generation of the function C from functions A and B: C = A + B.

Likewise, it can be easily shown that for the *series-interconnection* rule $y_a = y_b = y_c$ and $x_c = x_a + x_b$, so that the resultant spring c is characterized as follows:

For all $x_c > 0$,

$$y_c = C(x_c) \tag{5a}$$

where

$$C = (A^{-1} + B^{-1})^{-1} \tag{5b}$$

Thus, to obtain the graph of C for the series-interconnection rule, interchange the x and y axes of the graphs of A and B, "add" these two graphs, and then interchange the x and y axes of the resultant graph.

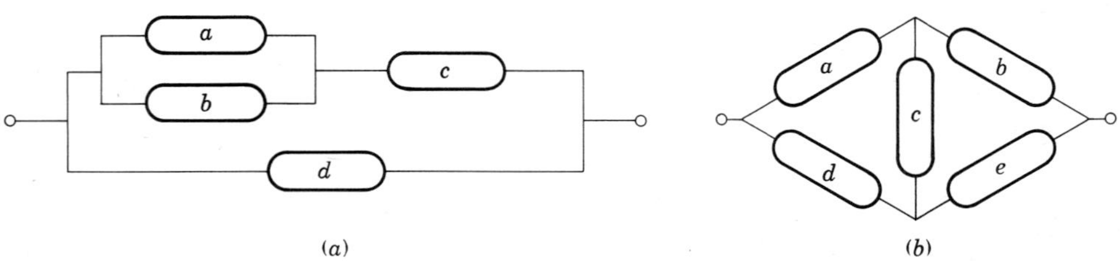

FIG. 2.1-6 Examples of two systems, one that is in Q and one that is not.

example, if it is important that the system consist of the minimum number of elementary springs, then Fig. 2.2-1*d* gives the answer. If instead it is required that the system consist of the maximum number of elementary springs which are connected in parallel, then Fig. 2.2-1*a* is the answer. We may want such a structure for reliability; i.e., if a spring breaks, we may wish minimum damage to the system—in the construction of Fig. 2.2-1*a*, if one elementary spring breaks, we have a spring constant $k=2$ which may still be adequate for our purposes. Alternatively, there may be *constraints* imposed on the synthesis. such as the requirement that the number of springs be kept less than three so as to save physical space. Accordingly, if it were desired that this constraint be met and the reliability be made "high," regardless of cost, Fig. 2.2-1*b* would be the answer.

As the above example might indicate, the task of synthesis is considerably more challenging and more difficult than that of analysis. In a few instances, these constraints and "goodness measures" can be meaningfully specified, and we are then able to follow synthesis techniques that result in a unique system. In most realistic cases, however, synthesis becomes synonymous with such intangible factors as experience, intuition, and ingenuity, which we do not know how to teach through a textbook. Consequently, this book is mainly concerned with the analysis of systems. Synthesis techniques are generally given whenever possible, provided that they are not unnecessarily complicated and that they possess some instructive value.

2.3 MEMORY AND MEMORYLESS SYSTEMS

There are numerous ways of classifying systems for purposes of study, analysis, and understanding. One such way, which forms a basis for this book, considers systems as consisting of two subsystems, one with and one without memory. These two subsystems are interconnected in a particular way or form called *the state model,* which we shall introduce in the following section. In order to establish the state model, however, it is desirable to understand what is meant by a memoryless and a memory subsystem (or system).

From an intuitive point of view, we expect a memoryless system to have a behavior which does not depend on what happened in the past, but rather on what happens "now." Conversely, a memory system should have a behavior which does depend on the past. Regardless of memory, we expect a system to interact with its environment (other systems, or ourselves) through two objects: an input u and an output y.† The variables u and y may represent physical "signals" such as voltages or currents, force, displacement, pressure, flow rate, or any other physical quantities by which we can represent information.

The output y will generally depend on the input u. We can thus write

$$y = F[u] \tag{1}$$

where F is a way of saying what the system does with input u in order to give

†Actually, when we say *input* we shall usually mean one or more input variables. Similarly with *output*.

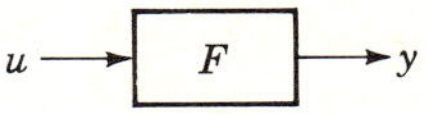

FIG. 2.3-1 *Illustration of a system.*

us output y. Equation (1) is represented in Fig. 2.3-1 (note arrows) and constitutes the starting point for our discussion.

Let us first consider systems without memory, starting with an example.

Example 2.3-1 *A Memoryless System—A Multiplier* Let the element set E of the class of *memoryless* explicit block diagram *systems* consist of three elements: *summers, squarers,* and *scalers,* defined and depicted in Fig. 2.3-2.

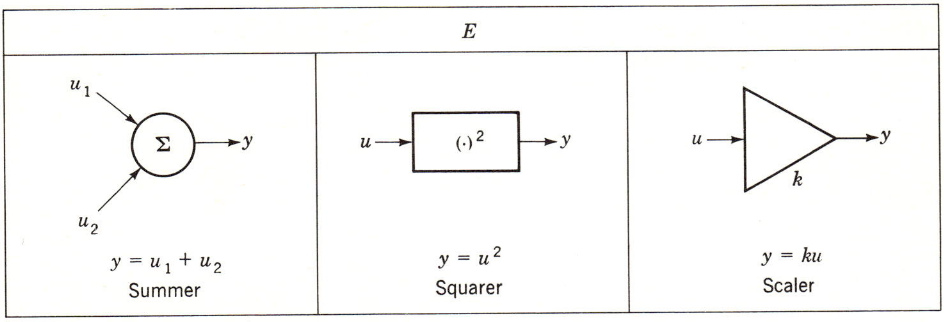

FIG. 2.3-2 *Summer, squarer, and scaler.*

Each of these elements has one (or two) input(s) and one output that take values from the set of real numbers. Physically, variables u and y for each element will normally represent a "voltage."† For example, 6.3 volts would normally be represented by real number 6.3, although it could be made to correspond to real number 12.6 if we so agreed. We are not concerned here with the physical nature of the information that is thus represented. It is in fact sufficient for our purposes to say that u and y, for each element, take real values.

As shown in Fig. 2.3-2, the summer output is related to the summer inputs by

$$y(t_1) = u_1(t_1) + u_2(t_1) \qquad \text{for any value } t_1 \text{ of time} \tag{2}$$

The squarer is characterized by

$$y(t_1) = [u(t_1)]^2 \qquad \text{for any value } t_1 \text{ of time} \tag{3}$$

and the scaler by

$$y(t_1) = ku(t_1) \qquad \text{for any value } t_1 \text{ of time} \tag{4}$$

where k is a given real number.

†These elements are normally made up of so-called operational amplifiers, resistors, and diodes, all of which are discussed in Chap. 4.

The rules of interconnection R for this system class are the same as those for class S in Example 2.1-1.

The desired system that we wish to construct from these elements is a *multiplier*, i.e., a system with two inputs u_1, u_2 and one output y, so that

$$y(t_1) = u_1(t_1)u_2(t_1) \tag{5}$$

The reader may find it challenging to stop reading any further at this point, and try to construct the above system from the given basic elements.

The construction of the multiplier is based on the identity

$$\tfrac{1}{4}[(u_1 + u_2)^2 - (u_1 - u_2)^2] = u_1 u_2 \tag{6}$$

as shown in Fig. 2.3-3. The reader may wish to trace through this system and make sure that the output y is indeed equal to the left side of Eq. (6), and hence to the product $u_1 u_2$. If, for example, at some time t_1, $u_1(t_1) = 3$ and $u_2(t_1) = 4$, it

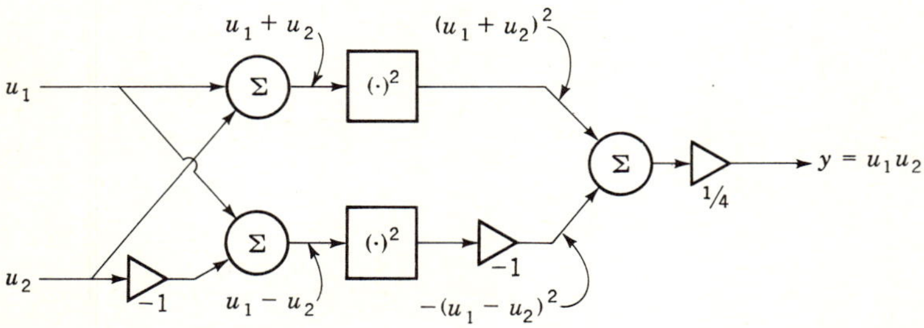

FIG. 2.3-3 Multiplier.

follows that $y(t_1) = (3)\,(4) = 12$, and the multiplier does what it is supposed to do. In "analog computers," multipliers are often made in this manner, because summers, squarers, and scalers are technologically easier to make out of network elements, whereas multipliers are generally not easily implemented.

We say that the system of Fig. 2.3-3 is *"memoryless," since at any time t_1, the value of the output $y(t_1)$ depends only on the values of the inputs $u_1(t_1)$ and $u_2(t_1)$ at the same time, and is absolutely independent of any past values (before t_1) that the inputs may have had.*

We are now ready to formalize the notion of a memoryless system.

DEFINITION 2.3-1 A system f with input u and output y is a *memoryless function*† *system* if and only if

For every t,

$$y(t) = f(u(t))$$

where t is time; $u(t)$ and $y(t)$ denote the *value* at time t of the input and output, respectively; and f is a function which assigns a value y to the output, given a value u of the input.

†We distinguish such systems with the name *function*, because as we shall see later in this section, there are memoryless systems (with loops) whose output values are not a function of the input values but stand in a mathematical relation to the former; i.e., it is possible for several (or no) output values to correspond to a given input value.

Throughout this book, the memoryless systems of interest to us, will be characterized by functions which are fixed and do not change with time. There are memoryless systems, however, where function f does vary with time in a known way, so that a distinction is usually made by referring to the former as *time-invariant* and to the latter as *time-varying*. In any event, if a function system is time-varying, it can always be converted to a time-invariant system by the addition of an extra input (time).

In memoryless systems, the past history of the input does not enter the process of determining the value y of the output, since whenever the input has a particular value u_1, the output will have the value $f(u_1)$ which can be completely determined, regardless of the time or times at which u has that value. We will therefore often write $y = f(u)$. Observe that although Definition 2.3-1 refers to *an* input and *an* output, it may be easily extended to accommodate several inputs and several outputs. Observe further that since we talk about the *value* that u or y attains, we must specify or assume the existence of *a set* U of possible input values and a set Y of possible output values. When we treat continuous memoryless systems as in Example 2.3-1, U and Y will be the set of real numbers or appropriate subsets. When we treat discrete systems, U and Y will each have a finite number of possible values. For example, if the input to a system is a push button, it may take on only one of two values (ON, OFF); if the output is a buzzer, it may also take on one of two values (buzz, no buzz). If the system is such that the buzzer buzzes if and only if the push button is OFF (why not?), then it is a memoryless system, since at any time the value of the input at that time determines the value of the output at the same time. We will encounter *hybrid* systems such as analog-to-digital and digital-to-analog converters, where one of the sets U, Y will be the real numbers and the other will be a finite set.

We proceed now with an example of a system with memory.

Example 2.3-2 *A System with Memory—An Electronic Combination Lock* In addition to the set E of elements of Example 2.3-1, consider the set E^* of elements shown in Fig. 2.3-4. This consists of a switch, an integrator, a delay, a comparator, and an AND gate. Their input-output behavior is defined as follows: *Switch:* If in position a, then $y_1 = u$, $y_2 = 0$; if in position b, then $y_2 = u$, $y_1 = 0$. *Integrator:* $y(t) = \int_{t_0}^{t} u(\tau)d\tau$, where t_0 is the last time at which the integrator was reset. *Delay:* $y(t) = u(t - \tau)$, where τ is a positive real number. *Comparator:* $\epsilon > 0$. If $u_1 - u_2 > \epsilon$, then $y_1 = 1$, $y_2 = y_3 = 0$; if $u_1 - u_2 < -\epsilon$, then $y_3 = 1$, $y_1 = y_2 = 0$; otherwise $y_2 = 1$, $y_1 = y_3 = 0$. AND *gate:* $y = 1$ if and only if $u_1 = u_2 = u_3 = 1$; otherwise $y = 0$.

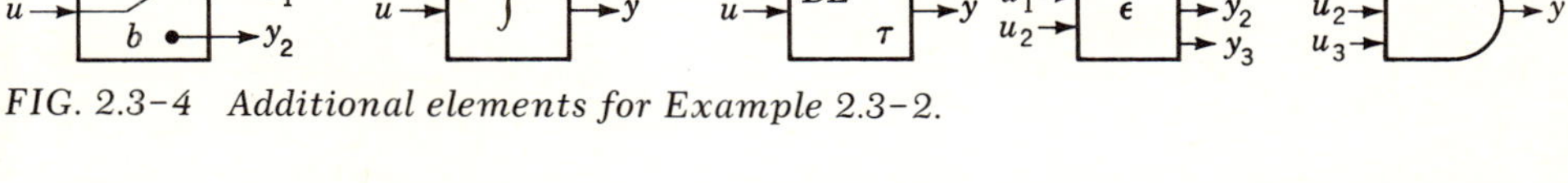

FIG. 2.3-4 Additional elements for Example 2.3-2.

Observe that the comparator and "AND gate" are memoryless, whereas the switch, the integrator, and the delay do not fit Definition 2.3-1; hence they are not memoryless. The switch makes y_1 or y_2 equal to u, depending on its position (which is also an input). The integrator has one input and one output, and may be "reset." Resetting at $t = t_0$, means that $y(t_0) = 0$. Thereafter

($t > t_0$), $y(t)$ is the time integral of the input. The delay has an output which is identical to the input delayed by τ units of time. The comparator assigns value 1 to y_1 if u_1 exceeds u_2 by more than ϵ. If u_2 exceeds u_1 by more than ϵ, then $y_3 = 1$, and if u_1 is about equal to u_2 (i.e., within the "zone" of width 2ϵ), then $y_2 = 1$; outputs which are not 1 are 0. Finally, the AND gate has an output which is 1 if and only if all three inputs are 1; otherwise the output is 0.

We shall construct with these basic elements an electronic "combination lock" as shown in Fig. 2.3-5.

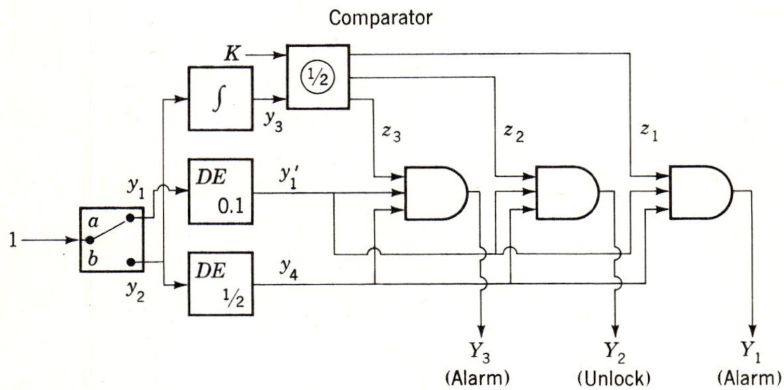

FIG. 2.3-5 An electronic combination lock.

Operation of this system is as follows: Let $K = 5$. The switch is normally in position a, hence $y_1 = 1$, $y_2 = 0$. Since the integrator and delay input (y_2) is zero and the outputs of both integrator (y_3) and delay (y_4) are initially zero, it follows that until the switch causes y_2 to change, $y_3 = y_4 = 0$. The comparator with inputs 5 and 0 causes $z_1 = 1$, $z_2 = z_3 = 0$. Finally, the three AND gates have each input $y_4 = 0$, which causes all system outputs Y_1, Y_2, Y_3 to be zero. These output values indicate that no action is to be taken by the system.

For the lock to operate properly (i.e., to unlock), the switch must be held in position b for a time $T = 5 \pm 1/2$ seconds (the $\pm 1/2$ indicates a tolerance, i.e., from 4.5 to 5.5 seconds). If it is held for a shorter or longer time, an alarm will ring as we will show below. First, let $T = 5$, starting at some time t_0. This causes $y_1(t) = 0$, $y_2(t) = 1$ for $t_0 < t \leq t_0 + 5$. Hence, the integrator output rises in time (with slope 1) until it reaches at $t = t_0 + 5$ the value $y(t_0 + 5) = 5$. Until that time, nothing happens at the system outputs, since $y_1' = 0$ "disables" all the AND gates (they have zero output). At $t = t_0 + 5$, however, the switch returns to position a and impresses a value $y_1(t_0 + 5) = 1$; hence $y_1'(t_0 + 5.1) = 1$. At time $t_0 + 5.1$, the comparator has outputs $z_2 = 1$, $z_1 = z_3 = 0$ since its two inputs are equal within 1/2 unit (indeed they are exactly equal). Moreover, the delayed $y_2(t)$ waveform which we called $y_4(t)$ is also 1 at $t = t_0 + 5.1$. Hence the AND gate with output Y_2 is the only AND gate that has all its inputs 1. Therefore,

$$Y_2(t_0 + 5.1) = 1$$

which is the electrical signal that causes the lock to unlock. The time signals discussed above are shown in Fig. 2.3-6.

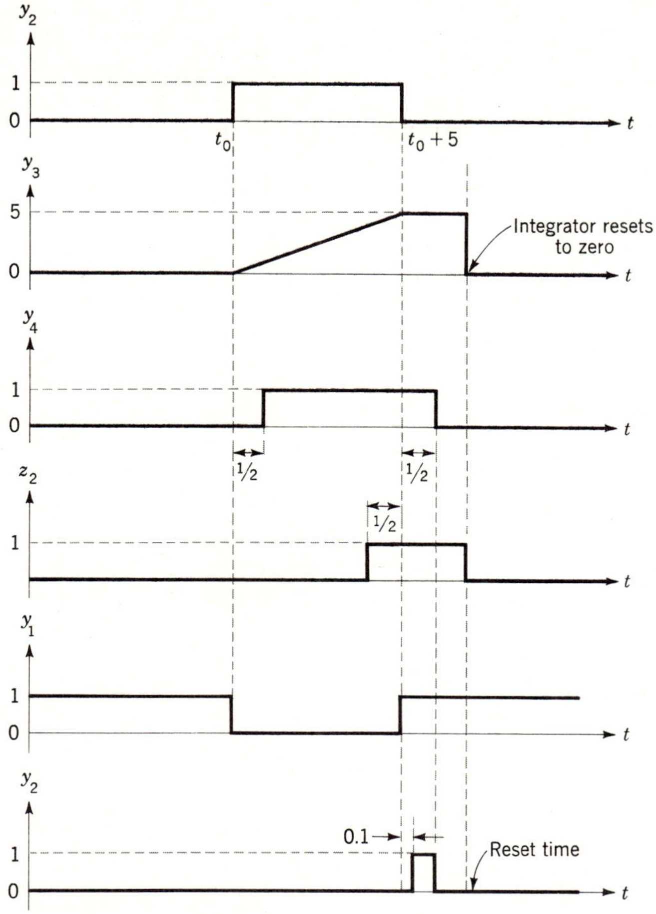

FIG. 2.3-6 Signals that unlock the lock.

When Y_2 becomes 1, it unlocks the lock and resets the system, i.e., returns the integrator output to zero with a slight delay. Observe that the lock will unlock if T is in the range $4.5 \leq T \leq 5.5$, since the comparator will have $z_2(t_0 + T) = 1$.

Suppose now that $T < 4.5$. First, if $0 \leq T < 1/2$ nothing will happen since $y_4(t_0 + T) = 0$. If, on the other hand, $1/2 \leq T < 4.5$, then at the moment $(t_0 + T + 1/2)$ when y_4 becomes 1, the comparator will have output $z_1 = 1$, $z_2 = z_3 = 0$; hence the system output Y_1 will become equal to 1. This is the signal that rings an alarm, indicating an unsuccessful attempt to unlock. Similarly, Y_3 will become 1 if $5.5 < T$.

To change the combination of this lock, it is only necessary to change the value of K. For any set value of K, the lock will have the following behavior. The gate will unlock only if 1/2 second after the switch has returned to a the value of y_3 is within $\pm 1/2$ unit from K. That is, the integrator must be allowed to integrate unity for at least $K - 1/2$ and at most $K + 1/2$ seconds. Hence by changing (internally) the value of K, the combination of the lock changes since the switch must be held at position b for $K \pm 1/2$ seconds. Any shorter or longer holding time will cause an alarm.

Quite clearly this system is not memoryless, since *its behavior at any time depends on the past history of the input.*

Earlier in the chapter, we dealt with series-parallel networks of springs, and we saw how the repeated use of interconnection rules simplifed analysis. Similar benefits are often present in memory systems, with the repetition or recursion occurring in time. This type of time recursion which forms a basis for the state model, is illustrated below:

Example 2.3-3 *A Reverberation Unit* Consider the task of building an electronic reverberation unit for a hi-fi system. Reverberation means the acoustical process of successive reflections experienced by sound when it hits reflective surfaces (a church usually has high or long reverberation because of flat, bare walls, whereas a room padded with carpets on floor, walls, and ceiling has very short or little reverberation). Our unit will cause artificial reverberation; that is, it will have an input to which we supply "pure" music without reverberation, and it will have an output from which music with reverberation will emerge. We will represent music as a "signal" $m(t)$, which for all time t takes on real values. (This could be a voltage for driving a loudspeaker, but we will not worry here about its exact physical nature.)

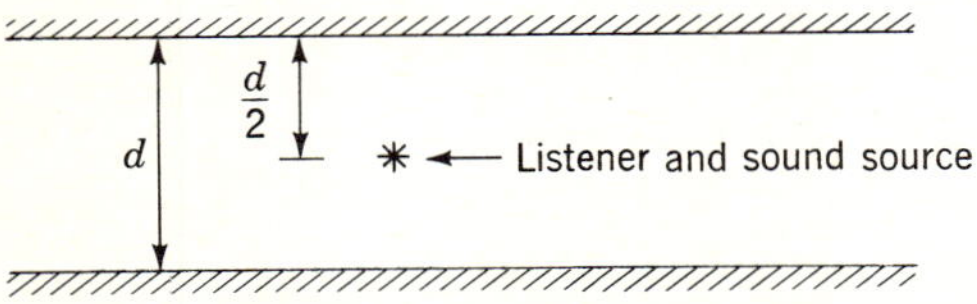

FIG. 2.3-7 Hallway used for reverberation test.

Next, we perform a small "thought" experiment in order to understand the nature of reverberation. In a hallway consisting of two walls d units apart (Fig. 2.3-7), we make a sudden sound of very short duration (say a pistol shot) at the location shown in Fig. 2.3-7. A listener located on the same position as the sound source hears the shot as it is fired; he also hears echoes caused by successive sound reflections at time intervals of d/c, where c is the speed of sound. What he hears is shown graphically in Fig. 2.3-8. Note that because of sound absorption, we expect that the echoes will die out, as time goes on. In fact, we assume each echo to be a fixed fraction, say 0.8, as loud as the previous echo; that is, we assume that 20 percent of the sound intensity is "absorbed" at the walls and in the intervening air, between any two successive echoes.

To construct a system that has the same input-output echo behavior as the room, observe that the presence of absorption and walls is equivalent to our taking the original "signal," delaying it by d/c, scaling it by 0.8, then delaying by $2d/c$ the original signal, scaling it by $(0.8)^2 = 0.64$, and so on until it is no longer

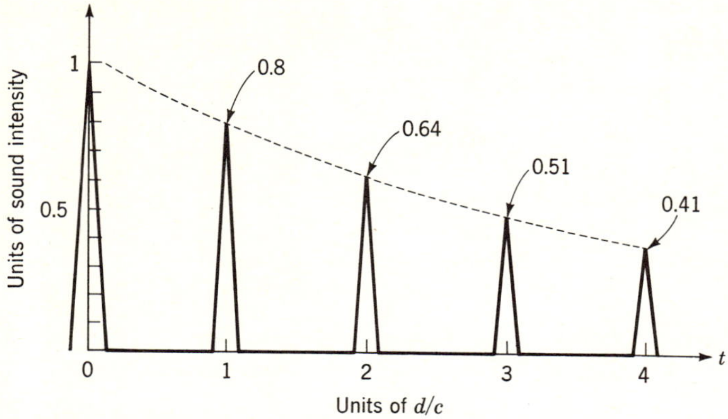

FIG. 2.3-8 Results of reverberation test.

audible (say after 20 echoes). All these delayed and scaled signals are then added to give the result of Fig. 2.3-8. This system, consisting of delays, scalers, and summers, is shown in Fig. 2.3-9.

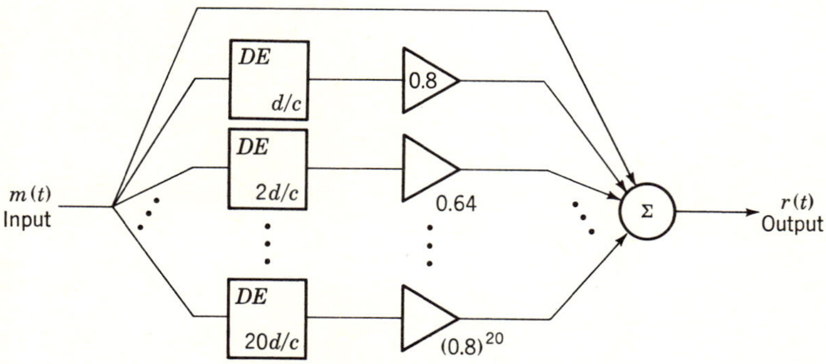

FIG. 2.3-9 First artificial reverberation unit.

Another way of accomplishing the same task with the same type of elements is shown in Fig. 2.3-10, as can be easily deduced by the reader.

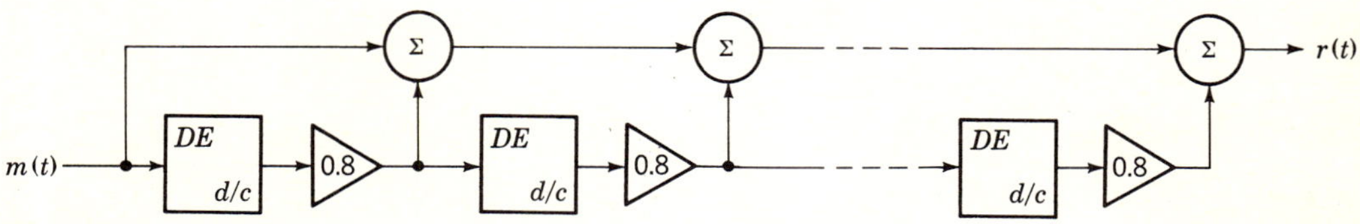

FIG. 2.3-10 Second artificial reverberation unit.

In either the system of Fig. 2.3-9 or in that of Fig. 2.3-10, if we supply as input $m(t)$ a "narrow" pulse, then the output $r(t)$ will be as shown in Fig. 2.3-8. What if, however, we supply for $m(t)$ a continuous signal representing music? The output $r(t)$ will consist of delayed and attenuated replicas of $m(t)$, all summed together, as we would expect to happen in the real hall.

Although both the above artificial systems seem to do the desired task of reverberation, we have not yet exploited the recursion in time of the reflection-attenuation process. Consider, for example, the system of Fig. 2.3-11 where we have taken the important step of allowing loops; that is, the interconnection rules are Rules 1, 3, and 4 of Table 2.1-1.

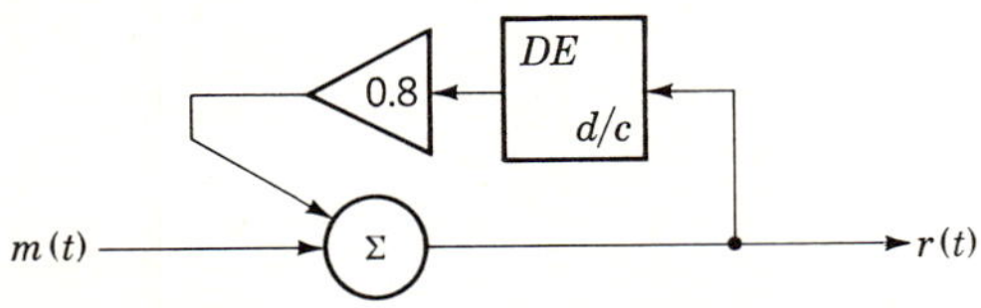

FIG. 2.3-11 Third artificial reverberation unit.

A little thought will reveal that this system behaves exactly as the system of Figs. 2.3-9 and 2.3-10. That is, given a narrow pulse for $m(t)$, $r(t)$ is as shown in Fig. 2.3-8. Moreover, this system is in a sense superior since it is not limited to a finite number of echoes (as are the systems of Figs. 2.3-9 and 2.3-10), but can "attenuate" and "reflect" forever, as is the case with the real hall. The economy of components achieved in this system is typical of many such recursive-in-time systems which we shall encounter quite often in the sequel.

Finally observe that none of these three artificial systems is memoryless, since the output at any time is generally a function of the past history of the input. Hence each is a memory system, as was the combination lock of the preceding example. We give next a formal definition of systems with memory.

DEFINITION 2.3-2 A system F with input u and output y is called a *memory system* if and only if

1. It is not memoryless, and
2. For some t, $y(t) = F(u[t_0,t])$.

Here, t is any value of time; $u[t_0,t]$, where $t_0 \leq t$, denotes the input *function* in the time interval $[t_0,t]$, i.e., *all* the values that the input takes in the interval $[t_0,t]$; $y(t)$ is the value of the output at time t, and F is a rule or function, which given *all* the values of input u in the interval $[t_0,t]$, assigns a value to output y.

Observe that in contrast to a memoryless system, a memory system must "know" more than the value of u at time t in order to generate $y(t)$; in fact, for at least some values of time, it must know the past history of the input, i.e., the values of u in the interval $[t_0,t]$, in order to generate an output value for y. The time t_0 serves to identify the memory span of the system. For some systems it may turn out that $t_0 = -\infty$; i.e., the behavior of these systems may be affected by what happened in the infinitely distant past; alternatively, it may be that the interval $t-t_0$ is infinitesimal, as is the case in Prob. 2.3-1. In any event, for a system to be memoryless, its output value must depend in some nontrivial way on the past history of the input. We say "nontrivial" because we wish to exclude contrived functions which, given values of u in the interval $[t_0,t]$, ignore all values other than the present value $u(t)$, thereby rendering the system memoryless.

Observe that the behavior of systems, whether memory or memoryless, does not depend on future values of the input; i.e., for all t, $y(t)$ is independent of $u(\tau)$ for $\tau > t$. Systems with this property are often called *causal*, for obvious reasons. It should also be noted that although we have been speaking of single-input, single-output systems, the above definition can be extended in a straightforward manner for the case of multiple-input multiple-output systems.

Finally, similarly to what we did in the case of memoryless systems, we distinguish a set U^* of input *functions* $u[t_1,t_0]$ and a set Y of output *values*, F being a function form U^* to Y. The only difference between this formulation and that of memoryless systems is that U^* is a set of *functions* (from time to a set of input values U), whereas in the memoryless case we are concerned only with the set of input values U. When we deal with the so-called continuous dynamic systems and networks, U^* will be a set of functions from a time interval of real values $[t_0,t]$ to the real numbers, and Y will be the set of real numbers. When we deal with finite-state machines, U^* will be the set of functions from a finite set of time values $t_0, t_1, t_2, \ldots, t_n$ to a finite set such as $\{0,1\}$, and Y will usually be the set $\{0,1\}$. On occasion we will encounter so-called hybrid systems with memory, where the inputs may be taken from a finite set and the outputs from the set of real numbers, or vice versa.

Memoryless Systems and Relations

Earlier in this section we mentioned the possibility of a memoryless system containing *loops* of memoryless elements. In such a case, the overall system input-output behavior is not necessarily characterized by a function, even though the constituent elements are all functions. Consider, for example, the system of Fig. 2.3-12 consisting of a squarer, summer, and scaler interconnected in a single loop.

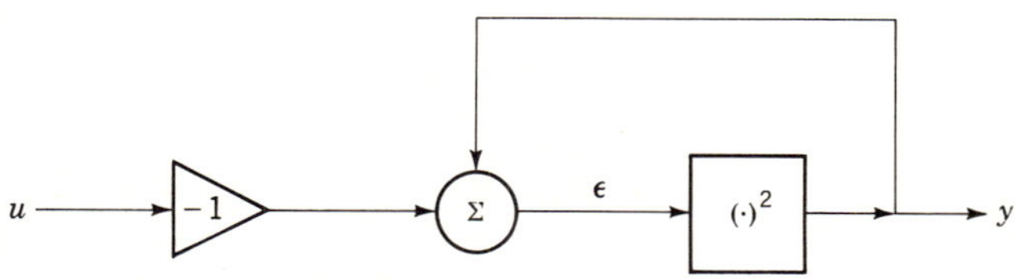

FIG. 2.3-12 A memoryless system with a loop.

We are interested in such a system for the simple reason that we can interconnect physical scalers, summers, and squarers in such a way, and we would therefore like to know the behavior of the resulting system. Letting ϵ be the summer output in Fig. 2.3-12, we interpret the interconnection by stating that the following two equations must be satisfied:

$$\epsilon = y - u$$

$$y = \epsilon^2$$

Combining these two equations yields the single equation relating y to u:

$$y = (y - u)^2 \tag{7}$$

We see that for $u \geq 0$, this equation has *two* real solutions for y; for example, if $u = 0$, then $y_1 = 1$ and $y_2 = 0$ are both solutions satisfying Eq. (7). We cannot therefore find a function or give a rule which given u, returns y; for what would such a function do if it is given for u the value 0? We conclude then that the above system is not a memoryless function system as dictated by Definition 2.3-1. Neither is it a memory system, for we cannot see how the past history of the input determines the value of y when $u = 0$. For example, suppose that at $t = 0$, $u = -1/4$; then there is only one solution: $y = 1/4$. If $u(t)$ now starts growing (Fig. 2.3-13*a*) as follows:

$$u(t) = -1/4 + t \qquad \text{for } t \geq 0 \tag{8}$$

then, as the reader may easily verify by solving Eqs. (7) and (8), either one of the following two solutions is possible (Fig. 2.3-13*b*):

$$y_1 = 1/4 + t + \sqrt{t} \qquad \text{for } t \geq 0$$

$$y_2 = 1/4 + t - \sqrt{t} \qquad \text{for } t \geq 0$$

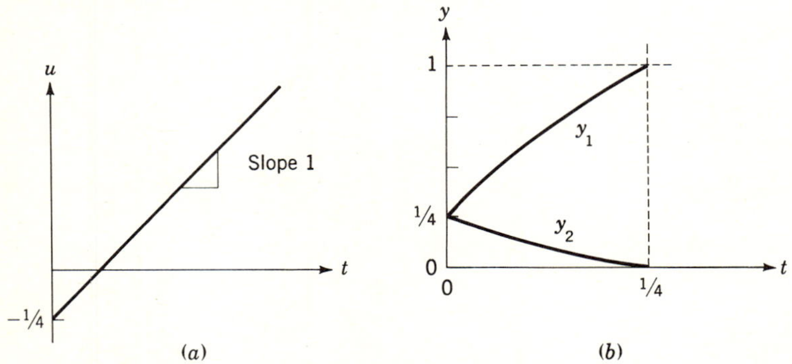

FIG. 2.3-13 Input-output graphs of the system of Fig. 2.3-12.

If no more information is known about the system of Fig. 2.3-12, then there is nothing we can say about which solution y_1 or y_2 will appear at the output. The *physical* system, represented by the model of Fig. 2.3-12 will exhibit *one* of these two waveforms at its output—depending on inadvertent presence of memory in the system (which we did not consider in the model) or/and in the presence of thermal agitation, mechanical vibration, or similar phenomena which we often call *noise* and which will "push" the output in one or the other of its two alternatives. As far as we are concerned, system models such as the one of Fig. 2.3-12, which can have other than a single output value for a given input value, but which do not depend on the past history of the input, will also be called memoryless systems and will be characterized by relations. As we shall see from the following definition, they include the memoryless function systems of Definition 2.3-1.

DEFINITION 2.3-3 A system R with input u and output y is *memoryless* if and only if

For every t,

$$u(t)Ry(t)$$

where t is time; $u(t)$ and $y(t)$ denote *values* at time t of input and output, respectively; and R is a *mathematical relation*, explained below.

As the reader can see, the only distinction between this definition and Definition 2.3-1 is the use of a relation instead of a function. In this context, a relation is formally defined as simply *a collection of pairs of values u, y;* a function, on the other hand, is a relation with the special property that every value u appears in only one such pair. In other words, a relation which is not a function can have a given value u appear in several u, y pairs. The notation u_1Ry_1 means that there is a pair u_1, y_1 in the relation; we also say u_1 is related to y_1. Thus, for example, the following collection of pairs is a function (hence also a relation):

u, y
0, 0
1, 2
3, 0
4, 5

whereas the following collection is a relation but not a function:

u, y
0, 0
1, 2
1, 3
3, 0
4, 2

since for $u = 1$, y can be either 2 or 3.

Relations can also be displayed in the form of a graph. Two such relations are shown in Fig. 2.3-14, where (a) is a function and (b) is not.

Presumably we could have given the relation for the circle of Fig. 2.3-14b by the tedious task of listing as many as possible of all the (infinite) pairs of points u, y which lie on the circle, but this is not necessary if we give *an equation or a rule for determining whether a pair of values is in the relation.*

Returning to the system of Fig. 2.3-12, we see that it is memoryless, for it can be characterized by the relation

$$y = (y - u)^2$$

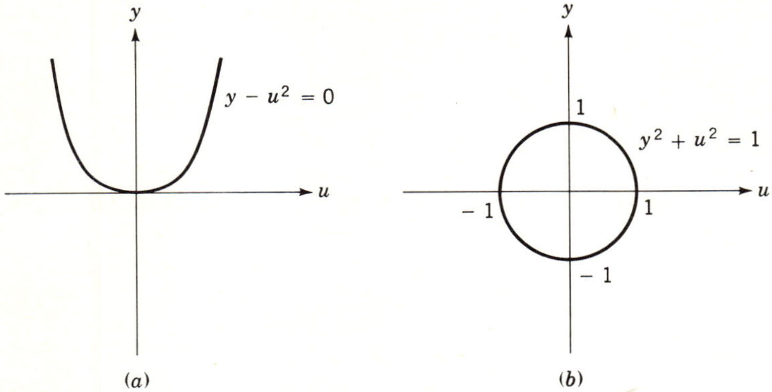

FIG. 2.3-14 Graphs of two relations.

where u and y are values of the input and output *at any time t*. That is, a pair u, y is in the relation if and only if it satisfies the above equation. The system is memoryless, since the past history of the input does not affect the output.

Throughout this book we shall be concerned primarily with memoryless function systems; we shall therefore drop the distinction "function," referring to the systems simply as memoryless, and clarifying when necessary whether they are characterized by relations and not by functions.

2.4 THE STATE MODEL

As we saw in the preceding section, a system has memory if the value $y(t)$ of its output at some time t depends on the past history of the input, i.e., on all the values $u[t_0,t]$ of the input in a time interval $[t_0,t]$. Although such a system can be thought of as being characterized by a rule F which given $u[t_0,t]$, returns $y(t)$, i.e.,

$$y(t) = F(u[t_0,t]) \tag{1}$$

often the rule is very complicated or impossible to state explicitly. For example, listing *all* possible input-output pairs is a hopeless task since there are infinitely many input functions $u[t_0,t]$ that could be applied to the system. The *state model* is a way of describing a system by separating it into two finite parts—one with and one without memory. This separation is performed in such a way that *the outputs of the memory part, the state variables, determine through their values at any given time all that is necessary for establishing the future input-output behavior of the system.* An economical state model characterization is one that has a small number of state variables and a simple memory part such as few integrators or flip-flops.

Not all systems can be characterized by a state model.† For example, systems using delays such as the reverberation unit of Fig. 2.3-11 cannot be

†That is, with a finite number of state variables.

generally put in state model form, since at any time, future system behavior is determined by what is stored in the delay at that time, i.e., an *infinite* number of values. Hence for that system we cannot find a finite number of state variables whose values determine future input-output behavior. The large majority of systems discussed in this book, however, can be put in state model form. These are dynamic systems made up of a finite number of integrators, electric networks consisting of a finite number of capacitors and inductors, and finite-state machines. They are often called *lumped systems*.

The state model characterization is used because it simplifies the description, and sometimes the analysis, of otherwise complicated dynamic systems, and because it provides us with useful insights into the operation of these systems. The following example illustrates how an integrator system is put in state model form.

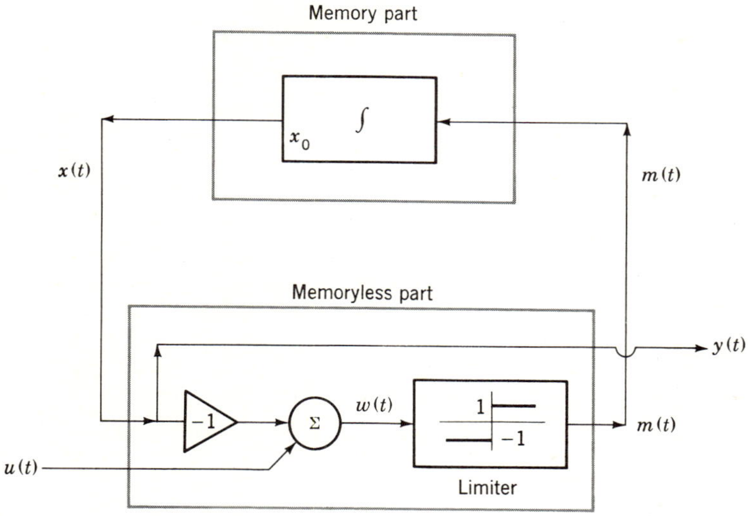

FIG. 2.4-1 A single integrator system.

Example 2.4-1 *A Single-integrator System* Consider the system of Fig. 2.4-1, consisting of only one memory element—an integrator—and three memoryless elements—a scaler, a summer, and a "limiter"; the limiter is characterized by the following input-output behavior:

$$\text{For all } t, \quad \begin{cases} m(t) = 1 & \text{if } w(t) \geq 0 \\ \\ m(t) = -1 & \text{if } w(t) < 0 \end{cases} \tag{2}$$

Observe that we have drawn the system in state model form, i.e., in two parts, one of which has memory and one of which does not. The state equations characterizing the system are the input-output equations of the two parts. Thus, the memoryless part, which has two inputs u and x, and two outputs m and y, is characterized by

$$\text{For all } t, \quad \begin{cases} y(t) = x(t) \\ m(t) = 1 & \text{if } u(t) - x(t) \geq 0 \\ m(t) = -1 & \text{if } u(t) - x(t) < 0 \end{cases} \tag{3}$$

The memory part, which has one input $m(t)$ and one output $x(t)$, is characterized by one equation:

For all $t \geq 0$,

$$x(t) = \int_0^t m(\tau)\, d\tau + x_0 \tag{4}$$

Note that in Fig. 2.4-1, we denote the initial value $x(0) = x_0$ of the integrator output by marking it within the integrator block.

Together, Eqs. (3) and (4) form the state characterization. The input-output behavior of the system is as follows: At any time t, the output $y(t)$ rises with unity slope $dy/dt = 1$ if $u(t) \geq y(t)$ and drops with unity slope $dy/dt = -1$ if $u(t) < y(t)$. A typical input-output behavior is shown in Fig. 2.4-2. Clearly,

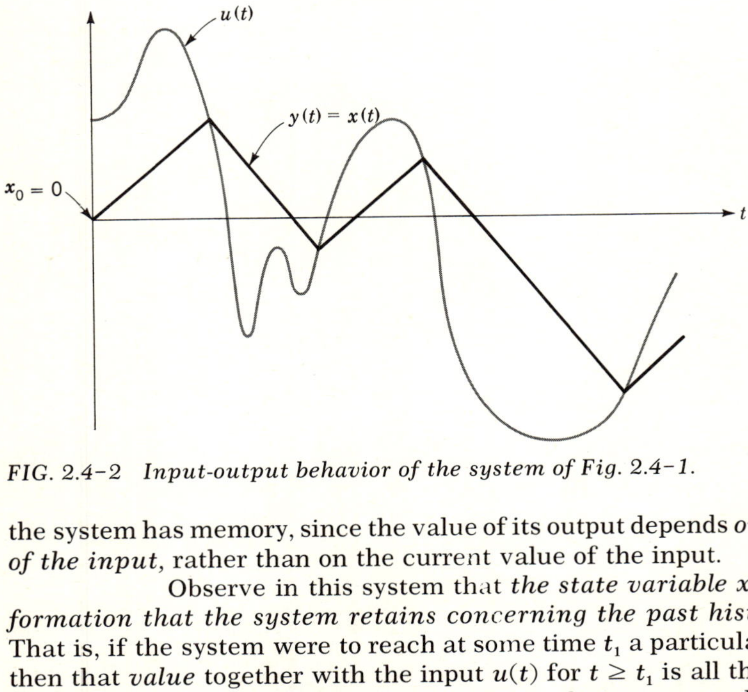

FIG. 2.4-2 Input-output behavior of the system of Fig. 2.4-1.

the system has memory, since the value of its output depends *on the past history of the input,* rather than on the current value of the input.

Observe in this system that *the state variable $x(t)$ is the only information that the system retains concerning the past history of the input.* That is, if the system were to reach at some time t_1 a particular value $x(t_1) = x_1$, then that *value* together with the input $u(t)$ for $t \geq t_1$ is all that is necessary to determine $y(t)$ for $t \geq t_1$. How the particular value x_1 was reached is not relevant for the future behavior of the system. Figure 2.4-3*a* and *b* show two different input functions u_1 and u_2, either one of which will result in $x(3) = -1$ if applied from $t = 0$, with $x(0) = 0$.

Observe in Fig. 2.4-3*a* and *b* that although the input functions u_1 and u_2 look very different in the interval [0,3], they nevertheless place the system

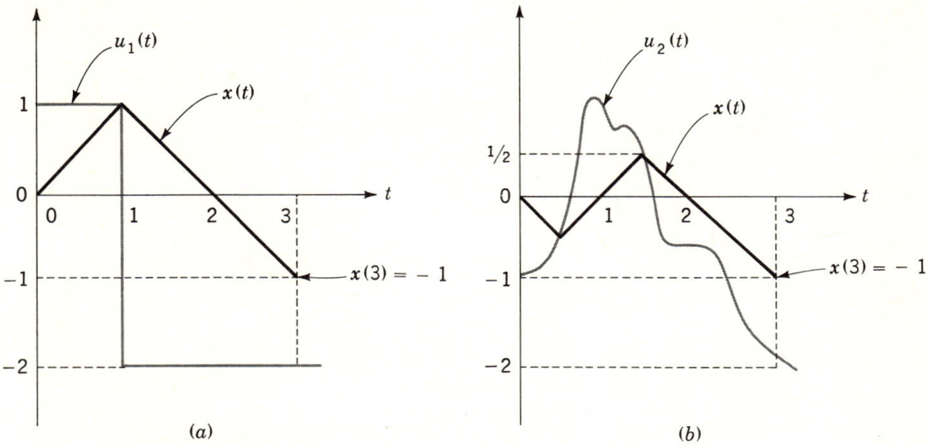

FIG. 2.4-3 Two different inputs resulting in the same state.

in *the same state* $x = -1$ at time $t = 3$. Thus, regardless of whether u_1 or u_2 was the past history of the input, the system remembers the same information or is in the *same state* at $t = 3$, i.e., $x(3) = -1$ in either case. This observation indicates that the state variables of a system do not generally "remember" all possible past input waveforms but only classes of waveforms. Indeed, very often an infinite number of input waveforms can lead a system to the same state, i.e., to the same value of $x(t)$. It is this ability of the state model *to classify as equivalent a number (usually infinite) of otherwise distinct input waveforms* that helps in the description and analysis of systems, as well as in the performance of a particular task by a particular system.

With this intuitive background, we are now in a position to formalize the state model representation of systems.

Generalization of the State Model

Generalizing from the above example we give the following definition.

DEFINITION 2.4-1 The *state characterization* or *state model* of a system consists of two parts or subsystems—the memory part and the memoryless part—which are interconnected as shown in Fig. 2.4-4.

The interconnection corresponds to the following equations:

Memoryless: $m = f(u,x)$ (5*a*)

$$y = g(u,x) \tag{5b}$$

Memory:

$$x(t) = F(m[t_0,t]) \qquad x(t_0) = x_0 \tag{5c}$$

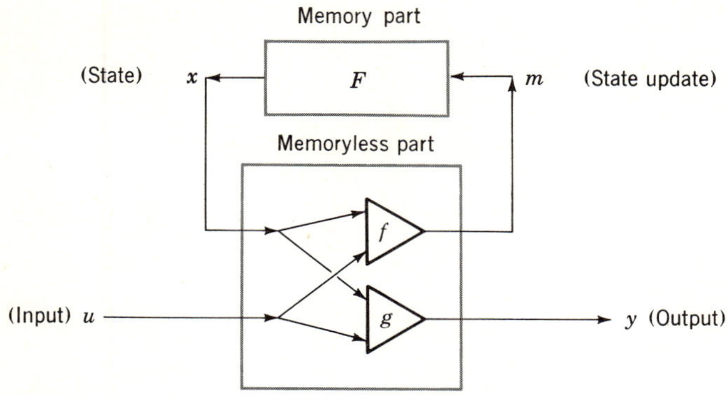

FIG. 2.4-4 General form of the state model.

where

f is a function which gives a value for the *state update* m, given values of the *input* u and *state* x.
g is a function which gives a value for the *output* y, given values of the input u and state x.
F is a rule† which gives a value for the state at any time t ($t > 0$), given x_0 and the values of the state update in the interval $[t_0,t]$, $t_0 < t$.

Observe the important property of F, specified above, which guarantees that knowledge of the initial state $x(t_0)$ at any time t_0, along with the future input behavior, $u[t_0,t]$, determines for any $t \geq t_0$ the output $y(t)$.

This definition is recursive, since it defines an arbitrary memory system in terms of another memory system such as F. The important distinction here is that the system of type F is a very *simple*, in fact an elementary, memory system such as an integrator, or as we shall see later a flip-flop, whereas the overall system defined by Eqs. (5) may be relatively complex and will generally perform tasks that cannot be performed by the elementary memory system alone.

As in our definitions of memoryless and memory systems (Definitions 2.3-1 and 2.3-2), each of the variables u, y, x, and m of Fig. 2.4-4 can and will generally represent *several* values. That is, there can be several system inputs $u_1, u_2, \ldots, u_k$; several system outputs $y_1, y_2, \ldots y_z$; several *state variables* $x_1, x_2, \ldots, x_p$; and several state-update variables $m_1, m_2, \ldots, m_p$; as shown in Fig. 2.4-5. Note that we do require the same number of state and state-update variables since there will be usually one elementary memory element associated with each state variable. Observe also that k, z, and p can each take any positive integer value. For example, we can have a system with three inputs ($k = 3$), two outputs ($z = 2$), one state variable ($p = 1$), and hence one state-update variable. However, in spite of the variety of state models possible through the choice of

†F is often called a *functional*, since it acts on functions—i.e., on $m[t_0t]$, instead of on a value $m(t)$.

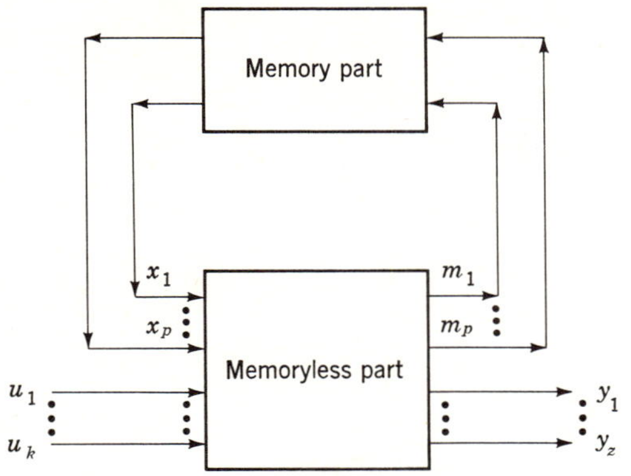

FIG. 2.4-5 The state model for systems with several variables.

k, z, and p, the basic ideas pertaining to state models can be discussed and explained in terms of the one variable model shown in Fig. 2.4-4.

Observe next that in order to make precise the notion of a state model we must specify the sets from which u, x, y, and m take their values. When all these sets are the real numbers, then we shall say that we are dealing with continuous dynamic systems.† When they are all finite sets, with values typically represented by groups of 0s and 1s, we shall say that we are dealing with finite-state machines or discrete dynamic systems. Still in other situations it will be possible for these sets to be mixed—some being the set of real numbers, the others being finite—in which case we shall say that we are dealing with hybrid dynamic systems. In all cases, if we remove the memory part of the state model, we will be left with memoryless continuous or memoryless discrete systems. The state model will accordingly be used as our main connective link among the large variety of systems that we shall study throughout this book.

It is natural to ask at this point what systems can be put in state model form. The obvious and general answer, of course, is: "every system for which variables x, u, y, and m, and functions f and g can be identified so that Eqs. (5) hold." A more useful answer can be given primarily in terms of interconnection rules as follows:

Rules for a System That Can Be Put in State Model Form

1. The system shall consist of a finite number of memoryless and memory *elements*—the latter being denoted as F_i. Each F_i must obey the conditions stated in Definition 2.4-1; i.e., its output value $x_i(t_0)$ and its input function $m_i[t_0,t]$, for any choice of t_0 and t ($t > t_0$), determine its output value $x_i(t)$ at time t.
2. The interconnection rules are as follows:
   (*a*) No two element outputs may be interconnected.

†Functions f and g will be generally continuous.

(*b*) At least k ($k \geq 0$) variables shall be designated as system inputs; no element output may be interconnected to such inputs.
(*c*) At least l ($l \geq 1$) variables chosen from element outputs and system inputs shall be designated as system outputs.
(*d*) Every element input shall be connected to either a system input or to an element output, and every element output shall be connected to some element input or to a system output.

3. The memoryless subsystem, formed by removing all memory elements, *must be* a function system (Definition 2.3-1). That is, even if the memoryless subsystem contains loops, it must be a memoryless system, which gives definite output values given values for all its inputs.

These rules are sufficient in the sense that a given system may not obey the rules and yet may by characterized by a state model. Practically, however, these rules cover the majority of interesting systems that can be put in state model form. Note that according to the rules such systems may have no inputs, in which case they are called *autonomous* (typical examples are oscillators and computer "clocks"), but they must have at least one output. Moreover, they may consist entirely of memory elements or entirely of memoryless elements or of both.

To establish that any system obeying the above rules can be put in state model form consider the following procedure:

1. First pull all the memory elements F_i out of the system, as shown in Fig. 2.4-6. The remaining system is memoryless and has as inputs the system inputs u_i and the memory-element outputs x_i. The outputs of this memoryless system are the system outputs y_i and the memory element inputs m_i.
2. By Rule 3, this memoryless system can be characterized by a function with inputs u_i x_j and outputs y_i, m_j. Split that function into two functions f and g, the former specifying the m_j and the latter specifying the y_i.
3. The system is now in state model form (Fig. 2.4-5) with state variables x_j, state-update variables m_j, inputs u_i, and outputs y_i. The reader should be

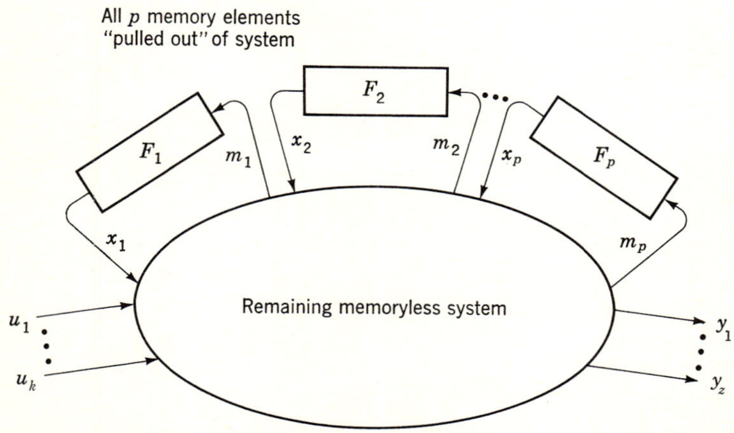

FIG. 2.4-6 Pulling all memory elements out of system.

able to convince himself that initial states $x_j(t_0)$ and inputs $u_i[t_0,t]$, for $t > t_0$, are adequate to establish output values $y_i(t)$, for every choice of t and t_0, $t > t_0$.

Note also that any loops which may be present among memory elements will pass through the memoryless part, thereby making the memory part loop-free.

Concluding Remarks

We have seen that the *state variables* of a system are a finite number of variables $x_1, x_2, \ldots, x_p$, such that knowledge of their values at any time t_1 is necessary and sufficient for determination of the future ($t > t_1$) system output(s), as well as future states, given the future system input(s). The *initial state* of a system is the value or set of values of the state variables at some time t_0 which is considered as a starting or initial time. Thus, the system of Fig. 2.4-3 has an initial state at time t_0 which is the initial value x_0 of the integrator output. As we shall see later, there are several possible choices for state variables for any given system.

We have also seen that not all systems can be characterized by a finite set of state variables, and in some cases state variables are either not known or difficult to identify. For a picturesque example consider the state variables characterizing the state of a used automobile at the time of its purchase. We are interested here in those variables such as condition of tires, engine, body, etc., whose values (good, bad, etc.) determine the future output (will the car last?), given the future input (normal driving). In practice, we always depend on the value of these state variables for our decision to buy or not to buy. Yet we are never sure that we can determine future behavior from inspection of the car's "state."

Systems may also be characterized in terms of their input-output behavior. In our foregoing car example, the input-output characterization entails knowledge of the input since the car's manufacture (how it was driven; did it have accidents, etc.). Such a knowledge could be used to determine future behavior, and the prognosis would most probably be quite accurate in such a case. To the best of our knowledge, neither the state nor the input-output characterization is more general, fundamental, or superior.

Throughout the book we shall use the state characterization as a guide. Thus, in Chaps. 3 to 5 we shall study the memoryless part of the state model with inputs and outputs taking on real-number values—the physical systems which fall in this category are the so-called memoryless block diagram systems and the resistive electric networks. In Chap. 6 we shall deal again with the memoryless part of the model, but this time inputs and outputs will take on values from a finite set—the physical systems in this category are logic or digital networks. In Chaps. 7 to 9 we shall examine systems with memory whose variables take on real-number values, which are characterized by either the state model or by their input-output behavior—the physical systems will be the continuous dynamic systems and the dynamic electric networks. Chapter 10 will examine finite-state machines, i.e., systems which are characterized by the entire state model whose variables take values from a finite set. Chapter 11 will concentrate on elementary continuous systems with delays, from the input-output point of view.

PROBLEMS

2.1-1 Construct three examples of systems that are and three examples of systems that are not in class S of Example 2.1-1. Repeat for class Q of Example 2.1-2.

2.1-2 Define the class of resistive electric networks; that is, find out what a resistive electric network is (by asking some knowledgeable person or by looking in Chap. 4), and then give E and R for that class.

2.1-3 One member of the class of ladder networks of springs is given in the figure. Note that an *odd* number of springs must be used.

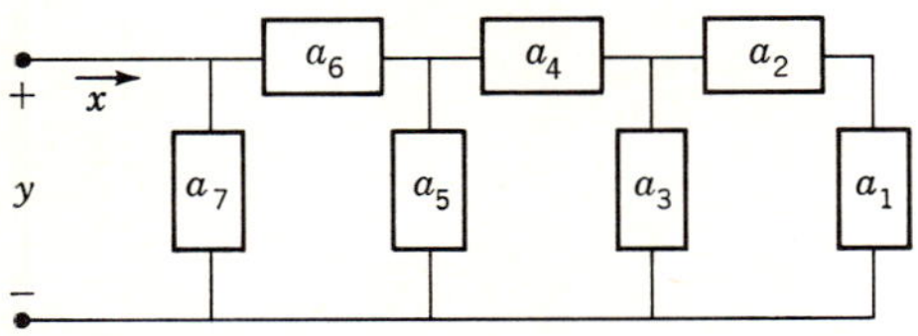

PROB. FIG. 2.1-3

(*a*) In analogy with the class of series-parallel spring networks defined in Chap. 2, give a recursive definition which specifies the class of ladder networks of springs to which this example belongs. You should specify an element set E and a set of rules R.

(*b*) Show that if each of the springs in a ladder of the type defined in (*a*) obeys a linear law,

$$y_i = k_i x_i \qquad i = 1, 2, \ldots, 2n+1$$

where y_i is the force and x_i is the displacement from equilibrium of the ith spring, then the entire network also obeys a linear law,

$$y = Kx$$

and determine K in terms of the individual k_i.
(*Hint:* Use a proof by induction.)

2.1-4 The ideas of Sec. 2.1 are quite general. In this problem we look at a "linguistic" application of these ideas.
A palindrome is a word which reads the same whether it is spelled left to right or right to left. More formally, let us define a palindrome on the alphabet {0,1} as follows:

Basis The following elementary words are each a palindrome: 0, 1, 00, 11.

Induction If S is a palindrome, then each of the words $1S1$, $0S0$ is also a palindrome.

Closure Only the words of the basis and those formed by a finite application of the above induction "rules" are palindromes.

(*a*) Determine which of the following words are palindromes according to the above definition:

1. 0011001100
2. 10101001101
3. (The *null* word consisting of no characters whatsoever.)

(*b*) Give the simplest basis and induction rules that will generate *all* the palindromes made up of the English letters P, S, E, O. That is, your rules should be able to construct *any* word made up of these letters, that reads the same forward or backward. We can think of six such words that have meaning; can you do better?

2.1-5 Consider three classes of systems S, $S1$, and $S2$. These all have the same interconnection rules R given in Table 2.1-1, but they differ in the gates that they include:

Class	*Gates Allowed*
S	AND, NOT, OR
$S1$	AND, NOT
$S2$	OR, NOT

Since class S includes class $S1$ as a special case, it is evident that for each system in class $S1$, there is a system in class S that is equivalent† to it. But is the converse true? Either prove it or find a system in class S for which there is no equivalent system in class $S1$. Repeat for class $S2$.

2.3-1 Consider the differentiator, i.e., a system with input $u(t)$ and output $y(t)$ such that for all time t,

$$y(t) = \frac{du}{dt}(t)$$

assuming that the derivative exists. Is this a memory or memoryless system? Justify your answer.

2.3-2 Using each of the integrator, delay, and differentiator elements (as many times as you wish) and any functions you wish, try to construct (approximately) each of the remaining two elements. For example, using delays, try to construct first a differentiator and then an integrator. Show your constructions as systems, and make any appropriate comments you deem necessary.

2.3-3 Consider the following as systems with the given inputs and outputs. State whether they are memory or memoryless, and identify the set of values U and Y that the input and output can take:

(*a*) A desk lamp where the input is a push button (press it once

†Two systems are *equivalent* if and only if they have identical input-output behavior under all possible inputs.

to turn the lamp on, press it again to turn it off) and the output is the light.

(*b*) A room light where the input is a toggle wall switch and the output is light.

(*c*) An electric typewriter, where the inputs are the keys and the output is a typewritten page.

(*d*) A cigarette vending machine where coins are inputs and cigarettes are outputs.

2.4-1 (*a*) What would be an appropriate set of state variables for the system shown in the figure.
(*Note: a, b, c, d* refer to the signals present in the lines which they are near.)

(*b*) Is there only one possible choice?

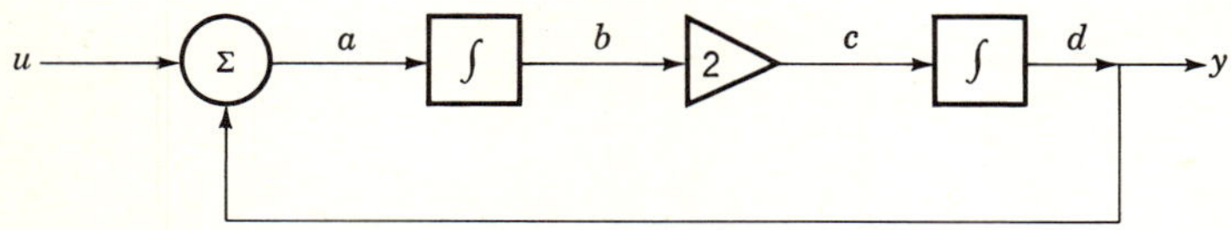

PROB. FIG. 2.4-1

2.4-2 A certain physical system consists of a water reservoir. There are two inputs: the angle u_1 of a faucet ($0 \leq u_1 \leq 2\pi$ radians) and a lever u_2 (u_2 either on or off) which actuates a drain. Water comes out of the faucet at $k_1 u_1$ cubic feet per minute (ft^3/min) where k_1 is a real number. Water drains at a rate k_2 when u_2 is on. The tank weighs as much as one cubic foot of water; call it one water weight unit (wwu); and it has a capacity of 1,000 wwu, i.e., no more water can come into the tank if it is full, regardless of what u_1 is.

(*a*) Draw a system diagram in state form, using an integrator and whatever functions are necessary so that you can relate the output, which is the weight y of tank and water at time t, to u_1 and u_2, which will generally change with time.

(*b*) Identify precisely your state variable x, the state update m, and the sets U, Y, X, and M for your system.

(*c*) Identify the functions f and g that relate input and state to state update and output, respectively.

2.4-3 Using unit-length ideal delays, summers, and scalers, design a linear system whose output is a rectangular pulse of length 3 when the input is a rectangular pulse of length 2.

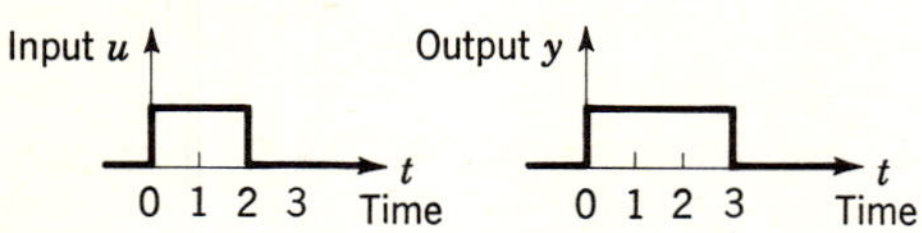

PROB. FIG. 2.4-3

chapter 3

Continuous Memoryless Systems

This chapter treats a class of continuous systems which are characterized as memoryless in the state model of Chap. 2, and which are also known as block diagram or function-block systems and are made up of elements such as scalers, summers, and multipliers. The main ideas of the chapter are (1) the distinction between explicit and implicit (with loops) systems and the formulation of equations that characterize these systems, (2) linearity and some of its important consequences, and (3) the input-output equivalence of systems which may have different internal structures. Also included are several electrical engineering examples which illustrate these ideas.

3.1 INTRODUCTION

This chapter is concerned with the memoryless portion of the general state model, that is, with systems whose output signals at any time t depend only on the values of the input signals at the same instant of time t. As stated in Definition 2.3-3, the input-output behavior of such memoryless systems is represented mathematically by functions or relations. In what follows we shall concentrate on systems which are described by *continuous* real-valued functions or relations, and we shall postpone until Chap. 6 our study of discrete logic systems whose input-output characteristics are described by bivalued (Boolean) functions.

We start with some basic definitions regarding function blocks and the rules of interconnecting these blocks to form memoryless block diagram systems. We then demonstrate that the resultant systems may be *explicit* or *implicit*, and we examine ways for characterizing them mathematically in terms of the mathematical description of the constitutent elements and the manner in which they are interconnected. We proceed to the special case where the functions characterizing the constituent elements are linear, and we examine the implications of this linearity condition on the mathematical characterization of linear memoryless systems. We conclude our examination of function-block systems with a discussion of *equivalence* and *approximate equivalence*, i.e., the relationship of structurally different systems having identical or approximately identical input-output behavior.

This chapter provides the basis for many subsequent chapters. For example, our study of function blocks leads us into the study of circuit elements and electric networks of Chap. 4. There, we will see that resistive networks are special physical systems that can perform memoryless mathematical operations by relating physical quantities such as current and voltage. We will also see how circuit elements are used to construct function-block systems. The implicit systems discussed in this chapter can be solved by iterative techniques, the subject of Chap. 5. The ideas of linearity and equivalence discussed in this chapter are used extensively throughout the book for continuous or discrete, and memoryless or memory systems.

3.2 DEFINITIONS

The class of *single-output continuous memoryless block diagram systems* is defined by an element set E and a set of interconnection rules R. The element set is as shown in Table 3.2-1. It consists of *function blocks*, each of which is described by a piecewise-continuous† function which is finite-valued for finite arguments. Associated with each function block is one *output* y, which "takes on" real values, and one or more *inputs* x_i that "take on" real values. The output and inputs are called *terminals* of the function block. When there are more than one input to the function block, say x_1 and x_2, we will sometimes denote the function, or the block, by $f(.\,,.)$ where the two dots indicate the

†f is piecewise-continuous if it has only a finite number of discontinuities, or "jumps."

TABLE 3.2-1 Element Set E

E	
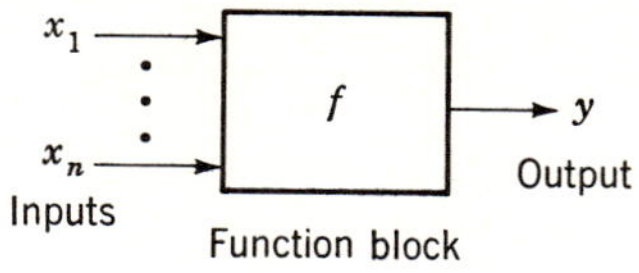	x_i and y are real numbers $y = f(x_1, \ldots, x_n)$

number and order of the arguments of f. In such a case, f will be a function from pairs of real numbers to real numbers. For three arguments we will use three dots, and so forth.

The set of interconnection rules R is as shown in Table 3.2-2. Note that in contrast to the interconnection rules of Table 2.1-1, these rules permit loops. In other words, it is possible to trace a closed path following arrows and passing through elements.

TABLE 3.2-2 Rules of Interconnection R

R
1. No two function-block outputs may be interconnected.
2. Every function-block input must either be connected to some function block output, or it shall be an input to the interconnected system.
3. Exactly one function block output will be the output of the interconnected system. Every other function block output must (and the above system output may) be interconnected to some function block input.

The mathematical interpretation of an interconnection is quite simple, as shown in Table 3.2-3. With E and R we can determine whether a system is in the class of continuous memoryless block diagram systems. For example, the systems of Fig. 3.2-1a and b are in the class, whereas the systems of Fig. 3.2-1c, d, and e are not in that class since they violate respectively Rules 1, 2, and 3 of R.

In Table 3.2-1 we saw that each function block is characterized by a function, which gives the value of the output of the block in terms of its input(s). Thus, the three function blocks of Fig. 3.2-2a are characterized by the following equations:

$$a = f(x) \tag{1a}$$

$$d = g(b,c) \tag{1b}$$

$$y = h(e,j,k) \tag{1c}$$

TABLE 3.2-3 Meaning of Interconnection

Whenever two function block terminals are interconnected, the variables representing these terminals are constrained to have identical values.

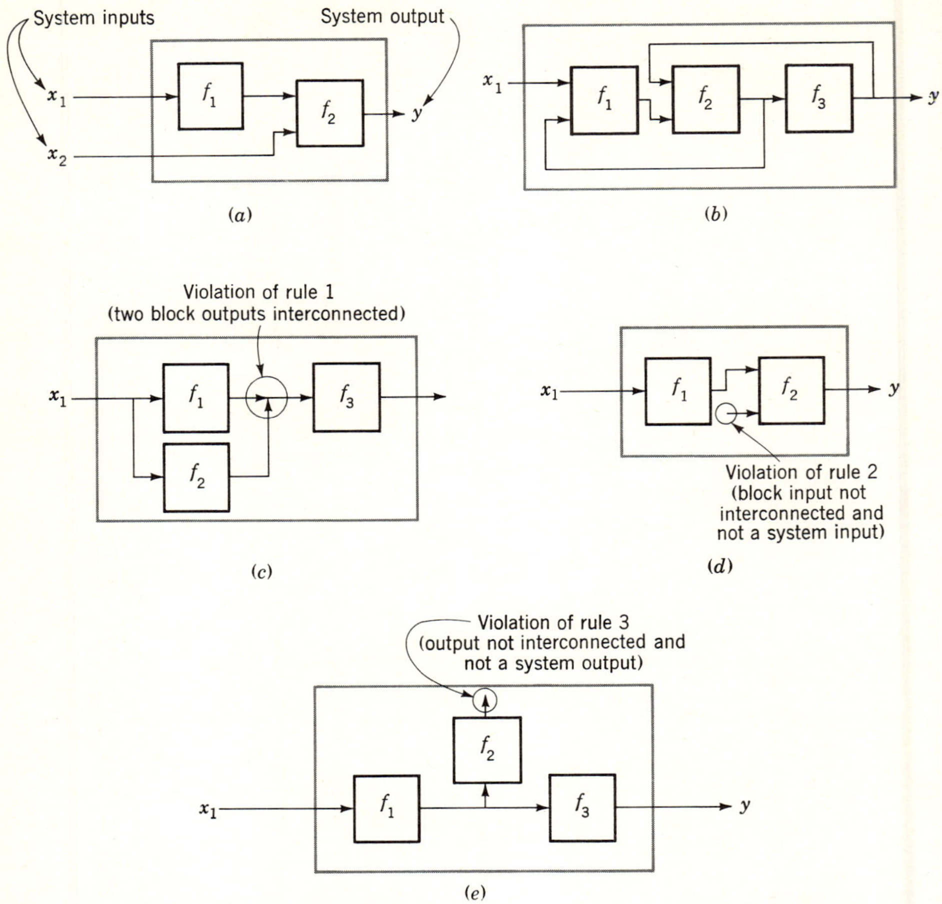

FIG. 3.2-1 Examples of systems that are and are not in the class of continuous memoryless block diagram systems.

When these blocks are interconnected, as in Fig. 3.2-2*b*, certain equality constraints are introduced by the interconnection process. They are summarized by the following equations:

$$a = b \quad \text{and} \quad a = e \tag{2a}$$

$$x = c \quad \text{and} \quad x = k \tag{2b}$$

$$d = j \tag{2c}$$

The first of these equations says that the value of the output of the f block is equal to the value of the top inputs of the g and h blocks. In engineering language, the output of the f block *is fed to* and acts as *an input to* the g and h blocks, respectively. Equations (1) and (2) may be combined by functional composition to give the system output y in terms of the system input x, through

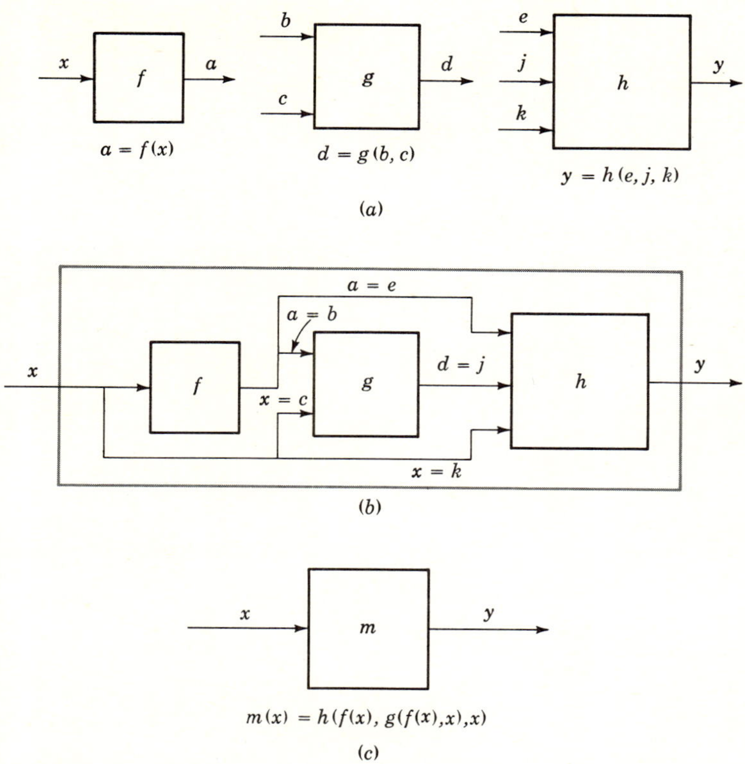

FIG. 3.2-2 Three distinct function blocks and an interconnected system.

elimination of a, b, c, d, e, j, and k. This yields

$$y = h(f(x),g(f(x),x),x) \tag{3}$$

Note that if we give a value for x, then since we know the functions f, g and h, all three arguments of h are specified, and so is the value of y.

If, for example, h were the product function, g the sum function, and f the squaring function, that is, if

$$h(e\,j,k) = ejk$$

$$g(b,c) = b + c$$

$$f(x) = x^2$$

then Eq. (3) would give for y the following:

$$y = (x^2)\,(x^2 + x)\,(x)$$

or (4)

$$y = x^5 + x^4$$

Several observations are now in order:

First, note that for every value of x, y is assigned a value by Eq. (3); hence we can write

$$y = m(x) \tag{5}$$

where m is a new function that can be constructed from h, f, and g. In our above example, $m(.)$ is the function that given a value for its argument, gives the sum of the fifth and fourth powers of that value as an answer. This means that the interconnected system of Fig. 3.2-2*b* may be represented by a single function block (Fig. 3.2-2*c*). This points out the recursive nature of the system class, since function block m could be used as an element of a larger system (perhaps in place of f in the system of Fig. 3.2-2*b*) and so forth, to an arbitrarily large number of recursion levels. As we discuss in Sec. 3.4, this step-by-step characterization of an interconnected system by one function relating inputs to output is always possible if the system is explicit (no loops), but not always if it is implicit (i.e., if it has loops).

Second, systems with more than one output can be constructed and analyzed, even though they are not possible according to E and R, simply by juxtaposing a number of single-output systems having common inputs, as shown in Fig. 3.2-3*a*. Alternatively, they may share some common blocks, as in Fig. 3.2-3*b*.

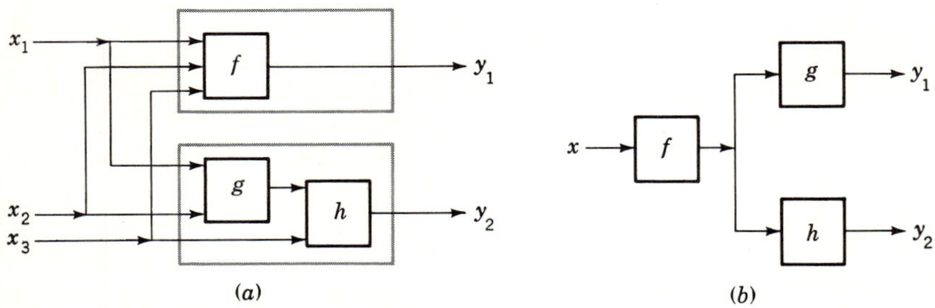

FIG. 3.2-3 Two-output systems.

Third, the function blocks that make up these systems are time-invariant or stationary; that is, functions f, g, h, etc., do not change with time, although the values of their input and output variables will generally change with time. In fact, time is unknown to the function blocks and to the systems of this chapter,† since it is only the value of the variables and the nature of the functions that determine the output value, regardless of the time at which these values are attained.

†This is an idealization. No physical system is truly memoryless, so that when actual function blocks are used, time does generally play an important role. However most physical systems can be analyzed by a decomposition into idealized memoryless and memory systems.

3.3 MODELING PHYSICAL DEVICES WITH FUNCTION BLOCKS

The function blocks and interconnected systems that we have been discussing are abstractions of physical entities which may have similar names. For example, there are physical engineering elements called *operational amplifiers* or *opamps* which have a physical input (a couple of wires) and a physical output (a couple of wires) as shown in Fig. 3.3-1*a*. As long as a voltage† x applied to the input of such a block varies slowly with time, the output voltage y will be given by the function f of Fig. 3.3-1*b*, i.e., $y = f(x)$.

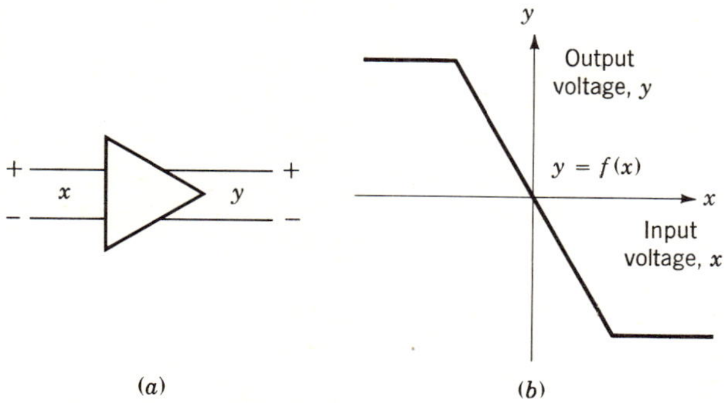

FIG. 3.3-1 An operational amplifier and its corresponding function for slow time variations.

Alternatively, a mechanical lever (Fig. 3.3-2*a*) can be represented by a function block, since the output y (lever output position) is related to the input (lever input position) by $y = kx$ where k is $-l_2/l_1$ (Fig. 3.3-2*b*). Note however that this modeling of the lever is not exact; some of the difficulties are explained below:

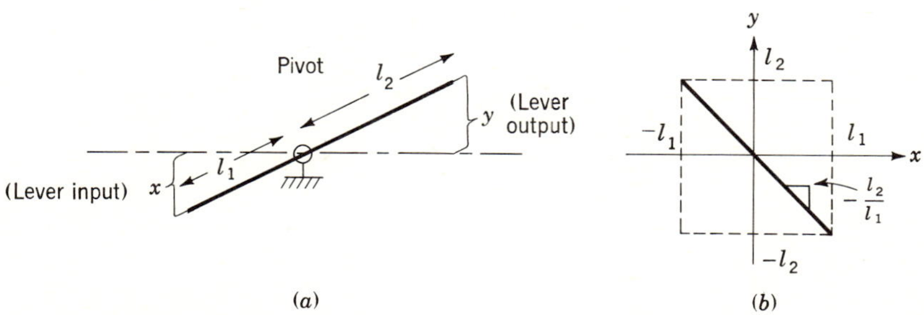

FIG. 3.3-2 A mechanical lever and its input-output behavior.

†Voltages and currents are defined in Chap. 4; knowledge of these quantities is not essential for studying this chapter.

1. The function is undefined for $x > l_1$ or $x < -l_1$, since the lever would have to "break" for x to attain such values.
2. A physical lever will bend under the load that it supports, and if the input x is caused to vary very rapidly, then the output y will no longer be given by kx (k constant) but will depend on the acceleration d^2x/dt^2 and the physical parameters of the lever (such as elasticity and inertia).
3. Numerous other phenomena such as fatigue, aging of the lever with repeated use, and different variations of l_2 and l_1 with temperature and humidity will make the input-output behavior of the lever different from the modeled behavior $y = kx$. In principle, we could try to account for all these effects with a more complicated model of the lever. With due care, however, we can use the simple model ($y = kx$) to predict with reasonable or acceptable accuracy what a very complicated system of interconnected levers will do.

Since electric networks have not yet been introduced in this book, a lever is much easier to understand at this stage. The above limitations, however, are present, regardless of the physical process that is modeled. For example, in electric networks, the breakage of the lever corresponds to burning, "poping," and smoking when certain values of voltage are exceeded; the effects of electric loading are very similar to the lever bending under mechanical load; and environmental changes are very often responsible for faulty operation of electronic equipment.

There is little that the student can do at this stage to guard against mismodeling. In due time, the intangibles called experience, background, general knowledge, and common sense will guide him to distinguish the relevant from the irrelevant in choosing and using a model. Meanwhile, we need only understand that the function blocks represent in an idealized and simplified way the behavior of a physical device.

The variables representing inputs and outputs correspond to physical quantities such as voltage, current, displacement, the contents of a computer register, velocity, pressure, and temperature. For the purposes of this chapter, however, it will be adequate to consider all variables as taking on values which are real numbers.

3.4 EXPLICIT AND IMPLICIT SYSTEMS

Explicit systems and *implicit* systems with a single output are constructed by the previously discussed rules R of Table 3.2-2, meaning that the former have no loops while the latter have at least one loop. We will make these ideas more precise, and independent of the "picture" of the system block diagram, much like a computer program would have to do in order to "understand" whether a system has or does not have loops.

We start by representing a block diagram system such as that of Fig. 3.2-2*b* by an *interconnection list* as follows:

h: f, g, x
f: x
g: f, x

Here, each function block occupies a line, and the name of the function is followed by a colon and by an ordered sequence of the names of the function block outputs or system inputs connected to that function block. For example, the first line says that function block h (of Fig. 3.2-2b) is connected to the output of function block f, to the output of function block g, and to the system input x, in that order.

We say that such a list is *sorted, if the lines can be rearranged, so that in every line, every name appearing to the right of a colon appears to the left of a colon somewhere above that line, or is a system input.* The above list is not sorted, because names f and g, appearing to the right of the colon on the first line, do not appear above that line. We can, however, sort the list by rearranging the lines as follows:

f: x
g: f, x
h: f, g, x

The list is now sorted, since in every line, every name to the right of a colon appears above that line to the left of a colon, or is a system input. In other words, a sorted list orders the function blocks in such a way that every input to a function block is either the output of some function block above it in the list or is a system input. It should be clear that every system block diagram constructed in accordance with Rules R of Table 3.2-2 can be represented by a list, although the list will not always be sortable.

Some examples of system block diagrams and their interconnection lists are shown in Fig. 3.4-1. Note that the list of the second of these systems cannot be sorted regardless of how the lines are rearranged. This unsortability of a list tells us that there is at least one loop in the system, and permits us to define explicit and implicit systems as follows:

DEFINITION 3.4-1 An *explicit* single-output memoryless system is constructed according to the rules of Table 3.2-2 and *can always* be described by a sorted interconnection list.

DEFINITION 3.4-2 An *implicit* single-output memoryless system is constructed according to the rules of Table 3.2-2, and *cannot* be described by a sorted interconnection list.

Implicit systems such as the one of Fig. 3.4-1b have at least one loop,

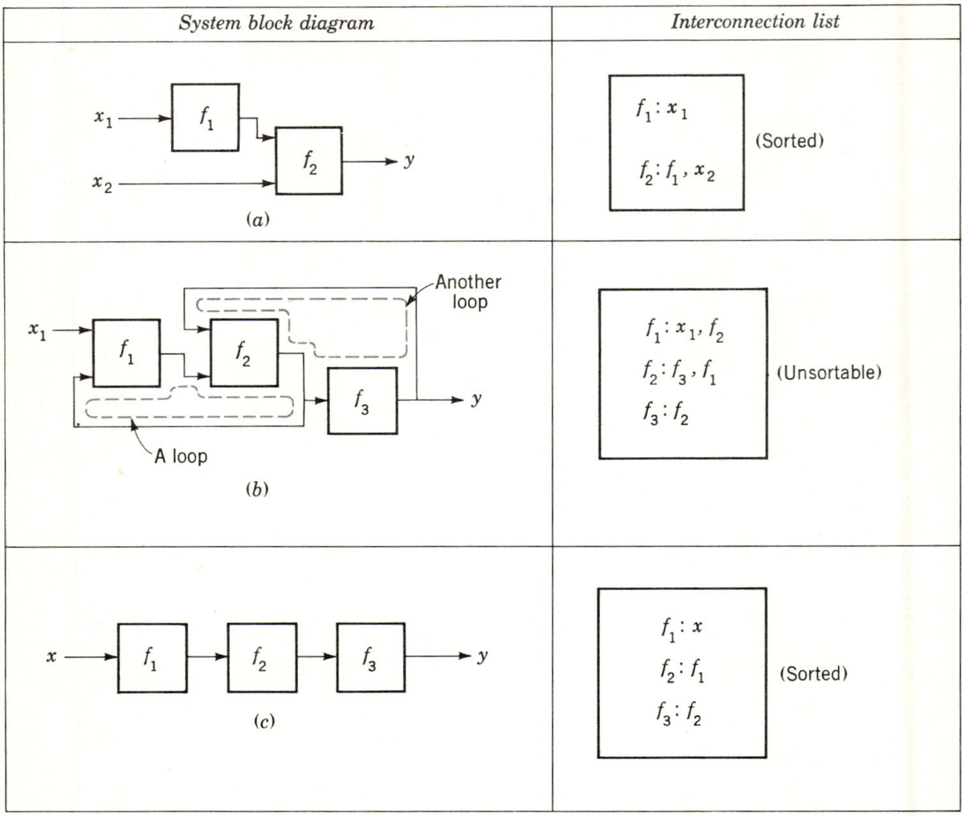

FIG. 3.4-1 Some systems and their interconnection lists.

i.e., a directed closed path which, starting from the output of one block A, and (possibly) passing through other blocks, returns as an input to A. It is the presence of such a loop that prevents the interconnection list of an implicit system from being sorted.

Observe that the above definitions can be extended in a straightforward manner to multiple-output systems. Observe further that in small or simple systems, it will be generally obvious, *by inspection of the system diagram*, whether a given system has or does not have loops, accordingly whether it is explicit or implicit. The formality of sorting lists gives us a picture-independent way of determining whether any given system is explicit, and becomes practically useful in treating very large systems† where visual inspection may lead to errors.

†Several computer programs that analyze block diagram systems start analysis by sorting a user-defined list which represents the given system.

3.5 FORMULATION AND SOLUTION OF SYSTEM EQUATIONS

Given a memoryless system, and the functions corresponding to its constituent function blocks, we wish to write equations, hopefully as few as necessary, that fully describe the input-output behavior of that system—this is the *formulation* task. After the equations have been written, we want to solve them; that is, we wish to find the value of the output, given values for the inputs —this is the *solution* task.

Explicit Systems

The formulation of equations for explicit systems is straightforward. We simply state the input-output equations for each function block in the order specified by the sorted list. Thus, for example, for the system of Fig. 3.5-1*a* with sorted list shown in Fig. 3.5-1*b* we give the following formulation:

$$a = f_1(x_1) \tag{1a}$$

$$b = f_2(a,x_2) \tag{1b}$$

$$y = f_3(a,b,x_2) \tag{1c}$$

Note that each line of Eqs. (1) corresponds to each line of the sorted list of Fig. 3.5-1*b*. Intermediate variables *a* and *b* were introduced to facilitate the formulation.

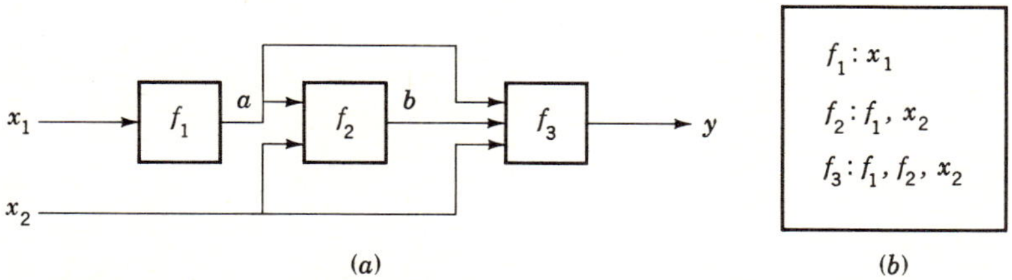

FIG. 3.5-1 An explicit system and its interconnection list.

The solution task is equally straightforward. Given values for the system inputs x_1 and x_2 we need only evaluate *in the order shown* the formulated equations (1). Thus, for example, if f_1 were the square function, f_2 the sum function, and f_3 the maximum function, i.e., if

$$\begin{aligned} f_1(x_1) &= x_1^2 \\ f_2(a,x_2) &= a + x_2 \\ f_3(a,b,x_2) &= \max(a,b,x_2) \end{aligned} \tag{2}$$

then the formulated equations (1) would become

$$
\begin{aligned}
a &= x_1^2 \\
b &= a + x_2 \\
y &= \max\,(a,b,x_2)
\end{aligned}
\tag{3}
$$

Thus, if $x_1 = -2$, $x_2 = 1$, we could substitute these values in Eqs. (3) to obtain

$$
\begin{aligned}
a &= (-2)^2 = 4 \\
b &= 4 + 1 = 5 \\
y &= \max\,(4,5,1) = 5
\end{aligned}
\tag{4}
$$

and the answer would be $y = 5$. In explicit systems, this approach of evaluating the output in the order of the list is always guaranteed to work, because an explicit system is described by a *sorted* list; hence, whenever we evaluate a function, we are guaranteed that all its arguments have been already evaluated. This is exactly the way in which several computer programs evaluate block diagram systems; i.e., they establish a "compiled" program which is obtained from the sorted list and which would read something like this, except, of course, that it would be in an appropriate computer language. Given x_1 and x_2:

1. Call subroutine that evaluates x_1^2, and store answer in a.
2. Call subroutine that evaluates $a + x_2$, and store answer in b.
3. Call subroutine that finds maximum of a, b, x_2, and store answer in y.

It should be clear to the reader that a computer program could be set up to do this solution task for *very large* and complicated systems with thousands of interconnections, as long as the systems are *explicit*.

Of course, it is not always desirable, or even necessary, to go through a formal list-sorting procedure in order to formulate and solve explicit-system equations. On the contrary, *the reader is urged to formulate and solve such equations by inspection of the system diagram, using as few variables as possible.*

Example 3.5-1 *The Multiplier of Example 2.3-1* Consider the multiplier of Fig. 2.3-3, which is reproduced in Fig. 3.5-2, with names assigned to each function block, and hence to each function-block output.

Formally, this system can be described by the list of Fig. 3.5-3*a*,

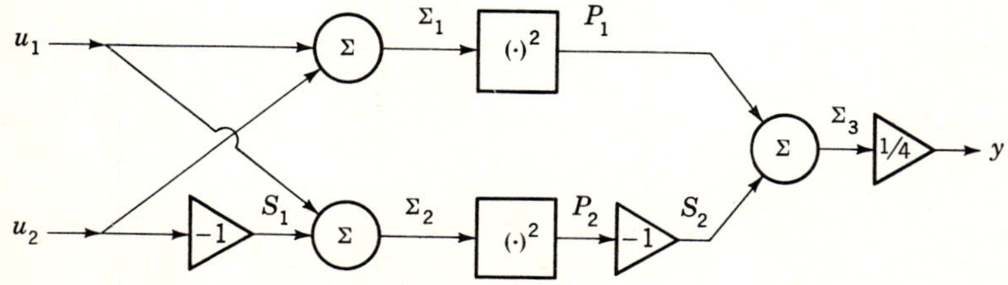

FIG. 3.5-2 The multiplier.

(a)	(b)	(c)	Row
Σ_1: u_1, u_2	Σ_1: u_1, u_2	$\Sigma_1 = u_1 + u_2$	1
P_1: Σ_1	P_1: Σ_1	$P_1 = (\Sigma_1)^2$	2
Σ_3: P_1, S_2	S_1: u_2	$S_1 = -u_2$	3
y: Σ_3	Σ_2: u_1, S_1	$\Sigma_2 = u_1 + S_1$	4
S_1: u_2	P_2: Σ_2	$P_2 = (\Sigma_2)^2$	5
Σ_2: u_1, S_1	S_2: P_2	$S_2 = -P_2$	6
P_2: Σ_2	Σ_3: P_1, S_2	$\Sigma_3 = P_1 + S_2$	7
S_2: P_2	y : Σ_3	$y = 1/4\,\Sigma_3$	8

FIG. 3.5-3 Lists describing multiplier.

which incidentally has as many rows as there are function blocks in Fig. 3.5-2; that list is not sorted, since the third line includes S_2 which does not appear above that line and is not a system input. The list of Fig. 3.5-3*a* can *be easily sorted,* as shown in Fig. 3.5-3*b;* hence the system is explicit. The actual functions, shown in Fig. 3.5-2, are then listed in Fig. 3.5-3*c* in the order of the sorted list. Finally, the expression for Σ_1 at the top row of Fig. 3.5-3*c* is substituted in the expression for P_1 in row 2, and, in general, substitution is carried downward in the list as shown below:

$$P_1 = (u_1 + u_2)^2 \qquad \text{(rows 1 and 2)}$$

$$S_2 = -(u_1 - u_2)^2 \qquad \text{(rows 3, 4, 5, and 6)}$$

$$\cdot y = 1/4\,(P_1 + S_2) \qquad \text{(rows 7, and 8)}$$

resulting in the desired expression:

$$y = 1/4[(u_1 + u_2)^2 - (u_1 - u_2)^2] \qquad \text{(all rows)}$$

This is the desired expression because it contains the least number of variables, and it relates output y to the two inputs u_1 and u_2.

Informally, the same result can be obtained by inspection of Fig. 3.5-2 where, starting from the output and progressing toward the inputs, our thinking may proceed along the following lines:

$y = 1/4$ of the sum of two variables, one of which is $(u_1 + u_2)^2$ and the other is $-(u_1 - u_2)^2$

Thus;

$$y = 1/4[(u_1 + u_2)^2 - (u_1 - u_2)^2]$$

Moreover, it is evident from a visual inspection of Fig. 3.5-2, that the system does not have any loops and hence is *explicit.*

Implicit Systems

Formally, the formulation of equations for implicit systems is accomplished by temporarily "breaking" the loops to obtain an explicit system, writing the equations for such a system, and then appending as many equations as there are loops, to take into account the fact that the loops are closed. This is best illustrated by an example:

Example 3.5-2 *Implicit-system Equations* Consider the system of Fig. 3.5-4*a*.

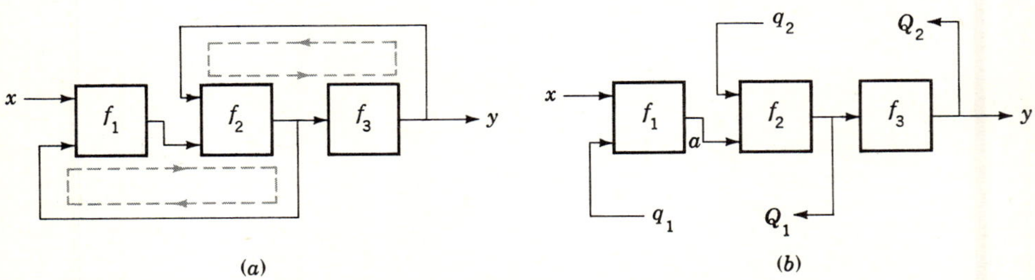

FIG. 3.5-4 An implicit system and the loopbreaking procedure.

The interconnection list of this system cannot be sorted; it is as follows:

f_1:	x, f_2
f_2:	f_3, f_1
f_3:	f_2

To formulate the system equations, we break the two loops of Fig. 3.5-4*a* as shown in Fig. 3.5-4*b*; i.e., we introduce temporary system inputs q_1 and q_2 in addition to x, and temporary intermediate variables Q_1 and Q_2 in addition to y. The list of this new system can be sorted as follows:

f_1:	x, q_1
f_2:	q_2, f_1
f_3:	f_2

where q_1 and q_2 are considered system inputs. As we did for explicit systems, we can now formulate equations for the system of Fig. 3.5-4*b*. They are:

$$\begin{aligned} a &= f_1(x,q_1) \\ Q_1 &= f_2(q_2,a) \\ Q_2 &= f_3(Q_1) \qquad \text{also} \qquad Q_2 = y \end{aligned} \tag{5}$$

Next, we "close the loops" with the following two constraints:

$$Q_1 = q_1$$
$$Q_2 = q_2 \tag{6}$$

Equations (5) and (6) can be combined, and intermediate variables other than the q_i can be eliminated to express the variables at the loop "breaks" in terms of themselves. Thus:

$$q_1 = f_2(q_2, f_1(x,q_1))$$
$$q_2 = f_3(q_1) \qquad \text{and} \qquad y = q_2 \tag{7}$$

Equations (7), or Eqs. (5) and (6), give the values of several variables, the q_i, in terms of themselves and the system inputs. We cannot evaluate the q_i until we know the q_i—a vicious circle that can be resolved by *iteration*, the method for solving such equations discussed in Chap. 5. The form of Eq. (7) which we will summarize here in terms of one variable as

$$q = f(q) \tag{8}$$

is known as an *implicit form*.† Until we treat this equation in the detail that it deserves, we remark here that it will normally be solved by assuming for q any initial value q_0, and thereafter calculating

$$q_1 = g(q_0)$$
$$q_2 = g(q_1) \tag{9}$$
$$\cdots\cdots$$
$$q_{i+1} = g(q_i)$$

where g is a function (the iteration rule) which takes into account f. When there are several loops, as in this example, then starting from any initial values for q_1 and q_2, say q_{10} and q_{20}, we will compute Q_1 and Q_2 by walking through Eq. (5); these values will be used in turn by the iteration rule to establish the next estimates q_{11}, q_{21}; and this process will continue until, and if, Eqs. (6) are satisfied either exactly or to a desired accuracy. The details of this iterative approach will be discussed in Chap. 5.

Before leaving this example, observe that *one* "break" at the output of f_2 would have been adequate for sorting the interconnection list. Letting the variable at that break be q would have resulted in the following single equation:

$$q = f_2(f_3(q), f_1(x,q))$$

which is, of course, in implicit form. Solution of this equation for q permits explicit solution for y, since

$$y = f_3(q)$$

Naturally, we would like to know, especially for large and complicated systems, the minimum number of breaks that permit sorting of the interconnection list.

†In contrast to the *explicit form* $y = f(x)$, which permits direct evaluation of y, given x.

Unfortunately, the only way we know of answering this question is by exhaustively trying all possible breaks.†

The following is a practical example of a *feedback system,* as systems with loops are often called. Given a function block f, and a linear difference sensor, this system approximates the inverse function f^{-1}. With this technique it is possible to construct such things as a "square-root" system, given a squarer block, or an "antilogarithm" system, given a logarithm block.

Example 3.5-3 *Use of an Implicit System to Approximate Inverse Functions* Block f in Fig. 3.5-5a behaves as a very sensitive linear difference sensor, so that

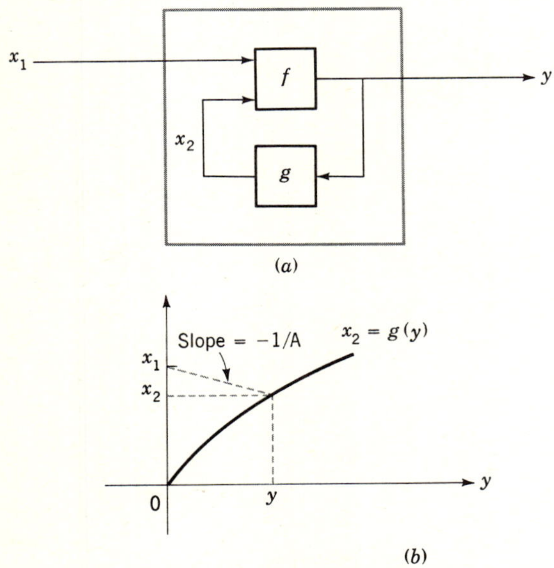

FIG. 3.5-5 Interconnection of systems f and g to form g^{-1}.

its output is equal to some large positive real number A times the difference of its two inputs. That is,

$$f(x_1,x_2) = (x_1 - x_2)A \qquad \text{and} \qquad A \gg 1 \tag{10}$$

Block g is characterized by function g, i.e., $x_2 = g(y)$. Formulating the equations that describe this system, we obtain

$$y = [x_1 - g(y)]A \tag{11}$$

Equation (11) may be rearranged in the form

$$x_1 = g(y) + \frac{y}{A} \tag{12}$$

†The problem is directly analogous to selecting the minimum number of prime implicants—a problem in Boolean function minimization which involves an exhaustive search.

which leads to the geometrical construction of Fig. 3.5-5*b*. The curve represents the functional relationship g between x_2 and y. To find the value of y associated with some particular value of the input variable x_1, we simply start at the point x_1 on the vertical axis in Fig. 3.5-3*b* and project a construction line of slope $-1/A$ until it intersects the curve. The horizontal coordinate of the intersection point is then the desired value of y. It is apparent from the construction of Fig. 3.5-5*b* that the choice of a large value for A makes x_2 nearly equal to x_1, so that the functional dependence of y upon x_1 is very nearly the inverse of function g. For example, if in some interval of values of y, the function $g(y)$ behaves like the square of y, then y may be expected to approximate the square root of x_1. Sensitive difference sensors such as f are called *operational amplifiers* or *opamps*. Values of A of over 10^6 are easily obtained. This approach is practical since, for example, a function block that squares its input may be more economical or easily constructed than a block that obtains the square root.

Some observations on implicit systems are in order. First, implicit systems are not necessarily function systems. That is, in some instances a solution may not exist, while in others more than one output value may be possible, given some values for the inputs. An example of the former is the system of

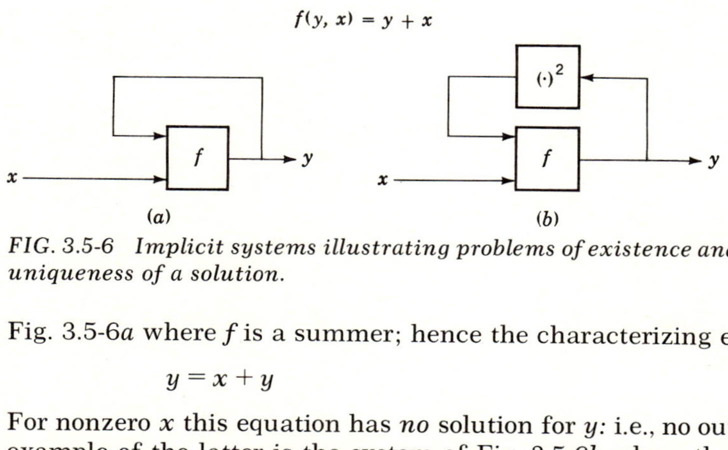

FIG. 3.5-6 Implicit systems illustrating problems of existence and uniqueness of a solution.

Fig. 3.5-6*a* where f is a summer; hence the characterizing equation is

$$y = x + y$$

For nonzero x this equation has *no* solution for y: i.e., no output is possible. An example of the latter is the system of Fig. 3.5-6*b* where the upper block represents the squaring function. Here, the input-output equation for the system is

$$y = y^2 + x$$

This equation is satisfied by *two* possible values of y; for example, if $x = 0$, $y = 0$ or $y = 1$. The above two systems illustrate two basic questions known as the *existence* and *uniqueness* of solutions to equations. These questions are treated in some detail in Chap. 5.

In terms of our definitions of memoryless systems, discussed in Chap. 2, we see that *implicit systems are generally characterized by mathematical relations* (see Definition 2.3-3) and not necessarily by functions. Thus, the mathematical relation characterizing the system of Fig. 3.5-6*a* is the null relation, that is, the relation which contains *no* input-output pairs, while the relation characterizing the system of Fig. 3.5-6*b* has the pairs (0,0) and (0,1) and hence is not a function.

Second, observe that *some implicit systems can give rise to explicit equation forms.* In other words, the existence of loops in the system diagram does not mean that the resulting equation will necessarily be in the implicit form, $q = f(q)$. For example, the system of Fig. 3.5-7, although implicit, is characterized by the explict form $y = {}^3\!/_7 x$, as the reader may easily verify.

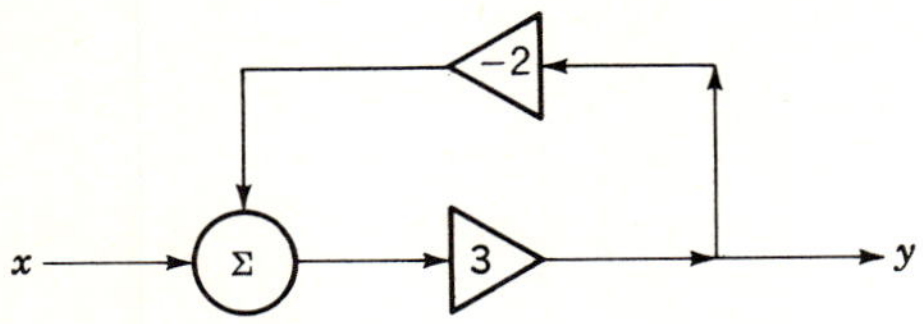

FIG. 3.5-7 An implicit system with an explicit solution.

Finally, observe, as in the case of explicit systems, that it will not be generally necessary to walk through the above formal procedure of loop breaking and list sorting in order to formulate equations. If the system is small, visual inspection will normally be adequate to establish the desired result. For example, in characterizing the system of Fig. 3.5-6*b*, our thinking might very well proceed as follows:

> $y = f$ of two variables, one being y^2, the other being x. Hence,
>
> $y = f(y^2, x)$

Nevertheless, our knowledge of the more formal procedures guarantees that we can handle more complex systems without resorting to intuition, or, alternatively, that we can write computer programs that perform the task of formulating system equations.

3.6 LINEARITY

A simple, yet powerful, constraint on memoryless continuous systems is *linearity*. We define linearity for a single-input single-output function system as follows:

DEFINITION 3.6-1 A system with input x and output $y = f(x)$ is *linear* if and only if $f(ax_1 + bx_2) = af(x_1) + bf(x_2)$, for all real values a, b, x_1, x_2.

This is an *operational* definition: i.e., it concerns the input-output operation. It is not a *structural* definition, since it says nothing about what function blocks make up the system under test. For example, the system of Fig. 3.6-1*a* is characterized by $y = 3(x + y)$, which can be further rearranged to give $y = -{}^3\!/_2 x$. Evidently, the test of linearity is met, since

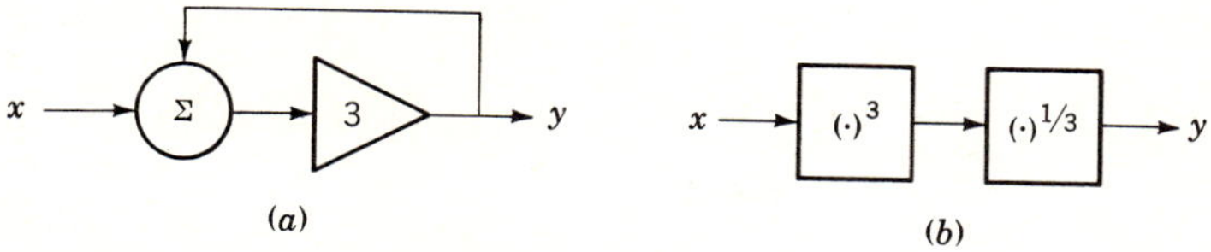

FIG. 3.6-1 Structurally and operationally linear systems.

$$-3/2(ax_1 + bx_2) = -\frac{3a}{2}x_1 - \frac{3b}{2}x_2 = a(-3/2x_1) + b(-3/2x_2)$$

The system of Fig. 3.6-1*b* is characterized by

$$y = (x^3)^{1/3}$$

which can be rearranged to give

$$y = f(x) = x$$

This system is also linear, since for all real values of a, b, x_1, and x_2;

$$f(ax_1 + bx_2) = af(x_1) + bf(x_2) = ax_1 + bx_2$$

Note, however, that the system of Fig. 3.6-1*a* is *structurally linear;* i.e., all its constituent function blocks obey the linearity constraint; while the system of Fig. 3.6-1*b* is not structurally linear $[(ax_1 + bx_2)^3 \neq ax_1^3 + bx_2^3]$.

Every system that is structurally linear is also operationally linear† (when its solution exists), but the converse is not true, as demonstrated by the system of Fig. 3.6-1*b*.

Linear systems with more than one input are defined by a straightforward extension of Definition 3.6-1, as illustrated here for a two-input system.

DEFINITION 3.6-2 A system with two inputs x_1, x_2 and output $y = f(x_1,x_2)$ is *linear* if and only if $f(aq_1 + bp_1, aq_2 + bp_2) = af(q_1,q_2) + bf(p_1,p_2)$ for all real values a, b, q_1, q_2, p_1, p_2.

If such a multiple-input system is linear, then it is also linear with respect to each input, when all other inputs are set to zero. This follows from Definition 3.6-2, for if we set the second input to zero, i.e., $p_2 = q_2 = 0$, we have

$$f_0(aq_1 + bp_1) = af_0(q_1) + bf_0(p_1) \tag{1}$$

where we denote by f_0 the single-variable function resulting from f, when its second argument is zero. That is, for all x,

$$f_0(x) = f(x,0) \tag{2}$$

Likewise, setting the first input to zero yields

$$f_1(aq_2 + bp_2) = af_1(q_2) + bf_1(p_2) \tag{3}$$

where for all x,

$$f_1(x) = f(0,x) \tag{4}$$

†Provided that the system has no other input that is a constant.

Note, however, that although Eqs. (1) through (4) are a consequence of Definition 3.6-2, *the converse is not true*. That is, a system could satisfy these equations but not the definition. Such a system, for example, could have an input-output equation,

$$y = x_1 \qquad \text{if } x_2 = 0$$

$$y = x_2 \qquad \text{if } x_1 = 0$$

$$y = x_1^2 + x_2^2 \qquad \text{otherwise}$$

We proceed now with an important consequence of linearity, which we state as a theorem.

THEOREM 3.6-1 A linear memoryless system f with inputs $x_1, \ldots, x_n$ and output y can always be characterized by the input-output relation

$$y = \sum_{i=1}^{n} a_i x_i$$

where the a_i are real constants.

PROOF: The proof, which will be demonstrated for $n = 2$, rests on Definition 3.6-2. Let $a = x_1$, $b = x_2$, $q_1 = p_2 = 1$, $q_2 = p_1 = 0$ in that definition. The result is:

For all x_1,x_2,

$$f(x_1,x_2) = x_1 f(1,0) + x_2 f(0,1)$$

Denoting $f(1,0)$ by a_1 and $f(0,1)$ by a_2 yields the desired result:

For all x_1,x_2

$$f(x_1,x_2) = a_1x_1 + a_2x_2 \tag{5}$$

If the system had more than two inputs, then Definition 3.6-2 would be appropriately extended, and the above process would have resulted, as it did in Eq. (5) above, in the desired result:

$$y = \sum_{i=1}^{n} a_i x_i \tag{6}$$

An obvious, yet important, consequence of linearity is given by the following corollary.

COROLLARY 3.6-1 Principle of Superposition The output of a single-output linear system with n inputs, $x_1, \ldots, x_n$ is equal to the sum of the outputs of n replicas of this system, with the kth replica having its kth input equal to x_k, and all other inputs set to zero.

That is, if the output y is related to, say, three inputs x_1, x_2, x_3 by $y = f(x_1,x_2,x_3)$, then the output is given by

$$y = f(x_1,0,0) + f(0,x_2,0) + f(0,0,x_3)$$

The proof is an immediate consequence of Theorem 3.6-1 and is left as an exercise for the reader.

Example 3.6-3 *Superposition* Consider for example the system of Fig.

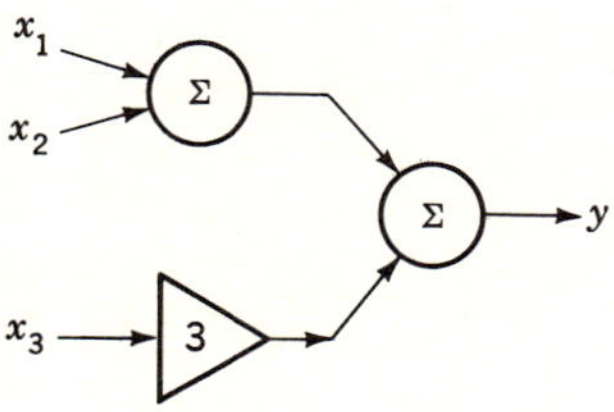

(a)

	x_1	x_2	x_3	y
	3	0	0	3
	0	−5	0	−5
	0	0	9	27
Superpose	3	−5	9	25

(b)

FIG. 3.6-2 Example of superposition.

3.6-2*a*. To find the output for $x_1 = 3$, $x_2 = -5$, $x_3 = 9$ we conduct three experiments. First, we set $x_1 = 3$, $x_2 = 0$, $x_3 = 0$. Looking at the system diagram of Fig. 3.6-2*a* we see that y is, by inspection, equal to 3. This we summarize on the top line of Fig. 3.6-2*b*. Next we set $x_1 = 0$, $x_2 = -5$, $x_3 = 0$. Again, by inspection, $y = -5$. Finally, we set $x_1 = 0$, $x_2 = 0$, $x_3 = 9$, and we obtain $y = 27$. Now we add (superpose) the three values of y to get the answer $y = 3 + (-5) + 27 = 25$, as shown in Fig. 3.6-2*b*.

While superposition may seem simple at this stage, it can become quite useful and interesting, as we shall see later.

Linearity is important since on the basis of Theorem 3.6-1 *it takes only n constants,* $a_1, \ldots, a_n$ *to fully characterize the input-output behavior of every n-input single-output linear system.* If linearity did not hold and nothing else were known about the system, i.e., if we had to "store" the input-output behavior for every set of values of $x_1, \ldots, x_n$, we would need a very large storage space.

For example, suppose that a system has six inputs ($n = 6$), and we are satisfied to approximate the value of each input by one of 10 "equally distant" values. Then the total number of input-value combinations is 10^6 or one million. We would therefore have to store *one million* values of y; and given any set of values for x_1 through x_6, we would have to search our storage for the closest stored values of these variables, in order to *approximate* y. If the system were linear however, we would only have to store *six* constants $a_1, \ldots, a_6$, and we could thereafter compute quite rapidly and *exactly* the output from Theorem 3.6-1 as

$$y = \sum_{i=1}^{6} a_i x_i \tag{7}$$

for any given values of $x_1, \ldots, x_6$.

The difference in these two approaches is indeed dramatic; no present-day computer can store a million real values in its primary memory; yet six constants can be adequately stored in a very small amount of space. The same is true with engineering block diagram systems: six scalers and a summer can synthesize the system of Eq. (7); yet it would take a million elementary blocks to store a million values if the other approach were followed.

When an n-input system is linear and has m outputs, that is, if it is constructed by putting together m n-input linear systems, then mn constants fully characterize the input-output behavior of the system. This set of constants is called a *matrix* and is discussed extensively in the second volume.

As the reader may undoubtedly suspect, the operations of summation and multiplication by a constant are desirable and useful for engineering system analysis and synthesis. In particular, digital computers are always equipped with a hardware adder, and very often with a hardware multiplier. Continuous blocks called summers and scalers are easily constructed from operational amplifiers and resistors, as we shall see in Chap. 4. Because they occur so frequently, they are represented by special blocks, as shown in Fig. 3.6-3. Here, the two-input summer (Fig. 3.6-3*a*) and scaler (Fig. 3.6-3*d*), which we have already encountered in Chap. 2, are really all that we require for forming any linear combination of variables. The differencer of Fig. 3.6-3*b* can be

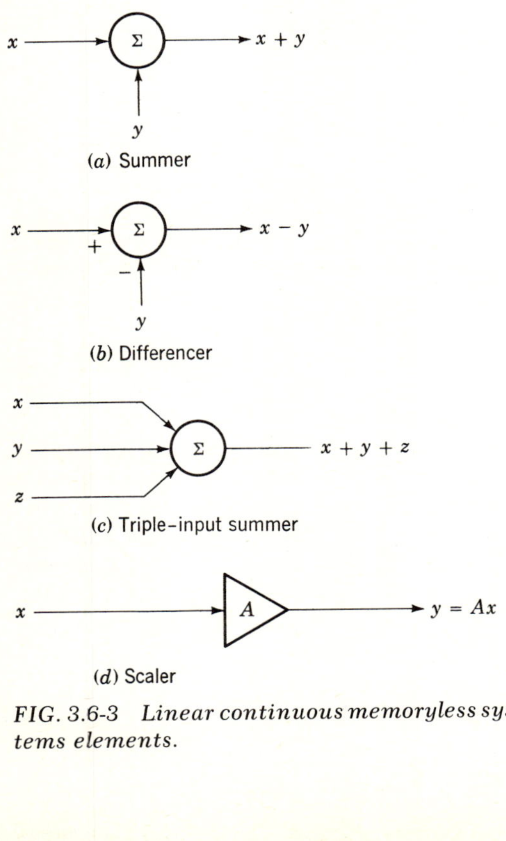

FIG. 3.6-3 Linear continuous memoryless systems elements.

made out of a summer and a negative scaler, while the three-input summer of Fig. 3.6-3*c* can be made out of two two-input summers.

We shall return to the ideas of linearity and linear systems in Chap. 4. We proceed now to another important system concept.

3.7 EQUIVALENCE AND APPROXIMATION

A powerful idea in the synthesis of systems is equivalence, defined as follows:

DEFINITION 3.7-1 Equivalence Two systems are *equivalent* if and only if for all possible input values their input-output behavior is identical.

The idea of equivalence is significant, because in engineering we often wish to simplify the structure of a system without changing its input-output behavior. Alternatively, equivalence is important in achieving so-called canonic realizations. For example, from Theorem 3.6-1 we deduced that every

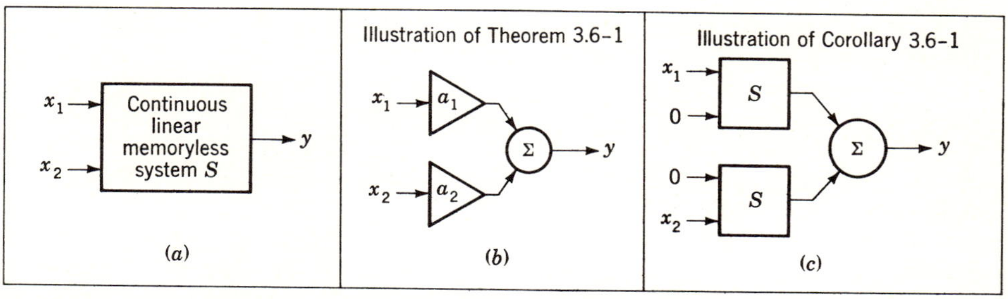

FIG. 3.7-1 Equivalence among two-input linear systems; all three systems are equivalent. (a) *Linear system S.* (b) *Canonic linear form of S.* (c) *Superposition used on system S.*

single-output linear system can be synthesized by two-input summers, and scalers in the "canonic form" of Eq. (6) of Sec. 3.6. Hence, we need only store a

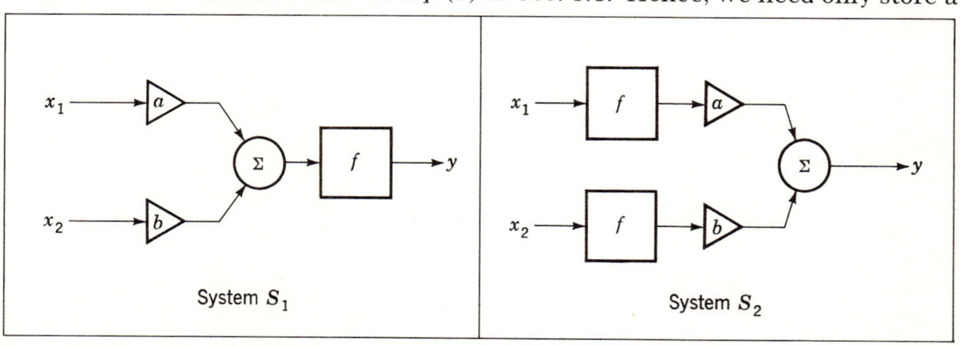

FIG. 3.7-2 Equivalence denoted by the definition of linearity: A single-input block f is linear if and only if systems S_1 and S_2 are equivalent.

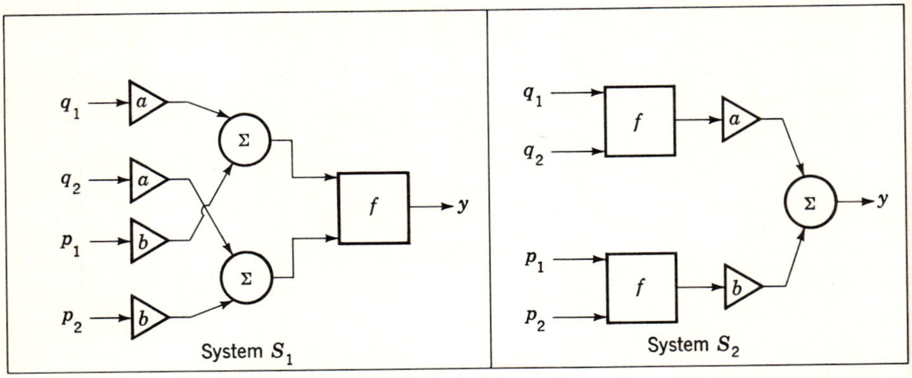

FIG. 3.7-3 Equivalence denoted by the definition of linearity: A two-input block f is linear if and only if systems S_1 and S_2 are equivalent.

large supply of these two types of elements, and we are guaranteed the ability to synthesize *any* desired linear continuous memoryless system, regardless of how large or complicated it may have to be.

From the discussion of the preceding section we deduce that every two-input linear system such as the one shown in Fig. 3.7-1*a* is (by Theorem 3.6-1) equivalent to the system of Fig. 3.7-1*b* and (by Corollary 3.6-1) to the system of Fig. 3.7-1*c*. Furthermore, from the basic definition of linearity, Definition 3.6-1, we see that every single-input linear system obeys the equivalence shown in Fig. 3.7-2. Similarly, every two-input linear system obeys the equivalence denoted by Definition 3.6-2 and shown in Fig. 3.7-3.

Equivalence, however, is by no means restricted to linear systems, since Definition 3.7-1 says nothing about linearity. In order to illustrate some helpful equivalences among nonlinear memoryless systems, we introduce

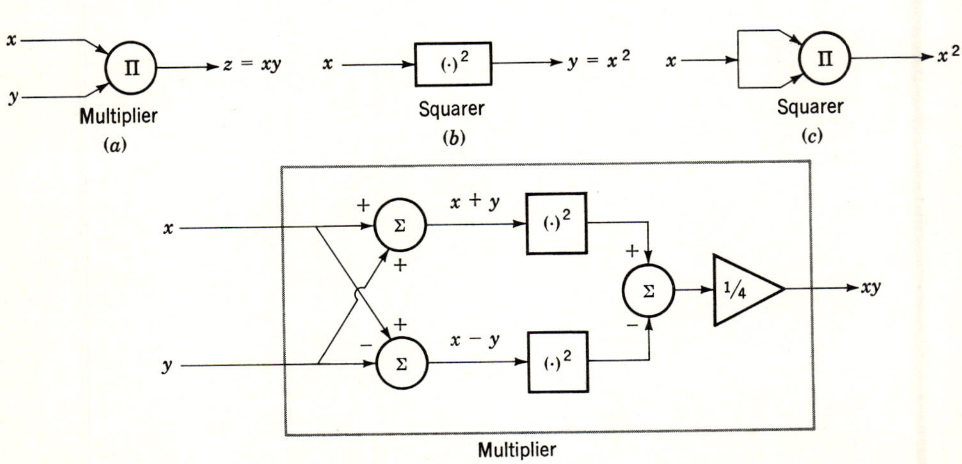

FIG. 3.7-4 Nonlinear continuous memoryless system elements.

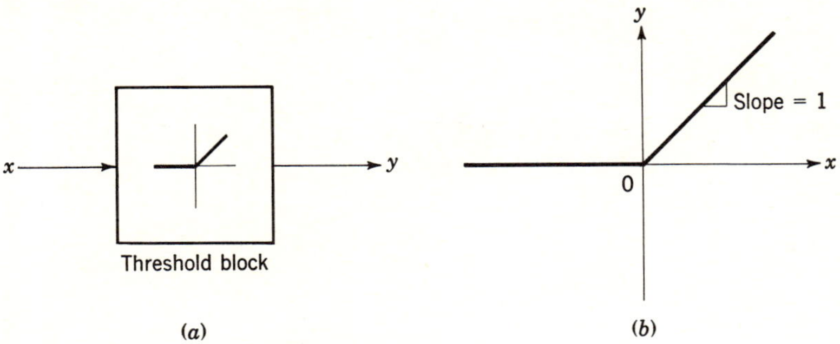

FIG. 3.7-5 The basic threshold block.

several additional basic elements, which do not obey the linearity constraint. These elements are the multiplier (Fig. 3.7-4*a*) and the squarer (Fig. 3.7-4*b*). As shown in Fig. 3.7-4*c*, a squarer is equivalent to a system which uses one multiplier. Alternatively, a multiplier is equivalent to the familiar system of Fig. 3.7-4*d*, which is made up of squarers, scalers, and summers.

Another useful nonlinear element is the threshold block of Fig. 3.7-5*a*. The input-output relationship of this block is defined as (Fig. 3.7-5*b*)

$$y = \begin{cases} 0 & \text{for } x < 0 \\ x & \text{for } x \geq 0 \end{cases} \tag{1}$$

Another way of saying the same thing is

$$y = \max(0, x) \tag{2}$$

where the function max (. , .) is understood to have a value equal to the largest of the values of its arguments.†

Example 3.7-1 *An Electronic Switch* A simple example of the utility of

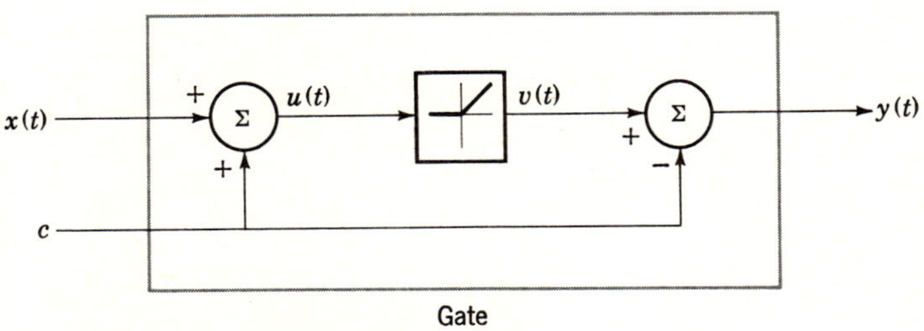

FIG. 3.7-6 An electronic switch.

the threshold block is shown in Fig. 3.7-6. The input-output relation of this

†More precisely $y = \max(a,b) = \begin{cases} a \text{ if } a \geq b \\ b \text{ if } a < b \end{cases}$

system is given by

$$y = \max(0, x + c) - c \tag{3}$$

For sufficiently large positive values of the parameter c, the input value appears unaltered at the output. For example, if

$$x = \sin(t) + 1 \qquad \text{and} \qquad c \geq 0$$

then from Eq. (3)

$$y = \sin(t) + 1$$

However, for sufficiently small values of c, it is $-c$, rather than the input signal x, that appears at the output; thus, if

$$x = \sin(t) + 1 \qquad \text{and} \qquad c \leq -2$$

then

$$y = 2$$

This system is useful since it behaves like an electronic switch or gate that "connects" x with y; yet has no moving parts that are sluggish, wear, or make noise. If, for example, the variable $x(t)$ represents the voice "signal" derived from a microphone, and if this signal ranges between v_1 and v_2 ($v_2 > v_1$), then by setting $c \geq v_1$, we allow the signal to "pass through" the system, and the electronic switch is "on." Alternatively, by setting c so that $c \leq -v_2$, we ensure that $y(t) = -c$, and the signal does not pass through to $y(t)$; i.e., the electronic switch is "off."

Approximation

An important engineering idea which is related to equivalence is system approximation. We define it as follows:

DEFINITION 3.7-2 System Approximation Two memoryless continuous systems are *approximately equivalent,* if under all possible identical inputs they have approximately equal outputs.

This definition is incomplete unless a way is specified for testing whether two outputs are "approximately equal." This is as it should be, for there are many possible ways of defining "approximately equal," each with its advantages, and limitations. For example, the outputs y_1, y_2 of two systems can be defined as approximately equal if

$$\max |y_1(x) - y_2(x)| < \epsilon \tag{4}$$

where x is the input to both systems, and ϵ is a positive real number that signifies the *maximum tolerable error*. Alternatively, we could define approximation of

$y_2(x)$ by $y_1(x)$ if

$$\max\,[y_1(x) - y_2(x)]^2 < \epsilon \tag{5}$$

or by

$$\int_{-\infty}^{\infty} [y_1(x) - y_2(x)]^2\,dx < \epsilon \tag{6}$$

We proceed now with two illustrations of approximate equivalence. First, we shall make an arbitrarily close approximation of a threshold block by an arbitrarily large number of multipliers:

Example 3.7-2 *Approximation of a Threshold Block with Multipliers* Consider the system of Fig. 3.7-7*a*. To establish $z(x)$ we write the equation

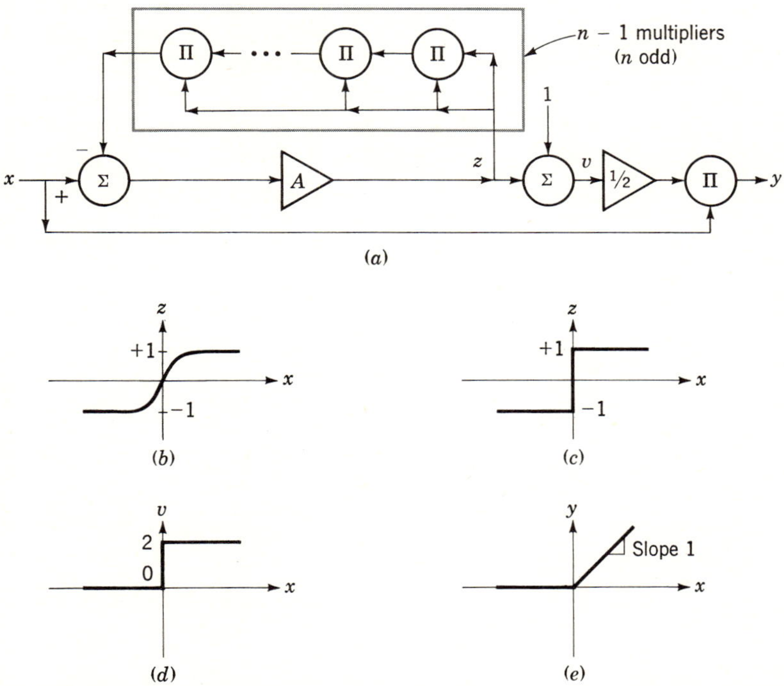

FIG. 3.7-7 Approximation of a threshold block with multipliers, summers, and scalers.

$$A(x - z^n) = z \tag{7}$$

where A is very large positive real constant, and n is a large *odd* integer. As we have already indicated in Example 3.5-3, the system acts as an inverter of the function $(\cdot)^n$, and we have for large A,

$$z \cong x^{1/n} \tag{8}$$

As n increases (staying always odd), the z versus x behavior, shown in Fig. 3.7-7*b*, approaches the z versus x behavior of Fig. 3.7-7*c*, to any desired accuracy. It follows from the system diagram (Fig. 3.7-7*a*) that v changes with x as shown in Fig. 3.7-7*d*, and y changes with x as shown in Fig. 3.7-7*e*. This is the desired characteristic of a threshold block, and we have thus shown how a system having an arbitrarily large number of multipliers can approximate to arbitrarily good accuracy a threshold block.

Very often, in communication, computation, and control systems it is desired to construct a system that, given x, responds with $f(x)$. For example, in an automatic landing system, x may be the vertical velocity of the airplane, and $f(x)$ the angle by which the elevators must be deflected to keep this vertical velocity constant, or decreasing with decreasing height. In fact, the general task of automatic control systems is to actuate some control parameters (such as velocity or temperature) in order to meet a certain desired performance (such as a specified trajectory or a constant temperature). This is a generalization of a system with input(s) x (the sensed parameters and desired performance), and output(s) y (the control parameters) so that $y=f(x)$. Sometimes f is implemented with "hardware," sometimes with "software"—the techniques in terms of system blocks are invariably similar. The following example discusses one of the many possible approaches for synthesizing a system such that $y=f(x)$, where f is given.

Example 3.7-3 *Synthesis of Piecewise-linear Functions* Consider the system of Fig. 3.7-8*a*. The constants a_i are ordered as follows:

$$0 < a_1 < a_1 + a_2 < a_1 + a_2 + a_3 < \cdots \tag{9}$$

For $x < a_1$, it is obvious from Fig. 3.7-8*a* that all the u_i will be negative; hence all the z_i will be zero; hence y will be zero.

For $a_1 \le x < a_1 + a_2$, u_1 will be nonnegative, and all the other u_i will be negative. Hence $y = z_1 = A_1 u_1 = A_1(x - a_1)$. As x increases further and exceeds the sums $a_1 + a_2$, $a_1 + a_2 + a_3$, . . . , the variables $z_2, z_3, \ldots$ become successively nonzero, resulting in the general form shown in Fig. 3.7-8*b*.

For $\sum_{i=1}^{k} a_i < x < \sum_{j=1}^{k+1} a_j$

$$\frac{dy}{dx} = \sum_{i=1}^{k} A_i \tag{10}$$

A function such as this is called a *piecewise-linear* function; it is generally represented (Fig. 3.7-8*c*) by pairs of *breakpoints* $x_i\ y_i$, and is evaluated as follows:

Assume that the x_i are ordered as follows:

$$x_1 < x_2 < x_3 < \cdots < x_n$$

then for $x = x_i$,

$$y_i = f(x_i)$$

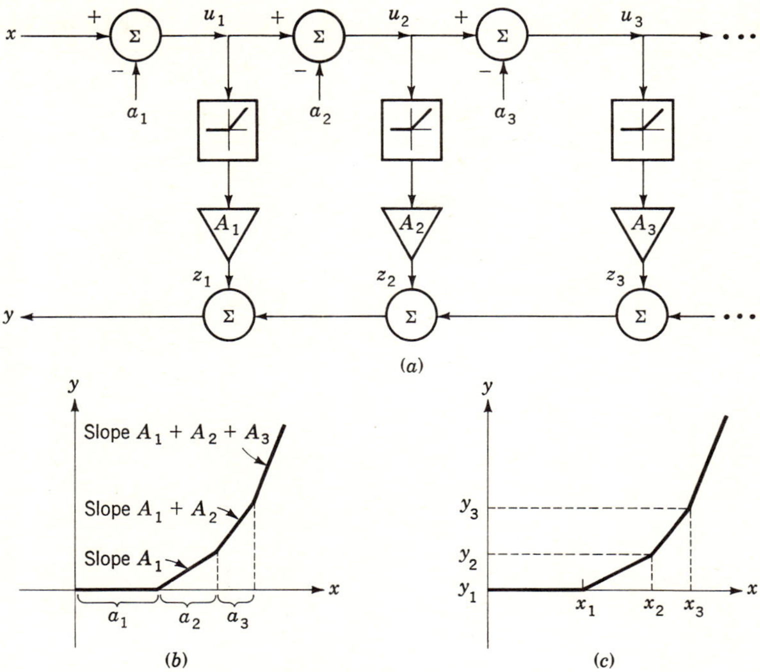

FIG. 3.7-8 Synthesis of piecewise-linear function.

and for $x_i < x < x_{i+1}$,

$$y = y_i + \frac{y_{i+1} - y_i}{x_{i+1} - x_i}(x - x_i) \tag{11}$$

That is, the graph of f connects with straight-line segments the adjacent breakpoints x_i, y_i. (Fig. 3.7-8*c*). The reader is invited to verify that for any given x_i, y_i, the constants a_i and A_i can be chosen so that Eqs. (10) and (11) are identical, i.e., so that the desired behavior of Fig. 3.7-8*c* is identical to the input-output behavior of Fig. 3.7-8*b*.

The system of Fig. 3.7-8*a* realizes exactly any desired piecewise-linear function. If a sufficiently large number of gates is used, then that system may be made *approximately equivalent to any continuous memoryless function block*. Thus, for example, we could synthesize the squarer, $y = x^2$, to an arbitrary degree of accuracy with the system of Fig. 3.7-8, hence with summers, scalers, and threshold blocks.

Observe now that in Fig. 3.7-4*c* we demonstrated how a multiplier can be used to realize a squarer, while in Fig. 3.7-4*d* we showed how squarers, scalers, and summers can be used to realize a multiplier. Moreover, observe that a threshold block can be approximated by multipliers, scalers, and summers (Fig. 3.7-7*a*). while a squarer can be approximated by threshold blocks, scalers, and summers as we discussed in the above paragraph. Hence, assuming that a large supply of scalers and summers is available, we see that either a supply of multipliers *or* squarers *or* threshold blocks is sufficient to realize approximately any system made up of scalers summers, multipliers, squarers, and threshold blocks.

PROBLEMS

3.4-1 For each of the systems shown in the figure write the interconnection list, and deduce whether that list is sortable. If so, give it in sorted form; if not, copy the diagram, and trace each loop.

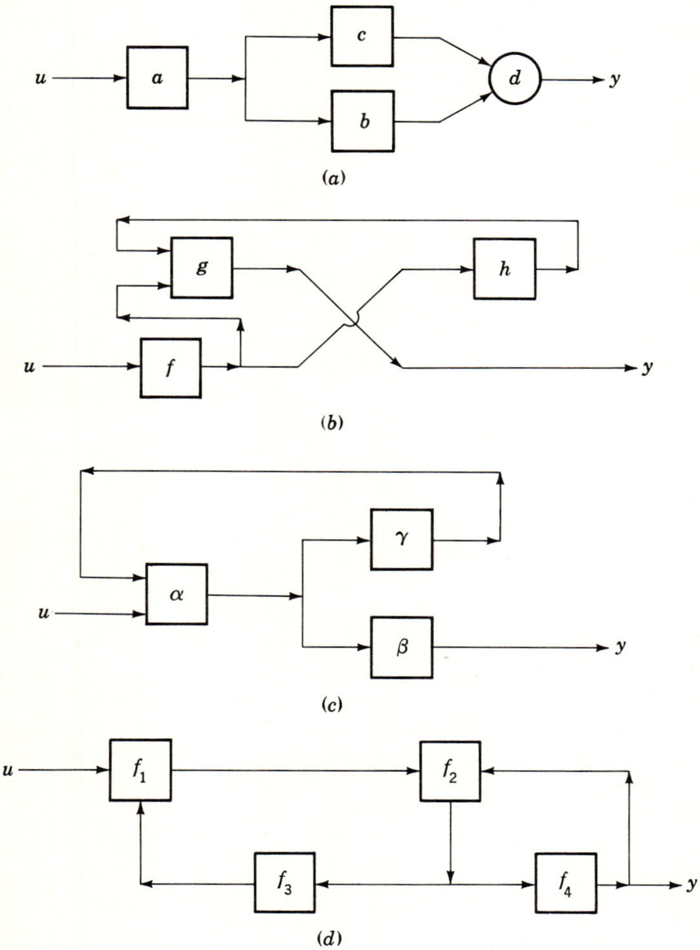

PROB. FIG. 3.4-1

3.4-2 Devise a simple algorithm which sorts a given interconnection list or stops when it discovers that the interconnection list cannot be sorted. Give your answer in terms of a flow chart.

3.5-1 Show that every explicit memoryless system with n inputs $u_1, \ldots, u_n$ and a single output y consisting of function blocks f_i can be characterized by an expression which contains only y on the left side of the equal sign and $u_1, \ldots, u_n$ as well as the f_i on the right side, and which *does not* contain any intermediate variables.

3.5-2 It is desired to build a system that will yield the ratio u_n/u_d of two inputs. The building blocks available are the summer, the multiplier, and the scaler. The proposed system is shown in the figure. Give the interconnec-

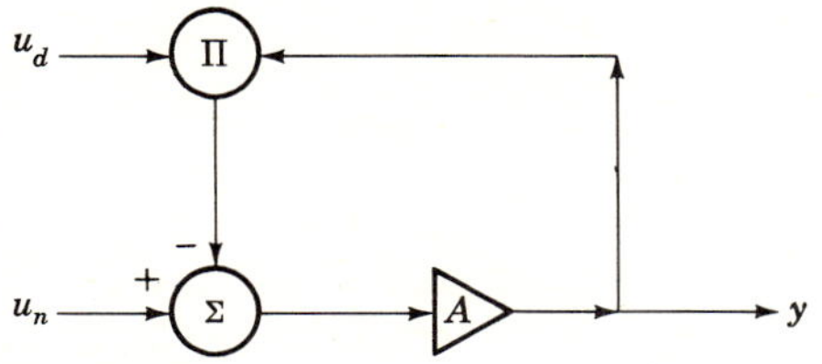

PROB. FIG. 3.5-2

tion list for this system, and deduce whether this list is sortable. If it is sortable, sort it; if not, identify the loop(s), and define new intermediate variable(s) $q_1, q_2, \ldots$ so that the list can be sorted. Express the equations in the form

$$q_i = \text{function of } (u_d, u_n, q_j)$$

$$y = \text{another function of } (u_d, u_n, q_j)$$

Solve this set of equations for y as a function of u_d and u_n, and comment about the circumstances under which this system would be a satisfactory "divider." What happens if you try to divide by zero?

3.5-3 Devise a simple algorithm which "breaks" *all* the loops of a given implicit-system interconnection list, by introducing intermediate variables. Give your answer in terms of a flow chart.

3.5-4 Give an example of a memoryless system with three loops, whose interconnection list can be sorted through introduction of only one intermediate variable.

3.5-5 In the system shown in the figure, $f(x_1, x_2) = A(x_1 - x_2)$, where $A = 10^6$.

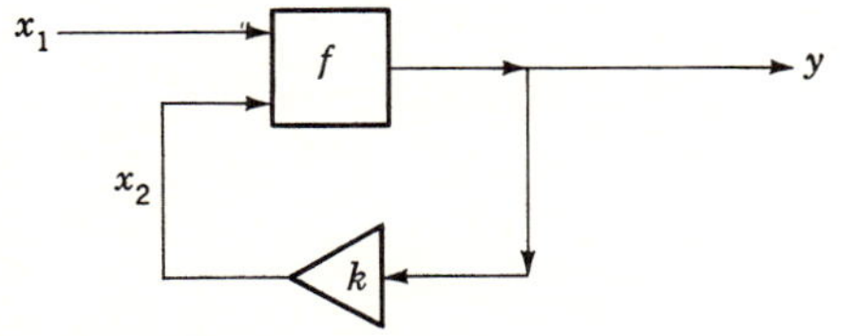

PROB. FIG. 3.5-5

(*a*) Give an *exact* explicit expression for $y(x_1)$, and then make this expression approximate if $k >> 10^{-6}$.

(*b*) Suppose that the "gain" A may vary with temperature up to ±10 percent of its value, while k may vary up to ±0.01 percent over the same temperature range. Up to how much can $y(x_1)$ vary as a consequence of such temperature changes if $0.1 < k < 1$?

(*c*) Because of physical limitations, scaler blocks with "gain" less than 1 can be made nearly temperature-independent, while scalers with gain larger than 1 will generally exhibit considerable variations. Assume that for $0 < k \leq 1$, variations of k and A are as in (*b*) above. Exactly which "scalers" can you construct, using the above feedback technique, so that their values will vary by no more than ±0.1 percent?

3.5-6 In the system shown in the figure, f is the linear difference sensor of the

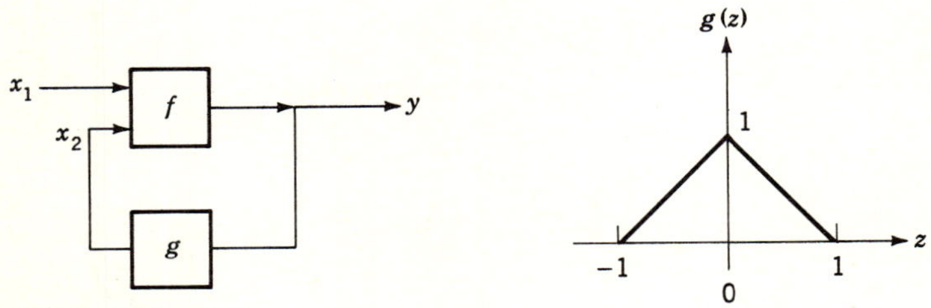

PROB. FIG. 3.5-6

preceding problem, with $A \to \infty$, and function g has the graph shown. Sketch the graph of y versus x_1. If this is a function system, give the function $y(x_1)$. If it is not, specify the relation which relates y to x_1.

3.6-1 Specify whether each of the systems shown in the figure is operationally linear, structurally linear, or nonlinear.

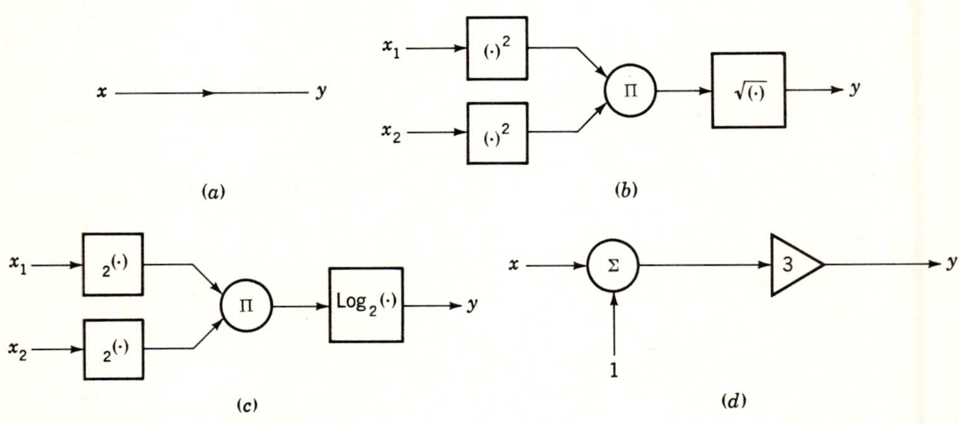

PROB. FIG. 3.6-1

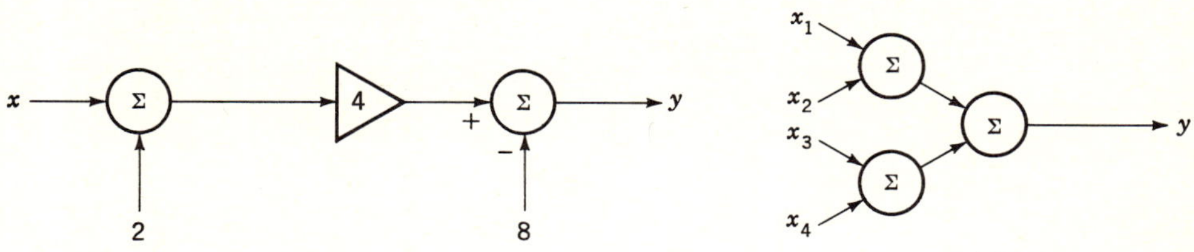

(e) (f)

3.6-2 Three experiments are conducted on a three-input *linear* system f with the following results:

$$f(1,1,1) = 6$$

$$f(0,1,1) = 3$$

$$f(0,0,1) = 1$$

Specify as simple a block diagram for f as possible.

3.6-3 (*a*) Show that if f_1 and f_2 are linear, then so is g.
(*b*) Give an *explicit* expression for $y(x)$, assuming that f_1 and f_2 are linear.

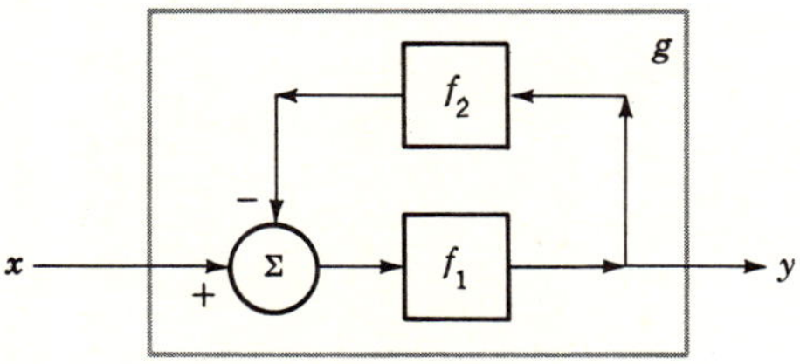

PROB. FIG. 3.6-3

3.7-1 State and justify whether the two systems shown in each of the pairs in the figure are equivalent, or not equivalent, where f_1, f_2, and f_3 are linear.

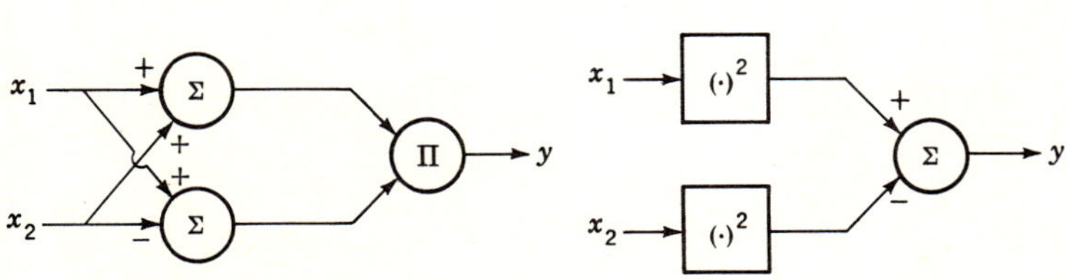

(*a*)

PROB. FIG. 3.7-1

(b)

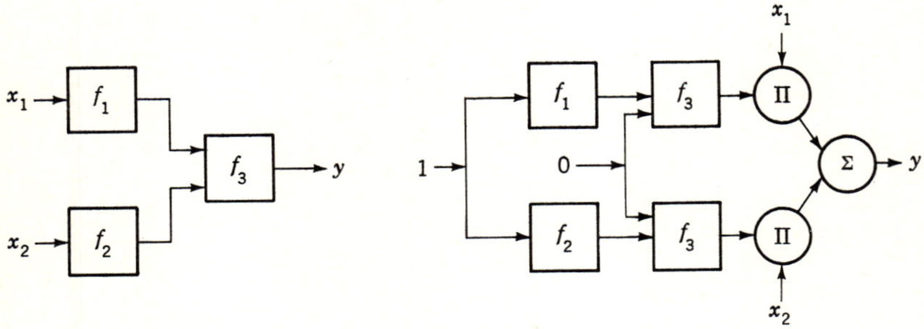

(c)

3.7-2 Using two threshold blocks and as many scalers and summers as you wish, design a system that has the input-output characteristic of the graph in the figure.

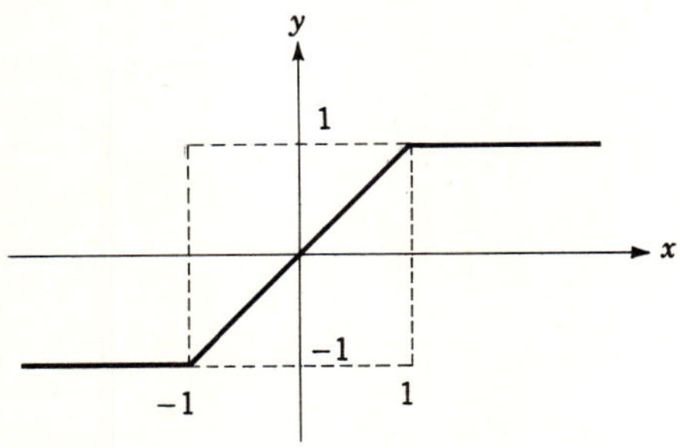

PROB. FIG. 3.7-2

chapter 4
Resistive Networks

Electric resistive networks are an important class of continuous memoryless systems. This chapter is concerned with the formulation of equations that characterize resistive networks and with the solution of these equations for certain special cases, in particular linear networks. The concepts and techniques associated with linearity and equivalence, which were discussed in Chap. 3, are equally important for resistive networks and are discussed in the context of several theorems and practical examples. A special class of nonlinear resistive networks—the diode networks—is discussed, along with ways in which resistive networks can be used to construct the function blocks of the preceding chapter.

4.1 INTRODUCTION

Resistive networks are a special class of electric networks. We are interested in studying such networks for several reasons:

In terms of the surrounding world, networks are used in today's technology for accomplishing a variety of tasks in computation,† communication, and control as well as in the generation and distribution of electric power. Numerous new concepts and systems are often implemented out of a relatively small number of different network elements such as resistors, capacitors, transistors, and magnetic cores. A good understanding of electric networks, however, is not restricted to the field of electronics; it is directly applicable to other diverse areas such as mechanics, acoustics, and hydraulics. For example, a good understanding of resistor-inductor-capacitor networks leads to a good understanding of vibrating strings and automobile suspension systems (and vice versa).

In terms of the methodology used in this book, resistive networks can be used to construct the function blocks that we have discussed in the previous chapter. Conversely, function blocks and the formulation of equations for function-block systems can be used to model or characterize the behavior of electric networks. The ideas of explicit and implicit equations, linearity, and equivalence carry over directly to resistive networks, permitting us to analyze and design a large number of useful systems with the same basic ideas. Resistive networks also represent the memoryless portion of our state model; in analyzing more complicated dynamic systems, we will find that we often need to decompose them into memory and memoryless subsystems, the memoryless portions being frequently resistive networks.

Finally, resistive networks are models of real physical electric circuits, as contrasted to function-block systems which are abstractions and generalizations of physical systems. Because of their physical nature, circuits introduce constraints (physical laws) which can be modeled by mathematical means. Thus, in contrast to pure mathematics, where any definition is permitted provided that it leads to something significant; in the case of networks we encounter the need for definitions, and, in fact, mathematics which must "bend a little" so as to accommodate the observations derived from the physical world. The knowledge of such things as time constants and the building up of an intuition or basic understanding of how a linear or nonlinear network behaves when a switch is closed have wider applicability than the specific networks to which they apply. In other words, this relationship of mathematics and physics, in terms of resistive networks, is important not only for treating the necessarily narrow field of networks but also for establishing an appropriate foundation which can be used in understanding and modeling new physical phenomena that technology inadvertently and progressively generates.

4.2 NETWORK STRUCTURES

We start with the class of *branch networks* which are defined by Table 4.2-1. Here the basis is a geometrical* object called an *element* with two

†For example, the logic gates and most of the so-called digital integrated circuits are modeled by resistive networks.

*For the time being we are only interested in the structure of branch networks, i.e., in the

terminals or *nodes*. We can construct a branch network by attaching to such an element, or to a more complicated branch network, a new two-terminal element by either one of the two interconnection rules. Rule I connects one node of the element being added to *any one* of the nodes of the branch network, thus creating a new network which has one more node than the old network. Rule II connects both nodes of the element being added to *any two* nodes of the old network, thus creating a new network with the same number of nodes as before.

Examples of branch networks are shown in Fig. 4.2-1*a* and *b*. The constructions of Fig. 4.2-1*c* and *d* are not branch networks: the former because it

TABLE 4.2-1 Definition of Branch Networks

Element set:

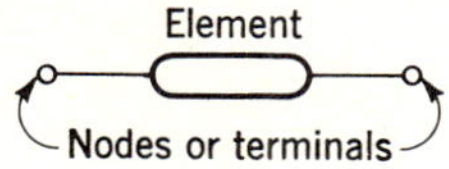

Rules of interconnection:

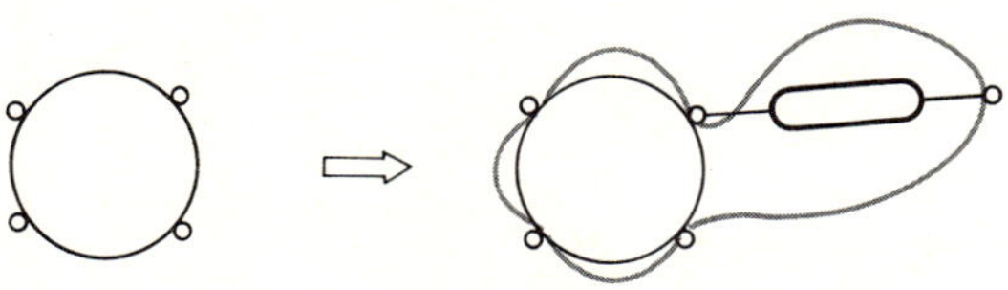

Rule I

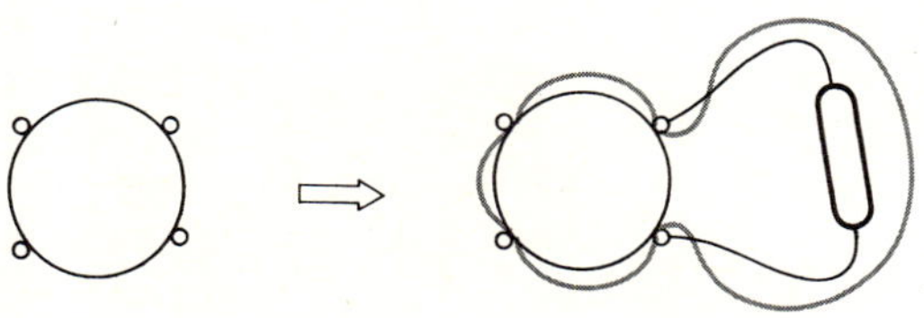

Rule II

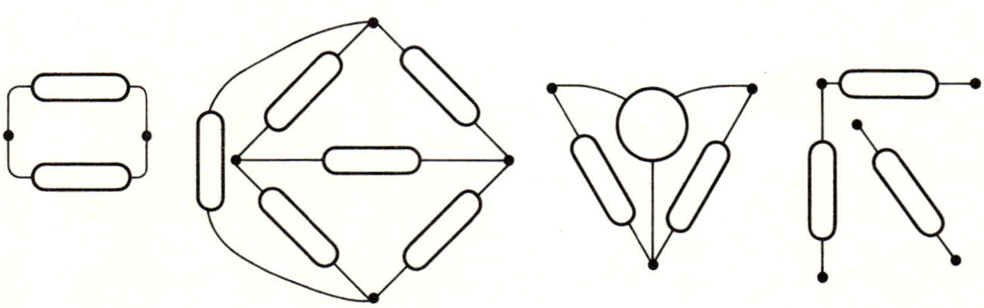

(*a*) (*b*) (*c*) (*d*)

FIG. 4.2-1 Examples of networks which can or cannot be generated by the rules of Table 4.2-1.

way the elements are to be interconnected. For this reason, it suffices to view elements and networks as geometrical or graphical objects that are drawn on paper.

contains a three-terminal element, and the latter because it is unconnected—neither network can be constructed from the rules of Table 4.2-1.

Observe that Table 4.2-1 is in no sense universal. The network structures of Fig. 4.2-1*c* and *d*, although not branch networks, can be quite useful in modeling physical circuits. For the purposes of this book, however, it is adequate that we study only branch networks, especially since in practice multiterminal elements such as the one of Fig. 4.2-1*c* are almost always modeled by branch networks; for example, a very common three-terminal element, the transistor shown in Fig. 4.2-2*a*, is very often replaced for modeling purposes by the branch network of Fig. 4.2-2*b*, where the special symbols used for the two-terminal elements are discussed later in the chapter.

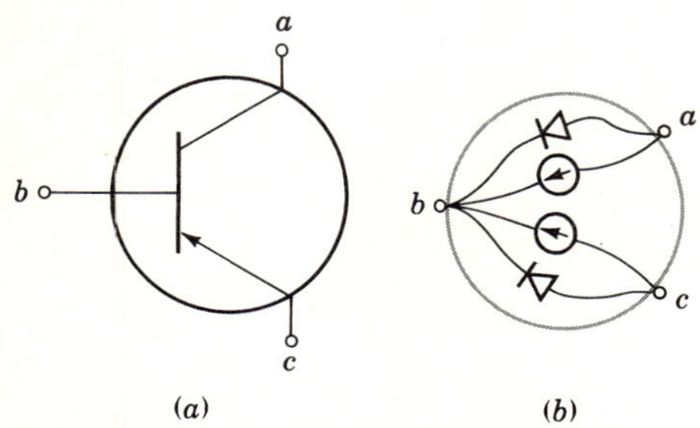

FIG. 4.2-2 Branch network representing a transistor.

4.3 THE CONSTRAINTS

Overview

An electric network as defined in the preceding section is a model or a representation of an electric circuit, which in turn is a structure made up from the electric interconnection of certain basic elements, such as *resistors, capacitors, inductors, voltage sources*, and *current sources*. Each basic element places a constraint between two quantities: for example, a linear resistor of resistance R relates the voltage v across it to the current i through it by $v = Ri$. Additional constraints are imposed by the network geometry, that is the way in which the network elements are interconnected; for example, if somewhere in the network only two elements are connected to one node, then the current through one element is forced, by this network structure, to equal the current through the other element.

The significance of networks is that very few types of basic elements with simple current-voltage relationships, when purposefully interconnected, make possible, through the above constraints, complex tasks such as logic operations in computers, communication of information in radio and television, and control in the guidance of space vehicles.

The Network Variables

In designing and analyzing electric circuits we deal with the following physical quantities: current i, voltage v, charge q, flux linkage λ, power p, energy w. The *network variables*, i, v, q, λ, p, and w are used to represent these quantities. In the following, we shall discuss briefly the meaning of each of the above variables and its relationship to the other variables. Table 4.3-1 summarizes this discussion and lists the appropriate units for these quantities in the so-called *mks* system of measurement.

TABLE 4.3-1 Electrical Quantities

Quantity	*Typical symbol*	*Unit*	*Illustration*	*Meaning*
Current	i	ampere, A	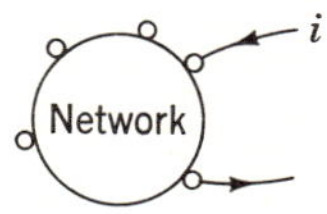	Number of unit charges flowing in direction of arrow, per unit time
Voltage or potential	v	volt, V	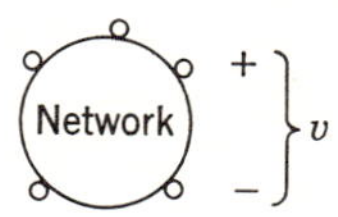	Work done upon, or expended by unit charge moving from (−) node to (+) node
Charge	q	coulomb, C	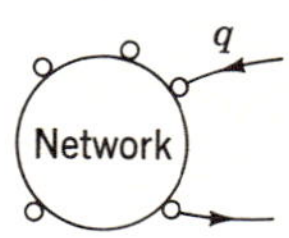	Total number of unit charges flowing in direction of arrow, since some time origin
Flux linkage	λ	weber, Wb	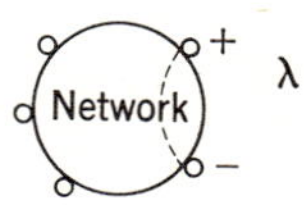	Magnetic flux linking circuit which terminates at the two nodes
Power	p	watt, W	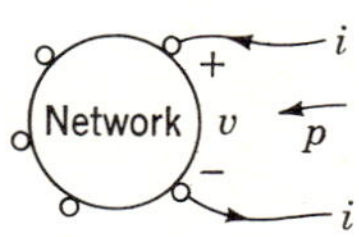	Electric power flowing between network and outside of network
Energy	w	joule, J	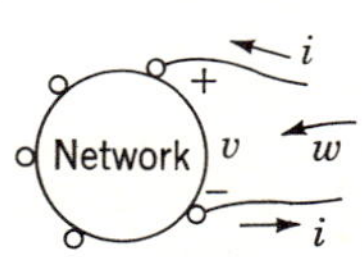	Electric energy flowing between network and outside of network

Current is said to flow *through* an element, or into (out of) a network terminal (from something which is outside the network). It is a measure of the amount of charge flowing per unit time through that terminal. The unit of current is the *ampere*, A, although in most circuits typical current flows are in mA (*milliamperes* or 10^{-3} A), μA (*microamperes* or 10^{-6} A), or nA (*nanoamperes* or 10^{-9} A). By way of a crude analogy, current which is the flow of a number of charged particles per unit time, can be thought of as corresponding to the flow of a number of liquid molecules through a pipe, i.e., to the mass-flow velocity of the fluid. An arrow is used to establish a reference direction for the current.

Voltage or *potential* or *voltage difference* is said to exist *across* an element or across exactly two nodes of a network. It is a measure of the work done upon or energy expended in moving a unit of charge from one node to the other. The unit of voltage is the *volt*, V, although, as in the case of current, we often speak of mV (millivolts) and μV (microvolts). In terms of our analogy with liquids flowing through pipes, voltage across an element corresponds to the pressure difference across a pipe. The + and − signs are appended to the two nodes to establish a reference direction for the voltage.

Charge, like current, flows *through* a branch or into (out of) a network. It is a measure of the net number of unit charges which flow through a terminal since some time origin. Hence, charge q and current i are related at any instant of time by

$$i = \frac{dq}{dt} \tag{1}$$

The unit of charge is the *coulomb*, C. In terms of our analogy, charge corresponds to the net amount of mass that has flown through a pipe since some time origin. As in the case of current, an arrow is used to establish a reference direction.

Flux linkage is to voltage, what charge is to current. That is, flux linkage λ is related to voltage v as follows:

$$v = \frac{d\lambda}{dt} \tag{2}$$

Flux linkage is said to exist across two nodes, as does voltage, and has as its unit the *weber*, Wb. This quantity stands for the magnetic flux linking a circuit† which terminates at the two nodes. We are hard pressed, however, to find in our crude analogy anything, other than the time integral of pressure, corresponding to flux linkage.

Power is a measure of the electric power flowing into (out of) a network. With respect to two terminals, as illustrated in Table 4.3-1, power is related to v and i as follows:

$$p = vi \tag{3}$$

The unit of power is the *watt*, W; although again we often speak of milliwatts, mW, and microwatts, μW; or in the case of power circuits, of kilowatts, kW (10^3 W), and megawatts, MW (10^6 W). In terms of our analogy, electric power corresponds to the mechanical power flowing into (out of) a hydraulic network.

Energy w is a measure of the total electric energy that has flowed into (out of) a network since some time origin. It is related to power p as follows:

†In this context, a circuit is a connected path.

$$p = \frac{dw}{dt} \tag{4}$$

The unit of electric energy is the *joule*, J. In our analogy, electric energy corresponds to the mechanical energy flowing into (out of) a hydraulic network.

Reference Directions

A two-terminal element within a network will incur at any instant of time a current or charge through and a voltage or flux linkage across its terminals. Similarly, electric power and energy will be flowing in or out of that element, from or into the rest of the network. In order to identify quantitatively these quantities, reference directions must be established, in the same sense that we need a coordinate center and a direction to specify the position or velocity of points on a line.

The reference direction for current and charge is indicated by an *arrow* along the path as shown in Fig. 4.3-1*a*. If we say that in Fig. 4.3-1*a*, $i = 2$,

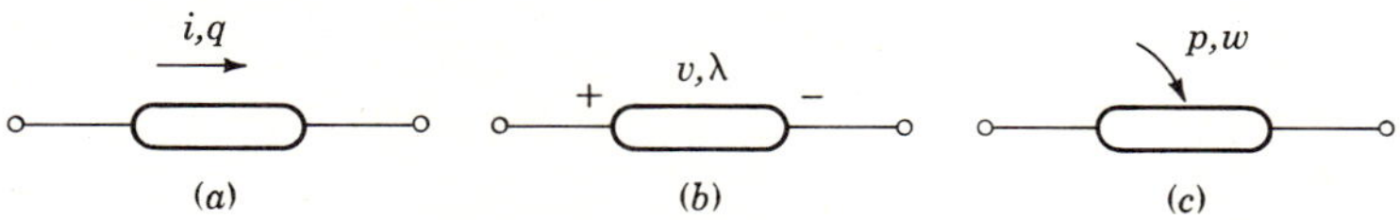

FIG. 4.3-1 Reference directions. (a) *Current and charge.* (b) *Voltage and flux linkage.* (c) *Power and energy.*

we mean that two units of current flow in the direction of the arrow; this corresponds to the flow of positive charges along the direction of the arrow. If the reference direction in Fig. 4.3-1*a* were reversed, but the current were the same as before, then we would say that $i = -2$. This is no different from changing the coordinate center from the lower left to the lower right corner of our paper—the horizontal position of a point on the paper would then change sign, as we changed coordinate systems. For voltage and flux linkage, the reference direction is indicated by + and − signs as shown in Fig. 4.3-1*b*. If we say that $v = 6$ in Fig. 4.3-1*b* we mean that six units of voltage are present across the two terminals in the direction indicated by the +, − signs. If the element is an ordinary resistor, then the charge flow causing or caused by this voltage will be positive from the "+" to the "−" terminal.

In the case of power and energy, the reference direction is established by an arrow going into or out of an element (or network). In Fig. 4.3-1*c*, $p = 4$ means that four units of electric power are entering the element (possibly to be stored or to be dissipated as heat).

With three reference directions to contend with, we could assign to every element in a network one of the possible eight combinations. Although this is often the case, a convention has been adopted for convenience, and we try to use it whenever possible. It is called the *passive convention* and is shown in Fig. 4.3-2.

Here the current reference is in the + to − direction of the voltage reference, and the power reference is into the element. If the element is passive,

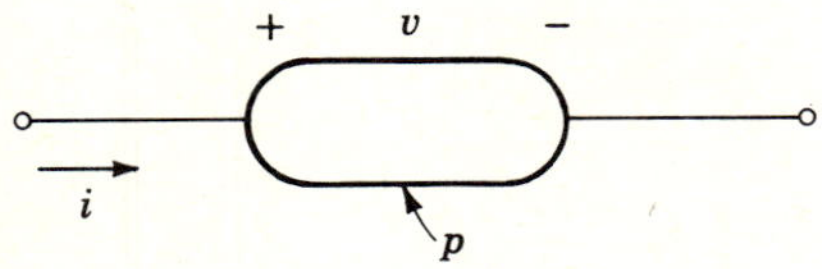

FIG. 4.3-2 The passive convention.

that is, if it generates no energy, then v and i will be both positive or both negative, and p, which is their product, will be positive.† In terms of our pipe analogy, the passive convention means that the reference direction of fluid flow is from the + to the − sign of the pressure reference. If the pressure across the pipe is positive, i.e., if the + side is at a higher pressure than the − side, then fluid will flow from the high to the low pressure, along the arrow; hence fluid flow will be positive and power will enter the pipe. If the − side is at a higher pressure than the + side, then the fluid will flow against the arrow; that is, both pressure and flow will be negative; yet power will again enter the device, and hence will be positive. The reader can easily appreciate the term passive convention, for if the device were instead of a "passive" pipe, an "active" pump, flow and pressure would always have opposite signs on the reference directions of Fig. 4.3-2, and power would always be negative, i.e., flowing out of the element, as is indeed the case with active electrical elements, such as batteries and generators.

Topological Constraints (KCL and KVL)

The interconnection of elements into networks results in structures whose electrical behavior is governed by *two types* of constraints—those due to the *geometry of the interconnection* and those due to the *nature of the elements*. This section deals with the former, and the next section deals with the latter.

Because of the physical law of conservation of charge, all current entering a network must balance the current leaving that network. This physical observation leads to our first axiom, known as the Kirchhoff current law.

Kirchhoff Current Law (KCL): The algebraic sum of the currents entering a network is zero.

Figure 4.3-3a shows a *subnetwork* N, i.e., a part of a larger network obtained by placing in an imaginary enclosure any number of elements of the larger network and showing the terminals 1, 2, 3, . . . , k which connect these elements with the rest of the network (not shown in Fig. 4.3-3a). The internal

†When dealing with sources (discussed later), we shall often reverse the reference directions of current and power in Fig. 4.3-2, so that positive current and positive voltage result in positive power along the new reference directions.

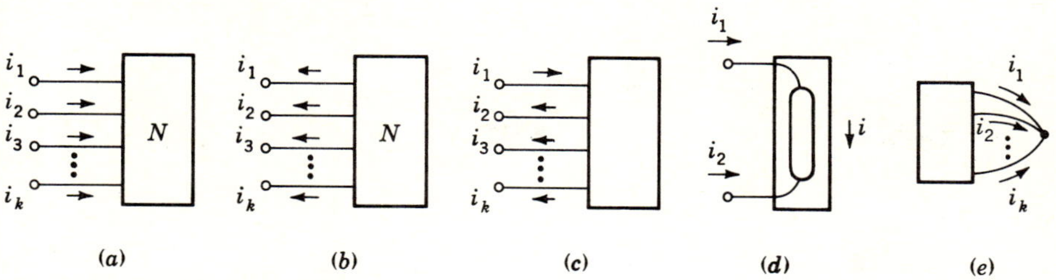

FIG. 4.3-3 Illustration of KCL.

structure of N is of no interest or consequence to this law. The law states that

$$\sum_{j=1}^{k} i_j = 0 \tag{5}$$

Evidently, Eq. (5) also holds for the reference directions of Fig. 4.3-3*b*, but *not* for Fig. 4.3-3*c*. Here i_1 would have to appear with a minus sign in Eq. (5) in order for Eq. (5) to hold. Figure 4.3-3*d* demonstrates that, as a special case of KCL, the current entering an element equals the current leaving that element: i.e.,

$$i_1 = i = -i_2 \tag{6}$$

Finally, Fig. 4.3-3*e* demonstrates another special case of KCL, i.e., the fact that all currents arriving at a node must balance, as indicated by Eq. (5). Clearly, a network with only *one* terminal is pretty restricted, since the current through that terminal would be, by KCL, always zero; this is why it takes at least two wires to do something that is electrically interesting.

In terms of our pipe analogy, KCL corresponds to the algebraic sum of all flows entering a pipe network being equal to zero. This is indeed true at any time, provided that the fluid is incompressible. Apparently, charge and current behave in such an "incompressible" fashion, for we cannot sneak up to the network and very rapidly change one current before the others have a chance to also change.† In other words, KCL holds at any instant of time.

Because of the physical law of uniqueness of the potential, *every terminal in a network has at any instant of time a unique potential or voltage with respect to some zero reference potential or "ground point."* The voltage across an element is the difference of the potentials at its two terminals. This physical observation leads to our second axiom, known as the Kirchhoff voltage law.

Kirchhoff Voltage Law (KVL): The algebraic sum of the voltage differences around any closed path in a network is zero.

---

†There is a note of caution here. Electric elements like resistors and capacitors behave as isolated network elements, and KCL holds only if the rate of change of the electric field in the network is not appreciable. This will always be the case for our purposes.

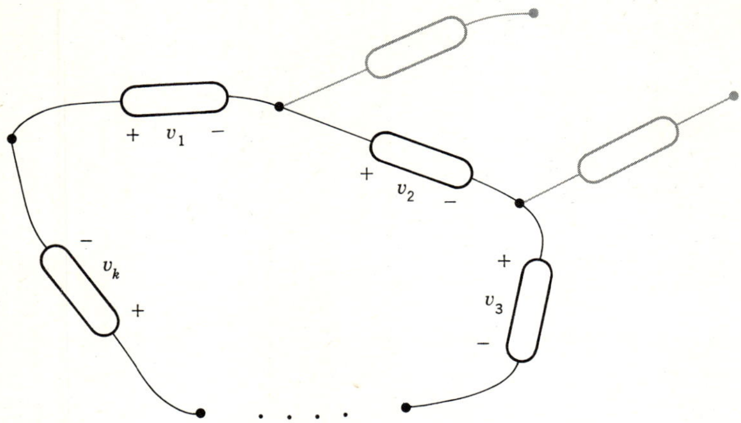

FIG. 4.3-4 Illustration of KVL.

In connection with the reference directions assigned to the elements of Fig. 4.3-4, KVL states that

$$\sum_{j=1}^{k} v_j = 0 \tag{7}$$

If all the reference directions were reversed, then Eq. (7) would still hold, as in the case of KCL. If, however, only one element reference direction were reversed, then the voltage variable corresponding to that element would have to appear negatively in Eq. (7) in order for that equation to hold.

In terms of our pipe analogy, KVL corresponds to the uniqueness of pressure. For example (Fig. 4.3-5), if the absolute pressures at several points of a

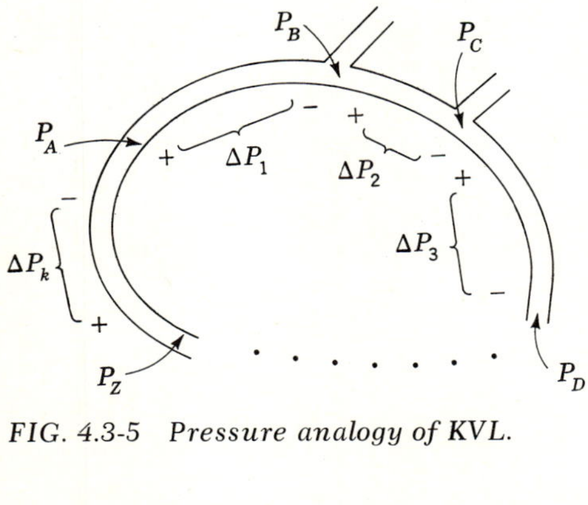

FIG. 4.3-5 Pressure analogy of KVL.

closed circuit of pipes are P_A, P_B, P_C, P_D, . . . , P_Z, then the algebraic sum of the pressure differences around the circuit will be zero.

$$\sum_{j=1}^{k} \Delta P_j = 0 \tag{8}$$

Equation (8) is obviously true, since if we substitute $P_A - P_B$ for ΔP_1, $P_B - P_C$ for ΔP_2, and so forth, we get

$$\sum_{j=1}^{k} \Delta P_j = (P_A - P_B) + (P_B - P_C) + (P_C - P_D) + \cdots + (P_Z - P_A) = 0 \quad (9)$$

That is, successive absolute pressure terms in Eq. (9) cancel one another, thereby confirming Eq. (8).

Finally, observe that KVL holds, only if in the network environment the rate of change of the magnetic field is not appreciable.

Example 4.3-1 Application of KCL to the network of Fig. 4.3-6*a* yields $i_1 + i_2 + i_3 = 0$. Hence, since $i_1 = 2$, $i_2 = -4$, and i_3 is not known, we can write

$$2 - 4 + i_3 = 0 \qquad \text{or} \qquad i_3 = 2 \quad (10)$$

If in the same network, different reference directions were chosen, as in Fig.

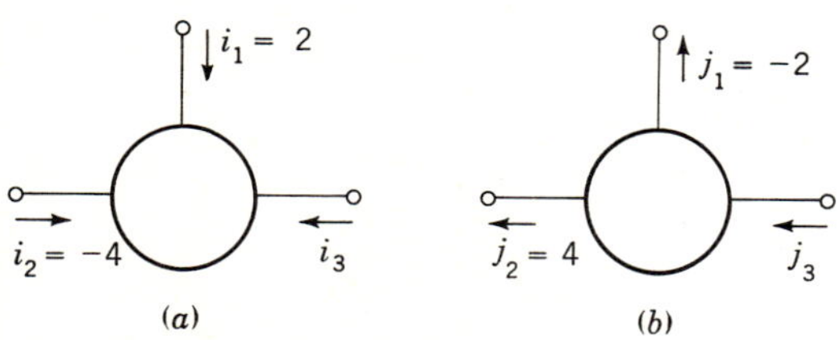

FIG. 4.3-6 KCL examples.

4.3-6*b*, then KCL would dictate

$$-j_1 - j_2 + j_3 = 0 \quad (11)$$

Note that in terms of these new reference directions $j_1 = -2$ and $j_2 = 4$, i.e., $j_1 = -i_1$ and $j_2 = -i_2$. Hence, substitution of these values in Eq. (11) yields

$$-(-2) - (4) + j_3 = 0 \qquad \text{or} \qquad j_3 = +2 \quad (12)$$

Here j_3 equals i_3, as it should, since the network and the values of the currents entering the network are unchanged, the reference direction for j_3 is identical to that for i_3, and the only changes are the reference directions for the other two currents—certainly, a change in the reference directions in a network should not affect what is going on in the network, any more than a change in the coordinate center of a geometric figure affects the figure.

Example 4.3-2 Application of KVL to the circuit of Fig. 4.3-7 yields

$$v_1 - v_2 + v_3 + v_4 = 0 \quad (13)$$

Substitution of the values for v_1, v_2, and v_3 given in Fig. 4.3-7 yields

$$2 - 3 + 4 + v_4 = 0 \qquad \text{or} \qquad v_4 = -3 \quad (14)$$

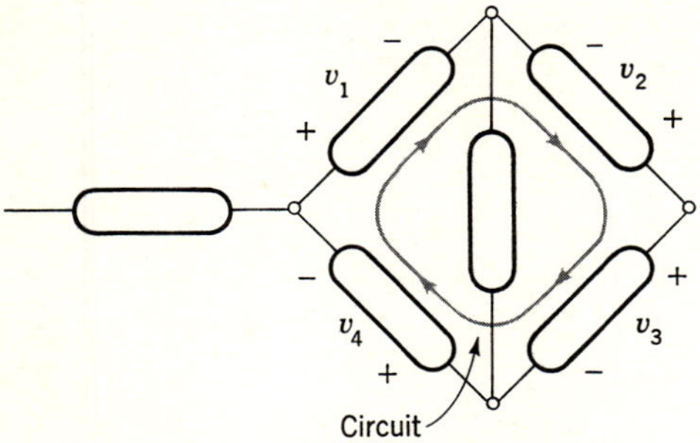

FIG. 4.3-7 KVL example: $v_1 = 2$, $v_2 = 3$, $v_3 = 4$.

Example 4.3-3 In the network of Fig. 4.3-8 there are four nodes, or junctions

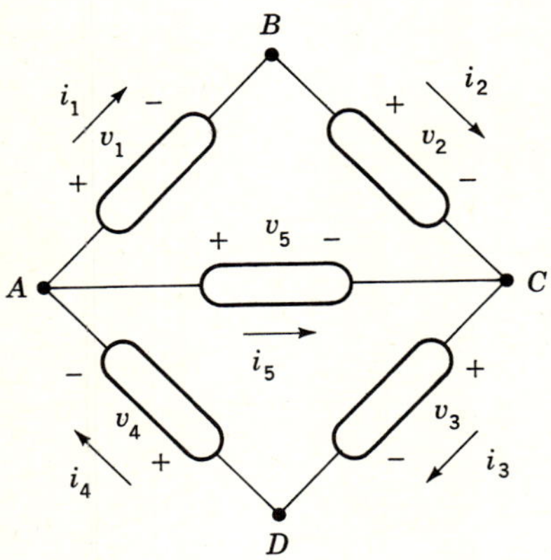

FIG. 4.3-8 Example of combined KVL and KCL use.

of elements; hence we can write for KCL four constraints as follows:

Node	*Equation*
A	$-i_1 - i_5 + i_4 = 0$
B	$i_1 - i_2 = 0$
C	$i_2 + i_5 - i_3 = 0$
D	$-i_4 + i_3 = 0$

(15)

Some observation will reveal that the system of Eq. (15) is overspecified; any three of the above equations can be used to derive the fourth.

Since in Fig. 4.3-8 we can identify three closed circuits, we can write three KVL equations as follows:

Circuit	*Equation*
1 2 5	$v_1 + v_2 - v_5 = 0$
1 2 3 4	$v_1 + v_2 + v_3 + v_4 = 0$
5 3 4	$v_5 + v_3 + v_4 = 0$

(16)

Here, the name of the circuit is formed by the subscripts of the elements in the circuit. Again, as in the case of Eqs. (15), Eqs. (16) are overspecified; any two equations can be used to derive the third.

The combined system of Eqs. (15) and (16), or their reduced forms consisting of any three KCL and any two KVL equations, constitutes the *topological constraints on the network* of Fig. 4.3-8. That is, regardless of what the five elements in the network are (such as resistors, voltage sources, inductors, capacitors, etc.), the above KVL and KCL constraints *must be met,* by virtue of the fact that the network elements are interconnected as in Fig. 4.3-8. When the element constraints are specified, then it is generally possible to solve for the currents and voltages of the network. For example, if the elements are all linear resistors, i.e.,

$$v_k = R_k i_k \qquad k = 1, 2, 3, 4, 5 \tag{17}$$

where R_k are positive real numbers, then solution of the combined KCL, KVL, and element constraints [Eqs. (15) to (17)] yields

$$v_k = i_k = 0 \qquad k = 1, 2, 3, 4, 5$$

that is, all currents and voltages in the network are zero.

We proceed now with a discussion of element constraints, and we shall return to a more systematic formulation and solution of all the combined constraints in Sec. 4.4.

Element Constraints

A network element, such as the one shown in Fig. 4.3-9*a* constrains the network variables associated with that element. Depending on the

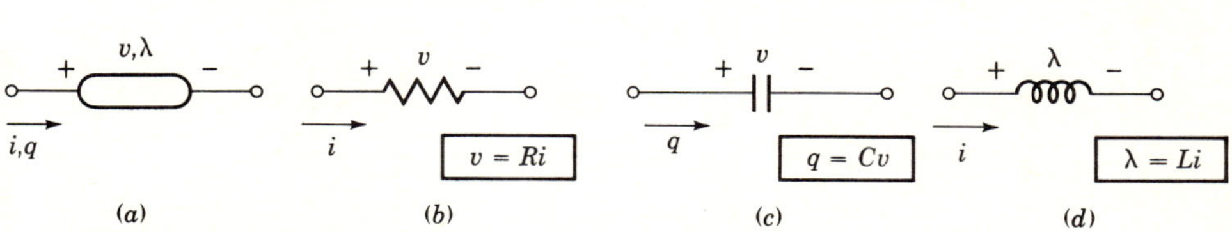

FIG. 4.3-9 Constraints of basic linear network elements.

variables that are constrained, the element is classified broadly as resistive, inductive, or capacitive. If the constraint is *linear,* then we have the three basic elements of Fig. 4.3-9*b, c,* and *d* called respectively *linear resistor, linear capacitor,* and *linear inductor.* As shown in the figure, a linear resistor constrains voltage and current to a constant ratio, the *resistance R;* a linear capacitor constrains charge and voltage to a constate ratio, the *capacitance C;* and a linear inductor constrains flux linkage and current to a constant ratio, the *inductance L.* Because of Eqs. (1) and (2), which state

$$i = \frac{dq}{dt} \qquad \text{and} \qquad v = \frac{d\lambda}{dt}$$

we can write for a linear capacitor

$$i = C\frac{dv}{dt} \tag{18}$$

and for a linear inductor

$$v = L\frac{di}{dt} \tag{19}$$

We shall deal with linear and nonlinear inductors and capacitors in Chap. 7. We limit our concern in this chapter to resistive elements, starting with the linear resistor. By common usage, the name *resistor* refers to a linear resistor with a positive real constant for R. Deviations from this type of constraint are identified with modifying adjectives, such as *nonlinear resistor, negative resistor,* and *time-varying resistor,* as we shall discuss later on.

In more detail, a linear resistor is a device which relates the current through it to the voltage across it (Fig. 4.3-9*b*) by:

At any time t

$$v(t) = Ri(t) \qquad \text{(Ohm's law)} \tag{20}$$

where the resistance R is a positive constant. The constant

$$G = \frac{1}{R} \tag{21}$$

is called the *conductance* of the resistor. The units of resistance are called *ohms*, Ω, while the units of conductance are called *mhos*, † ℧. Dimensionally,

$$\begin{aligned} &\text{(Volts)} = \text{(ohms)} \times \text{(amperes)} \qquad v = Ri \\ &\text{(Amperes)} = \text{(mhos)} \times \text{(volts)} \qquad i = Gv \end{aligned} \tag{22}$$

Graphically (Fig. 4.3-10), a linear resistor can be described by a straight line in the v-i plane, passing through the origin and having slope R or G, depending on whether i or v is along the horizontal axis.

Since, by Eq. (3) the power entering a resistor is given by

$$p = vi$$

†mho is ohm spelled backward.

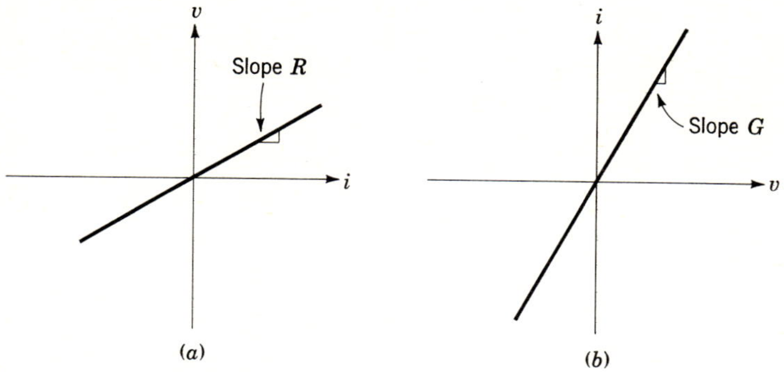

FIG. 4.3-10 Linear resistor i-v constraint.

it follows that

$$p = Ri^2 = Gv^2 \tag{23}$$

Since R (hence G) is positive, and i^2 and v^2 are always non-negative, it follows that the power entering a resistor is *always* non-negative. That is, a resistor can only consume power, and can never return it to the rest of the network to which it is connected. This power is dissipated in the form of heat. Sometimes it is a nuisance, because fans must be used to remove the generated heat from equipment such as computers; at other times, this "resistive" heat is generated intentionally as in toasters and electric heaters.

Besides linear resistors, there are many useful resistive network elements whose characteristics can be modeled by a curve (not a straight line) in the *i-v* plane. These devices can be further subdivided into two classes:

1. Current-controlled resistors
2. Voltage-controlled resistors

A *current-controlled* resistor is described by an equation of the form

$$v = r(i) \tag{24}$$

where $r(.)$ is a function and is described by a curve in the *i-v* plane, as shown in Fig. 4.3-11*a*. A *voltage-controlled* resistor is described by an equation of the form

$$i = g(v) \tag{25}$$

where g(.) is a function,† and is defined by a curve in the *v-i* plane as illustrated in Fig. 4.3-11*b*.

If the slope of the $r(.)$ or of the $g(.)$ graph is negative for some range of the current or voltage, then we say that the resistor exhibits *negative resistance* or *negative conductance.*

Diodes are a class of electronic devices that conduct current more easily in one direction than in the other. The *tunnel diode*, a semiconductor device developed by the Japanese researcher Esaki, exhibits a voltage-controlled

†The $g(.)$ and $r(.)$ functions are generally continuous.

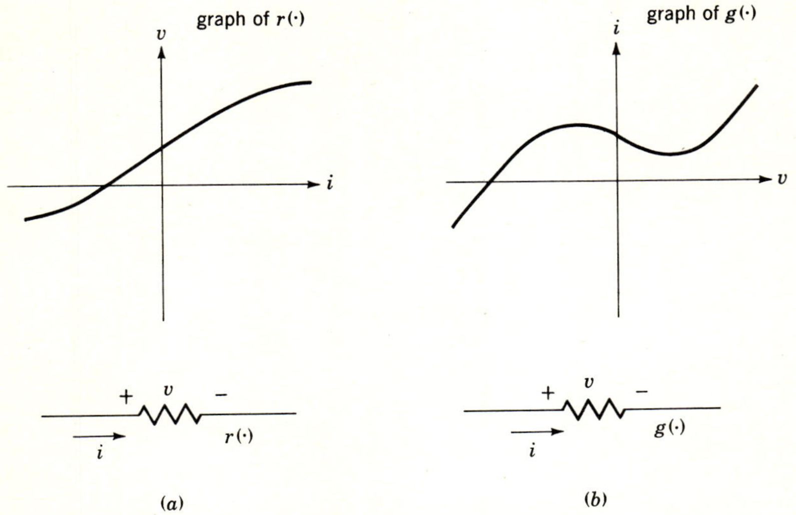

FIG. 4.3-11 Typical characteristics of: (a) *a current-controlled resistor;* (b) *a voltage-controlled resistor.*

negative resistance attributable to a quantum-electrical effect known as "charge tunneling." *Gas diodes,* an example of which is the neon-filled glow lamp, sometimes used as an indicator light, have current-controlled negative resistance behavior, as a consequence of the ionization process within the gas.

The common semiconductor junction diode is a two-terminal element that behaves like a large conductance for one polarity of applied voltage and like a small conductance for the opposite polarity of applied voltage. (For very rapid changes of terminal voltage or current, a resistive model is inadequate, and the model must be augmented to account for the effects of energy storage.) The *ideal-diode* symbol in Fig. 4.3-12*a* has the associated idealized characteristic indicated in Fig. 4.3-12*b*. The mathematical description of the idealized diode characteristic is given by

For all $i \geq 0$

$$v = 0 \tag{26}$$

For all $v \leq 0$

$$i = 0$$

The idealized characteristic is to be interpreted as a limiting form approached by the semi-ideal characteristic in Fig. 4.3-12*c*, as the forward conductance G_f and the back conductance G_b are allowed to become large and small, respectively. The actual characteristic of a semiconductor junction diode is shown in Fig. 4.3-12*d*. The mathematical form for this characteristic is

$$i = i_0(e^{v/v_0} - 1) \tag{27}$$

For germanium at room temperature, v_0 might be something like 25 mV. On

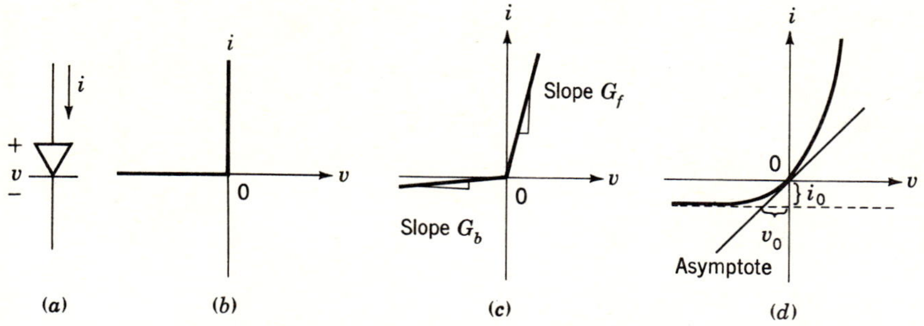

FIG. 4.3-12 The diode element. (a) *Network symbol for the ideal diode.* (b) *Ideal diode i versus v characteristic.* (c) *A nonideal characteristic.* (d) *Theoretical characteristic of semiconductor junction.*

scales of current and voltage for which i_0 and v_0 are small, the theoretical junction characteristic approaches the appearance of the idealized diode characteristic. As we shall see in Sec. 4.7, the ideal diode is a primitive network element that can be used to synthesize many interesting nonlinear functions.

Besides linear and nonlinear resistors, there is a class of useful elements called *sources.* Sources are circuit elements which are used to provide excitation waveforms or inputs to the network. Some common sources are batteries, which provide a constant voltage, and the "plug in the wall" which generates a time-varying voltage,

$$v(t) = 110\sqrt{2}\,\sin 2\pi 60t$$

From an elementary point of view, an *ideal voltage source* with source voltage v_s can be interpreted as a resistive element with the property that the voltage v across it is given by

$$v = v_s$$

independent of the current i through it.

An *ideal current source* with source current i_s is an element with the property that the current i through it is given by

$$i = i_s$$

independent of the voltage v across it.

The element symbols and current-voltage characteristics associated with independent voltage and current sources are shown in Fig. 4.3-13. The battery symbol (*a*) denotes a constant source voltage, while the more general voltage source symbol (*b*) may be used to represent either a constant voltage or a voltage having some prescribed variation with time. As indicated by the characteristic (*c*), the voltage is specified, but the current is determined by the properties of the remainder of the network to which the voltage source is connected. Similar observations hold for the current source (*d*) and its characteristic (*e*). The networks (*f*) and (*g*) are more realistic models of actual sources, and are entirely equivalent with respect to their terminal behavior, which is described

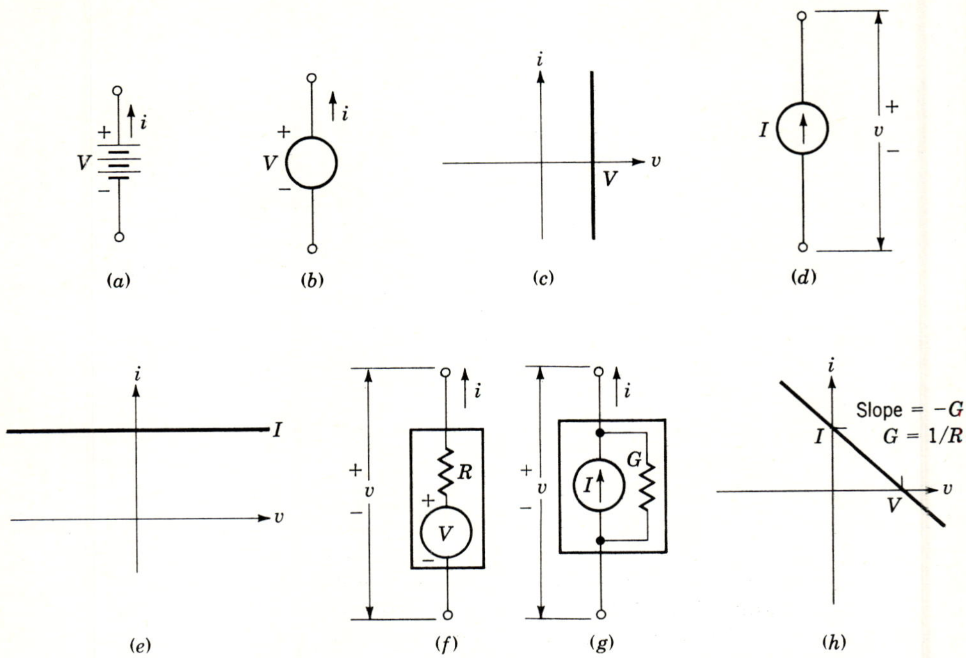

FIG. 4.3-13 Independent sources. (a) *Battery.* (b) *Voltage source.* (c) *Voltage-source constraint.* (d) *Current source.* (e) *Current-source constraint.* (f) *Voltage source with internal resistance R.* (g) *Current source with internal conductance G.* (h) *Source constraint.*

by the characteristic (*h*). Either network may be made to approach the behavior of either an ideal voltage source or an ideal current source by appropriate choice of the parameters V, I, R, or G. For example, the network in Fig. 4.3-13*f* approaches an ideal current source if we allow both V and R to become large, while holding the quotient V/R fixed.

The next element to be added to our list of basic network elements is the *dependent source.* A dependent source is a voltage or current constraint at one pair of terminals that is proportional to the current or voltage applied or appearing at some other pair of terminals. Hence there are four different kinds of dependent sources (current or voltage, controlled by either current or voltage)†; and for each kind the two terminal pairs may be separated as in Fig. 4.3-14*a*, or they may have a common terminal as in Fig. 4.3-14*b* and *c*. Two of the four kinds of dependent sources are displayed in Fig. 4.3-14. A third kind is shown in Fig. 4.3-15*a*, together with its "output terminal-pair characteristics" in Fig. 4.3-15*b*. The output characteristics are a family of curves associated with different values of the input "control" variable, which in this case is the current i_1. For the dependent source in Fig. 4.3-14*a*, the mathematical description is

†A voltage source whose voltage depends on some current is called a *current-controlled voltage source.* Likewise, we can have a *voltage-controlled voltage source.*

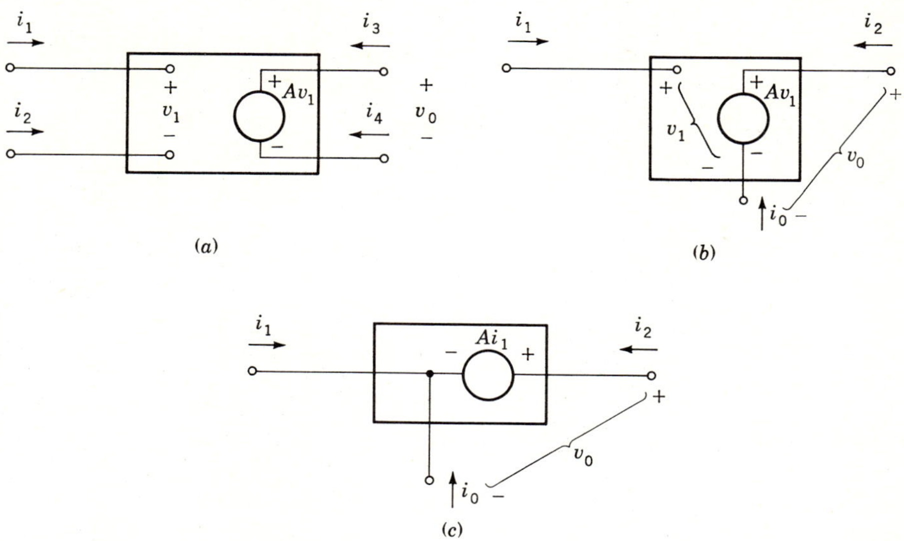

FIG. 4.3-14 *Dependent voltage sources.* (a) *General voltage-controlled voltage source.* (b) *Three-terminal voltage-controlled voltage source.* (c) *Three-terminal current-controlled voltage source.*

$$i_1 = i_2 = 0$$
$$i_3 + i_4 = 0$$
$$v_0 = Av_1 \tag{28}$$

where A is constant; for the dependent source of Fig. 4.3-14*b*,

$$\begin{aligned} i_1 &= 0 \\ i_2 + i_0 &= 0 \\ v_0 &= Av_1 \end{aligned} \tag{29}$$

where A is constant; and for that of Fig. 4.3-14*c*,

$$\begin{aligned} i_1 + i_2 + i_0 &= 0 \\ v_0 &= Ai_1 \end{aligned} \tag{30}$$

where A is constant.

Finally, for the *current-controlled current source* of Fig. 4.3-15*a*.

$$\begin{aligned} v_1 &= 0 \\ i_2 &= Bi \end{aligned} \tag{31}$$

where B is constant. Identification of similar constraint equations for a *voltage-controlled current source* remains a suggested exercise for the reader.

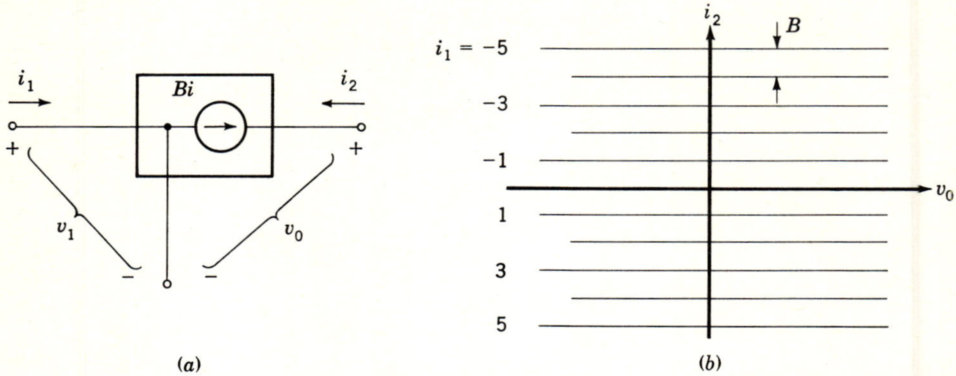

FIG. 4.3-15 (a) *A current-controlled current source.* (b) *The associated family of i_2 versus v_0 curves, with i_1 the parameter.*

Example 4.3-4 *Rectification* A simple network which forms the key to the conversion of alternating to direct current is shown in Fig. 4.3-16*a*. The source waveform is given by

$$v_s(t) = 100 \sin 2\pi t \tag{32}$$

and is shown in Fig. 4.3-16*b*. The diode characteristic is as in Fig. 4.3-12*c* with

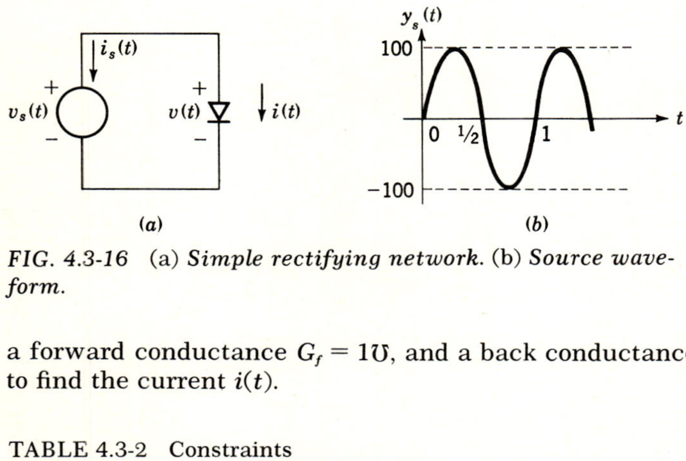

FIG. 4.3-16 (a) *Simple rectifying network.* (b) *Source waveform.*

a forward conductance $G_f = 1℧$, and a back conductance $G_b = 0.01℧$. We wish to find the current $i(t)$.

TABLE 4.3-2 Constraints

Topological		*Element*	
KVL	*KCL*	*Diode*	*Source*
$v_s(t) = v(t)$	$i_s(t) = -i(t)$	For $v \geq 0$: $i(t) = v(t)$ For $v \leq 0$: $i(t) = 0.01v(t)$	$v_s(t) = 100 \sin 2\pi t$

The topological constraints (KVL and KCL) and the element constraints (diode and source) are shown in Table 4.3-2.

Evidently,

For $v_s(t) \geq 0$,

$$i(t) = v(t) = v_s(t) = 100 \sin 2\pi t$$

For $v_s(t) \leq 0$,

$$i(t) = 0.01v(t) = 0.01v_s(t) = \sin 2\pi t$$

and the answer therefore is

$$\left.\begin{array}{l} \text{For } n \leq t \leq n + \tfrac{1}{2} \\ \qquad i(t) = 100 \sin 2\pi t \\ \text{for } n + \tfrac{1}{2} \leq t \leq n + 1 \\ \qquad i(t) = \sin 2\pi t \end{array}\right\} \quad n = 0, 1, 2, \ldots \tag{34}$$

which is shown graphically in Fig. 4.3-17. Observe that with this waveform for

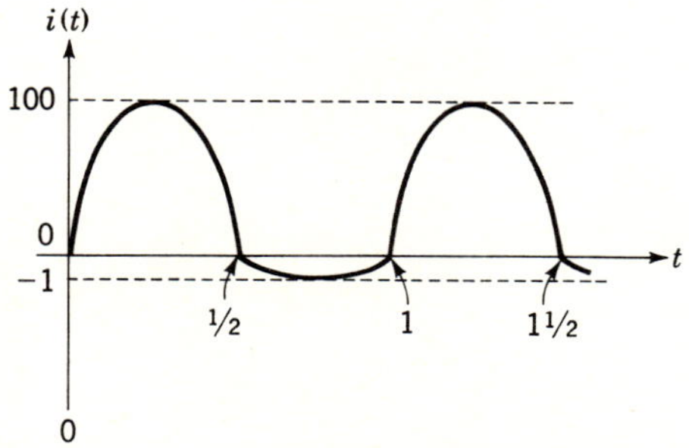

FIG. 4.3-17 The current waveform i(t).

$i(t)$, *all* constraints are satisfied. Observe specifically that the source current is given by

$$i_s = i(t)$$

This current does not disturb the source voltage, since the voltage $v_s(t)$ of the latter is given by Eq. (32) and is independent of the current passing through the source.

Thus, the alternating source voltage which is equally positive and negative has given rise, through the diode, to a current which is almost exclusively positive. As the ratio G_f/G_b increases, i.e., as the diode approaches more and more ideal characteristics, the current waveform of Fig. 4.3-17 approaches more and more half of the time a positive sinusoid, and the rest of the time the value zero. In conventional "battery-charger" circuits for radios and other appliances, a capacitor is often added to this network, as we shall see in Chap. 6, in order to average out the peaks and valleys of Fig. 4.3-17.

4.4 FORMULATION AND SOLUTION OF RESISTIVE NETWORK EQUATIONS

The purpose of this section is to show how the topological and element constraints of Sec. 4.3 lead to an orderly formulation of network equations, and how these equations are, in turn, solved for the various network variables. The objective of formulation is to establish a set of equations which reflect all the constraints *without unnecessary overspecification,* and the objective of solution is to get values or explicit expressions for the network variables of interest. Since we are dealing with resistive networks, where element constraints are algebraic, and since KVL and KCL are linear algebraic equations, we expect that formulation will lead to a system of algebraic equations. Moreover these equations will be linear if the constraints are linear. Of the many possible ways of formulating resistive network equations, we shall concentrate in this section on the so-called *node-voltage* method of formulation. We start, accordingly with a review of node voltages.

Node Voltages

The uniqueness of potential permits us to assign a voltage to every node of a network with respect to a common, reference, or datum node. This datum node, often called *ground,* and symbolized by three parallel lines, may be external to the network, as in Fig. 4.4-1*a,* or part of the network, as in Fig. 4.4-1*b*.

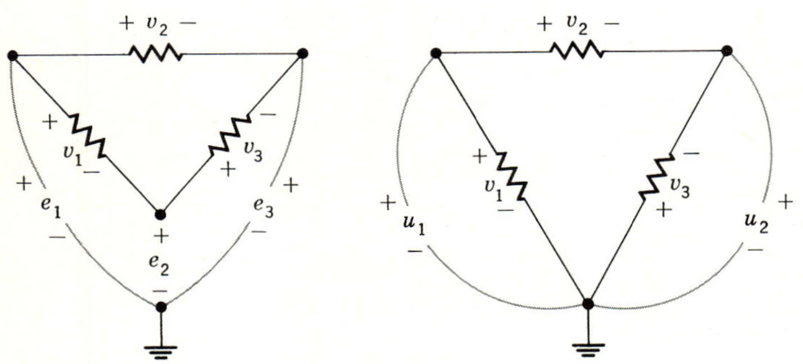

FIG. 4.4-1 Illustration of node voltages.

In Fig. 4.4-1*a*, the node voltages are e_1, e_2, and e_3. The *branch voltages* v_1, v_2, and v_3, i.e., the voltages across each element, are related to these node voltages by KVL as follows:

$$\begin{aligned} v_1 &= e_1 - e_2 \\ v_2 &= e_1 - e_3 \\ v_3 &= e_2 - e_3 \end{aligned} \tag{1}$$

In other words, the circuit or loop to which KVL applies is formed by the branch under consideration and the datum node, so that every branch voltage equals the difference between the node voltages on either side of the branch.† By the same reasoning, the three branch voltages of Fig. 4.4-1*b* are related to the two node voltages u_1 and u_2 as follows:

$$\begin{aligned} v_1 &= u_1 \\ v_2 &= u_1 - u_2 \\ v_3 &= -u_2 \end{aligned} \tag{2}$$

The sufficiency of two node voltages to specify all the branch voltages of the three-node network of Fig. 4.4-1 illustrates the more general fact that *n node voltages are sufficient to specify all the branch voltages of an* $(n+1)$*-node network,* if the datum node is a node in the network. As a matter of fact, $n-1$ node voltages *are not* sufficient to specify all the branch voltages of an $(n+1)$-node network. Specifically, it is not possible to express voltages of branches connected to the node, whose voltage with respect to ground is not assumed known. Thus, n node voltages are both *necessary* and *sufficient* to specify all the branch voltages of any $(n+1)$-node network. The general procedure which is illustrated in Fig. 4.4-2*a* is as follows:

1. Assign one of the network nodes to be ground, and assign node potentials $e_1, e_2, \ldots, e_n$ to the remaining nodes, with respect to ground.
2. As shown in Fig. 4.4-2*a*, express every branch voltage v_k as the difference of the two node voltages e_i and e_j associated with that branch (element).

Finally, observe that the specification of branch voltages in terms of node voltages guarantees satisfaction of KVL everywhere in the network, since the use of node voltages asserts the uniqueness of potential at each network node.

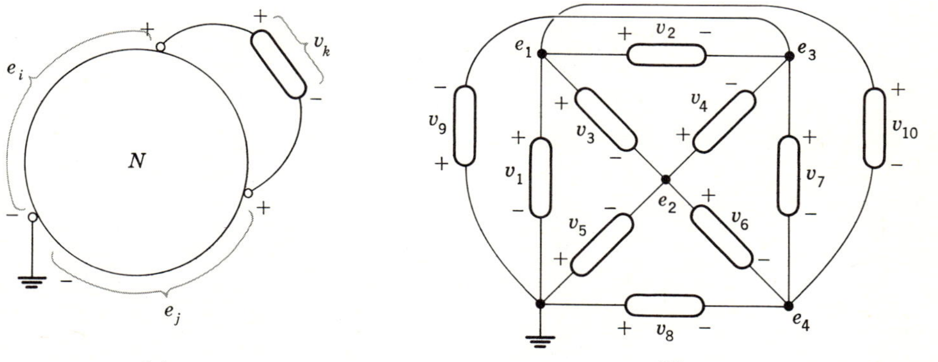

FIG. 4.4-2 (a) *Relationship between node and branch voltages.* (b) *The full-graph network of five nodes.*

†In terms of the fluid analogies of Sec. 4.3, this is analogous to saying that the pressure across a pipe equals the difference of the pressures at the two ends of the pipe, with respect to *any* reference pressure (datum).

Example 4.4-1 *A Full-graph Network* The five-node network of Fig. 4.4-2*b* is called a full-graph network because every node is connected to every other node through an element. Following the procedure outlined in this section, we chose the lower left node in Fig. 4.4-2*b* as the ground node, and identify e_1, e_2, e_3, and e_4 as the voltages of the remaining four nodes with respect to ground (it is assumed but not shown here that all the reference directions of these node voltages have their minus signs at the ground node). A full-graph network of n nodes has $\binom{n}{2}$ branches,† i.e., in this case $\binom{5}{2}$ which is (5) (4)/2 = 10.

These 10 branch voltages are related to the four node voltages as follows:

$$\begin{aligned} v_1 &= e_1 & v_6 &= e_2 - e_4 \\ v_2 &= e_1 - e_3 & v_7 &= e_3 - e_4 \\ v_3 &= e_1 - e_2 & v_8 &= -e_4 \\ v_4 &= e_2 - e_3 & v_9 &= -e_3 \\ v_5 &= -e_2 & v_{10} &= e_1 - e_4 \end{aligned} \tag{3}$$

Observe the economy achieved in being able to express any one of 10 branch voltages through only 4 node voltages. If this were a 100-node full-graph network, then this economy would be even more impressive since 99 node voltages would be adequate to express any one of 4,950 branch voltages.

Note, finally that KVL is guaranteed to hold everywhere.* For example in the circuit formed by elements 1, 2, 7, and 8 we have

$$v_1 + v_8 - v_7 - v_2 = 0 \tag{4}$$

Substituting Eq. (3) into the left side of Eq. (4) yields

$$e_1 + (-e_4) - (e_3 - e_4) - (e_1 - e_3) \tag{5}$$

which is identically equal to zero, as expected.

General Node-voltage Equations

Formulation of the network equations entails satisfaction of KVL, KCL, and the element constraints. The node-voltage method of formulating network equations achieves this objective for networks consisting of voltage-controlled resistors and current sources as follows:

1. Node voltages are assigned, and are used to express all branch voltages. *This step satisfies KVL,* everywhere in the network.
2. The element constraints are used to specify all the branch (element) currents

†All the combinations of n nodes taken two at a time.

*Because, as we discussed earlier, every node has a unique potential, the assigned node voltage.

through resistors in terms of the branch voltages, and hence in terms of the node voltages. The current through current sources is specified independently of branch voltages, as given. *This step assures satisfaction of all the element constraints,* and specification of all branch currents.

3. KCL is applied at every node, other than the reference node, to relate the element currents of step 2, above. *This step guarantees satisfaction of KCL* everywhere in the network.

The method will be justified after the following example:

Example 4.4-2 *Node-voltage Analysis of a Simple Network* Consider the three-node network of Fig. 4.4-3 with node voltages, branch voltages, branch currents, and resistor constraints as shown in that figure.

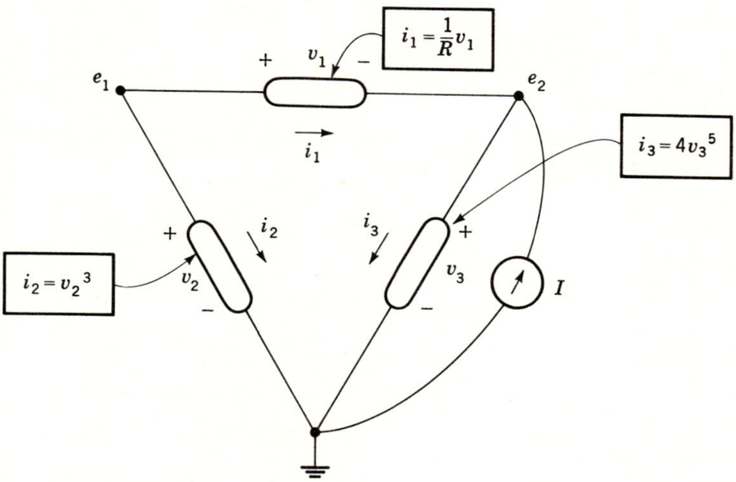

FIG. 4.4-3 A simple network.

Step 1 of the node-voltage method leads to the following relations which assure satisfaction of KVL:

$$
\begin{aligned}
v_1 &= e_1 - e_2 \\
v_2 &= e_1 \\
v_3 &= e_2
\end{aligned}
\tag{6}
$$

Step 2 involves the following relations:

$$
\begin{aligned}
i_1 &= \frac{1}{R} v_1 = \frac{1}{R}(e_1 - e_2) \\
i_2 &= v_2{}^3 = e_1{}^3 \\
i_3 &= 4v_3{}^5 = 4e_2{}^5
\end{aligned}
\tag{7}
$$

which along with the current I of the current source† ensure satisfaction of all the element constraints.

Finally, step 3 leads to the following KCL relations:

$$-i_1 - i_2 = 0$$

$$i_1 - i_3 + I = 0$$

which, using Eqs. (7), are rewritten in the final form:

$$\begin{aligned} -\frac{1}{R}(e_1 - e_2) - e_1{}^3 = 0 \\ \frac{1}{R}(e_1 - e_2) - 4e_2{}^5 + I = 0 \end{aligned} \tag{8}$$

Equations (8) are the desired result, for they relate the two unknown node voltages to the known current-source value I, *taking into account KVL, KCL, and the element constraints.* Solution of Eqs. (8) enables us to know the value of *every* network variable. For example, if $I = 129$, and $R = 1$, then Eqs. (8) become

$$\begin{aligned} e_1 - e_2 + e_1{}^3 = 0 \\ e_1 - e_2 - 4e_2{}^5 = -129 \end{aligned} \tag{9}$$

whose solution is $e_1 = 1$, $e_2 = 2$. From Eqs. (6), the branch voltages are

$$v_1 = 1 - 2 = -1$$

$$v_2 = 1$$

$$v_3 = 2$$

From Eqs. (7), the element currents are

$$i_1 = \frac{-1}{1} = -1$$

$$i_2 = (1)^3 = 1$$

$$i_3 = 4(2)^5 = 128$$

All these values are also shown in Fig. 4.4-4, from where it can be easily ascertained that KVL, KCL, and all the element constraints are satisfied.

Let us now look a little more carefully at the node-voltage method of formulation. The first step, i.e., the assignment of n node voltages is by construction possible for any given $(n + 1)$-node network, as has been already indicated. This step assures satisfaction of KVL and poses no conditions on the network topology, or on the element constraints.

The second step, i.e., the conversion of each element voltage into an element current poses the constraint that the elements be *voltage-controlled resistors, or current sources.* This ensures that the currents through every

†Which, by definition of an independent current source, is independent of the voltage e_2, across that source.

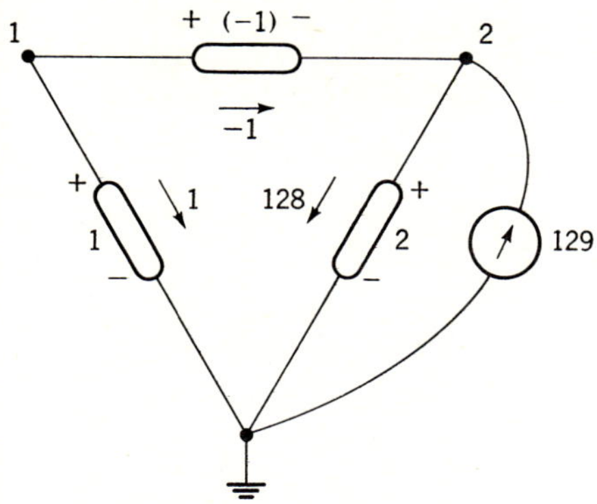

FIG. 4.4-4 The network of Fig. 4.4-3 solved.

element are known or can be expressed in terms of the node voltages. This step satisfies *all* element constraints, and imposes no further restrictions on the types of elements or on network topology.

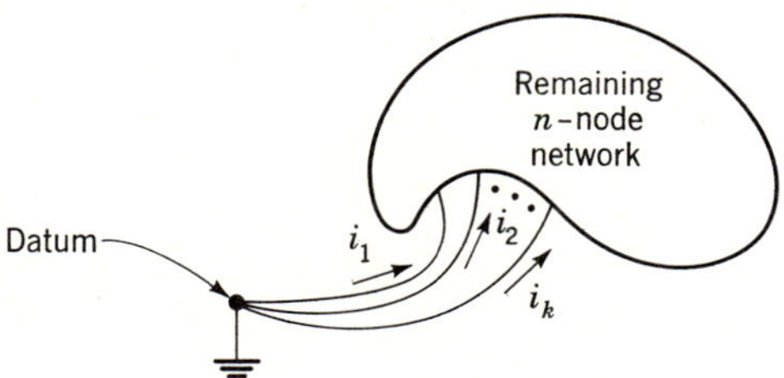

FIG. 4.4-5 Satisfaction of KCL in an $(n+1)$ node network.

The final step, i.e., the writing of n KCL equations at each of the network nodes, except datum, ensures satisfaction of KCL everywhere in the network as will be shown by using Fig. 4.4-5. First, observe that if KCL is satisfied at every one of the n other-than-datum nodes, then KCL is also satisfied for all the currents entering the n-node network which contains these nodes. This assertion is easy to prove by induction on the number of nodes contained in a subnetwork, and the reader is invited to develop a proof. Hence, it is true that

$$\sum_{j=1}^{k} i_j = 0 \tag{10}$$

But the sum of Eq. (10) is precisely the sum of currents *leaving* the $(n + 1)$st datum node. Hence, KCL is also satisfied at that node. From the above, it follows that Step 3 ensures satisfaction of KCL everywhere in the network. Observe that writing these KCL equations imposes no constraints on the network topology or on the elements making up the network. Hence we conclude that

THEOREM 4.4-1 The formulation of node-voltage equations is always possible for networks consisting of voltage-controlled resistors and independent current sources.

Observe that trouble may arise if there is a network node to which current sources are connected, so that the sum of their currents does not satisfy KCL, as shown in Fig. 4.4-6*a*. Writing the nodal equation at that node would

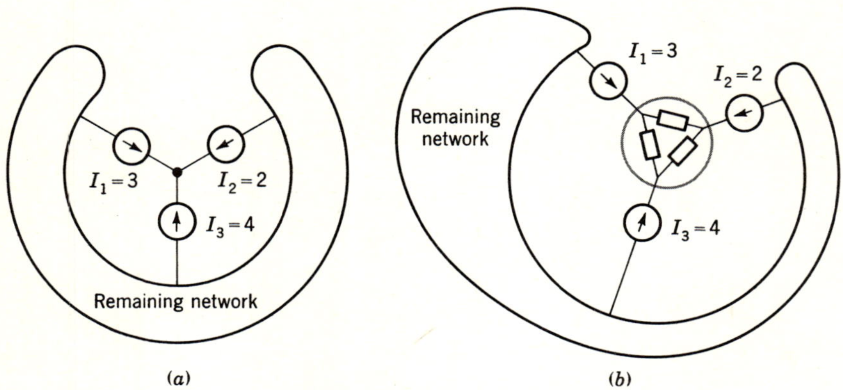

FIG. 4.4-6 Violation of KCL by current sources.

result in

$$I_1 + I_2 + I_3 = 0$$

which is not possible with the values assigned to these currents in Fig. 4.4-6*a*. Hence, *although formulation is possible, solution is not.* Observe further that it is not essential for the current sources to be connected at a common node in order to violate KCL. Figure 4.4-6*b* illustrates such a case, where the sum of the currents entering the encircled network does not satisfy KCL.

Whenever there are voltage sources or dependent sources in a network, the above formulation must be augmented with appropriate additional constraints. Although in principle this can always be done, the resulting equations exceed in number the number of nodes in the network, thereby increasing the solution effort (computer time for the solution of N linear equations grows as N^3). For this reason, we discuss in the following section formulation procedures for three cases of linear networks as follows:

1. Linear resistive networks with current sources
2. As in case 1 above, with the addition of voltage sources
3. As in case 2 above, with the addition of all four types of dependent sources.

Although these procedures are not conceptually any more fundamental than the unstructured use of all network constraints, they nevertheless guarantee formulation of N or fewer equations for any $(N + 1)$-node network in one of the three classes listed above.

Linear Node-voltage Equations: Current Sources

A common and special case of networks consisting of voltage-controlled resistors and independent current sources is the class of networks consisting of *linear resistors* and *independent current sources*. We start with an example.

Example 4.4-3 *Node-voltage Analysis of Linear Networks* Consider the network of Fig. 4.4-7 consisting of linear resistors of conductance G_{ij} and current

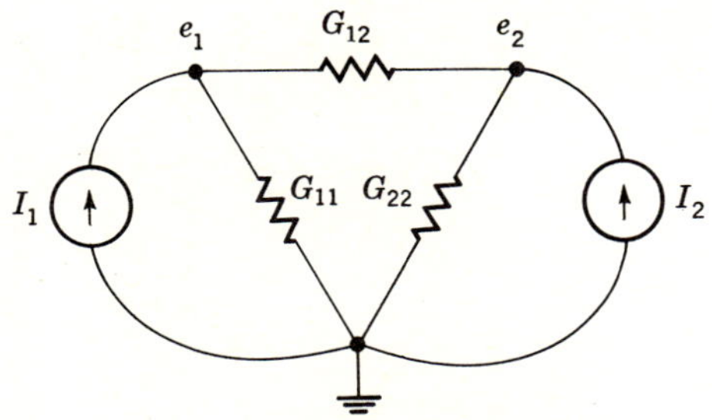

FIG. 4.4-7 Example of node-voltage analysis of linear networks.

sources I_1 and I_2. Execution of steps 1, 2, and 3 of the node-voltage formulation method leads to the following equations:

$$I_1 + G_{12}(e_2 - e_1) - G_{11}\, e_1 = 0 \tag{11}$$

$$I_2 + G_{12}(e_1 - e_2) - G_{22}\, e_2 = 0$$

These may be rearranged as follows:

$$(G_{11} + G_{12})e_1 - G_{12}e_2 = I_1 \tag{12}$$

$$-G_{12}e_1 + (G_{22} + G_{12})e_2 = I_2$$

Equations (12) in the above example reflect the following general facts which apply to the formulation of node-voltage equations to a linear network of any size:

1. In the first equation the coefficient of e_1 is the sum of all conductances connected to node 1. In the second equation the coefficient of e_2 is the sum of all conductances connected to node 2. *In general, in the jth equation the coefficient of e_j is the sum of all conductances connected to the jth node.*
2. In the first equation the coefficient of e_2 is (with a minus sign) the sum of all conductances connected between nodes 1 and 2. In the second equation the coefficient of e_1 is (with a minus sign) the sum of all conductances connected between nodes 2 and 1. *In general, in the jth equation the coefficient of e_k ($k \neq j$) is (with a minus sign) the sum of all conductances connected between nodes j and k.*
3. In the first equation the right-hand term is I_1, the value of the current entering node 1 from a current source. In the second equation the right-hand term is I_2, the value of the current entering node 2 from the current source connected to that node. *In general, the right-hand term of the jth equation is the sum of the currents entering the jth node from all current sources connected to that node.*

The above observations and their generalizations are immediate consequences of the node-voltage method of formulating network equations, and of the linearity of the resistors. They are summarized in a set of rules, or algorithm, that can be easily implemented by a computer program (Table 4.4-1).

TABLE 4.4-1 Node-voltage Method of Formulating Linear Network Equations†

Class of Networks Consisting of Linear Resistors and Current Sources

1. Choose one network node as datum, and assign node voltages $e_1, e_2, \ldots, e_n$ to the remaining nodes.
2. Write the n desired equations. The jth equation ($j = 1, 2, \ldots, n$) is written as follows:

$$\left(\sum_k G_{jk}\right)e_j - \sum_{k \neq j} G_{jk}\, e_k = I_j$$

where

$$k = 1, 2, \ldots, n \text{ and}$$

$\sum_k G_{jk}$ is the sum of *all* conductances connected to node j.

G_{jk} is the conductance (or sum of conductances) connected between nodes j and k. Note that G_{jj} denotes the conductance between node j and ground.

I_j is the sum of current-source currents entering node j.

Linear Node-voltage Equations; Voltage Sources

So far we have been dealing with networks consisting of linear resistors and current sources. We will now extend the node-voltage method of formulation to networks which contain voltage sources in addition to the foregoing.

Example 4.4-4 *Node-voltage Equations: Voltage Sources* Consider the network of Fig. 4.4-8 which is made up of three linear resistors, a current source, and a voltage source. If we assign a current i flowing through the voltage source and node voltages e_1 and e_2 as before, we can formulate conventional node voltage equations as follows:

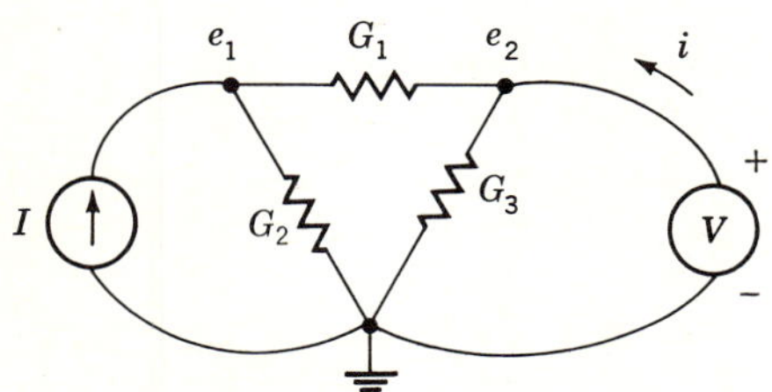

FIG. 4.4-8 A simple network.

†In this and in subsequent formulation approaches, we shall assume that the network under consideration has n + 1 nodes.

$$\begin{aligned} (G_1 + G_2)e_1 - G_1e_2 &= I \\ -G_1e_1 + (G_1 + G_3)e_2 &= i \end{aligned} \tag{13}$$

These equations are not satisfactory, since they include i which is a third unknown, and they do not take into account the value V of the voltage source. If we append to Eqs. (13), however, the equation

$$e_2 = V \tag{14}$$

then we have a system of three equations and three unknowns, which constitutes an adequate formulation of the network equations.

In general, every voltage source that is connected between two nodes of a network introduces a new unknown—the current through the voltage source—and provides us with a new equation which constrains the difference of the node voltages on either side of the voltage source to equal the value of that source. This is illustrated, in general, for the $(n + 1)$-node network of Fig. 4.4-9.

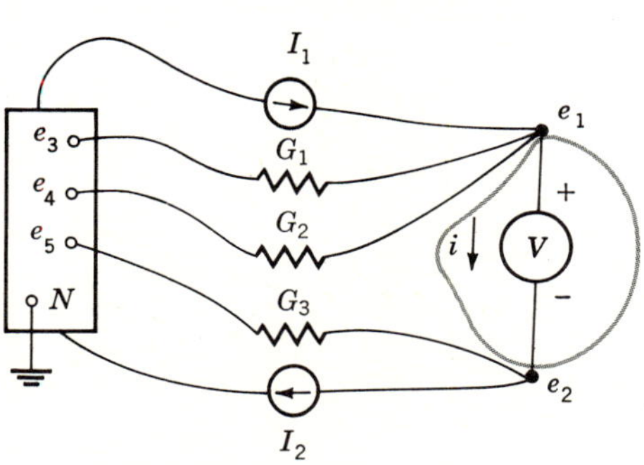

FIG. 4.4-9 The general case.

Here, writing regular node-voltage equations yields the following at nodes 1 and 2:

At node 1

$$(G_1 + G_2)e_1 - G_1e_3 - G_2e_4 = I_1 - i \tag{15}$$

At node 2

$$G_3e_2 - G_3e_5 = -I_2 + i$$

The n node node-voltage equations which contain the above two, have $n + 1$ unknowns—the node voltages $e_1, e_2, \ldots, e_n$, and the voltage-source current i. They are augmented by one more equation which relates e_1, e_2, and V by KCL, as follows:

$$e_1 - e_2 = V \tag{16}$$

thus resulting in $n + 1$ equations in $n + 1$ unknowns.

Often, the currents through the voltage sources of a network are of no interest, and we wish to eliminate them at the formulation stage so that the total number of equations is small. This elimination can be easily done by a topological approach. Specifically, instead of writing one KCL equation at node 1 and another at node 2, we shall write only one KCL equation, avoiding the voltage-source current i. This is done by (1) using $e_2 + V$ in place of e_1, and (2) "enclosing" the voltage source by a line, as shown in Fig. 4.4-9, and then writing one KCL equation for all currents entering the encircled network. This approach yields in addition to the $n - 2$ equations (for all nodes other than 1 and 2), the following equation:

$$I_1 + G_1[e_3 - (e_2 + V)] + G_2[(e_4 - (e_2 + V)] + G_3(e_5 - e_2) - I_2 = 0 \tag{17}$$

which can be rearranged as follows:

$$(G_1 + G_2)(e_2 + V) + G_3e_2 - G_1e_3 - G_2e_4 - G_3e_5 = I_1 - I_2 \tag{18}$$

Hence, the total number of network equations is now $n - 1$, and the total number of unknowns is also $n - 1$ ($e_2, e_3, \ldots, e_n$), since $e_1 = e_2 + V$.

This topological approach may be used in any $(N + 1)$-node network by enclosing voltage sources, and renaming node voltages in terms of the voltage sources. It will always give rise to N or fewer equations. More systematically the approach is as shown in Table 4.4-2.

TABLE 4.4-2 Node-voltage Method of Formulating Linear Network Equations

Class of Networks Consisting of Linear Resistors, Independent Current Sources, and Independent Voltage Sources

1. Enclose *each* group of voltage sources that form a connected path in a subnetwork, along with any other elements connected between voltage-source nodes of that group.
2. Assign a node voltage e_i to exactly one node of each such subnetwork and to each free node, i.e., a node that is not in a subnetwork. Express the voltages of all other nodes within a subnetwork in terms of the node voltage e_i assigned to that group and the voltage-source values V_i. If any node of a subnetwork is the network datum, use this node as a reference for that group, with node voltage zero.
3. Write a KCL equation for every one of the free nodes, and for every one of the subnetworks. Do not write a KCL equation for the subnetwork, if any exists, that contains the network ground. In writing these equations, use current-source values, and express all other branch currents, using the conductance values and the node voltages assigned by step 2, above, to the two nodes of each such conductance.

This approach is illustrated in Fig. 4.4-10. Here, voltage sources V_2, V_3, and V_4, along with all resistors connected between their respective nodes, are enclosed in *one* subnetwork N_2 (because these sources are connected), and source V_1 is enclosed by another subnetwork N_1. Next, a node voltage, e_1, is assigned to one node of N_1, and the other node of N_1 is related to e_1 through KVL and V_1. The two upper nodes of N_2 are related to the bottom node which is datum (or 0) by KVL; that is, the node on the left is $V_3 - V_2$, and the node on the right is $V_3 + V_4$. Finally, e_2 is assigned to the only other node which is a free node since it is not in a subnetwork.

Two node-voltage equations will be written, one for N_1, and one for node 2. No equation is written for N_2 because N_2 includes ground. Observe that there are two unknowns, e_1 and e_2. Hence the number of equations matches the number of unknowns. The two KCL equations are:

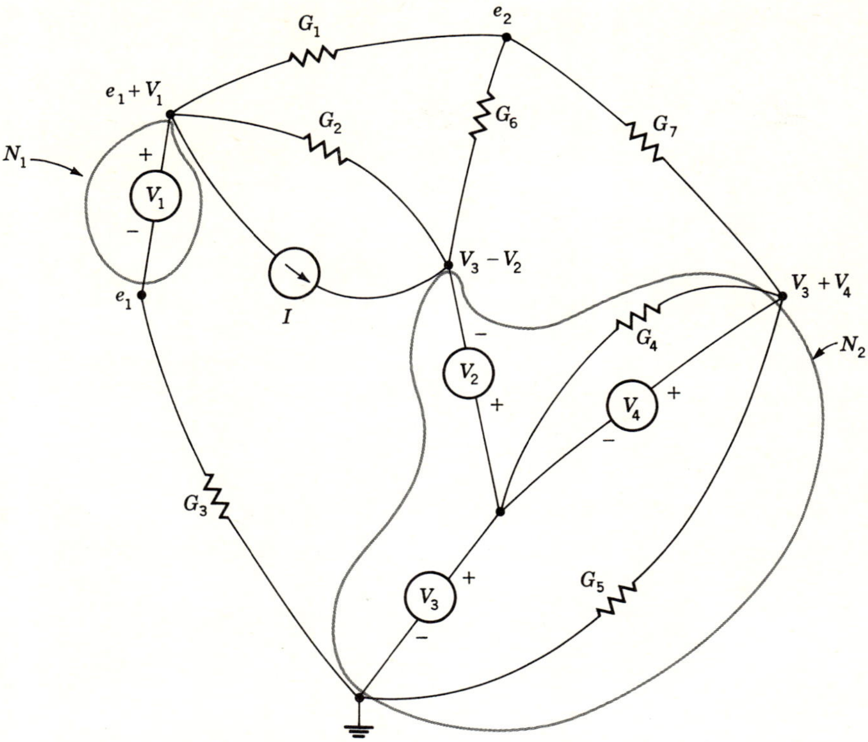

FIG. 4.4-10 Illustration of the topological approach.

For N_1

$$G_3e_1 + G_1[(e_1 + V_1) - e_2] + G_2[(e_1 + V_1) - (V_3 - V_2)] = -I \tag{19}$$

For node 2

$$G_1[e_2 - (e_1 + V_1)] + G_6[e_2 - (V_3 - V_2)] + G_7[e_2 - (V_3 + V_4)] = 0$$

Observe that having solved the above for e_1 and e_2, we know, or can obtain by simple addition or multiplication, *all* the currents and voltages in the network.

We will give next a plausibility argument for the general validity of the above approach. First of all, the formation of subnetworks encloses *all* the voltage-source currents, so that these currents will not appear in the formulated equations. Second, because each subnetwork contains connected sources, it is always possible, using KVL, to rename all nodes in the subnetwork in terms of one node voltage and the enclosed voltage-source values. (Note that trouble could arise here if the voltage sources are connected in a circular path or loop, and their values, which are independently supplied, violate KVL. We shall discuss this matter, which is equivalent to KCL violation by current sources connected at a node, later in the section.) Thus, we have a node voltage associated with each subnetwork and with each *free node,* i.e., a node which is not in a subnetwork. A special case arises when the subnetwork includes ground, for then the node voltage used for reference purposes within that network is ground, which is known and has zero value. Finally, when KCL equations are written, we are guaranteed by the above approach that there will be as many KCL equations as there are subnetworks plus free nodes, i.e., as many KCL equations as there are unknowns.† Moreover, in writing these KCL equations, we will encounter either known values of current sources or currents which can be expressed in terms of known conductances and voltages assigned at network nodes. In the special case of a subnetwork including ground, no KCL equation need be written for that network, since satisfaction of KCL at every other free node and subnetwork ensures satisfaction of KCL at the network which excludes the subnetwork under discussion; this, however, ensures satisfaction of KCL at that subnetwork, since the currents entering the latter are the currents leaving the former.

Linear Node-voltage Equations: Dependent Sources

This subsection discusses how node-voltage equations are written for networks which contain dependent sources, in addition to resistors, current sources, and voltage sources.

Example 4.4-5 *A Transistor Amplifier* A transistor is a three-terminal de-

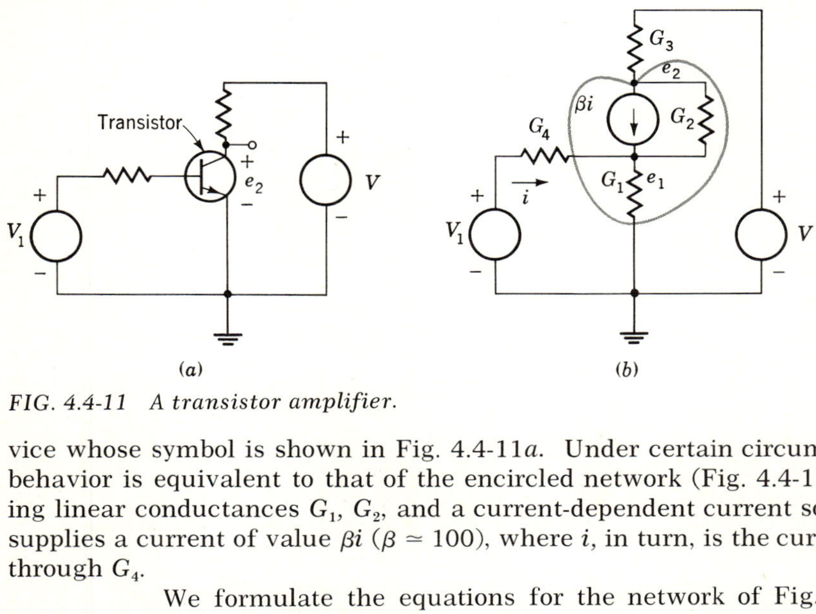

FIG. 4.4-11 A transistor amplifier.

vice whose symbol is shown in Fig. 4.4-11*a*. Under certain circumstances, its behavior is equivalent to that of the encircled network (Fig. 4.4-11*b*) containing linear conductances G_1, G_2, and a current-dependent current source which supplies a current of value βi ($\beta \simeq 100$), where i, in turn, is the current flowing through G_4.

We formulate the equations for the network of Fig. 4.4-11*b* as though the dependent source were an independent current source. This leads to

†If a subnetwork includes ground, then the number of unknowns and the number of equations are one less than that stated above.

the following equations:

At node 1

$$G_4(e_1 - V_1) + G_1 e_1 + G_2(e_1 - e_2) = \beta i \tag{20}$$

At node 2

$$G_3(e_2 - V) + G_2(e_2 - e_1) = -\beta i$$

The unknown i is then related to the voltage across G_4 as follows:

$$i = G_4(V_1 - e_1) \tag{21}$$

Substitution of Eqs. (20) into Eq. (21) yields the desired system of two equations in two unknowns,

$$\begin{aligned} G_4(\beta + 1)(e_1 - V_1) + G_1 e_1 + G_2(e_1 - e_2) = 0 \\ G_3(e_2 - V) + G_2(e_2 - e_1) - \beta G_4(e_1 - V_1) = 0 \end{aligned} \tag{22}$$

which may be further rearranged by placing unknowns on the left and known quantities on the right, as follows:

$$\begin{aligned} [G_4(\beta + 1) + G_1 + G_2]e_1 - G_2 e_2 = G_4(\beta + 1)V_1 \\ -(G_2 + \beta G_4)e_1 + (G_3 + G_2)e_2 = G_3 V - \beta G_4 V_1 \end{aligned} \tag{23}$$

For typical values of $V = 10$, $G_3 = G_4 = 10^{-3}$, $G_2 = 10^{-6}$, $G_1 = 0.02$, and $\beta = 100$, these equations yield

$$e_2 \simeq 10 - 17V_1 \tag{24}$$

That is, a small input voltage V_1 is *amplified* by a factor of 17 before it appears as an output voltage e_2 which is (1) in the opposite direction from V_1 and (2) offset by 10 V from zero. This principle of amplification is widely used in communication, computation, and control for magnifying weak signals.

Returning to the more general case of node-voltage equations with dependent sources, we observe in the above example that we first treated the dependent source as an independent source in order to formulate the node-voltage equations, and we then related the current on which the source depends

TABLE 4.4-3 Node-voltage Method of Formulating Linear Network Equations

Class of Networks Consisting of Linear Resistors, Independent Current Sources, Independent Voltage Sources, and All Four Types of Dependent Sources

1. Temporarily, consider dependent current sources as independent current sources, and dependent voltage sources as independent voltage sources, and formulate node-voltage equations for the resulting network which now has only resistors and "independent sources."
2. In the equations formulated from step 1, above, replace the values of the temporarily independent sources by the branch currents and voltages on which they depend (and their multiplying factors), and express these currents or voltages in terms of node-voltage and conductance values.

to the conductance through which it flows and to the node voltages on either side of that conductance. This, in fact, will be our general approach which we state in Table 4.4-3.

This approach does indeed formulate the necessary equations, since step 1 yields the equations that would result if the network consisted only of resistor and independent sources, and step 2 introduces all the dependences of the sources. The latter can always be expressed in terms of conductance values and node voltages, if the conductances of the network are linear.

Nonlinear Node-voltage Equations

The preceding formulation approaches which deal with classes of linear networks can be extended to nonlinear networks in a rather straightforward way.

Specifically, for the third class of networks, consisting of conductances, both types of independent sources, and all four types of dependent sources, node-voltage formulation is always possible if

1. Step 1 of Table 4.4-3 can be carried out. This in turn is always possible if each resistor current can be expressed in terms of the node voltages of the nodes to which the resistor is connected. This is possible *if the nonlinear resistors are voltage-controlled.*
2. The dependent sources will be normally linear, i.e., of the form $b=Ka$, where b is the value of a voltage or current source, and a is a voltage or current on which b depends through the proportionality constant K. These dependent sources may also be nonlinear, i.e., of the form

   $$b=f(a)$$

   where f is a known function, without affecting the formulation approach.

The second of the above two conditions ensures that step 2 of Table 4.4-3 can be carried out. Hence, both the above conditions are sufficient for formulating

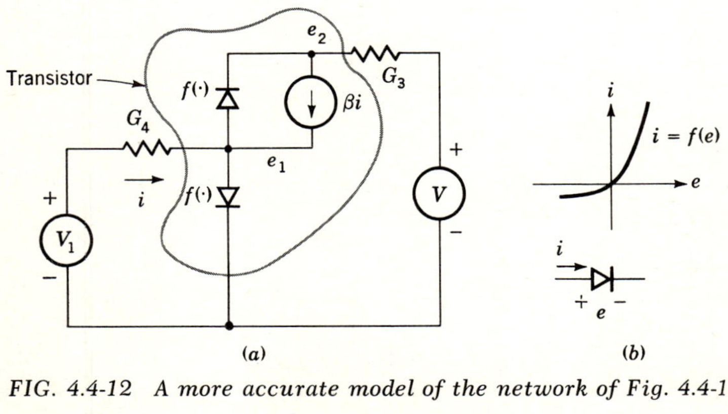

*FIG. 4.4-12 A more accurate model of the network of Fig. 4.4-11*a.

node-voltage equations. The following example illustrates such a formulation for a practical circuit.

Example 4.4-6 *Transistor Amplifier (continued).* The transistor network of Fig. 4.4-11*a* can be modeled more accurately as shown in Fig. 4.4-12*a*.

Recall that the diode is a nonlinear resistor with the characteristics (note reference directions) shown in Fig. 4.4-12*b*. Proceeding as before, we formulate the node-voltage equations by first letting the dependent source be independent, by writing KCL equations for nodes 1 and 2, and then taking account of the dependence and letting $i = (V_1 - e_1)g_4$. The result is

$$(e_1 - V_1)G_4 + f(e_1) + f(e_1 - e_2) = G_4\beta(V_1 - e_1) \tag{25a}$$

$$(e_2 - V)G_3 - f(e_1 - e_2) = -G_4\beta(V_1 - e_1) \tag{25b}$$

These equations differ from Eqs. (22) in that they use function $f(.)$ instead of conductances G_1 and G_4 to express the current through the two diodes in terms of node voltages. Observe also that the diode current in Equation (25b) is

$$-f(e_1 - e_2) \qquad \text{and not} \qquad f(e_2 - e_1)$$

These two terms are vastly different, as can be seen from the graph of $f(.)$ in Fig. 4.4-12*b*, and should not be confused with each other. We will study the solution of such equations in Chap. 5; for the time being, we wish to observe that when Eqs. (25) are solved, they give rise to a relationship between V_1 and e_2, as

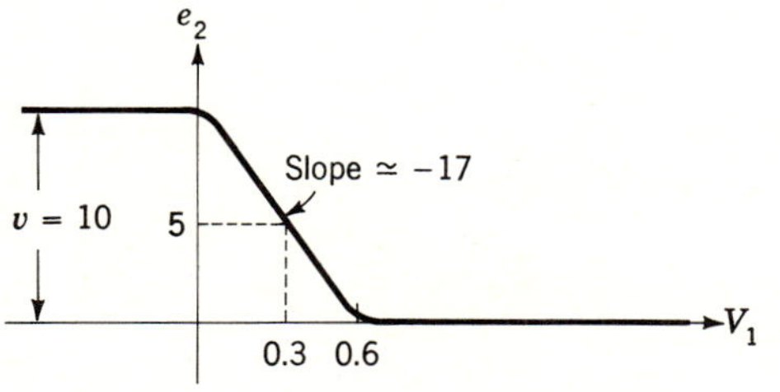

FIG. 4.4-13 Input-output relationship.

shown in Fig. 4.4-13 for the typical values used in Example 4.4-5. Observe that the slope at $V_1 = 0.3$ is about -17, which is the value found in Example 4.4-5. This characteristic is similar to that discussed in Sec. 3.3.

Observe further that *amplification* is accomplished by supplying small V_1-versus-time input signals (which may represent music or speech) about $V_1 = 0.3$ V; the *output* e_2 will then equal the input signal amplified by a factor of -17. If however, the input signal is large enough so that V_1 falls outside the interval $[0, 0.6]$, then the output no longer represents faithfully the input, and we have the familiar case of *distortion*, or as it is called, a case of *saturation* (if $V_1 > 0.6$) or a case of *cutoff* (if $V_1 < 0$).

The following example illustrates the node-voltage method of formulation for a network which is not practical, but rather illustrates the points made in formally stating the method.

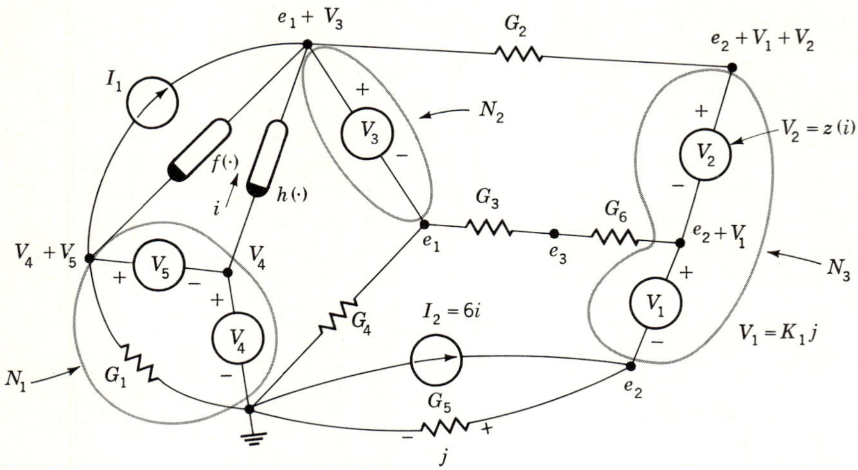

FIG. 4.4-14 Illustration of node-voltage formulation.

Example 4.4-7 Consider the network of Fig. 4.4-14 which contains linear conductances G_1, G_2, G_3, G_4, G_5, G_6; nonlinear conductances $f(.)$ and $h(.)$; independent current source I_1; independent voltage sources V_3, V_4, V_5; current-dependent current source I_2; current-dependent voltage source V_2†; and voltage-dependent voltage source V_1. Formulation is done as follows:

1. Three subnetworks consisting of connected voltage sources are identified, and reference voltages 0, e_1, e_2 are assigned within the subnetworks. Voltage e_3 is assigned to the free node. All other nodes in each subnetwork are assigned voltages in terms of 0, e_1, e_2 and the voltage source values V_1, V_2, V_3, V_4, V_5.
2. Three KCL equations are written for the subnetworks containing e_1, e_2 and for the free node e_3. They are as follows:

For N_2

$$G_4e_1 + G_3(e_1 - e_3) + G_2[(e_1 + V_3) - (e_2 + V_1 + V_2)] \\ -f((V_4 + V_5) - (e_1 + V_3)) - h(V_4 - (e_1 + V_3)) = I_1 \qquad (26)$$

For Node 3,

$$G_3(e_3 - e_1) + G_6[e_3 - (e_2 + V_1)] = 0 \qquad (27)$$

For N_3,

$$G_2[(e_2 + V_1 + V_2) - (e_1 + V_3)] + G_6[(e_2 + V_1) - e_3] + G_5(e_2) = I_2 \qquad (28)$$

Next, the source values I_2, V_1, and V_2 which were so far treated as independent, are replaced as follows:

†Note that the dependence is nonlinear, i.e., $V_2 = z(i)$.

$$
\begin{aligned}
I_2 &= 6i = 6h(V_4 - (e_1 + V_3)) \\
V_1 &= K_1 j = K_1 e_2 \\
V_2 &= z(i) = z(h(V_4 - (e_1 + V_3)))
\end{aligned}
\tag{29}
$$

Substitution of Eqs. (29) into Eqs. (26) to (28) yields respectively the desired equations (30) to (32):

$$
G_4 e_1 + G_3(e_1 - e_3) + G_2[(e_1 + V_3) - (e_2 + K_1 e_2 + z(h(V_4 - (e_1 + V_3))))] \\
- f((V_4 + V_5) - (e_1 + V_3)) - h(V_4 - (e_1 + V_3)) = I_1 \tag{30}
$$

$$
G_3(e_3 - e_1) + G_6[e_3 - (e_2 + K_1 e_2)] = 0 \tag{31}
$$

$$
G_2[(e_2 + K_1 e_2 + z(h(V_4 - (e_1 + V_3)))) - (e_1 + V_3)] \\
+ G_6[(e_2 + K_1 e_2) - e_3] + G_5(e_2) = 6h(V_4 - (e_1 + V_3)) \tag{32}
$$

Admittedly, these are complicated-looking equations; yet they constitute a system of *three equations in three unknowns*, which can be tackled by an iterative computer method, as we shall see in subsequent chapters.

Solution of Node-voltage Equations: Existence and Uniqueness

Although formulation of network equations is a very important part of network analysis, it is the solution of these equations that completes the analysis task and provides information about network behavior.

The three approaches that have been discussed for formulating node-voltage equations have as their end result a system of k equations in k unknowns. When the network is linear, these equations are also linear, and their solution can be carried out by any one of the well-known methods such as Cramer's rule, successive unknown elimination, and superposition (which is discussed in the following section). When the equations are nonlinear, then the solution is carried out by an orderly trial-and-error approach known as *iteration*, which is discussed extensively in subsequent chapters.

Two important issues, however, involved in the solution of these equations are the questions of existence and uniqueness, that we encountered in Chap. 3. Recall that given a set of equations, a solution may or may not exist (existence); if a solution exists, it may be the only one (uniqueness), or there may be several solutions. Clearly, if we know that a solution does not exist, we can save a great deal of our effort or/and computer time by avoiding a solution attempt. Furthermore, knowing that the solution, if it exists, is unique can also save computer time that might have otherwise been expended in looking for other solutions. It is therefore desirable for us to know *before* attempting analysis or even formulation, whether a solution exists and whether if it exists, it is unique. Unfortunately, such knowledge cannot be obtained for all possible networks. For a large class of networks, however, we can guarantee existence and uniqueness of a solution, prior to formulation and solution as follows:

THEOREM 4.4-2 Existence and uniqueness of a solution is guaranteed for networks which consist of (1) resistors, whose voltage-current characteristics are continuous, monotonically increasing with bounded nonzero slopes (when the slopes exist); and (2) independent current and voltage sources whose interconnection does not violate either KCL or KVL.

This theorem is presented here without proof, since its proof is beyond the scope of the book. Observe that the characteristics of Fig. 4.4-15*a, b, c* meet the conditions of Theorem 4.4-2, while those of Fig. 4.4-15*d, e,* and *f* do not.

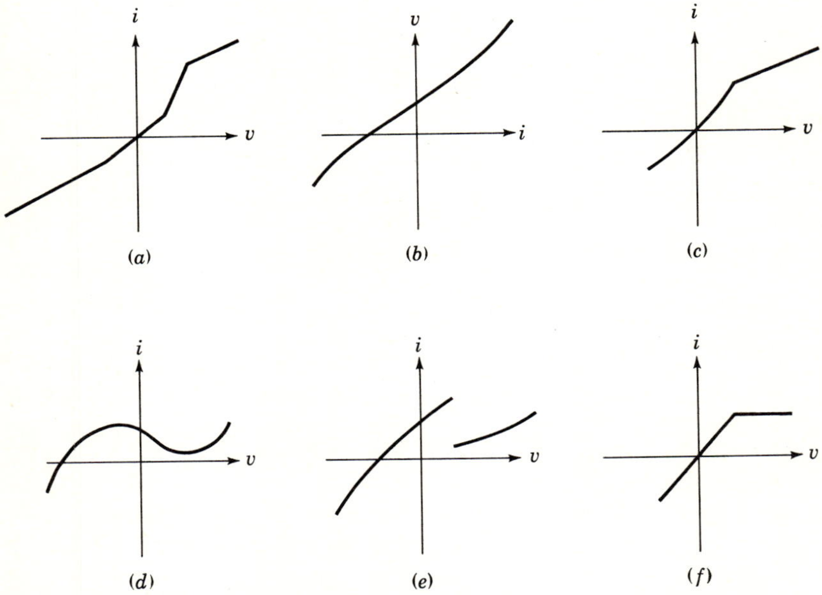

FIG. 4.4-15 Examples of acceptable and unacceptable resistor characteristics for existence and uniqueness of solution.

Furthermore, observe that (1) independent current sources are not allowed to violate KCL, as shown in Fig. 4.4-6*a* or *b*, and (2) independent voltage sources are not allowed to violate KVL; i.e., they cannot lie in a closed loop or path, such that the values of the sources violate KVL.

Finally observe that linear networks *without dependent sources* obey the conditions of this theorem and hence are guaranteed to have a unique solution.

4.5 LINEARITY

When a resistive network is made up of linear conductances and dependent and independent sources, it is characterized by a system of linear equations; hence it obeys the principle of superposition discussed in Sec. 3.6 This section looks at some of the consequences and implications of linearity as applied to resistive networks.

Linearity of Node-voltage Equations

We start with the following theorem:

THEOREM 4.5-1 Networks consisting of linear conductances and dependent and independent voltage or/and current sources† give rise to a system of linear node-voltage equations.

PROOF: The node-voltage method of formulating network equations gives rise to a system of equations, with *each* equation having the form

$$i_1 + i_2 + \cdots + i_k = 0 \tag{1}$$

Here, each i_j term corresponds to a current and is expressed in one of the following ways:

1. $i_j = G(e_k - e_m)$ for a linear conductance G having at its terminals unknown node voltages e_k and e_m
2. $i_j = I$ for an independent current source
3. $i_j = G\left[\left(e_m + \sum_i h_i V_i\right) - \left(e_k + \sum_p q_p V_p\right)\right]$

   if independent voltage sources V_i are present in the network. Here the h_i and q_p constants are either +1 or −1, and the two parenthesized terms are the voltages at the conductance terminals expressed in terms of network node voltages and voltage-source values. If there are dependent voltage sources in the network, then the corresponding V_i, above, will be replaced by terms of the form Kq or $K(u_1 - u_2)$, discussed below, according to whether the V_i depend on current or voltage.
4. $i_j = Kq$ for a current-dependent current source, where K is a constant and q represents a current somewhere in the network, i.e., a term which is, recursively, of the same form as i_j (including this form of i_j)
5. $i_j = K(u_1 - u_2)$ for a voltage-dependent current source, where K is a constant, and u_1, u_2 are either unknown node voltages e_i, e_j or more generally terms of the form $\left(e_i + \sum_j h_j V_j\right)$, when there are voltage sources in the network.

From the above, it follows that Eq. (1) is of the form

$$a_1 e_1 + \cdots + a_k e_k + b_1 V_1 + \cdots + b_m V_m + c_1 I_1 + \cdots + c_n I_n = 0 \tag{2}$$

where the e_i are unknown node voltages; the V_i and I_i are independent voltage- and current-source values; and the coefficient a_j, b_j, c_j are real numbers which depend on conductance values, dependent-source constants, and network topology.

Equation (2) may be rearranged with the unknown e_i terms on the left side and the known V_i and I_i terms on the right. The system of equations then is linear and has the following form:

$$\begin{aligned} a_{11}e_1 + a_{12}e_2 + \cdots + a_{1n}e_n &= b_{11}y_1 + b_{12}y_2 + \cdots + b_{1k}y_k \\ a_{21}e_1 + a_{22}e_2 + \cdots + a_{2n}e_n &= b_{21}y_1 + b_{22}y_2 + \cdots + b_{2k}y_k \\ a_{n1}e_1 + a_{n2}e_2 + \cdots + a_{nn}e_n &= b_{n1}y_1 + b_{n2}y_2 + \cdots + b_{nk}y_k \end{aligned} \tag{3}$$

†As before, these sources cannot be interconnected so as to violate KCL or KVL.

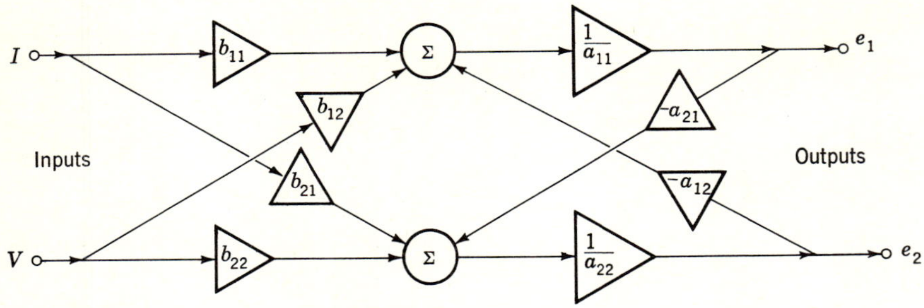

FIG. 4.5-1 Block diagram representation of a network.

where the terms $y_1, \ldots, y_k$ are the independent sources [the I_i and V_i terms of Eq. (2)]. Q.E.D.

Thus, if a network has two unknown node voltages and any number of dependent sources but one independent current source and one independent voltage source, it will give rise to a system of equations of the form

$$a_{11}e_1 + a_{12}e_2 = b_{11}I + b_{12}V \tag{4}$$

$$a_{21}e_1 + a_{22}e_2 = b_{21}I + b_{22}V \tag{5}$$

Here, I and V may be considered as the *inputs* of a block diagram system, and e_1 and e_2 as the *outputs* of that system. Expressing from Eq. (4) e_1 in terms of e_2, I, and V, and expressing from Eq. (5) e_2 in terms of e_1, I, and V, we obtain the block diagram system of Fig. 4.5-1, which models the above network. Observe that (1) this memoryless system is *implicit*, and (2) any network for which equations can be formulated, can be modeled by a block diagram system of this type. The above system is evidently linear, because it obeys Definition 3.6-1, and is characterized by a system of linear equations. Consequently superposition, which was discussed in Sec. 3.6, is applicable.

Superposition

Recall from Chap. 3 that in a linear system the output resulting from several inputs which are simultaneously applied is the sum of the outputs caused by each input acting independently while all other inputs are set to zero. This is the principle of superposition which was shown to be a consequence of a system's linearity. Often, when this principle is applied to networks, it yields topological simplifications which facilitate analysis. We shall demonstrate superposition in Example 4.5-2, after examining in Example 4.5-1 several simple networks which appear often as a part of more complex networks.

Example 4.5-1 *Some Simple Networks* The *voltage divider* shown in Fig. 4.5-2*a* is a network consisting of two linear resistors with resistance R_1 and R_2. From Fig. 4.5-2*a* the following node equation is formulated:

$$\frac{e}{R_1} + \frac{e - V}{R_2} = 0$$

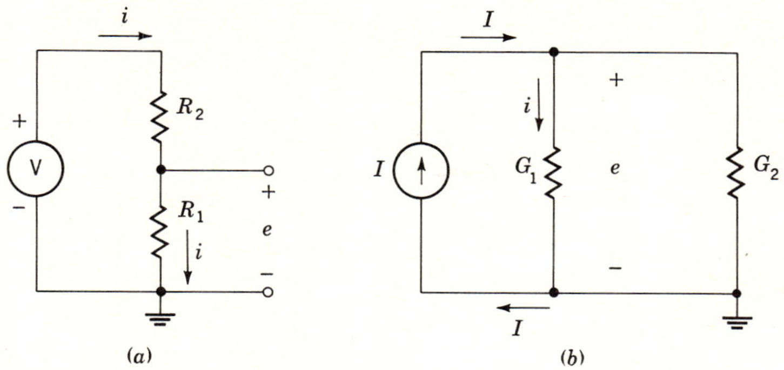

FIG. 4.5-2 Some simple networks. (a) *Voltage divider.* (b) *Current divider.*

which when rearranged yields

$$e = \frac{R_1}{R_1 + R_2} V \tag{6}$$

This network is called a *divider* because it multiplies the input voltage V by a constant which is always less than unity. For example, if $R_1 = R_2$, then Eq. (6) yields $e = \frac{1}{2}V$. Observe that by KCL the current i flowing in R_1 equals the current flowing through R_2, and by KVL the voltage across R_2 is $V - e$.

Next consider the *current divider* of Fig. 4.5-2*b* with linear conductances G_1, G_2. Here current source I is considered to be the input, and current i the output. The simple node-voltage equation for this network is

$$(G_1 + G_2)e = I$$

which when rearranged yields

$$e = \frac{I}{G_1 + G_2}$$

The desired current i equals $G_1 e$; hence

$$i = \frac{G_1}{G_1 + G_2} I \tag{7}$$

That is, output current i is less than input current I, by a multiplying factor which is of the same form as Eq. (6), except that in Eq. (7) conductance values replace the resistance values of Eq. (6). Observe further that in Fig. 4.5-2*b* the same voltage appears across both conductances.

Example 4.5-2 *Illustration of Superposition* Consider the network of Fig. 4.5-3. We wish to find v_0 in terms of V, I, and the resistance values. Considering v_0 as the output of our linear system and V and I as the two inputs. We shall first find the output $v_0 = v_{0V}$ caused by the voltage source, with $I = 0$. Setting $I = 0$, however, means that regardless of the voltage across the current source, no current shall pass through it. This is "topologically" equivalent to replacing the

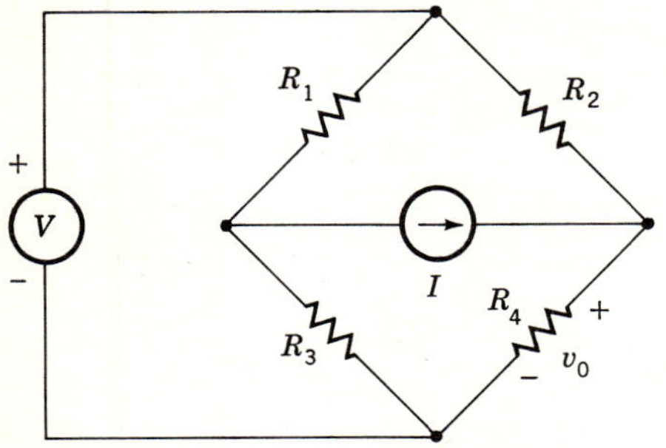

FIG. 4.5-3 Illustration of superposition.

current source with an *open circuit,* i.e., a conductance of zero value, which is equivalent to removing the current source from the network, as shown in Fig. 4.5-4*a*.

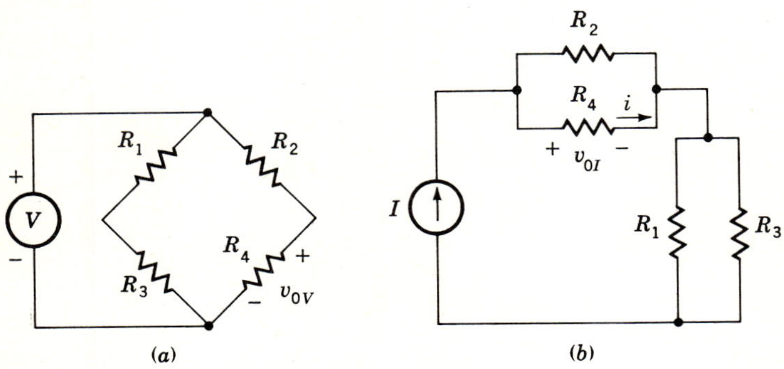

FIG. 4.5-4 Applying superposition.

Observe that R_2 and R_4 form a voltage divider; hence from Eq. (6) it follows that

$$v_{OV} = \frac{R_4}{R_2 + R_4} V \tag{8}$$

Next, we find the output v_{OI} caused by the current source, with $V = 0$. Setting $V = 0$ means that the voltage across the voltage source will be zero, regardless of the current passing through it. This is topologically equivalent to replacing the voltage source with a *short circuit,* i.e., a resistance of zero value, which in turn is equivalent to short-circuiting the voltage source, as shown in Fig. 4.5-4*b* (with some geometrical rearrangement of the network).

Observe that R_2 and R_4 form a current divider; hence use of $G_2 = 1/R_2$, $G_4 = 1/R_4$, and Eq. (7) yields for the current i through R_4 (see Fig. 4.5-4*b*) the following:

$$i = \frac{G_4}{G_2 + G_4} I \tag{9}$$

But the voltage v_{0I} is simply equal to $R_4 i$; hence

$$v_{0I} = \frac{1}{G_2 + G_4} I$$

which can be rewritten as

$$v_{0I} = \frac{R_2 R_4}{R_2 + R_4} I \tag{10}$$

From the principle of superposition, the desired output v_0 is given by

$$v_0 = v_{0V} + v_{0I} \tag{11}$$

Hence, substituting Eqs. (8) and (10) into Eq. (11) yields the desired result

$$v_0 = \frac{R_4}{R_2 + R_4} V + \frac{R_2 R_4}{R_2 + R_4} I$$

which can be rewritten as

$$v_0 = \frac{R_4}{R_2 + R_4} (V + R_2 I) \tag{12}$$

Observe that this input-output relationship turned out to be independent of R_3 and R_1.

The principle of superposition will be used again later in this chapter as well as in later chapters after the introduction of dynamic networks.

Homogeneity

Homogeneity is a special case of linearity. It simply states that *scaling the input by a constant scales the output by the same constant.* Recall that if x is the input and y the output of a linear system, then linearity dictates that

If

$$x = ax_1 + bx_2 \tag{13}$$

then

$$y = ay_1 + by_2$$

where a and b are real numbers, y_1 is the output when x_1 is the input, and y_2 is the output when x_2 is the input. For the special case $b = 0$, Eqs. (13) reduce to the statement of *homogeneity:*

If

$$x = ax_1$$

then

$$y = ay_1$$

Following our discussion of Chap. 3 for multiple-input multiple-output systems, observe that if the system has several inputs with values $x_1, x_2, x_3, \ldots, x_n$, which give rise to output values $y_1, y_2, \ldots, y_n$; then scaling all inputs by a, i.e., assigning them values $ax_1, ax_2, ax_3, \ldots, ax_n$ results in output values $ay_1, ay_2, \ldots, ay_n$, which are scaled in the same way.

The following examples illustrate homogeneity:

Example 4.5-3 The voltage divider of Fig. 4.5-2*a* obeys homogeneity, since, if by Eq. (6) it relates an input value v_1 to a corresponding output value e_1, by

$$e_1 = \frac{R_1}{R_1 + R_2} v_1$$

then scaling the input, i.e., letting $V = av_1$, yields by Eq. (6) an output

$$e = \frac{R_1}{R_1 + R_2} av_1$$

which equals ae_1.

Example 4.5-4 The node-voltage equation for the "star" network of Fig. 4.5-5

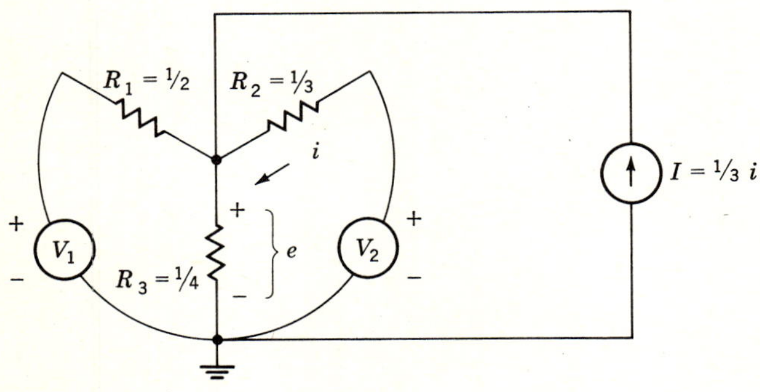

FIG. 4.5-5 A "star" network.

is

$$\left(\frac{1}{R_1} + \frac{1}{R_2} + \frac{1}{R_3}\right)e - \frac{1}{R_1}V_1 - \frac{1}{R_2}V_2 = \frac{1}{3}\frac{V_2 - e}{R_2}$$

Substitution of the resistor values shown in Fig. 4.5-5 yields

$$9e - 2V_1 - 3V_2 = V_2 - e$$

or

$$e = 0.2V_1 + 0.4V_2 \tag{14}$$

If, for example,

$$V_1 = 1 \qquad \text{and} \qquad V_2 = 2$$

then, by Eq. (14),

$$e = 1$$

If both inputs are multiplied by 10, i.e., if

$$V_1 = 10 \qquad \text{and} \qquad V_2 = 20$$

then, by Eq. (14),

$$e = 10$$

which confirms that homogeniety holds.

Note that the dependent source is not considered to be an input, since it cannot be independently assigned a value, as is the case for V_1 and V_2.

In the following example, homogeneity is used to solve a problem which might otherwise entail considerably more work.

Example 4.5-5 *A Ladder Network* Consider the *ladder* network of Fig. 4.5-6.

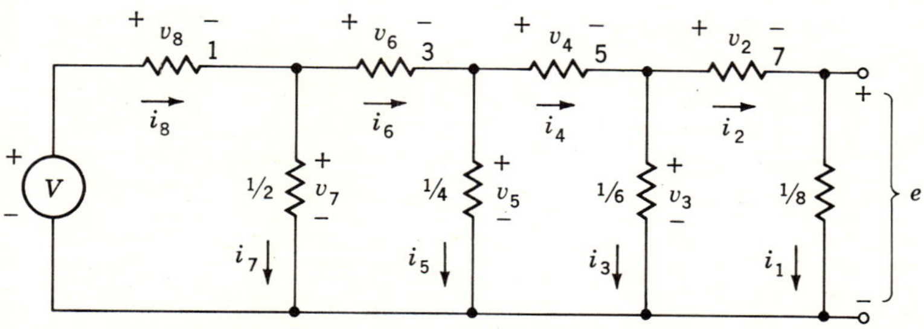

FIG. 4.5-6 A ladder network. Shown values are resistances.

We wish to find e as a function of V.

Since the network is linear, it follows that homogeneity holds, and the desired result will be of the form

$$e = KV \tag{15}$$

To find K, all we need is *a pair* of values e, V. Instead of solving the network with four node equations, we *assume* that $e = 1$, and work backward using KCL and KVL as follows:

If $e = 1$, then by Ohm's law $i_1 = 8$.
By KCL, $i_2 = i_1$; hence $i_2 = 8$.
By Ohm's law, $v_2 = 7i_2$; hence $v_2 = 56$.
By KVL, $v_3 = e + v_2$; hence $v_3 = 57$.
By Ohm's law, $i_3 = 6v_3$; hence $i_3 = 342$.

The process continues, yielding the following successive results:

$$i_4 = 350 \qquad v_7 = 24{,}541$$
$$v_4 = 1{,}750 \qquad i_7 = 49{,}082$$
$$v_5 = 1{,}807 \qquad i_8 = 56{,}660$$
$$i_5 = 7{,}228 \qquad v_8 = 56{,}660$$

$$i_6 = 7{,}578 \qquad V = 81{,}201$$

$$v_6 = 22{,}734$$

Hence it takes $V = 81{,}201$ to give $e = 1$. We conclude then from Eq. (15) that $K = 1/81{,}201$.

4.6 EQUIVALENCE

General Considerations

The idea of block diagram system equivalence that was introduced in Sec. 3.7 is extended in this section to the notion of *network equivalence*. The ability to identify and understand network equivalences is helpful in the design and analysis of electronic circuits, since relatively complex networks can often be replaced by relatively simple equivalent networks. The experienced designer relies heavily on this ability to identify equivalences, and it is primarily through the use of such complex-to-simple-circuit equivalences that he is able to examine an apparently intricate circuit and predict or improve upon its behavior.

In order to define the notion of network equivalence, we must identify a set of nodes which we shall call the *external nodes* or *terminals* of a given network. The reason for identifying this set is that the network will be connected to other networks, or will communicate, in general, with its environment *only through these terminals*. The voltages at other "internal" nodes, or the currents through network elements, will not interest us as long as the network behaves in a specified or desirable manner with respect to its terminals.

DEFINITION 4.6-1 Network Equivalence Two networks N_1 and N_2 are equivalent if the terminals of N_1 can be put in one-to-one correspondence with the terminals of N_2, such that the relation of terminal currents and terminal voltages of N_1 is identical to the relation of terminal currents and terminal voltages of N_2.

Consider, for example, the two three-terminal networks shown in

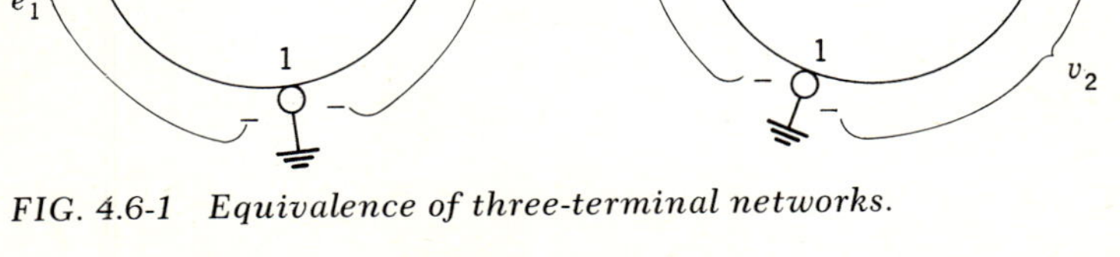

FIG. 4.6-1 Equivalence of three-terminal networks.

Fig. 4.6-1. By KCL and KVL, two voltages and two currents are necessary and sufficient to specify the voltage across any two external nodes, and the current entering any external node of a three-terminal network. Let these voltage and current pairs be specified as shown in Fig. 4.6-1. In applying our definition to the two networks of Fig. 4.6-1, observe that we have established a correspondence of external nodes which we have indicated by numbering the terminals with integers 1, 2, and 3. Applying the definition to N_1 and N_2, we obtain the following:

Networks N_1 and N_2 are equivalent if e_1, e_2, i_1, and i_2 are related in exactly the same way that v_1, v_2, j_1, and j_2 are related.

If, for example, e_1 and e_2 were independently specified by voltage sources, this equivalence would mean that

$$\text{Whenever } v_1 = e_1 \text{ and } v_2 = e_2 \qquad \text{then } j_1 = i_1 \text{ and } j_2 = i_2 \tag{1}$$

Alternatively, if currents i_1 and i_2 were independently specified by current sources connected to N_1, equivalence would dictate that for all values of i_1 and i_2,

$$\text{Whenever } i_1 = j_1 \text{ and } i_2 = j_2 \qquad \text{then } v_1 = e_1 \text{ and } v_2 = e_2 \tag{2}$$

In the simpler case of *two-terminal* networks,† shown in Fig. 4.6-2,

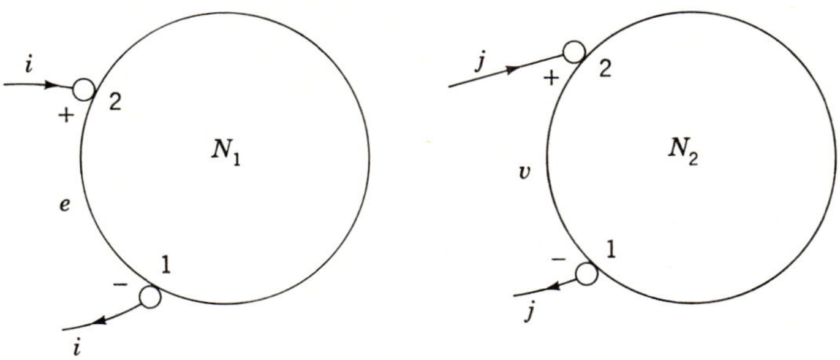

FIG. 4.6-2 Equivalence of two-terminal networks.

equivalence dictates that

For all e,

$$\text{if } v = e \qquad \text{then } j = i \tag{3}$$

or for all i,

$$\text{if } j = i \qquad \text{then } v = e \tag{4}$$

We proceed now with some common two-terminal network equivalences.

Series and Parallel Resistors

Consider the *series interconnection* of two linear resistors R_1 and R_2, as shown in Fig. 4.6-3*a*. This network is equivalent to one consisting of a

†Where voltage is a function of current or current is a function of voltage.

single linear resistor R where

$$R = R_1 + R_2 \tag{5}$$

as shown in Fig. 4.6-3b. To prove this let us analyze both networks.

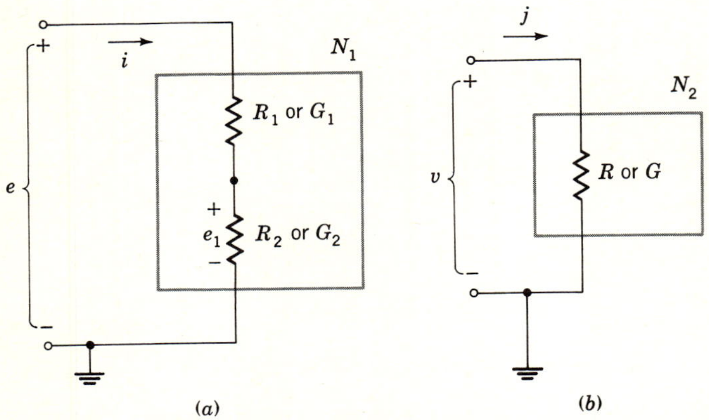

FIG. 4.6-3 In order for networks N_1 and N_2 to be equivalent, it is necessary and sufficient that $R = R_1 + R_2$ or $G = G_1G_2/(G_1 + G_2)$.

For network N_1, KCL requires that the same current i flow through both resistors. Thus, for any node voltages v_1 and v_2 the following relations hold:

$$\begin{aligned} e - e_1 &= R_1 i \\ e_1 &= R_2 i \end{aligned} \tag{6}$$

Eliminating the node voltage e_1, we obtain, for all e and i,

$$e = (R_1 + R_2)i \tag{7}$$

For network N_2, Ohm's law states that, for all v and j,

$$v = Rj \tag{8}$$

In order for the e-i relation to be identical to the v-j relation, it is necessary and sufficient from Eqs. (7) and (8) that

$$R = R_1 + R_2$$

or

$$G = \frac{G_1 G_2}{G_1 + G_2} \tag{9}$$

A similar argument can be used to show that the networks of Fig. 4.6-4a and b are equivalent.

Consider next the *parallel connection* of two linear resistors as shown in Fig. 4.6-5a.

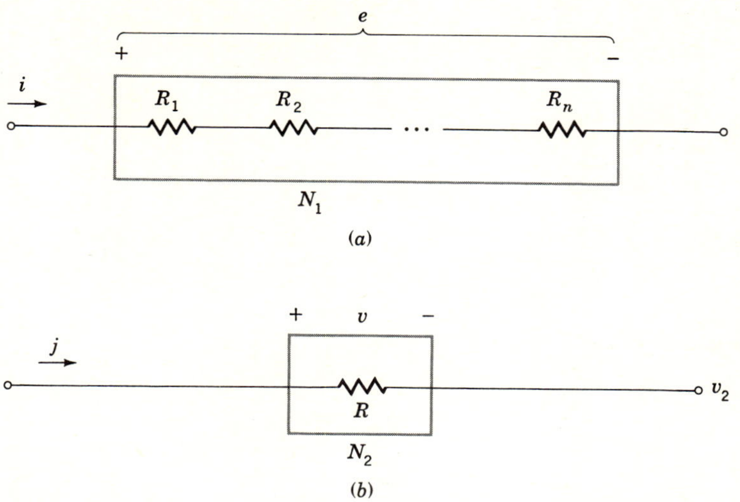

FIG. 4.6-4 The networks N_1 and N_2 are equivalent if and only if $R = R_1 + R_2 + \cdots + R_n$.

By KCL, we have for network N_1

$$i = i_1 + i_2 \tag{10}$$

By Ohm's law, we have

$$\begin{aligned} i_1 &= G_1 e \\ i_2 &= G_2 e \end{aligned} \tag{11}$$

Hence,

$$i = (G_1 + G_2)e \tag{12}$$

For network N_2 of Fig. 4.6-5*b*, we have

$$j = Gv \tag{13}$$

In order for N_1 and N_2 to be equivalent, it is necessary and sufficient that

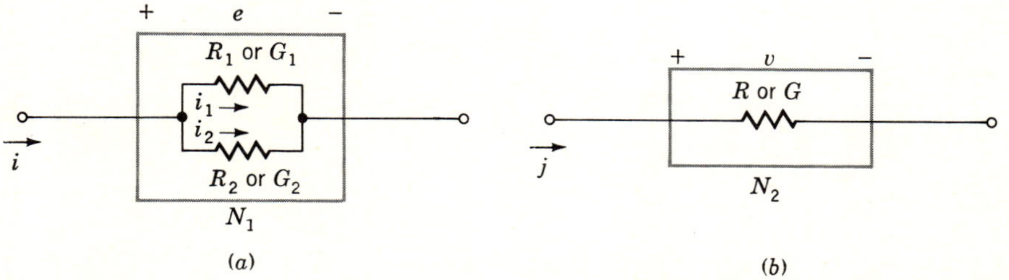

FIG. 4.6-5 The networks N_1 and N_2 are equivalent if and only if $G = G_1 + G_2$ or $R = R_1R_2/(R_1 + R_2)$.

$$G = G_1 + G_2 \tag{14}$$

or

$$R = \frac{R_1 R_2}{R_1 + R_2} \tag{15}$$

A similar argument can be used to find the equivalent resistance of an arbitrary number of linear resistors connected in parallel, as indicated in Fig. 4.6-6.

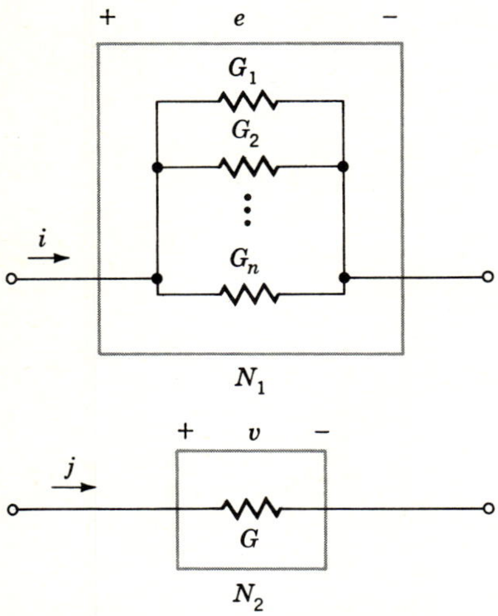

FIG. 4.6-6 The networks N_1 and N_2 are equivalent if and only if $G = G_1 + G_2 + \cdots + G_n$.

The above relationships are used in the following example to illustrate how equivalence can simplify analysis.

Example 4.6-7 *The DAC Ladder* A common resistive network used in DACs (Digital to Analog Converters) is the resistive ladder of Fig. 4.6-7, where horizontally drawn resistors have value 1 Ω and vertically drawn resistors have value 2 Ω (except for R_1).

Starting from the right side of Fig. 4.6-7, we can combine series resistors R_1 and R_2 into a single resistor R, whose value is, by Eq. (9), $R = R_1 + R_2 = 2$. Following this step, we have the network of Fig. 4.6-8*a*. Observe that since N_1 is equivalent to N_2, with respect to the terminals a and b of N_1 and N_2, the overall network of Fig. 4.6-8*a* exhibits the same $e - i$ behavior as the network of Fig. 4.6-7.

Consider next the two parallel resistors of network N_3 in Fig. 4.6-8*a*. With respect to the behavior at terminals c and d, N_3 is, by Eq. (15), equivalent to N_4 of Fig. 4.6-8*b*. The rest of the network, N_5, "does not know" the difference between N_3 in Fig. 4.6-8*a* and N_4 in Fig. 4.6-8*b*, since N_5 communicates with N_3 or N_4 through terminals c and d, and the current-voltage behavior of

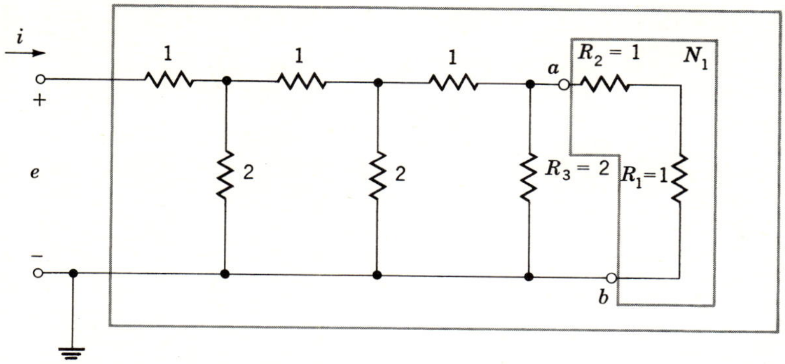

FIG. 4.6-7 *The DAC ladder.*

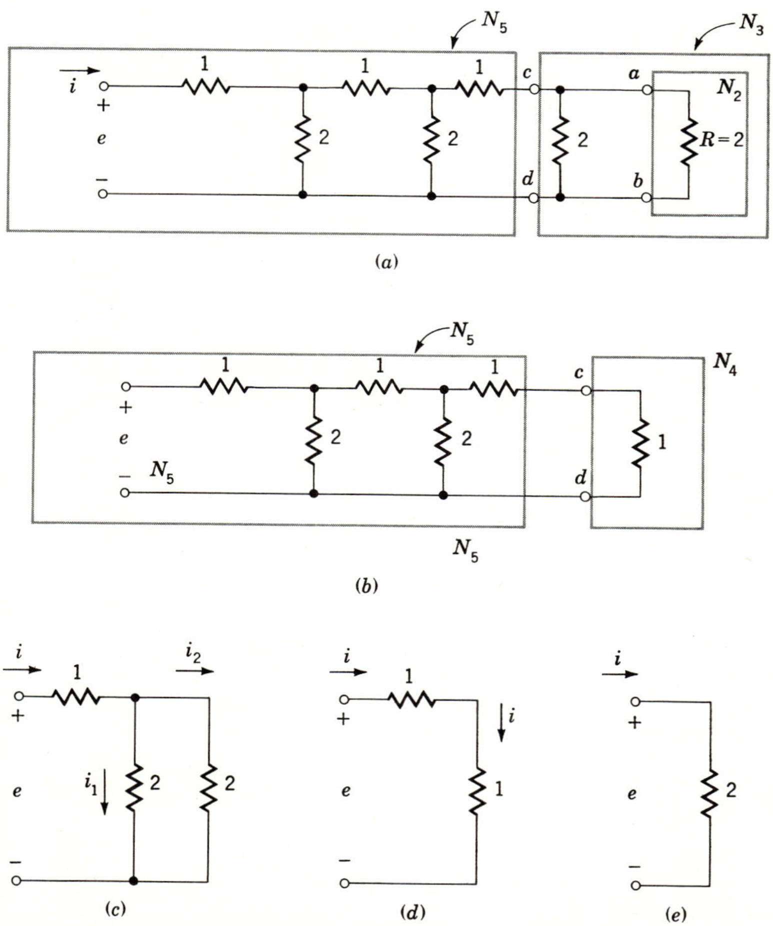

FIG. 4.6-8 *Use of equivalence to simplify the network of Fig. 4.6-7 into the network of Fig. 4.6-8e.*

N_3 is identical to that of N_4 at these two terminals. It follows then that the networks of Fig. 4.6-8*a* and *b* are equivalent in their *e-i* behavior.

Observe, however, that the network of Fig. 4.6-8*b* is identical to the original network, except that is has one less "stage." It follows that repeated use of the foregoing series and parallel reductions will give rise to equivalent networks, shown in Figs. 4.6-8*c*, *d*, and *e* for the last three steps of the process. Thus, the network of Fig. 4.6-7 is equivalent to a resistor of value 2 Ω.

Use of the above steps *in reverse* leads to complete analysis of the network of Fig. 4.6-7. In particular, from Fig. 4.6-8*e*, it follows that

$$i = \tfrac{1}{2}e \tag{16}$$

From Fig. 4.6-8*c* and our knowledge of current dividers (see previous section), it follows that

$$i_1 = i_2 = \tfrac{1}{2}i = \tfrac{1}{4}e \tag{17}$$

Proceeding accordingly, we find that in the network of Fig. 4.6-7 the current entering any node other than datum from the left divides into two *equal* currents, one flowing to the right, the other vertically down to datum. Moreover, since we now know all currents through vertically drawn resistors, we can establish by Ohm's law all node voltages. All these results are shown in Fig. 4.6-9. The network has been therefore analyzed.

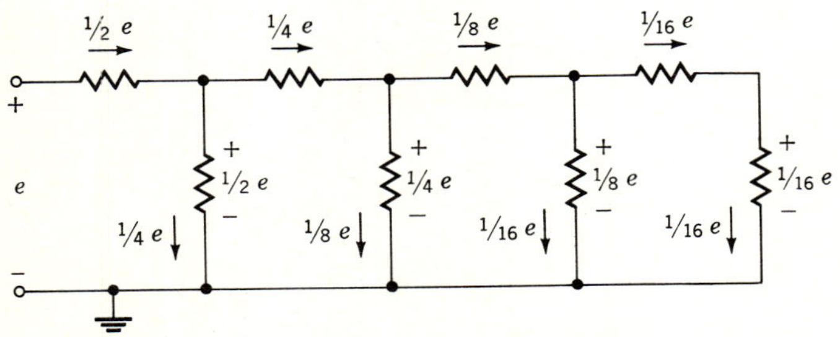

FIG. 4.6-9 The network of Fig. 4.6-7 analyzed.

The use of the ladder in a DAC hinges on the idea of using or not using in a summing network, currents or voltages of successive stages of the ladder, according to whether the corresponding "bit" of a binary number is 1 or 0. For example, if the given binary number is 1101, then a summing network could be made to add the first, second, and fourth of the horizontally flowing currents in Fig. 4.6-9. That is

$$i = (\tfrac{1}{2} + \tfrac{1}{4} + \tfrac{1}{16})e$$

which for $e = 16$ gives the value

$$i = 13$$

This in turn is the integer value (analog) of the binary (digital) number 1101, thus indicating that digital-to-analog conversion does indeed take place.

Nonlinear Network Equivalences

Although equivalence is by no means restricted to linear networks, it is generally difficult to establish explicit relationships between the elements of equivalent nonlinear networks if the networks are more complex than series-parallel. For series or parallel connections of nonlinear resistors, the equivalences are as follows:

Consider the two nonlinear resistors of Fig. 4.6-10a which are

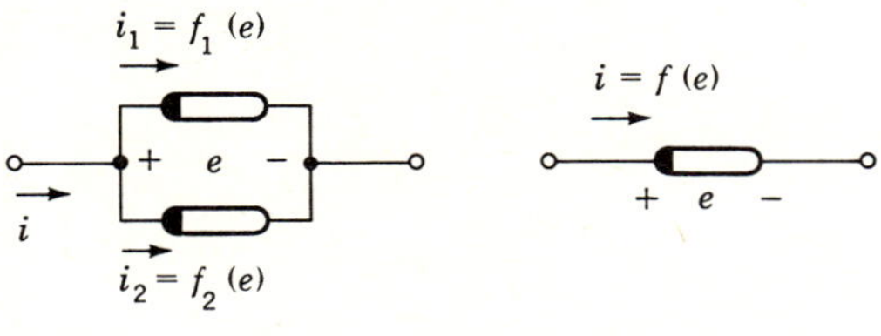

FIG. 4.6-10 Equivalence of nonlinear resistors connected in parallel.

voltage-controlled, i.e.:

For all e,

$$\begin{aligned} i_1 &= f_1(e) \\ i_2 &= f_2(e) \end{aligned} \tag{18}$$

where f_1 and f_2 are given continuous functions. By KCL it follows that

$$i = i_1 + i_2$$

Hence, for all e,

$$i = f_1(e) + f_2(e) \tag{19}$$

Since, given any value e, we are guaranteed to obtain by Eq. (19) a value i, it follows that there exists a function f which we can construct from knowledge of f_1 and f_2 so that

For all e,

$$i = f(e) \tag{20}$$

Hence, the network of Fig. 4.6-10a is with respect to e and i equivalent to the network of Fig. 4.6-10b, where the single nonlinear voltage-controlled resistor has a voltage-current characteristic given by Eq. (20).

For example, if

$$f_1(x) = \sin^2 x,$$

and

$$f_2(x) = \cos^2 x$$

then

$$f(x) = 1$$

That is, the equivalent network is a current source of value 1.

Alternatively, if f_1 and f_2 are the piecewise-linear functions of

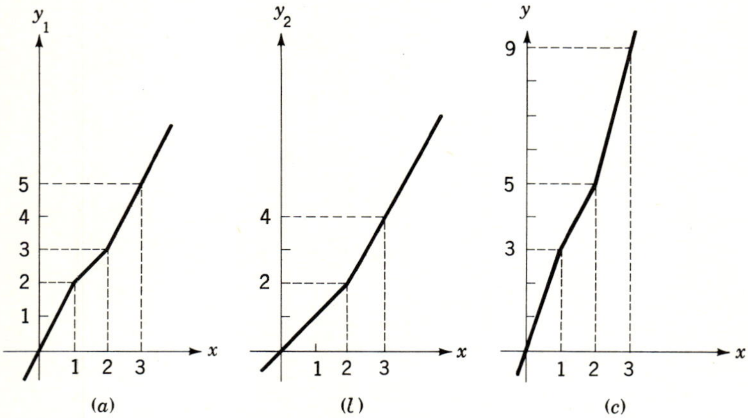

FIG. 4.6-11 Addition of piecewise-linear functions: (a) $y_1 = f_1(x)$; (b) $y_1 = f_2(x)$; (c) $y = f(x)$.

Figs. 4.6-11*a*, and *b*, then the desired function f is obtained, by adding at each value of x, the corresponding values of y_1 and y_2. The result is function f, shown in Fig. 4.6-11*c*. Hence, a nonlinear resistor with the voltage-current characteristic of Fig. 4.6-11*c* is equivalent to two parallel resistors with the voltage-current characteristics of Fig. 4.6-11*a* and *b*, respectively.

EXERCISE 4.6-1 Show that the voltage e across two nonlinear current-controlled resistors which are connected *in series* is related to the current i flowing through the resistors, by a function $e = h(i)$, where h is given by

$$h = h_1 + h_2$$

and h_1 and h_2 are the current-voltage characteristics of each resistor, i.e.,

$$e_1 = h_1(i_1) \qquad \text{and} \qquad e_2 = h_2(i_2)$$

EXERCISE 4.6-2 Show that if two resistors are simultaneously current- and voltage-controlled, then their parallel voltage-current characteristic is given by

$$f = (h_1^{-1} + h_2^{-1})^{-1}$$

where f, h_1, and h_2 are as in the above examples, and f^{-1} means the inverse function of f.

Thévenin and Norton Equivalent Circuits

In this section we wish to establish two-types of very simple circuits which are equivalent to any *two-terminal linear* network.

From our development of superposition in Sec. 3.6 it follows that in a network consisting of linear resistors, dependent sources, n independent voltage sources, and m independent current sources, any node voltage e is given by

$$e = \sum_{i=1}^{n} a_i V_i + \sum_{j=1}^{m} b_j I_j \tag{21}$$

where

1. V_i are the values of the independent voltage sources.
2. I_j are the values of the independent current sources.
3. a_i and b_j are constants whose values depend *only* on the conductances of the network, the multiplying factors of dependent sources, and on network topology.

We shall consider this equation in the network of Fig. 4.6-12, where we have

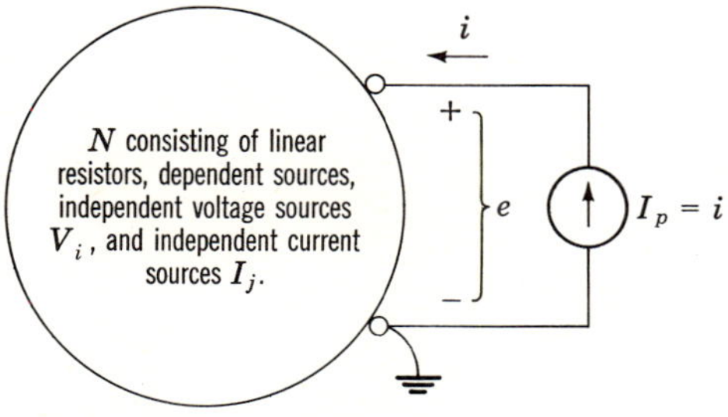

FIG. 4.6-12 A two-terminal network.

chosen one terminal as datum, and we have connected a two-terminal network N to a current source of value i. For the network consisting of N *and the current source*, Eq. (21) can be rewritten as follows:

$$e = \sum_{i=1}^{n} a_i V_i + \sum_{\substack{j=1 \\ j \neq p}}^{m} b_j I_j + b_p I_p \tag{22}$$

where we have pulled out of the current-source sum of Eq. (21) the current source I_p, which is external to N. Observe that the two sum terms of Eq. (22) are now independent of e and I_p (that is, i); hence we shall denote them with a single variable V_{oc}.

In other words we set

$$V_{oc} = \sum_{i=1}^{n} a_i V_i + \sum_{\substack{j=1 \\ j \neq p}}^{m} b_j I_j \tag{23}$$

which combined with Eq. (22) and the fact that $i = I_p$ yields

$$e = V_{oc} + b_p i \tag{24}$$

If we now define a new constant,

$$G_{\text{int}} = \frac{1}{b_p} \tag{25}$$

we can rewrite† Eq. (24) as follows:

$$i = G_{\text{int}}(e - V_{oc}) \tag{26}$$

This result is significant for it describes the terminal behavior of a two-terminal network made up of a single voltage source of value V_{oc} in series with a resistor of conductance G_{int}, as shown in Fig. 4.6-13*a*.

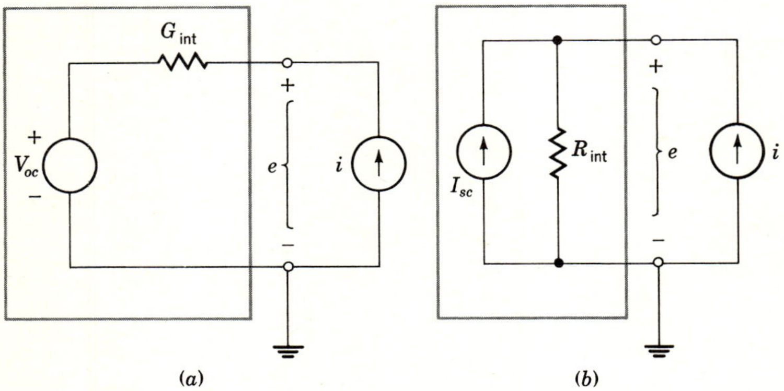

FIG. 4.6-13 The (a) *Thévenin and* (b) *Norton equivalent networks.*

Therefore, the network of Fig. 4.6-13*a* is *equivalent* to the network of Fig. 4.6-12 with respect to the two terminals shown in these figures. That is, if we set in the network of Fig. 4.6-13*a*, V_{oc} and G_{int} according to Eqs. (23) and (25), respectively, then the network of Fig. 4.6-13*a* behaves identically to the network of Fig. 4.6-12, when connected to other networks or sources through the two terminals under consideration. The simple circuit of Fig. 4.6-13*a*, which is equivalent to any linear two-terminal network N, is called the *Thévenin equivalent network.*

Observe next that if we set $i = 0$ in Fig. 4.6-12 or Fig. 4.6-13*a*, i.e., if there is an "open circuit" at the terminals, Eq. (26) reduces to

$$e = V_{oc}$$

whence the subscripts of V_{oc}. That is, V_{oc} is the voltage that appears across the terminals when the terminals are not connected to any networks or sources.

†Note that G_{int} will exist as long as $b_p \neq 0$. If $b_p = 0$, then e does not depend on i, and the network is equivalent to a voltage source of value V_{oc}.

Observe further, that if we set all independent voltage and current sources to zero, i.e., if

For $i = 1, \ldots, n,$

$$V_i = 0 \tag{27}$$

and for $j = 1, \ldots, p-1, p+1, \ldots, m,$

$$I_j = 0$$

then Eq. (23) yields $V_{oc} = 0$, and Eq. (26) becomes

$$G_{\text{int}}e = i \tag{28}$$

That is, G_{int} is the equivalent (internal) conductance "seen" at the two terminals, when all *independent* sources of N are set to zero, i.e., when all independent current sources of N are replaced by open circuits, and all independent voltage sources of N are replaced by short circuits.†

The *Norton equivalent* network (Fig. 4.6-13*b*) is obtained in the same manner, by rewriting Eq. (26) as follows:

$$e = \frac{1}{G_{\text{int}}}(i + G_{\text{int}}V_{oc}) \tag{29}$$

which can be written in the desired form

$$e = R_{\text{int}}(i + I_{sc}) \tag{30}$$

where

$$R_{\text{int}} = \frac{1}{G_{\text{int}}}$$

and

$$I_{sc} = G_{\text{int}}V_{oc} \tag{31}$$

Observe that Eq. (30) is the equation describing the Norton equivalent circuit of Fig. 4.6-13*b*. Hence, the original network, as well as the Thévenin equivalent of that network, is equivalent to the above Norton equivalent circuit.

If we let $e = 0$ regardless of the value of i, i.e., if we short-circuit the terminals of the network, then from Eqs. (26) and (31) we have

$$i = I_{sc}$$

That is, I_{sc} is the current flowing through the network (notice direction) when the terminals are short-circuited. As before, since $R_{\text{int}} = 1/G_{\text{int}}$, it follows that R_{int} is the resistance seen at the two terminals when all internal *independent* voltage sources are short-circuited and all internal current sources are opened.

We shall apply these two equivalences in the following example.

Example 4.6-2 *Thévenin and Norton Equivalent Circuits* Consider the network N of Fig. 4.6-14*a*. We wish to find the Thévenin and Norton equivalent circuits of N.

†The dependent sources are left intact, since they will generally affect the coefficient b_p.

First we will establish V_{oc} by superposition; that is, we shall compute the voltages V_{oc1}, V_{oc2}, and V_{oc3} caused at the terminals by sources I_1, I_2, and V_3 acting alone, with the terminals open in all three cases. By superpositon, V_{oc} will then be given by

$$V_{oc} = V_{oc1} + V_{oc2} + V_{oc3} \tag{32}$$

To find V_{oc1}, we set $I_2 = V_3 = 0$; i.e., we replace I_2 with an open circuit and V_3 with a short circuit, as shown in Fig. 4.6-14*b*. This circuit can be solved

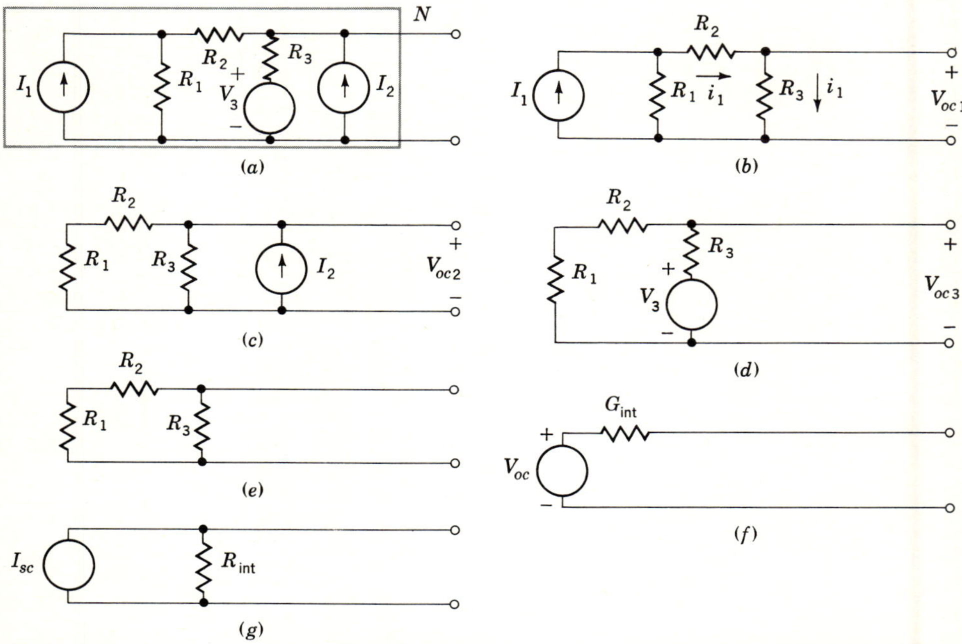

FIG. 4.6-14 Use of Thévenin and Norton equivalences. (a) *The network.* (b) V_{oc} *due to source 1.* (c) V_{oc} *due to source 2.* (d) V_{oc} *due to source 3.* (e) *Evaluation of* R_{int}. (f) *Thévenin equivalent.* (g) *Norton equivalent.*

by inspection, since the current I_1 will divide, as in a current divider, into two parts, one flowing through R_1, and the other through the series equivalent of R_2 and R_3, i.e., through a resistor of value $R_2 + R_3$. Thus, the current i_1 of Fig. 4.6-14*b* is given by

$$i_1 = \frac{R_1}{R_1 + R_2 + R_3} I_1$$

and since $V_{oc1} = i_1 R_3$, we have

$$V_{oc1} = \frac{R_1 R_3}{R_1 + R_2 + R_3} I_1 \tag{33}$$

To compute V_{oc2}, we open I_1 and short-circuit V_3 as shown in Fig.

4.6-14*c*. I_2 now flows through a network whose equivalent resistance is the parallel combination of R_3 on one hand, and the equivalent resistance of R_1 and R_2 in series, on the other. That is,

$$V_{oc2} = \frac{R_3(R_1 + R_2)}{R_3 + R_1 + R_2} I_2 \tag{34}$$

To compute V_{oc3}, we open I_1 and I_2, as shown in Fig. 4.6-14*d*. Now we have a voltage divider with resistances R_3 and $R_1 + R_2$. Hence, by inspection

$$V_{oc3} = \frac{R_1 + R_2}{R_1 + R_2 + R_3} V_3 \tag{35}$$

Application of Eq. (32) to Eqs. (33) to (35) yields

$$V_{oc} = \frac{R_1R_3I_1 + R_3(R_1 + R_2)I_2 + (R_1 + R_2)V_3}{R_1 + R_2 + R_3} \tag{36}$$

Next, we compute the internal resistance, by setting to zero I_1, I_2, and V_3, as shown in Fig. 4.6-14*e*. This resistance is the parallel combination of R_3 with $R_1 + R_2$. Hence

$$R_{\text{int}} = \frac{R_3(R_1 + R_2)}{R_1 + R_2 + R_3} \tag{37}$$

or

$$G_{\text{int}} = \frac{R_1 + R_2 + R_3}{R_3(R_1 + R_2)} \tag{38}$$

Finally, since, by Eq. (31) $I_{sc} = G_{\text{int}}V_{oc}$, we have, from Eqs. (36) and (38),

$$I_{sc} = \frac{R_1R_3I_1 + R_3(R_1 + R_2)I_2 + (R_1 + R_2)V_3}{R_3(R_1 + R_2)} \tag{39}$$

The equivalences have, therefore, been established, since Eqs. (36) through (39) give all the desired quantities for the equivalent circuits of Fig. 4.6-14*f* and *g*.

Example 4.6-3 *Thévenin Equivalent: Dependent Sources* The network of Fig. 4.6-15 contains an independent source and a dependent source. We seek the Thévenin equivalent at terminals *a* and *b*.

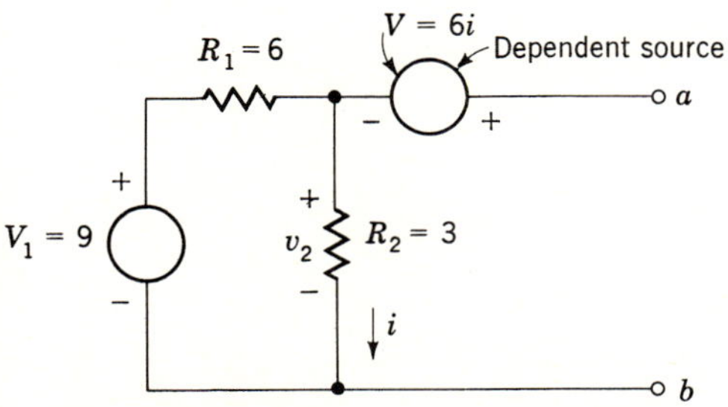

FIG. 4.6-15 A network with a dependent source.

To compute V_{oc}, observe that since no current flows through the dependent source, R_2 and R_1 form a voltage divider. Hence

$$v_2 = \frac{R_2}{R_1 + R_2} V_1 = 3$$

Hence $i = v_2/R_2 = 1$ and $V = 6i = 6$. Since $V_{oc} = v_2 + V$, we have

$$V_{oc} = 9$$

To compute G_{int}, we short-circuit V_1 and obtain the network of Fig. 4.6-16.

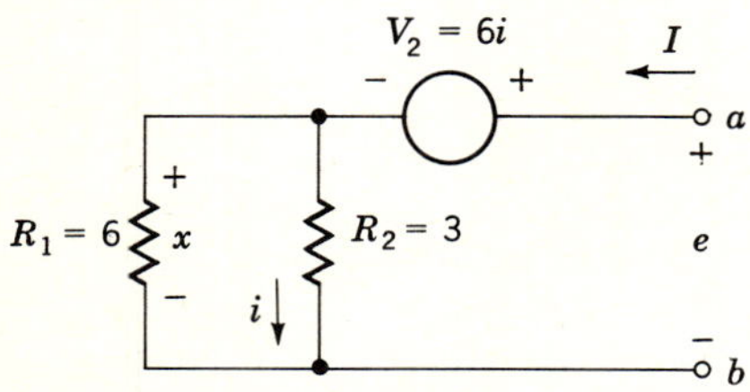

FIG. 4.6-16 Finding G_{int}.

Here (Fig. 4.6-16), current I divides through the current divider formed by R_1 and R_2, so that

$$i = \frac{R_1}{R_1 + R_2} I = \tfrac{2}{3} I$$

Hence the voltage x is by Ohm's law, $3i$ or $2I$, and e is, by KVL,

$$e = 2I + 6i = 2I + 4I = 6I$$

Since $G_{\text{int}} = I/e$, it follows that

$$G_{\text{int}} = \frac{1}{6} \qquad \text{or} \qquad R_{\text{int}} = 6$$

The resulting Thévenin circuit is shown in Fig. 4.6-17.

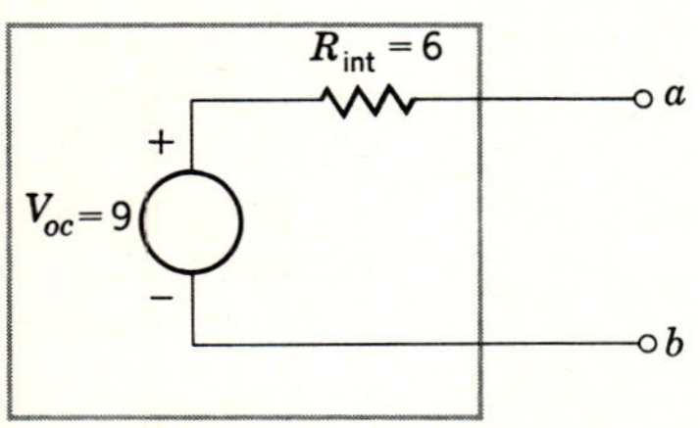

FIG. 4.6-17 The Thévenin equivalent.

As an example of the usefulness of this equivalent circuit, consider attaching a resistor of resistance 3 at the terminals a, and b of Fig. 4.6-15. This is equivalent to attaching the same resistor at the terminals a and b of Fig. 4.6-17. The latter, however, leads by inspection of Fig. 4.6-17 to a total resistance of 9, and hence a current of value 1, flowing through that resistor. Arriving at this result from the network of Fig. 4.6-15 would have necessitated a considerably greater effort.

4.7 BLOCK DIAGRAM SYSTEMS FROM RESISTIVE NETWORKS

General Considerations

In this section we discuss how simple resistive networks can implement some of the memoryless block diagram elements of Chap. 3. In the process of showing this, we shall encounter several simple networks which appear frequently within more complex networks. A good understanding of the former can therefore lead to a greater facility in analyzing and synthesizing the latter.

Realization of Scalers

Perhaps the simplest form of a scaler (introduced in Fig. 3.6-3) is a resistor of resistance R, since the current through the resistor may be considered as "input" x, and the voltage across the resistor as "output" y, resulting in

$$y = Rx \tag{1}$$

If the input and output variables are to be of the same physical nature, e.g., both voltages, then the voltage divider of Ex. 4.5-1 provides an adequate realization (Fig. 4.7-1) of the form

$$y = Kx \tag{2}$$

Here K cannot exceed unity since

$$K = \frac{R_1}{R_1 + R_2}$$

where R_1 and R_2 are always positive.

To obtain scalers of value greater than unity, dependent sources

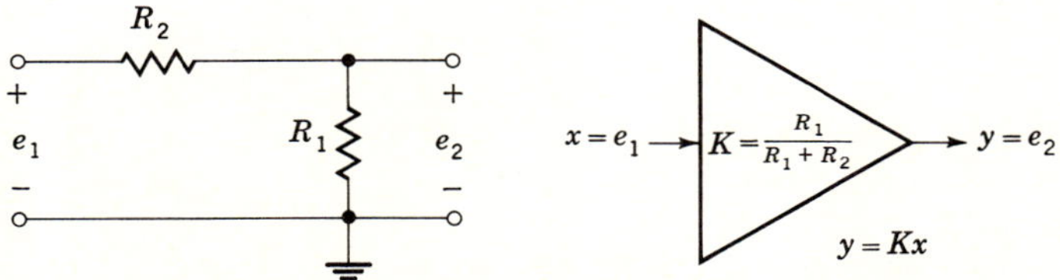

FIG. 4.7-1 Realization of a scaler by a voltage divider.

must be used. These, in turn, can be realized, as we have seen, by transistors. A common transistor circuit which is generally used for this purpose is called an *operational amplifier* or *opamp*. It is symbolized in Fig. 4.7-2*a*. Its behavior is

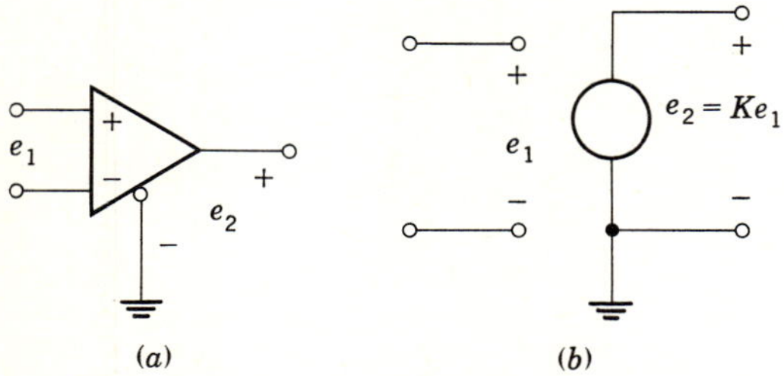

FIG. 4.7-2 The operational amplifier. (a) *Opamp symbol.* (b) *Opamp model.*

approximately equivalent (for slow in time variations of voltages and currents) to the dependent source of Fig. 4.7-2*b*. The gain K is a very large number of the order of 10^6.

The opamp is used in analog computers as a scaler by connecting an input resistor R_1 and a *feedback resistor* R_2, as shown in Fig. 4.7-3*a*, and grounding its + terminal.

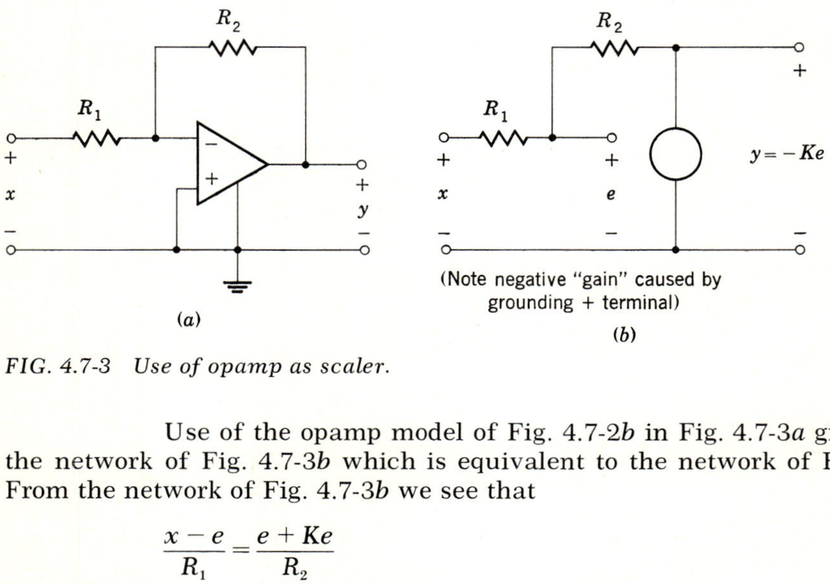

FIG. 4.7-3 Use of opamp as scaler.

Use of the opamp model of Fig. 4.7-2*b* in Fig. 4.7-3*a* gives rise to the network of Fig. 4.7-3*b* which is equivalent to the network of Fig. 4.7-3*a*. From the network of Fig. 4.7-3*b* we see that

$$\frac{x-e}{R_1}=\frac{e+Ke}{R_2}$$

Rearranging in terms of y and x yields

$$y=-\left(\frac{R_2}{R_1+(R_1+R_2)/K}\right)x$$

which for $K \to \infty$, as is the case here, gives

$$y \simeq -\frac{R_2}{R_1}x \tag{3}$$

The approximation error here is of the order of $(1/K)\ (R_2/R_1)$, or 10^{-6}, which is considerably less than the typical variation† of resistance values with temperature and with aging. Hence, for all practical purposes, Eq. (3) is quite exact.

Realization of Summers

The operation of summing is realizable in its most elementary form by a simple linear conductance network as shown in Fig. 4.7-4. By KCL,

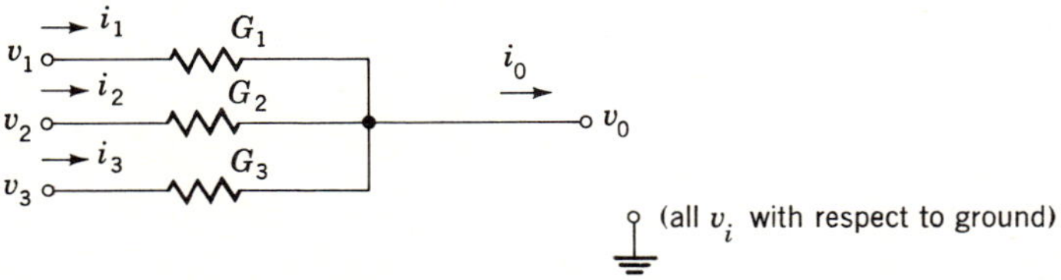

FIG. 4.7-4 The conductance "summer."

$$i_0 = (v_1 - v_0)G_1 + (v_2 - v_0)G_2 + (v_3 - v_0)G_3 \tag{4}$$

which may be rearranged in the form

$$v_0 = \frac{v_1G_1 + v_2G_2 + v_3G_3}{G_1 + G_2 + G_3} - \frac{i_0}{G_1 + G_2 + G_3} \tag{5}$$

If we assume that i_0 is negligible (suppose, for example, that v_1, v_2, and v_3 represent applied source voltages, whereas v_0 is to be measured by a high-resistance meter or is to be connected to an opamp that does not draw appreciable current), then Eq. (5) reduces to the form of a "weighted average,"

$$v_0 = \frac{v_1G_1 + v_2G_2 + v_3G_3}{G_1 + G_2 + G_3} \tag{6}$$

A weighted average is a special case of the more general linear function, of several variables, having the property that if all of these variables have the same value v, then the value of the function is also v. If the weighting coefficients are all the same, i.e., if $G_1 = G_2 = G_3$, then the weighted average reduced to the "ordinary" average,

$$v_0 = \frac{v_1 + v_2 + v_3}{3} \tag{7}$$

Thus with all the G_i equal, the network of Fig. 4.7-4 computes the average of a number of independent variables. Observe that summing and averaging are

†0.1 to 1 percent.

trivally related by a scale factor equal to the number of independent variables; hence, in order to realize a summer, this network must be followed by a scaler.

The notion that the voltage at any node of a conductance network is the weighted average of the voltages at all neighboring nodes† allows us to formulate node-voltage equations of a conductance branch network. For example, in Fig. 4.7-5, we see that the voltage v_1 is the weighted average of the

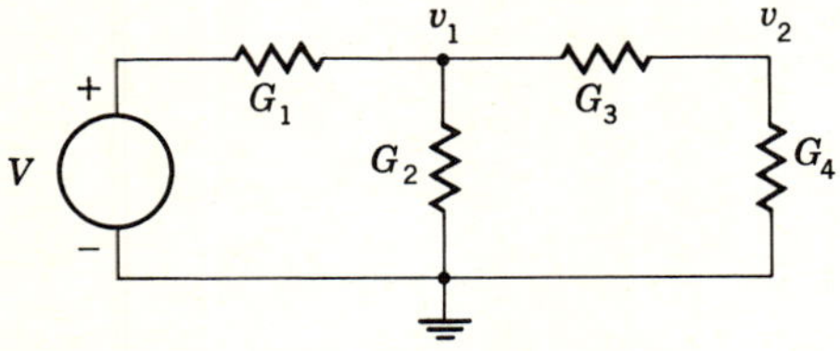

FIG. 4.7-5 A conductance network. Conductance summing tells us that $V > v_1 > v_2$.

three neighboring node voltages V, 0, and v_2. Thus,

$$v_1 = \frac{VG_1 + (0)G_2 + v_2G_3}{G_1 + G_2 + G_3} \tag{8}$$

and

$$v_2 = \frac{v_1G_3 + (0)G_4}{G_3 + G_4} \tag{9}$$

Rearranging yields

$$v_1 = \frac{VG_1 + v_2G_3}{G_1 + G_2 + G_3} \tag{10}$$

and

$$v_2 = \frac{v_1G_3}{G_3 + G_4} \tag{11}$$

Equations (10) and (11), which are the node-voltage equations, can of course be solved to find the values of v_1 and v_2. Observe, however, that the interpretation of a node voltage as a weighted average of neighboring node voltages can help our intuitive understanding of the distribution of node voltages in a conductance network. In particular, in the network of Fig. 4.7-5, if V and the conductances are positive, it follows (from the "weighted average" notion) that v_1 and v_2 are both positive, and that v_1 is greater than v_2 but less than V.

Resistor-Diode Networks and the Threshold Block

The foregoing analysis of the conductance summer may be used to explain the behavior of the resistor-diode networks of Fig. 4.7-6. As we shall

†That is, nodes connected through a conductance to the node under discussion.

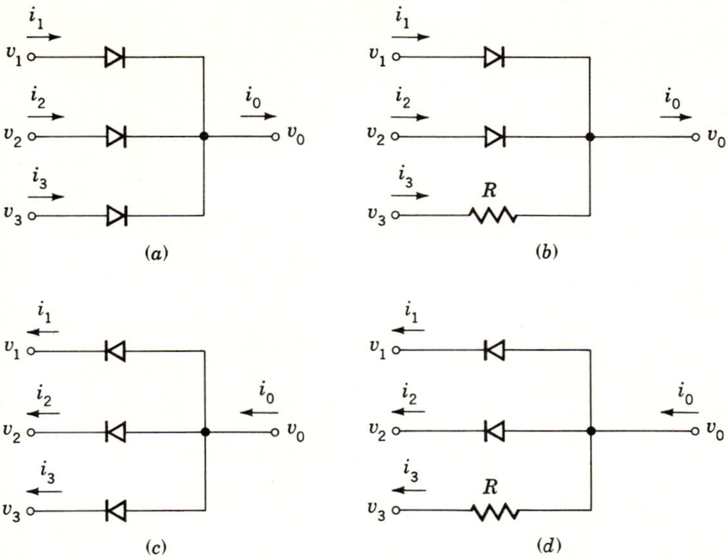

FIG. 4.7-6 Diode gates. (a, b) *These networks realize the "max" function, i.e.,* $v_0 = \max(v_1, v_2, v_3)$. (c, d) *These networks realize the "min" function, i.e.,* $v_0 = \min(v_1, v_2, v_3)$.

see, these networks represent physical realizations of the "max" function which was introduced in Sec. 3.7. To analyze the network of Fig. 4.7-6*a* let us assume that the three node voltages v_1, v_2, v_3 are unequal and that v_1 is the largest of the three, i.e.,

$$v_1 > v_2 \qquad v_1 > v_3$$

Let us think in terms of a diode model, which is a large linear conductance, G_f if the diode happens to be carrying positive current in the forward direction (along arrow), and a small linear conductance G_b if the current is directed backward through the diode. In order to decide whether to use G_f or G_b for the conductance of a diode branch in the network, let us tentatively assume that v_0 is less than v_1 but greater than either of the two other voltages, v_2 and v_3; this trial assumption can later be tested for validity. Under this assumption, the current i_1 is positive, and the currents i_2 and i_3 are negative. Hence

$$v_0 = \frac{v_1 G_f + v_2 G_b + v_3 G_b}{G_f + 2G_b} - \frac{i_0}{G_f + 2G_b} \tag{12}$$

Now, to approach the behavior of an ideal diode, we let G_f and G_b become very large and very small, respectively; it is apparent from Eq. (12) that the voltage v_0 will then approach the voltage v_1. That is,

$$\lim_{\substack{G_f \to \infty \\ G_b \to 0}} v_0 = v_1$$

Our trial assumption that v_0 exceeds v_2 and v_3 (but not v_1) is therefore justified. (Although we shall not go through the details here, it is true that any other trial assumption would have led, in the limit of large G_f, to a contradiction.) Evidently v_0 assumes the largest of the values v_1, v_2, v_3; hence we can write for the network of Fig. 4.7-6*a* the following equation:

$$v_0 = \max\,(v_1,v_2,v_3) \tag{13}$$

This result, is, of course, not restricted to three independent variables, and hence three diodes, but can be extended to an arbitrary number.

By a slight modification of the foregoing arguments, it is possible to demonstrate that the max function is also realizable by means of the network of Fig. 4.7-6*b*, in which any one, *and only one*, of the diodes has been replaced by a resistance R (here the value of R should lie in an intermediate range of values much greater than the forward resistance $1/G_f$, and much less than the back resistance $1/G_b$ of the diode. As before, it is also necessary in this case to assume that the current i_0 is negligible. The derivation of this result is left as an exercise for the reader.

With the diodes reversed, as in Fig. 4.7-6*c* and *d*, the max diode networks become min networks. That is,

$$v_0 = \min\,(v_1,v_2,v_3) \tag{14}$$

In either max or min function, the number of independent variables can, in principle, be increased by including a larger number of diodes in the network. In practice, because physical diodes are not ideal, the number of diodes in such networks is usually limited to four or five. Such networks are commonly called "gates" because the value of one of the input voltages may be used to gate the flow of signals through the network from the other inputs to the output.

A simple application of the diode gate network is illustrated in Fig. 4.7-7. The network of Fig. 4.7-7*a* is recognizable as a diode gate having two inputs, namely v_1 and zero (ground), and output v_2. It follows from Eq. (13) that $v_2 = \max\,(0, v_1)$; hence if the input v_1 is a sinusoid, the output is the so-called "half-wave rectification" of the input voltage waveform which we encountered in Ex. 4.3-4 and which is shown again in Fig. 4.7-7*b*.

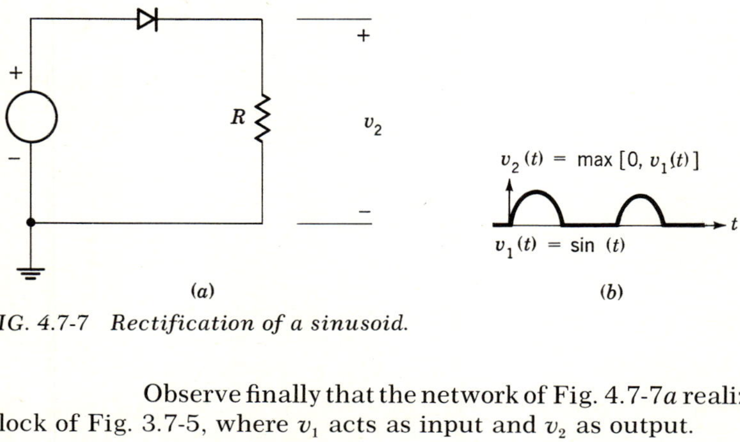

FIG. 4.7-7 Rectification of a sinusoid.

Observe finally that the network of Fig. 4.7-7*a* realizes the threshold block of Fig. 3.7-5, where v_1 acts as input and v_2 as output.

Realization of Monotonically Increasing Piecewise-linear Functions

Consider the network of Fig. 4.7-8. The voltage-source values

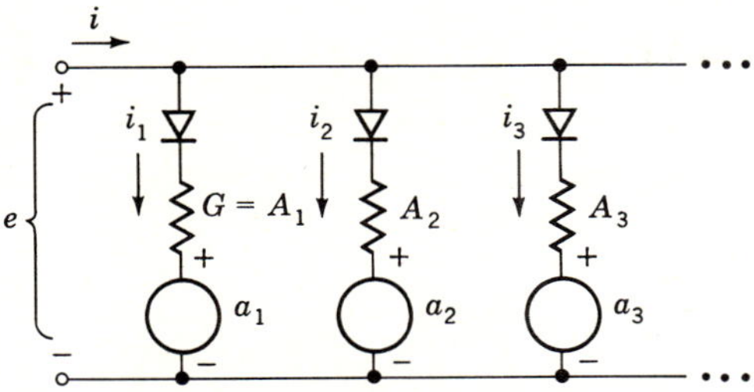

FIG. 4.7-8 Network for the realization of piecewise-linear characteristics.

$a_1, a_2, \ldots$ are ordered. That is,

$$a_1 < a_2 < a_3 < \cdots$$

The conductances have values $A_1, A_2, \ldots$ and the diodes are ideal.

The current i_1, flowing through the leftmost diode, is by a straightforward extension of the above discussion.

$$i_1 = \max\,[0, A_1(e - a_1)]$$

Similarly,

$$i_2 = \max\,[0, A_2(e - a_2)] \tag{15}$$

and

$$i_3 = \max\,[0, A_3(e - a_3)]$$

From Eqs. (15) it follows that the current i, which is given by

$$i = i_1 + i_2 + i_3 + \cdots \tag{16}$$

is related to e by the graph of Fig. 4.7-9.

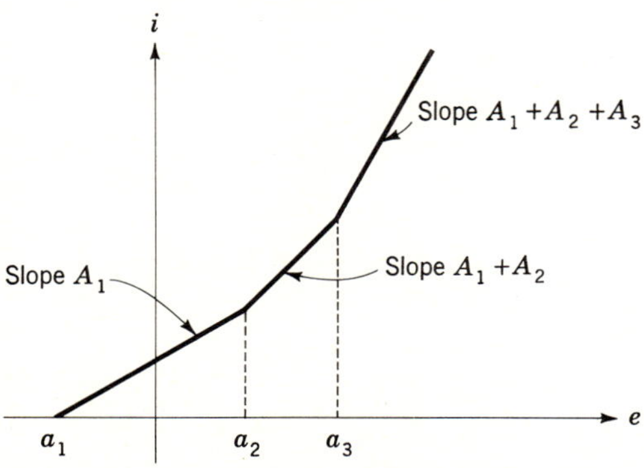

FIG. 4.7-9 e-i characteristic of the network of Fig. 4.7-8.

It is easy to see that by appropriately choosing the constants a_i and A_i, it is possible to construct any desired *monotonically increasing* piecewise-linear characteristic.

Observe finally that this circuit behavior is similar, but not identical, to the input-output characteristic of the block diagram system of Fig. 3.7-8.

In this section we have seen how resistive networks can be used to realize several block diagram system elements. We proceed now with the fundamentals of solving by computer the equations formulated from *nonlinear* networks and block diagram systems.

PROBLEMS

4.2-1 Determine whether each of the networks shown in the figure is or is not a branch network, and briefly justify your conclusion.

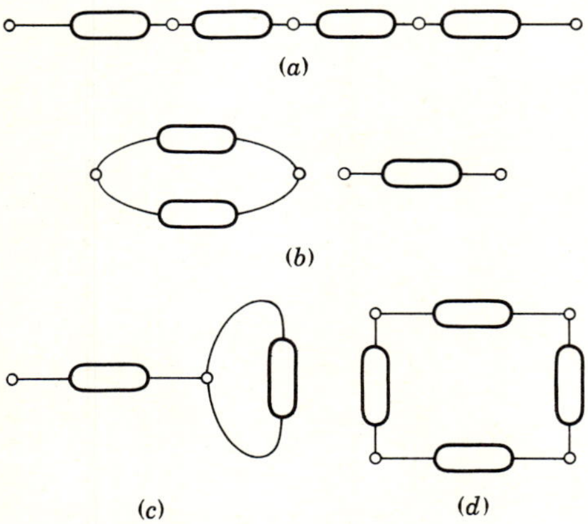

PROB. FIG. 4.2-1

4.3-1 In each of the networks shown in the figure find the unknown voltages and currents so that KVL and KCL are satisfied: (*a*) Find i. (*b*) Find v_1, v_2, v_3. (*c*) Find i_1, i_2

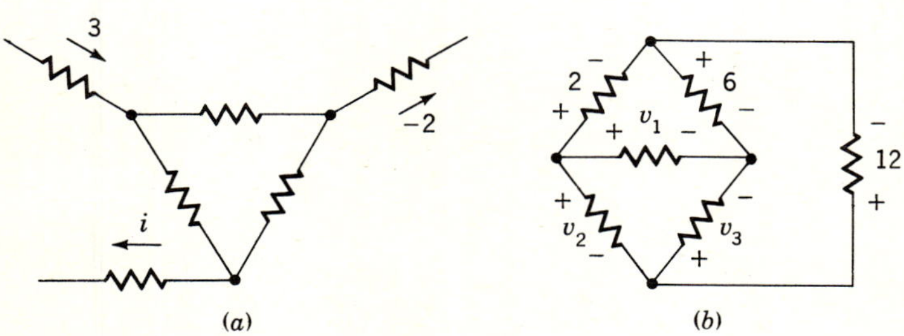

PROB. FIG. 4.3-1

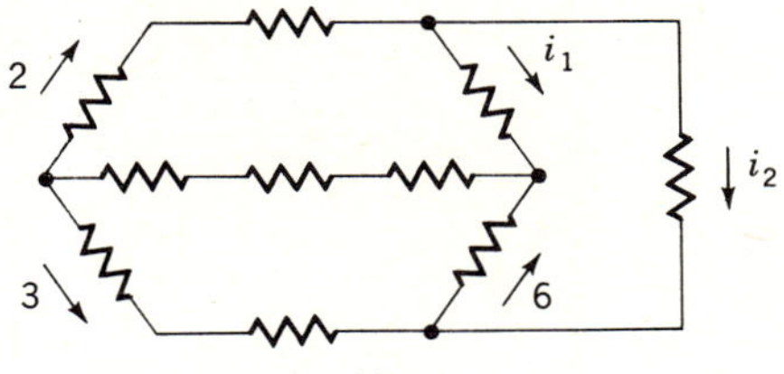

(c)

PROB. FIG. 4.3-1 (con't)

4.3-2 First write a KVL equation for *every* closed circuit, and a KCL equation for *every* node. Then write the smallest number of (independent) equations from which all the above equations can be derived.

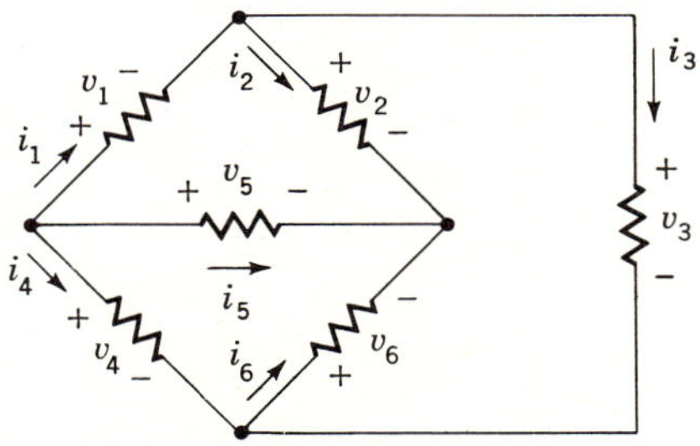

PROB. FIG. 4.3-2

4.4-1 Write node-voltage equations for the network shown in the figure.

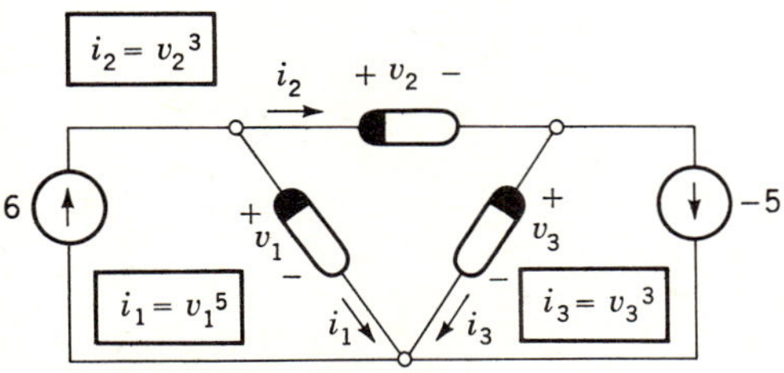

PROB. FIG. 4.4-1

4.4-2 Formulate and solve node-voltage equations for the network shown in the figure. (Indicated resistor values are conductances.)

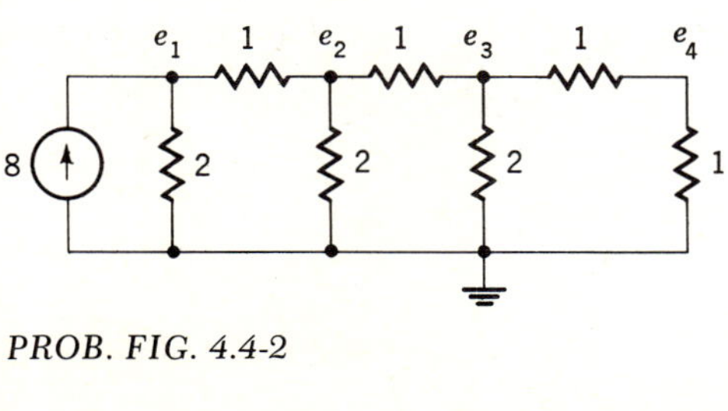

PROB. FIG. 4.4-2

4.4-3 The set of equations

$$5v_a - 2v_b - v_c = 2$$

$$-2v_a + 4v_b - v_c = 1$$

$$-v_a - v_b + 5v_c = 0$$

has been written for a network on a node-voltage basis. Determine a network with appropriate sources that will lead to this set of equations.

4.4-4 Write node-voltage equations for the network shown in the figure.

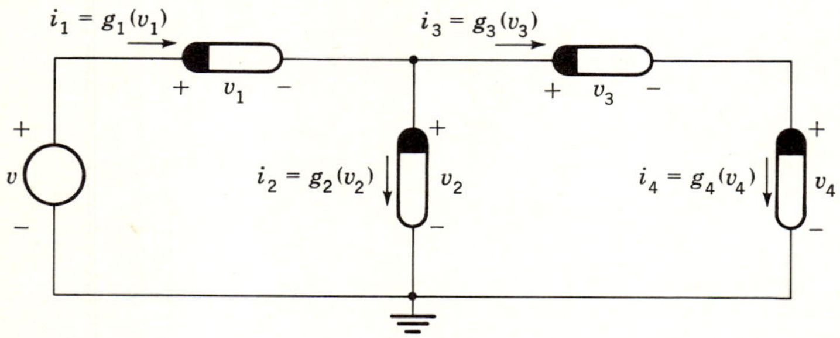

PROB. FIG. 4.4-4

4.4-5 Let nonlinear elements E_1, E_2, and E_3 be defined by the characteristics shown in part *a*. For each of the connections in parts *b* and *c* determine all solutions for the voltage *v*, which is indicated.

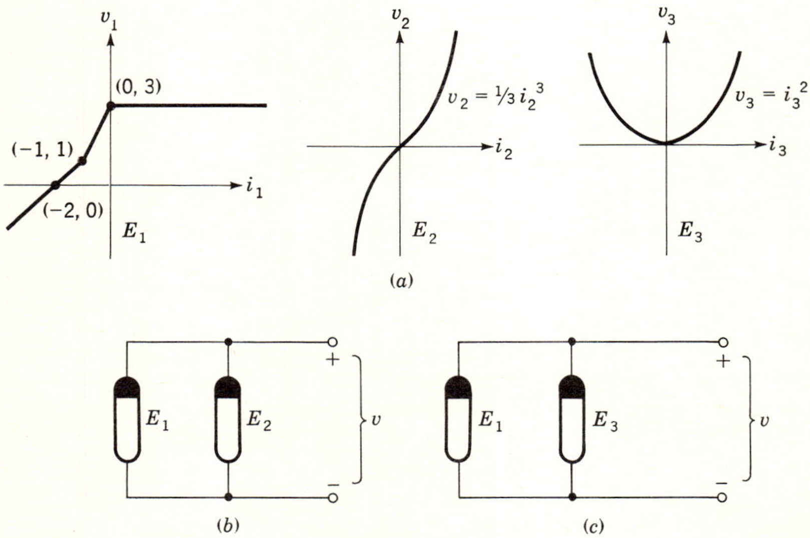

PROB. FIG. 4.4-5

4.4-6 The transistor model to be used in the practical amplifier (part *a*) is shown in part *b*. Note that each diode has characteristic $i = f(v)$, and there is a dependent current source. (All resistor values are conductances.) Formulate node-voltage equations for this amplifier.

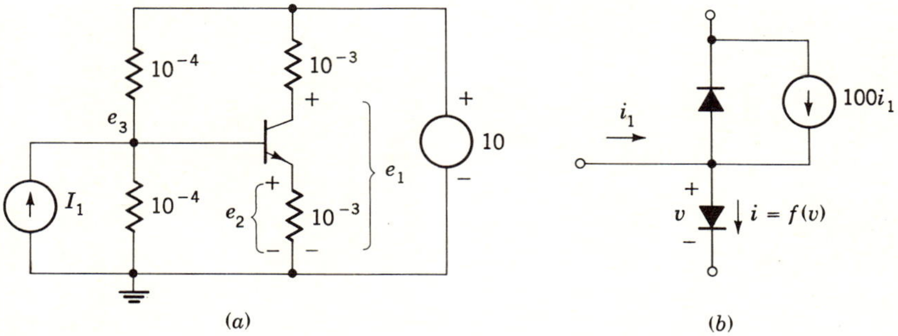

PROB. FIG. 4.4-6

4.5-1 (*a*) Determine R so that $i = 2/3$.
(*b*) Determine R so that $i = 2/3$.
(*c*) Determine R so that $v = 2/3$.
(*d*) Determine R so that $v = 2/3$.
(All quantities are in volts, amperes, or ohms.)

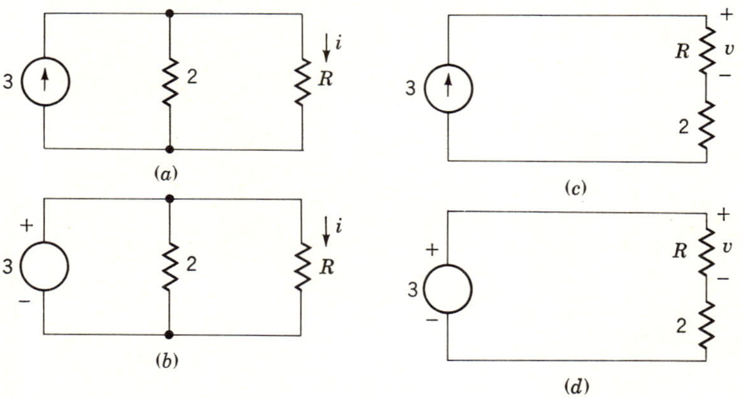

PROB. FIG. 4.5-1

4.5-2 The "black box" consists of an ideal internal voltage source v_s in series with an internal resistance R_s. The black box is designed to supply a very small current i to a specimen in a biological experiment. The specimen is represented by a resistance R which may vary from 10 to 500 Ω during the experiment. What fixed values of v_s and R_s would you suggest so that the black box will approximate a current source in this experiment, with i remaining within 1 percent of one mA throughout the experiment?

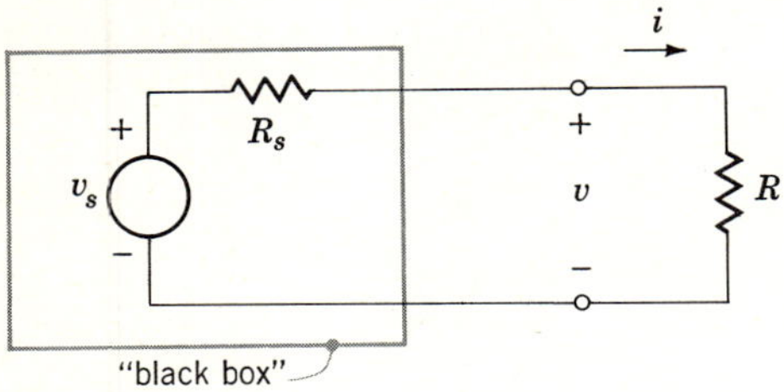

PROB. FIG. 4.5-2

4.5-3 The linear network shown in the figure contains an independent current source i_1 and an independent voltage source v_2.

(*a*) Calculate v_1 and i_2 in terms of R_1, R_2, i_1, and v_2.

(*b*) Redraw the network with source v_2 set to zero. Compute v_1 and i_2 under this condition.

(*c*) Redraw with i_1 set to zero and with v_2 restored to the same (generally) nonzero value as in (*a*). Compute v_1 and i_2 under this condition.

(*d*) Show that "superposition applies," that is, that the value of v_1 obtained in (*a*) is the sum of the values obtained in (*b*) and (*c*), and similarly for i_2, provided the value of i_1 is the same in (*a*) and (*b*), and the value of v_2 is the same in (*a*) and (*c*).

(*e*) Is a short circuit or zero resistance equivalent to a voltage source of zero value?

(*f*) Is an open circuit or zero conductance (infinite resistance) equivalent to a current source of zero value?

(*g*) Using scalers and summers, construct a system model for the above network, having inputs i_1 and v_2 and outputs v_1 and i_2.

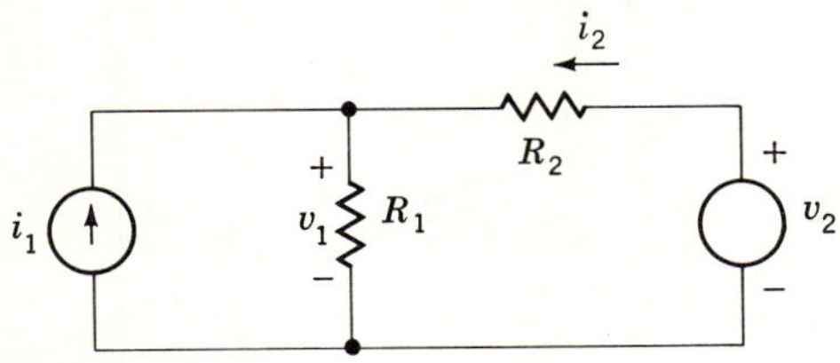

PROB. FIG. 4.5-3

4.5-4 Consider the network shown in the figure.

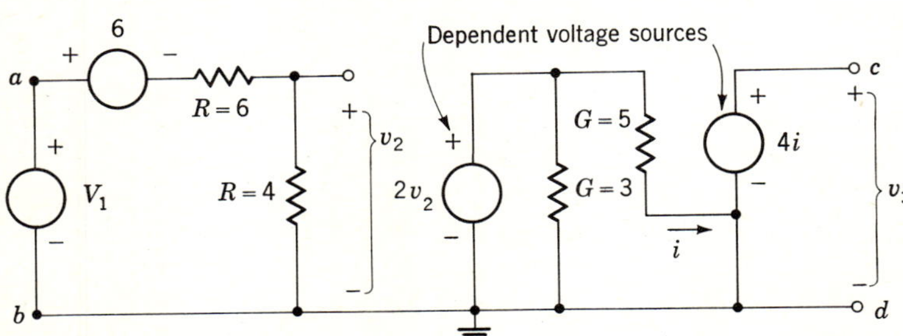

PROB. FIG. 4.5-4

(*a*) Express v_3 in terms of V_1 by showing a sequence of simple calculations that establish first v_2 in terms of V_1, then i in terms of v_2, and then v_3 in terms of i. (Note that the rightmost source is a dependent voltage source with voltage $4i$.)

(*b*) Remove the leftmost voltage source, and then connect node *c* to node *a*, and node *d* to node *b*. Now, using your results of part (*a*) above, solve "instantly" for the voltage between the terminals *c* and *d*.

4.5-5 Find the value of v in the network shown in the figure.

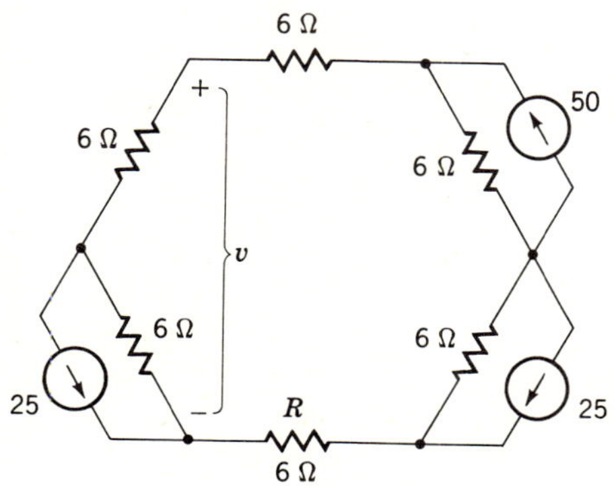

PROB. FIG. 4.5-5

4.5-6 Consider the network of the preceding problem. Suppose that the power dissipated in the resistor *R* is required. Two ways of computing this power are suggested:

(*a*) Use superposition; i.e., first calculate the power supplied by each source, and then add the results.

(*b*) Calculate the entire power due to all sources at once. One method is wrong. Explain which is incorrect and why, and use the correct method to find the power dissipated in *R*.

4.6-1 (*a*) Sketch the v-i characteristics for each of the networks I, II, and III, labeling and dimensioning intercepts and slopes.

(*b*) What relationship among R, i_a, and v_b causes the v-i characteristics for the networks of I and II to be the same?

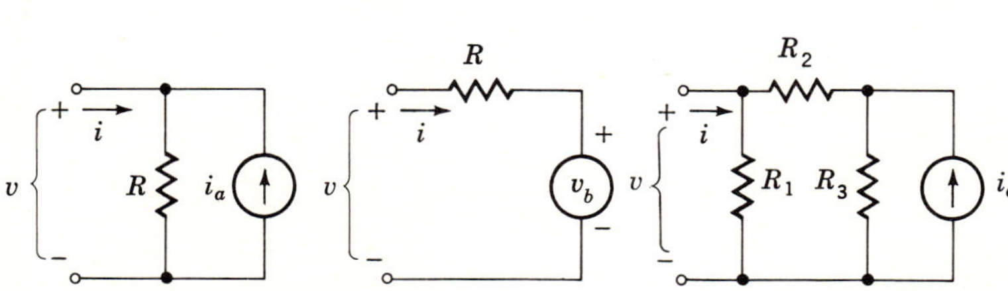

PROB. FIG. 4.6-1

(*c*) Determine R and i_a for network I so that its v-i characteristic is the same as that of network III.

4.6-2 For the network shown plot i versus v. Use the ideal model for the diode.

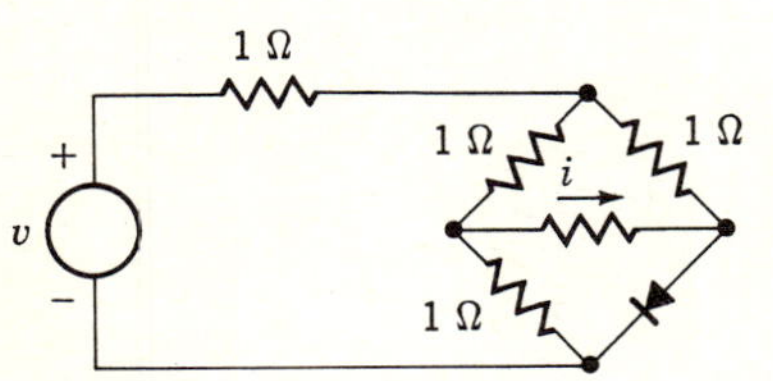

PROB. FIG. 4.6-2

4.6-3 (*a*) Write a set of algebraic equations relating the node voltages e_1, e_2, and e_3 in the network shown. Find the value of the current i_d through the dependent voltage source.

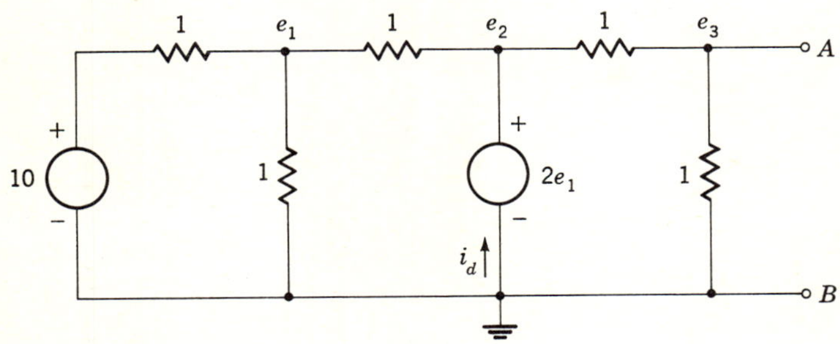

PROB. FIG. 4.6-3

(*b*) Determine the Thévenin equivalent network between the terminals A and B.

4.6-4 (*a*) Determine the voltages at nodes b, c, and d relative to the volt-

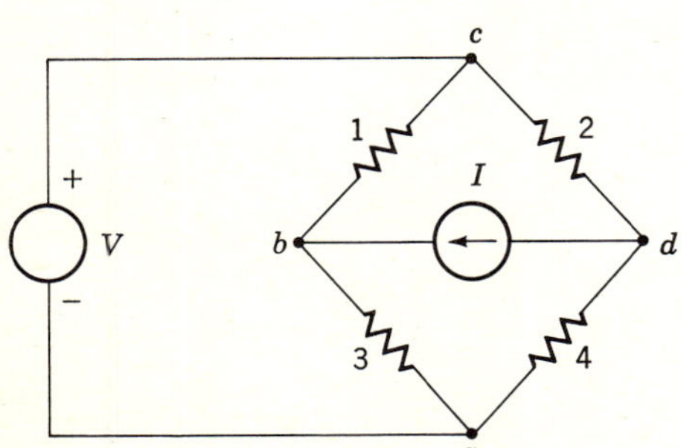

PROB. FIG. 4.6-4

age at node a by using node-voltage analysis and assuming $V=5$ and $I=-1$. (All resistance values are in ohms.)

(*b*) Solve the network using the results of (*a*): that is, sketch the network, labeling *all* the branch voltages and currents with their polarities when $V=5$ and $I=-1$. Include the branches with sources.

(*c*) Determine the voltage at node b relative to that at node a when

$$V=0 \quad \text{and} \quad I=1 \tag{1}$$

and

$$V=1 \quad \text{and} \quad I=0 \tag{2}$$

Using these results and superposition, determine the voltage at node b relative to that at node a when $V=5$ and $I=-1$.

(*d*) Determine the Thévenin equivalent of the network "seen" by the 3-Ω resistor (and therefore not including this resistor) connecting nodes a and b.

(*e*) Determine the voltage of node b relative to node a, using the equivalent network of (*d*) and the 3-Ω resistor.

4.6-5 Network A contains current sources, voltage sources, and linear resistances. Using the results of experiments a and b, find v in experiment c.

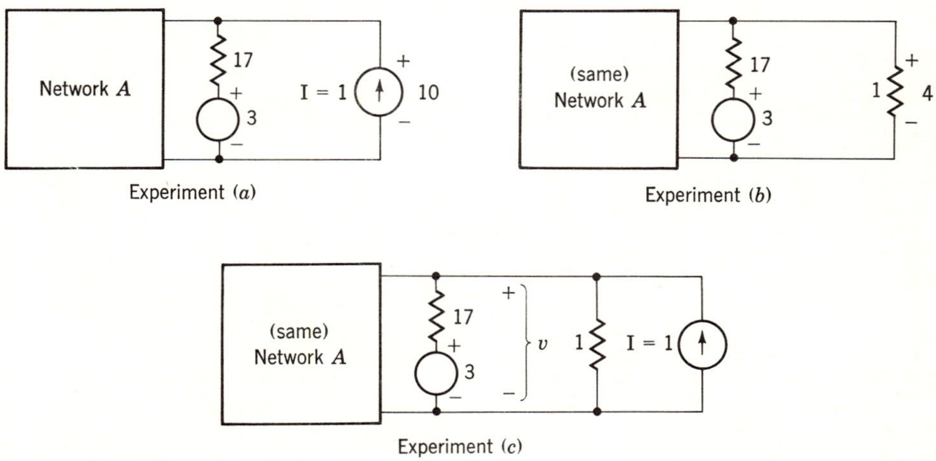

PROB. FIG. 4.6-5

4.6-6 The v_a-i_a characteristic for network A is shown in part a. An experiment is performed on network B, which is made up of *linear resistors* and sources, and the results are:

(i) When the terminals are short-circuited ($v_b=0$), it is found that $i_b=10$ amperes.

(ii) When the terminals are open ($i_b=0$), it is found that $v_b=5$ volts.

Sketch the v-i characteristic for the network obtained by connecting networks A and B in parallel as shown in part b. Clearly indicate your reasoning.

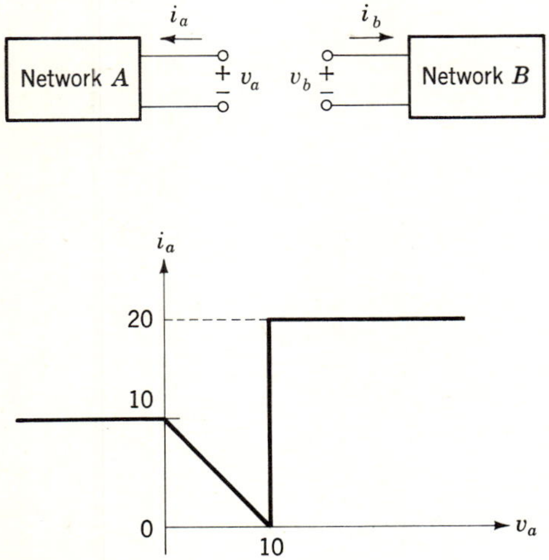

PROB. FIG. 4.6-6a

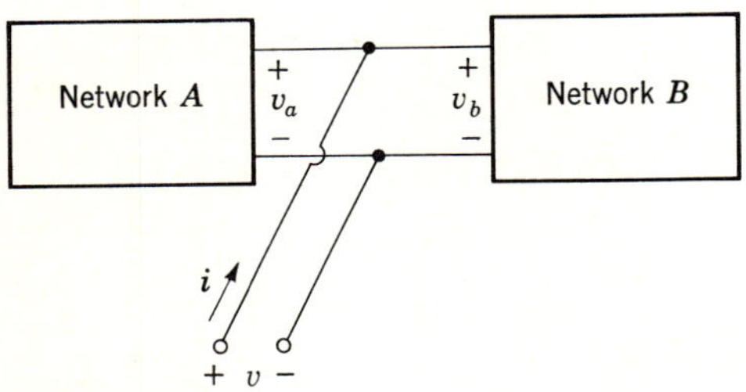

PROB. FIG. 4.6-6b

4.6-7 This problem deals with maximum power transfer. Consider the network shown in part (a).

(a) Determine the power dissipated in the resistor R_1 and in the resistor R_2. [$R_1 > 0$ (source resistor); $R_2 > 0$ (load resistor).]

(b) Suppose now that R_1 is fixed; how would you choose the resistor R_2 so as to maximize the power dissipated in R_2?

Another way of maximizing the power delivered to a load resistor is to use an ideal transformer. An ideal transformer is represented by the symbol of part (b). Mathematically, an ideal transformer is specified by the relationships

$$\frac{v_1(t)}{N_1} = \frac{v_2(t)}{N_2} \qquad N_1 i_1(t) + N_2 i_2(t) = 0$$

Suppose now that one is given a network with the Thévenin equivalent of

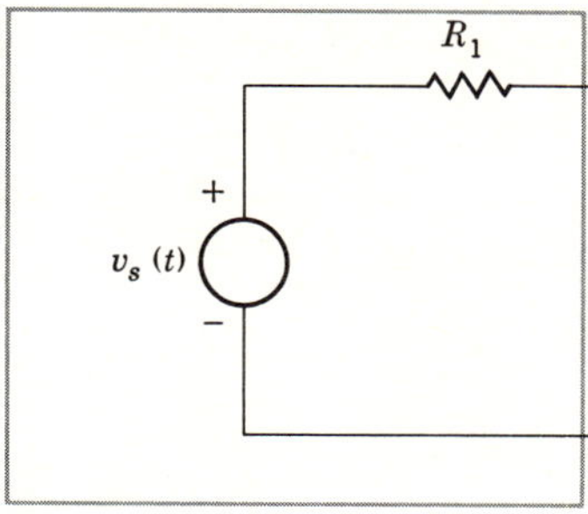

PROB. FIG. 4.6-7a

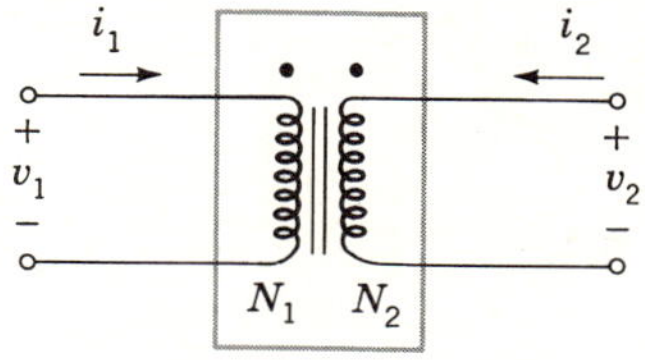

PROB. FIG. 4.6-7b

N_1, N_2

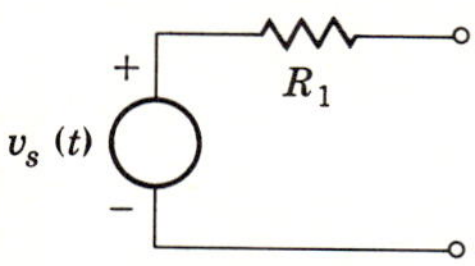

PROB. FIG. 4.6-7c

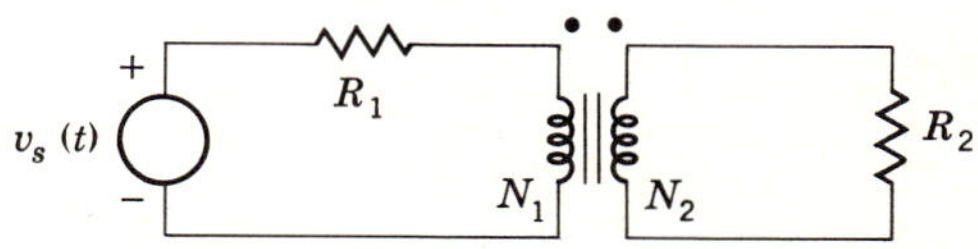

PROB. FIG. 4.6-7d

part (*c*) and a fixed load resistor R_2. These networks can be coupled by a transistor as shown in part (*d*).

(*c*) Determine the turns ratio N_1/N_2 so that the power delivered to R_2 is maximized when R_1 and R_2 are fixed. Show your reasoning.

4.7-1 In network *M* all resistors are linear with values in thousands of ohms (i.e., 1 kΩ = 1,000 Ω). Each of the four switches $s_1, \ldots, s_4$ is either closed ($s_i = 1$) or open ($s_i = 0$).

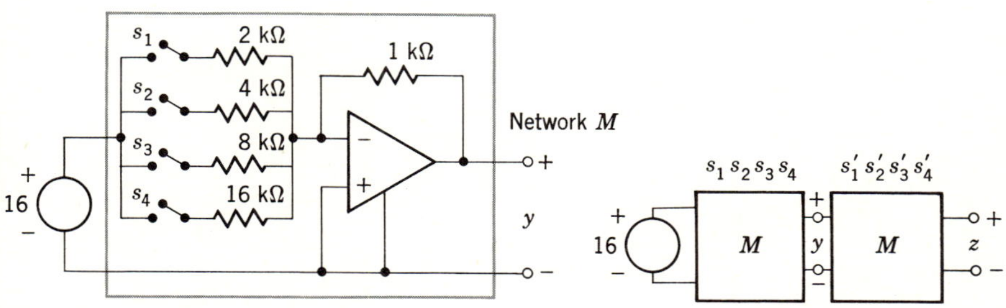

PROB. FIG. 4.7-1

(*a*) Give an equation for the voltage *y* in terms of the variables s_1, s_2, s_3, s_4. What task does this circuit perform?

(*b*) Consider now the cascade of two *M* networks as shown in the figure. Relate *z* to $s_1, \ldots, s_4$ and $s'_1, \ldots, s'_4$. What task does this circuit perform?

4.7-2 In the problem shown in the figure, V_1, V_2, and V_3 are either 0 or 3 V. Using resistors and diodes, design network *N* so that

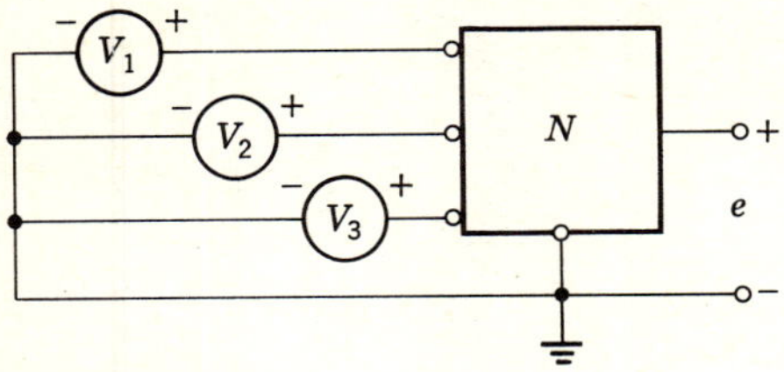

PROB. FIG. 4.7-2

(*a*) $e = 3$ if $V_1 = V_2 = V_3 = 3$; and $e = 0$ otherwise. (This is an AND gate.)

(*b*) $e = 0$ if $V_1 = V_2 = V_3 = 0$; and $e = 3$ otherwise. (This is an OR gate.)

(*Hint:* Use min, max diode circuits.)

chapter 5

Iteration

The implicit memoryless systems of Chap. 3 and the nonlinear resistive networks of Chap. 4 are generally characterized by nonlinear implicit equations of the form $x = f(x)$ and $f(x) = 0$. This chapter is concerned with the computational solution of such equations by an orderly trial-and-error approach called iteration. Several well-known iteration techniques—Picard's and Newton's methods—are discussed from the point of view of the underlying mathematical ideas; particular attention is given to the existence and uniqueness of a solution to $x = f(x)$, and to the convergence, convergence speed, and accuracy of the iteration rule that solves such an equation. Although our discussion is in terms of a single unknown, the results can be generalized to an arbitrary number of variables.

5.1 INTRODUCTION

In the preceding two chapters we learned how to formulate equations for memoryless block diagram systems and resistive networks. Analytical solution of these equations is possible when they are linear or in very rare instances when they are nonlinear. For example, if the formulated equation is $ay + b = x$ [where x is an excitation or input, y is a desired output, and a, b are constants ($a \neq 0$)], we can solve analytically for y in the form $y = (x - b)/a$. Therefore, given any value for x, we can evaluate in a direct and straightforward way the corresponding value of y. Similarly, if the formulated equation is $e^{ay+b} = x^2$, we can still solve for y in the form $y = (1/a)(\ln x^2 - b)$. Both the above cases have in common the fact that y can be expressed in terms of x in an explicit form which permits direct evaluation of y, given a value for x.

Very often, especially in the case of implicit nonlinear block diagram systems and nonlinear resistive networks, the formulated equations cannot be solved in such a manner. For example, if the equation is $\sin y + y = x + a$, then we cannot express y explicitly in terms of x. It is primarily in such cases that we resort to an orderly trial-and-error approach called *iteration* through an appropriate digital computer program.

Although in this chapter we shall study specific iteration methods such as Picard's and Newton's, our intent is not to catalog the large number of known iteration methods. Instead, we are concerned with basic mathematical concepts such as existence and uniqueness of a solution and convergence of iteration, which will permit us to evaluate a proposed iteration method, or alternatively to design one which may be better suited to our special needs. Accordingly, in this chapter we shall examine equations in *one* unknown; this approach facilitates explanation of the underlying fundamentals, and forms a foundation for the study of computational methods on many variables.

5.2 ITERATION

Suppose that the equation to be solved is of the form

$$f(x,u) = 0$$

where x is the desired unknown for a given value of u. Iteration involves the following steps:

1. Guess a value for x.
2. If the proposed value for x is "sufficiently close" to the exact solution [e.g., if $|f(x,u)| < 0.0001$], then Stop; otherwise continue with Step 3, below.
3. Propose a new value for x, on the basis of a well-defined rule [e.g., $x_{new} = x_{old} + f(x_{old}, u)$], and continue with Step 2, above.

Normally, Steps 2 and 3 will be repeated several times until a solution is found; when such a solution has been found, we say that the iteration process has *converged*. On less fortunate occasions, Steps 2 and 3 can be repeated many times with no apparent convergence to a solution; we then say that the iteration process *does not converge*. When it does converge, we are often interested in "how fast" (how many repetitions or iterations of Steps 2 and 3 are involved) and how accurately the solution is obtained; we refer to these properties as *convergence speed* and *convergence error*, respectively. Finally, we would like to know what will happen if the problem which is to be solved by iteration has either no solution or several solutions; we are therefore concerned with the *existence* and *uniqueness* of solution(s) to that problem.

In this section, we shall study the above concepts both abstractly, and in terms of applications to simple network and system problems. We start with a simple example from network analysis so that we may better understand the process of iteration and the questions that it raises.

Example 5.2-1 *Iterative Solution of a Nonlinear Network* Consider the nonlinear network of Fig. 5.2-1, where we wish to find the voltage e with respect to ground. Observe that the three voltage-controlled resistors have their currents i_j equal to the first, fifth, and third powers of their respective voltages v_j, as shown in Fig. 5.2-1.

We begin by writing a nodal equation for this network:

$$\sum_1^3 i_j = (6-e) + (-1-e)^5 + (4-e)^3 = 0 \tag{1}$$

Observe that we know of no explicit analytical ways for solving this equation for

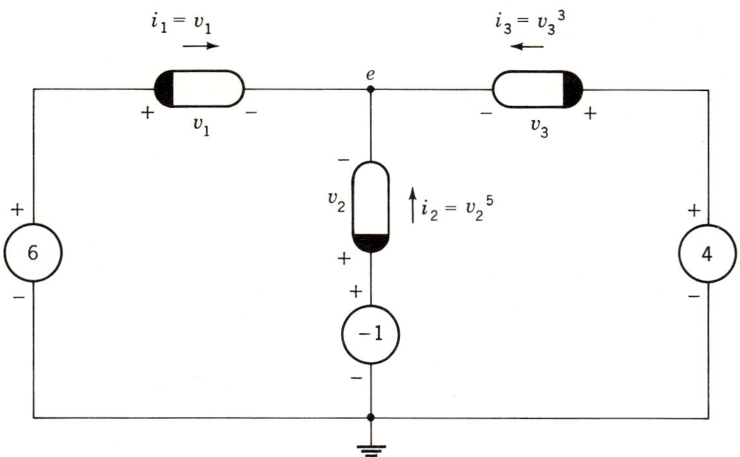

FIG. 5.2-1 Network to be solved by iteration.

e and must therefore resort to iteration. We propose accordingly an iteration process based on the following informal reasoning: Suppose that we try for the unknown of Eq. (1) an arbitrary value, say e_0; then the term $\Sigma_1^3 i_j$ of Eq. (1) will be

1. Positive if $e_0 < e$.
2. Zero if $e_0 = e$.
3. Negative if $e_0 > e$.

To improve our "guess" e_0, we seek a rule that yields a new value e_1, which in Case 1 above is greater than e_0, in Case 2 equal to e_0, and in Case 3 less than e_0. Furthermore, we feel that this rule should make a correction that is monotonically related to the term $\Sigma_1^3 i_j$; that is, the bigger the magnitude of this term, the bigger should be the corresponding correction.

The following simple rule satisfies all the above requirements:

$$e_1 = e_0 + \alpha \sum_1^3 i_j \tag{2}$$

where α is a positive constant ($\alpha \neq 0$).

Unless we are very lucky, our new guess, e_1, will not be the correct solution, although it will be, we hope, closer than e_0 to the desired solution. We will then replace e_0 by e_1, and e_1 by e_2, in Eq. (2), in order to obtain what, we hope, will be a better guess, e_2. Hence, our *rule* or *algorithm* for the entire iteration process becomes as follows:

1. Choose an arbitrary value e_0. Let $k = 0$.
2. Compute $\Sigma_1^3 i_j$ using Eq. (1), with $e = e_k$. If $|\Sigma_1^3 i_j| < 0.001$, then Stop; the desired solution is e_k. Otherwise go to Step 3.
3. Compute $e_{k+1} = e_k + \alpha \Sigma_1^3 i_j$. Increase k by 1 and go to Step 2, above.

We will now test this algorithm on a computer, following an "experimental" rather than a theoretical approach, so that we may identify and motivate the questions that we will treat from a theoretical point of view in the following section. From Eqs. (1) and (2) we see that our iteration rule is as follows:

$$e_{k+1} = e_k + \alpha[(6 - e_k) + (-1 - e_k)^5 + (4 - e_k)^3] \tag{3}$$

As a first experiment, we let $\alpha = 0.005$ and $e_0 = 0$. The results of the iteration process are shown in Table 5.2-1 where the leftmost column marks the *iteration index*, ($k = 0, 1, 2, \ldots$) and the remaining columns correspond to the values of e_k, i_1, i_2, i_3, and $\Sigma_1^3 i_j$, respectively. Note that the meaning of aEk in the computer printout† is $a10^k$; for example, at $k = 0$ the table states $i_2 = -1$, $i_3 = 64$, and Σ_1^3 69. Observe that it takes 17 iterations for the algorithm to stop, and that the final value ($e_{17} = 0.999994$) results in $|\Sigma_1^3 i_j| = 0.00065$ which is indeed less than the stipulated 0.001. Indeed, the exact solution is $e = 1$, so that e_{17} is for all practical purposes the desired solution.

In our second experiment we let $\alpha = 0.01$ and $e_0 = 0$. The results, shown in the same format in Table 5.2-2, exhibit *faster convergence*, since the algorithm terminates in six iterations. We observe also that the term $\Sigma_1^3 i_j$ becomes alternately positive and negative for $k \geq 3$.

In the third experiment, where $\alpha = 0.015$ and $e_0 = 0$ (Table 5.2-3) the *convergence rate* has decreased, and it takes 19 iterations for the algorithm to terminate. Here the term $\Sigma_1^3 i_j$ changes sign as it decreases in magnitude, with increasing k.

†APL running on IBM 360-67.

TABLE 5.2-1 Iteration with $\alpha = 0.005$ and $e_0 = 0$

K	E	I1	I2	I3	SUM
	0.000000000E0	6.000000000E0	¯1.000000000E0	6.400000000E1	6.900000000E1
1	3.450000000E¯1	5.655000000E0	¯4.401608601E0	4.882723637E1	5.008062777E1
2	5.954031389E¯1	5.404596861E0	¯1.033599310E1	3.946363478E1	3.453223854E1
3	7.680643316E¯1	5.231935668E0	¯1.727787448E1	3.375888722E1	2.171294841E1
4	8.766290736E¯1	5.123370926E0	¯2.327513463E1	3.046987623E1	1.231811253E1
5	9.382196363E¯1	5.061780364E0	¯2.735362859E1	2.870265694E1	6.410808721E0
6	9.702736799E¯1	5.029726320E0	¯2.969154379E1	2.781058980E1	3.148772323E0
7	9.860175415E¯1	5.013982458E0	¯3.089693509E1	2.737928870E1	1.496336067E0
8	9.934992219E¯1	5.006500778E0	¯3.148330759E1	2.717590163E1	6.990948177E¯1
9	9.969946959E¯1	5.003005304E0	¯3.176029714E1	2.708122452E1	3.239326891E¯1
10	9.986143594E¯1	5.001385641E0	¯3.188930224E1	2.703742958E1	1.495129755E¯1
11	9.993619243E¯1	5.000638076E0	¯3.194898650E1	2.701723171E1	6.888328335E¯2
12	9.997063407E¯1	5.000293659E0	¯3.197651415E1	2.700792958E1	3.170908498E¯2
13	9.998648861E¯1	5.000135114E0	¯3.198919235E1	2.700364824E1	1.459100473E¯2
14	9.999378411E¯1	5.000062159E0	¯3.199502760E1	2.700167832E1	6.712883921E¯3
15	9.999714055E¯1	5.000028594E0	¯3.199771251E1	2.700077206E1	3.088142867E¯3
16	9.999868463E¯1	5.000013154E0	¯3.199894771E1	2.700035515E1	1.420591486E¯3
17	9.999939492E¯1	5.000006051E0	¯3.199951594E1	2.700016337E1	6.534817685E¯4

TABLE 5.2-2 Iteration with $\alpha = 0.01$ and $e_0 = 0$

K	ANSWER E	I1	I2	I3	SUM
0	0.000000000E0	6.000000000E0	¯1.000000000E0	6.400000000E1	6.900000000E1
1	6.900000000E¯1	5.310000000E0	¯1.378584918E1	3.626469100E1	2.778884182E1
2	9.678884182E¯1	5.032111582E0	¯2.951225187E1	2.787632621E1	3.396185920E0
3	1.001850277E0	4.998149723E0	¯3.214829632E1	2.695007332E1	¯2.000732843E¯1
4	9.998495445E¯1	5.000150455E0	¯3.198796537E1	2.700406250E1	1.624758586E¯2
5	1.000012020E0	4.999987980E0	¯3.200096164E1	2.699967545E1	¯1.298210051E¯3
6	9.999990383E¯1	5.000000962E0	¯3.199992306E1	2.700002597E1	1.038669972E¯4

TABLE 5.2-3 Iteration with $\alpha = 0.015$ and $e_0 = 0$

K	E	I1	I2	I3	SUM
0	0.000000000E0	6.000000000E0	¯1.000000000E0	6.400000000E1	6.900000000E1
1	1.035000000E0	4.965000000E0	¯3.489973006E1	2.606598213E1	¯3.868747934E0
2	9.769687810E¯1	5.023031219E0	¯3.019945159E1	2.762662906E1	2.450208696E0
3	1.013721911E0	4.986278089E0	¯3.311291989E1	2.663120043E1	¯1.495441373E0
4	9.912902908E¯1	5.008709709E0	¯3.130926562E1	2.723584554E1	9.352896291E¯1
5	1.005319635E0	4.994680365E0	¯3.242784073E1	2.685662438E1	¯5.765359852E¯1
6	9.966715955E¯1	5.003328404E0	¯3.173461243E1	2.708996666E1	3.586826387E¯1
7	1.002051835E0	4.997948165E0	¯3.216448395E1	2.694463833E1	¯2.218974552E¯1
8	9.987233733E¯1	5.001276627E0	¯3.189800016E1	2.703448359E1	1.377600597E¯1
9	1.000789774E0	4.999210226E0	¯3.206323185E1	2.697868171E1	¯8.533991418E¯2
10	9.995096754E¯1	5.000490325E0	¯3.196079326E1	2.701324093E1	5.293798791E¯2
11	1.000303745E0	4.999696255E0	¯3.202430700E1	2.699179971E1	¯3.281103934E¯2
12	9.998115797E¯1	5.000188420E0	¯3.198492921E1	2.700508767E1	2.034687570E¯2
13	1.000116783E0	4.999883217E0	¯3.200934372E1	2.699684699E1	¯1.261351093E¯2
14	9.999275801E¯1	5.000072420E0	¯3.199420683E1	2.700195538E1	7.820972801E¯3
15	1.000044895E0	4.999955105E0	¯3.200359174E1	2.699878786E1	¯4.848773890E¯3
16	9.999721631E¯1	5.000027837E0	¯3.199777311E1	2.700075160E1	3.006327902E¯3
17	1.000017258E0	4.999982742E0	¯3.200138067E1	2.699953404E1	¯1.863889430E¯3
18	9.999892997E¯1	5.000010700E0	¯3.199914399E1	2.700028891E1	1.155624464E¯3
19	1.000006634E0	4.999993366E0	¯3.200053073E1	2.699982088E1	¯7.164821633E¯4

TABLE 5.2-4 Iteration with $\alpha = 0.03$ and $e_0 = 0$

```
      ALPHA←.03
      ITERATE

   E[0] = 0
   E[1] = 2.07
   E[2] = ¯5.777555245
   E[3] = 97.28860667
   E[4] = ¯275217112.6
   E[5] = 4.73694582E40
DOMAIN ERROR
```

Finally, (Table 5.2-4), letting $\alpha = 0.03$, gives rise to a rapidly growing e_k, which by the 5th iteration reaches the value $e_5 = 4.7 \times 10^{40}$, indicating no intention of converging on the solution. The format of this table is different but self-explanatory.†

Several questions arise from this example: Will the process converge for any initial value e_0? For the given value e_0, will the process converge if α is made larger (smaller)? How do our computational errors (e.g., rounding off of decimal digits) affect convergence? How is the convergence speed related to the value of α? Are there better iteration rules, and how can we construct them? These and other related questions motivate our study of iteration from a more theoretical point of view. We shall start this study by formalizing below the form of the iteration rule, and by establishing the class of problems which we shall solve by iteration.

Observe that Eq. (3) establishes a value for e_{k+1} on the basis of the previous value e_k, and when convergence has been achieved, i.e., when, at the limit, $e_{k+1} = e_k$, Eq. (3) assumes the form

$$\alpha[(6 - e_k) + (-1 - e_k)^5 + (4 - e_k)^3] = 0$$

which is the equation that we wanted to solve, with the solution "plugged in."

We can generalize these two observations as follows: Suppose that the *problem to be solved is a nonlinear equation* [such as Eq. (1)] of the form

$$\boxed{\text{Find } x \text{ so that } F(x) = 0} \tag{4}$$

which has a solution, say $x = s$. Suppose also that this solution can be found by an *iteration rule* [such as Eq. (3)] of the general form

$$\boxed{x_{k+1} = f(x_k)} \tag{5}$$

where $f(.)$ is (another) function. As was the case in our example, in order for the rule to yield the solution s, we expect that $x_{k+1} = x_k = s$; hence

$$s = f(s) \tag{6}$$

That is, *the equation resulting from the iteration rule when $x_{k+1} = x_k$ must have the same solution as the equation to be solved.* Besides satisfying Eq. (6), the rule, and therefore the function $f(.)$, must satisfy certain additional criteria that will ensure convergence, as we shall see in the following section. For the time being, however, we observe that for a given equation (4), *many* iteration rules, such as Eq. (5), can be chosen with the property (6). That is, we can think of many ways for obtaining the function $f(.)$ from the function $F(.)$ while satisfying Eq. (6). For example, in Eq. (3), which is in the form of Eq. (5), we can choose any one of an infinite number of values for α, while satisfying Eq. (6). Alternatively, we could construct an iteration rule of a different form as follows:

$$e_{k+1} = e_k + \alpha[6 - e_k + (-1 - e_k)^5 + (4 - e_k)^3]^3$$

In this case, $f(e) = e + \alpha[6 - e + (-1 - e)^5 + (4 - e)^3]^3$. Finding a solution for the equation $e = f(e)$ corresponds to solving the equation of the network of Fig. 5.2-1.

†The computer comment "DOMAIN ERROR" indicates that e_6 is so large that it cannot be stored in the computer.

We see then that the task of converting the given equation which is to be solved into the form $x = f(x)$, and hence the task of establishing an iteration rule, is subject to choice, which will be usually made on the basis of the problem class under consideration and the problem solver's understanding of the iteration process. We proceed now under the assumption that the given equation has been converted to the form $x = f(x)$, solution of which we shall seek through the iteration rule $x_{k+1} = f(x_k)$.

5.3 ITERATIVE SOLUTION OF $x = f(x)$: PICARD'S ALGORITHM

We have already motivated by our network example the need for solving nonlinear algebraic equations. In this section we shall present a simple iterative rule, often called *Picard's algorithm* which can be used to obtain the solution of the equation

$$x = f(x) \tag{1}$$

What we mean by a solution of Eq. (1) is a real number s (provided that such a number exists†) with the property that

$$s = f(s) \tag{2}$$

The Picard algorithm is defined as follows:

Step 1: Choose an initial value x_0 for the solution; set the *iteration index* $k = 0$.
Step 2: If $|f(x_k) - x_k| \leq \epsilon$, Stop ($\epsilon > 0$). The solution is then, within the maximum tolerable error ϵ, $s \simeq x_k$. Otherwise, go to Step 3, below.
Step 3: Compute $x_{k+1} = f(x_k)$. Increase k by 1 and go to Step 2, above.

Observe that in this, as in every iterative algorithm, s is never computed exactly but rather to within a certain accuracy as dictated by the value of ϵ. The smaller the value of ϵ, the more accurate the computation. On the other hand, the smaller ϵ, the larger the number of iterations.

It is easy to see that this algorithm is simple to implement on a digital computer. Hence, it is certainly attractive from that point of view. Nonetheless, before we blindly use such an algorithm, we must pause and consider certain fundamental questions:

1. How do we know that Eq. (1) has a solution? Certainly, if no solution exists it seems futile to use the algorithm in the first place.
2. Suppose that Eq. (1) has a solution. Is the solution unique? For, if Eq. (1) has more than one solution, then we may be interested in finding all the solutions.
3. Suppose that Eq. (1) has a unique solution s. Is then Picard's algorithm guaranteed to converge? If so, how fast? In other words, can we estimate how many iterations will be conducted before the algorithm terminates?

All the above questions are of fundamental importance not only to Picard's algorithm but also to any iterative algorithm. It is for this reason that we shall spend quite a bit of time and care in analyzing them.

†For example, the equation $x^2 + 1 = 0$ has no *real* solution.

Existence of Solutions

In considering the solution(s) of Eq. (1) let us suppose that x takes on values in the interval $[a,b]$, i.e.,

$$a \leq x \leq b \tag{3}$$

Furthermore, suppose that

For all $x(a \leq x \leq b)$,

$$a \leq f(x) \leq b \tag{4}$$

Thus, we assume that for each value of x in the real interval $[a,b]$, the value of the function $f(x)$ is also in that interval. The graphical interpretation of this

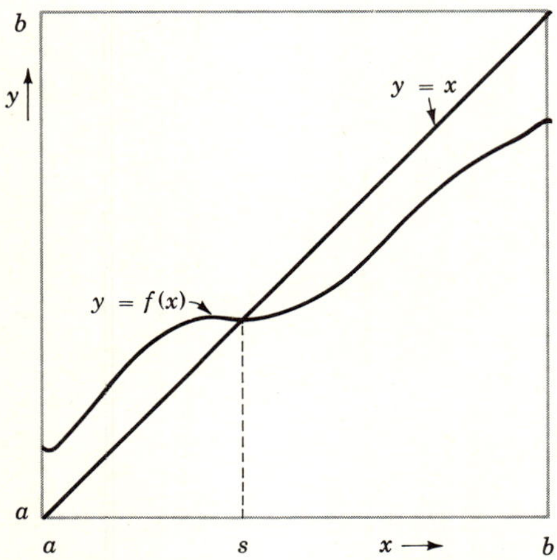

FIG. 5.3-1 Graphs of $y = f(x)$ and $y = x$.

assumption is illustrated in Fig. 5.3-1; the graph of $f(.)$ lies in a square whose side is b − a.†

In the special case illustrated in Fig. 5.3-1 the solution s of the equation $x = f(x)$ exists and is, in fact, unique. Clearly, a solution s will exist if and only if the graph of the function $f(.)$ intersects the straight line $y = x$. Inspection of Fig. 5.3-2a reveals that if $f(x)$ is *not* continuous, then $x = f(x)$ *may not* have a solution. If, however, $f(x)$ is constrained to be continuous, then at least one solution exists. This can be demonstrated graphically, since the graph of $f(x)$ must start somewhere on the left edge and must terminate somewhere on the right edge of the square; hence, it must intersect $y = x$, at least once, though such an intersection may be at an end point of $[a,b]$. Therefore, the conditions

†These assumptions do not restrict the generality of our approach.

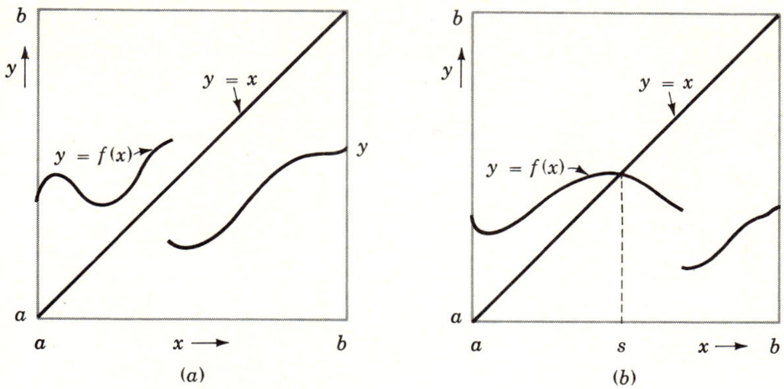

FIG. 5.3-2 Discontinuities of f.

$$\boxed{\begin{array}{l} \text{For all } x(a \le x \le b),\ a \le f(x) \le b \\ f(x) \text{ continuous} \end{array}} \qquad (5)$$

are *sufficient* for the existence of a solution. They are *not necessary*, since $f(x)$ could be as in Fig. 5.3-2*b*, i.e., discontinuous, while possessing a solution.

Uniqueness of Solution

Next, we examine uniqueness. Specifically, we wish to find in addition to Eq. (5), constraints on the class of functions $f(.)$, so that the equation $x = f(x)$ has only one solution.

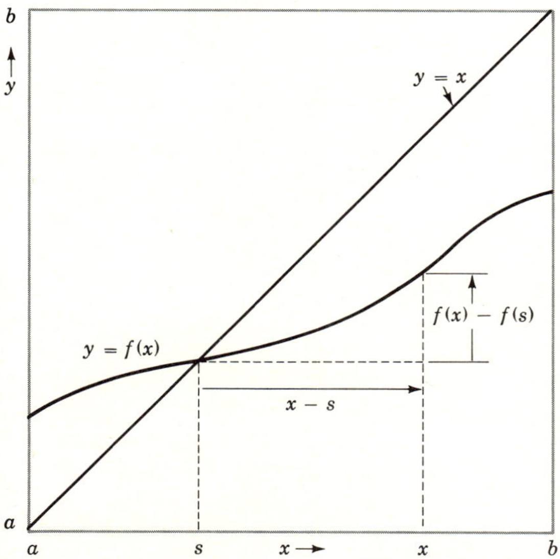

FIG. 5.3-3 Toward necessary and sufficient conditions for uniqueness.

Suppose that s, as shown in Fig. 5.3-3, is a solution. In order to guarantee uniqueness we restrict the function $f(.)$ to obey the following conditions:

For $x > s$,

$$f(x) - f(s) < x - s \tag{6a}$$

For $x < s$,

$$f(x) - f(s) > x - s \tag{6b}$$

That is, we require $f(x)$ to be "below" the line $y = x$, when x is to the right of s, and "above" that line when x is to the left of s. Multiplying Eqs. (6) by the sign of $x - s$ we obtain

For $x > s$,

$$f(x) - f(s) < |x - s| \tag{7a}$$

For $x < s$,

$$f(s) - f(x) < |x - s| \tag{7b}$$

Equation (7) is easily shown to be both *necessary and sufficient* for the uniqueness of a solution of Eq. (1). Equation (7) depends, however, on the value of s which is *a priori* unknown. Since we cannot legitimately restrict $f(.)$ in such a manner, we proceed to the next available alternative by changing conditions (7) as follows:

For all x_1, x_2,†

$$|f(x_1) - f(x_2)| < |x_1 - x_2| \tag{8}$$

Evidently, if Eq. (8) holds, then Eqs. (7) hold (by setting $x_2 = s$). The converse, however, is not true. Equations (8) and (5) are therefore *sufficient* but *not* necessary for uniqueness.

Equation (8) can also be stated in the form

For all x_1, x_2,

$$\boxed{|f(x_1) - f(x_2)| \leq L|x_1 - x_2| \qquad L < 1} \tag{9}$$

which is known as a *Lipschitz condition* on f; L is known as the *Lipschitz constant*. Alternatively, f is said to be a *contraction* mapping. In graphical terms, the Lipschitz condition (9) states that a straight line connecting any two points on the graph of $f(x)$ has a slope magnitude which is less than unity.

To prove that Eq. (9) is *sufficient* for uniqueness, we use a contradiction argument. Suppose that Eqs. (5) and (9) hold, and suppose that $x = f(x)$ has more than one solution. There are then, because existence is guaranteed by

†Hereafter, "for all x_1, x_2" shall be interpreted as "for all x_1, x_2 in the interval $[a,b]$."

Eq. (5), at least two solutions, say s_1 and s_2. Therefore

$$s_1 = f(s_1)$$

and (10)

$$s_2 = f(s_2)$$

Subtracting these two equations and taking absolute values yields

$$|s_1 - s_2| = |f(s_1) - f(s_2)| \tag{11}$$

which violates Eq. (9) for the values $x_1 = s_1$, $x_2 = s_2$, thus leading to a contradiction. Hence, we conclude that Eq. (9) implies uniqueness.

Note that the Lipschitz condition (9) can be related to the derivative $f'(x)$ of the function $f(x)$, as follows. Suppose that $f(x)$ is differentiable. By the *mean value theorem*,† there exists for any x_1, x_2 a value x such that

For all x_1, x_2,

$$f(x_2) - f(x_1) = f'(\hat{x})(x_2 - x_1) \tag{12}$$

Taking absolute values, we obtain

$$|f(x_2) - f(x_1)| = |f'(\hat{x})(x_2 - x_1)| \tag{13}$$

Further rearrangement yields

$$|f(x_2) - f(x_1)| = |f'(\hat{x})||x_2 - x_1| \tag{14}$$

If in the above equation, the derivative magnitude is less than unity, i.e., if

For all x,

$$|f'(x)| \leq L \qquad \text{where } L < 1 \tag{15}$$

then we obtain the Lipschitz condition (9). Thus we have shown that *a differentiable function with slope magnitude less than unity implies the Lipschitz condition.* The converse, however, is not true; that is, *the Lipschitz condition does not imply the differentiability* of $f(.)$, although it does imply that $|f'(x)| < 1$ wherever that derivative exists. The reader may convince himself of the validity of this statement by inspection of Fig. 5.3-4, where $f(.)$ meets the Lipschitz condition but is not differentiable at the "corner." Thus, for discussing uniqueness, *the Lipschitz condition is less restrictive than the differentiability and bounded-slope-magnitude condition.*

Summarizing, we have shown that the following conditions:

For all $x(a \leq x \leq b)$, $a \leq f(x) \leq b$
$f(.)$ continuous
For all x_1, x_2, $\lvert f(x_1) - f(x_2)\rvert \leq L\lvert x_1 - x_2\rvert \qquad L < 1$

(16)

†The mean value theorem states: Suppose $h(x)$ is continuous for $a \leq x \leq b$, and $h'(x)$ exists for $a < x < b$. Then there is an $\hat{x}$ $(a < \hat{x} < b)$, such that

$$h(b) = h(a) + h'(\hat{x})(b - a)$$

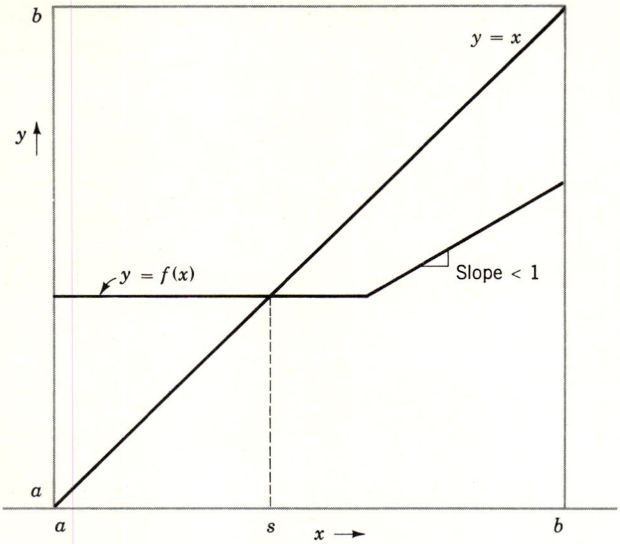

FIG. 5.3-4 A nondifferentiable f(.) that meets the Lipschitz condition.

guarantee both the *existence* and the *uniqueness* of a solution to the equation $f(x)=x$.

These conditions, however, are only sufficient, since, for example, the function $f(.)$ illustrated in Fig. 5.3-5 violates *all* the conditions in Eq. (16); yet the equation $x=f(x)$ has a unique solution.

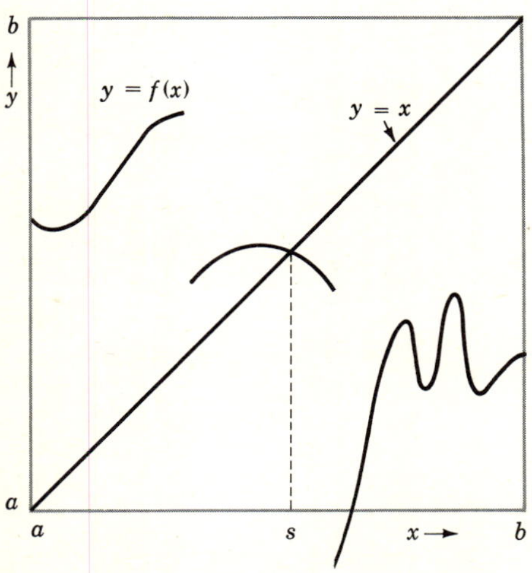

FIG. 5.3-5 Graph of a function that violates all the conditions in Eq. (16) but has a unique solution.

Convergence of the Algorithm $x_{k+1} = f(x_k)$

So far, we have been discussing properties of $f(.)$, which ensure that the equation $f(x) = s$ possesses a unique solution. In this section we shall be concerned with the convergence of the iteration rule $x_{k+1} = f(x_k)$ when $f(.)$ has the above properties.

First, we will establish a "distance" between the value x_k, computed at the kth iteration, and the solution s, in terms of results obtained at the $(k-1)$st iteration. Observe that the kth value x_k is given in terms of the $(k-1)$st value x_{k-1} by

$$x_k = f(x_{k-1}) \tag{17}$$

We assume that the conditions that guarantee the existence and uniqueness of a solution s hold; by definition of a solution we have

$$s = f(s) \tag{18}$$

Subtracting Eq. (18) from Eq. (17) yields

$$x_k - s = f(x_{k-1}) - f(s) \tag{19}$$

From Eq. (19) and through the use of the Lipschitz condition (9) we deduce that

$$|x_k - s| = |f(x_{k-1}) - f(s)| \leq L|x_{k-1} - s| \qquad L < 1 \tag{20}$$

Following an identical reasoning and replacing k with $k - 1$, we deduce that

$$|x_{k-1} - s| = |f(x_{k-2}) - f(s)| \leq L|x_{k-2} - s|; \qquad L < 1 \tag{21}$$

From Eqs. (20) and (21) we now obtain the inequality

$$|x_k - s| \leq L^2|x_{k-2} - s| \tag{22}$$

Repeated substitution of x_{k-i-1} for x_{k-i}, until $i = k - 1$ yields

$$|x_k - s| \leq L^k|x_0 - s| \tag{23}$$

By Eqs. (16), $|x_0 - s| \leq b - a$; hence $|x_0 - s|$ is bounded. Since $L < 1$, we conclude from Eq. (23) that

$$\lim_{k\to\infty} |x_k - s| = 0 \tag{24}$$

or, equivalently,

$$\lim_{k\to\infty} x_k = s \tag{25}$$

and we say that the sequence established by the iteration rule converges to s with increasing k.

We pause here in order to clarify and formalize the meaning of *convergence*.

DEFINITION 5.3-1 A sequence $x_0, x_1, x_2, \ldots$ converges† on s if for every $\epsilon > 0$,

†We also say that s is the limit of $x_0, x_1, x_2, \ldots$, and we write $x_1 \to s$.

there is an integer N such that for $k > N$, $|x_k - s| < \epsilon$. A sequence diverges when it does not converge.

In other words, convergence ensures that we can always find a *large enough k* so that x_k comes *arbitrarily* close to s, and stays arbitrarily close thereafter.

Returning to the Picard iteration rule $x_{k+1} = f(x_k)$, we see that if the sequence $x_0, x_1, x_2, \ldots$ converges on s, then the algorithm *will indeed terminate in a finite number of iterations*, because of the test performed in Step 2 of the rule. We shall often, under these conditions, say that *the algorithm or iteration rule converges.* Since $L < 1$ and $x_0 - s$ is finite, we conclude by inspection of Eq. (23) that the sequence $x_0, x_1, x_2, \ldots$ converges on s if $f(.)$ satisfies conditions (16).

It follows, therefore, that the conditions of Eq. (16) are sufficient not only for the *existence* and *uniqueness* of a solution, but also for the *convergence* of the Picard algorithm on this solution.

In graphical terms, the Picard algorithm corresponds to the "tra-

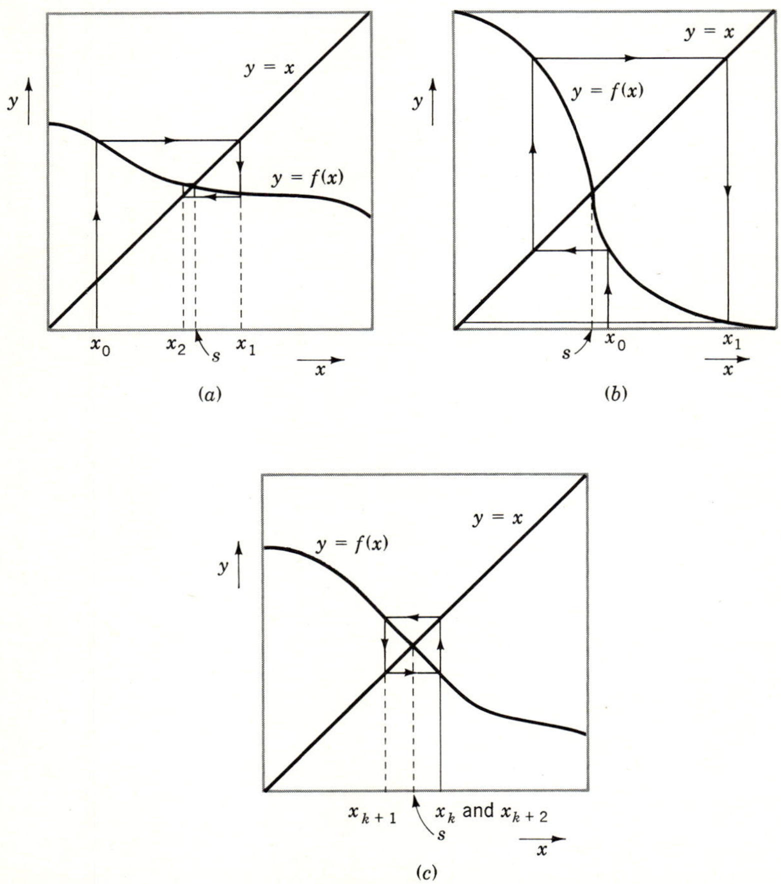

FIG. 5.3-6 *Trajectories of the Picard algorithm.*

jectories" of Fig. 5.3-6*a*, *b*, and *c*, where each trajectory connects consecutive x_k's. That is, a vertical line is drawn from x_0 up until it intersects the graph of $f(.)$. The "height" of this line from the line $y = 0$ is $f(x_0)$, which equals x_1. To transfer the distance x_1 to the horizontal axis we draw a line of zero slope from the above intersection until it intersects the diagonal line $y = x$. A vertical line is then drawn from that intersection until it intersects the graph of $f(.)$, and a horizontal line of zero slope is drawn from the latter intersection to the diagonal, thereby establishing x_2. This process of successively drawing vertical lines to the graph of $f(.)$ and horizontal lines to the diagonal, corresponds to the successive computations of $x_0, x_1, x_2, \ldots$ by Picard's algorithm. Observe that in Fig. 5.3-6*a* the algorithm converges to the solution, whereas in Fig. 5.3-6*b* and *c*, where conditions (16) are obviously *not* met, the algorithm diverges.

Speed of Convergence

Having established conditions that guarantee convergence of Picard's algorithm, we now wish to establish the number of iterations required for the iteration rule to converge; that is, we wish to study the speed of convergence of the Picard algorithm. We start with an appropriate definition.

DEFINITION 5.3-2 The convergence time† m of a convergent iteration rule is the least upper bound (lub) of the number of iterations k required to reduce the distance $|x_k - s|$ to 1/e of its initial value $|x_0 - s|$.

Assuming that $f(.)$ satisfies Eq. (16), we know that the algorithm $x_{k+1} = f(x_k)$ will indeed converge to the unique solution s of $x = f(x)$. Let us now recall that we have already established, in Eq. (23), how the error at the kth step, $x_k - s$, is related to the initial error, $x_0 - s$; the relation is

$$|x_k - s| \leq L^k |x_0 - s| \qquad L < 1 \tag{26}$$

Notice that we do not have an *a priori* idea as to how many times we have an equality or a strict inequality in Eq. (26) until we apply the algorithm. Some thought, however, should convince the reader that the more often we have a strict inequality in Eq. (26), the faster the convergence. On the other hand, the *worst case* (from the point of view of convergence) occurs when equality holds at each value of k. In other words, if

For all k,

$$|x_k - s| = L^k |x_0 - s| \tag{27}$$

then we have the *slowest* rate of convergence.

Observe that Eq. (27) will hold if $f(.)$ is a straight line: i.e.,

$$f(x) = Lx + c \qquad L < 1 \tag{28}$$

where c is an arbitrary constant. Applying Picard's algorithm to this function

†Strictly speaking, the number is not time but a number of iterations, which in turn is proportional to computing time.

yields

$$x_{k+1} = f(x_k) = Lx_k + c \tag{29}$$

from where it is easy to verify that Eq. (27) does indeed hold.

To compute the time of convergence m according to Definition 5.3-3, we require that for $k = m$,

$$|x_m - s| = \frac{1}{e}|x_0 - s| \tag{30}$$

But from Eq. (27) we have

$$|x_m - s| = L^m|x_0 - s| \tag{31}$$

Hence, the Lipschitz constant L and the time of convergence m are related by

$$L^m = \frac{1}{e} \tag{32}$$

or, equivalently,

$$m = -\frac{1}{\ln L} \tag{33}$$

Note that when the Lipschitz constant $L \to 1$, then $m \to \infty$, which means that the convergence rate is extremely slow. As $L \to 0$, then $m \to 0$, and hence the speed of convergence is fast. Thus, as a rule of thumb, the smaller the value of the Lipschitz constant L, the faster the convergence of the Picard algorithm.

From a graphical point of view, if Eq. (27) holds, then the error sequence $\epsilon_0, \epsilon_1, \epsilon_2, \ldots$, where $\epsilon_k \triangleq |x_k - s|$, converges exponentially to zero as shown in Fig. 5.3-7. Equation (27) can then be written as

$$\epsilon_k = L^k\epsilon_0 = \epsilon_0 e^{-k/m} \tag{34}$$

Because of the above exponential relationship between the error ϵ_k and k, the number of correct decimal places obtained after k iterations is

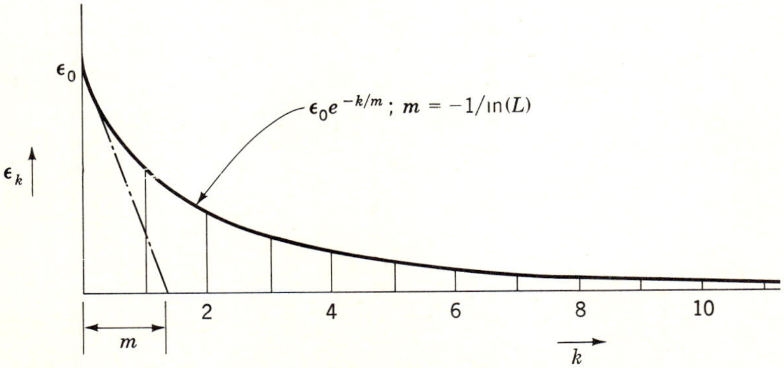

FIG. 5.3-7 Convergence of the error sequence.

approximately *proportional* to k. Hence, this type of convergence is often known as *linear convergence* or *first-order convergence.* This name is also attributed to the fact that the error at the kth iteration is, in the worst case, *proportional* to the error at the $(k-1)$st iteration: i.e., $|x_k - s| \leq L|x_{k-1} - s|$. An obvious improvement in convergence speed would result if this relationship were of order higher than linear, i.e., if

$$|x_k - s| \leq \alpha |x_{k-1} - s|^p \tag{35}$$

where $\alpha < 1$, and p is an integer greater than 1. For $p = 2$, the corresponding class of iteration rules is said to possess *quadratic convergence,* a subject which we shall examine in a subsequent section.

Graphical Interpretation of Contraction

Earlier in the section we remarked that when a function $f(.)$ meets the Lipschitz condition

$$|f(x_1) - f(x_2)| \leq L|x_1 - x_2| \qquad L < 1 \qquad \text{(9) repeated}$$

that function is also called a *contraction* mapping. The graphical interpretation of contraction gives additional insight into the uniqueness of a solution to $x = f(x)$; and the convergence of Picard's algorithm.

Observe first that we can think of the term $|a - b|$, where a and b are real numbers, as a *distance* between a and b; in particular, if the points a and b are marked on a line of real numbers, then $|a - b|$ corresponds to the distance between these two points on that line (Fig. 5.3-8a).

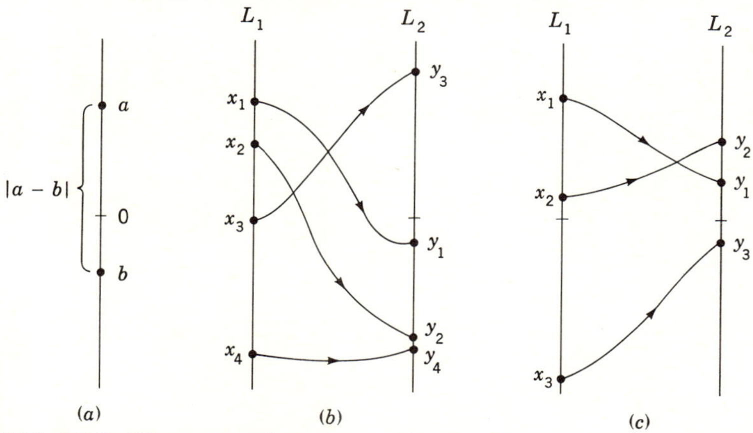

FIG. 5.3-8 Distance, a function, and a contraction.

Observe next (Fig. 5.3-8b) that a function f can be viewed as mapping every point x on line L_1 into a point $f(x)$ on line L_2; in Fig. 5.3-8b we indicate such a mapping by connecting the four points x_1, x_2, x_3, x_4 with their images y_1, y_2, y_3, y_4. Observe finally (Fig. 5.3-8c) that if a function f meets the Lipschitz condition (9), then the distance between *any* two points on line L_1, is *guaranteed* to be greater than the distance between the images of these points on line L_2 [this is exactly what Eq. (9) "says"]. For example, in Fig. 5.3-8c $|x_1 - x_2| > |y_1 - y_2|$. Thus, a function which satisfies Eq. (9) causes the distance of two points to *contract* under the mapping; hence the name contraction.

It is easy to see now why satisfaction of the Lipschitz condition ensures uniqueness of a solution to the equation $x = f(x)$. Let s_1 and s_2 be solutions to this equation. Then they satisfy $s_1 = f(s_1)$ and $s_2 = f_2(s)$, and it follows that they are their own images, as shown in Fig. 5.3-9. But this is impossible if

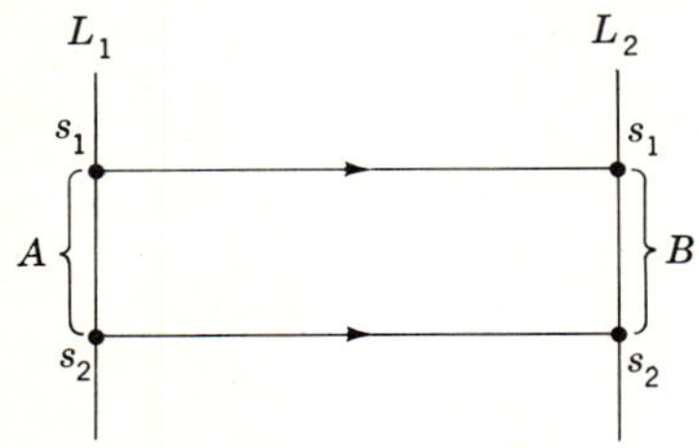

FIG. 5.3-9 Uniqueness of solution to $x = f(x)$. Note $A = B$.

f is a contraction, since the distance between s_1 and s_2 must contract, and as Fig. 5.3-9 shows, it does not. Hence we conclude that if f is a contraction and there is a solution, it must be unique.

Now let us consider Picard's algorithm. In terms of our graphical approach, the successive iterations of the algorithm

$$x_1 = f(x_0)$$

$$x_2 = f(x_1)$$

correspond to successive mappings first of point x_0 lying on L_1, to its image x_1 lying on line L_2, then of point x_1 to x_2 lying on line L_3, and so forth, as shown in Fig. 5.3-10.

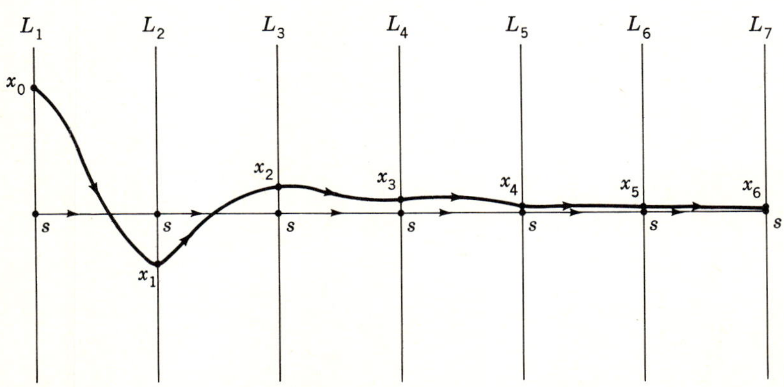

FIG. 5.3-10 How contraction leads to convergence.

If s is the solution to $x = f(x)$, then s will certainly map to itself under every mapping from line L_k to line L_{k+1}, thereby resulting in the straight line of Fig. 5.3-10. That convergence of the algorithm is an inevitable consequence of contraction follows by inspection of Fig. 5.3-10. In particular, the distance between the two points x_k and s *contracts* with increasing k; hence x_k gets progressively closer to the solution, and the algorithm converges. Observe also that if L is made smaller, the contraction per iteration is greater, and the algorithm converges faster.

An Old Example in a New Light

Let us now return to our motivating network example of Sec. 5.2, and let us probe a little deeper into some questions that were raised at that time on the basis of what has been established in this section.

Recall that in that example we wanted to determine the voltage e at the common node of three resistors ($i_1 = v_1$, $i_2 = v_2^5$, $i_3 = v_3^3$), the other sides of which are connected respectively to 6, −1, and 4 V. Equation (1) of Sec. 5.2 which represents the nodal equation for that circuit states that $F(e) = 0$, i.e.,

$$F(e) = (6 - e) + (-1 - e)^5 + (4 - e)^3 = 0 \tag{36}$$

Observe from Eq. (36) that $F(e)$ is continuous and decreases monotonically, as e increases, with a slope $F'(e)$ which is, for all e, strictly negative. Hence, the graph of $F(e)$ plotted versus e *must* intersect the line of zero ordinate in, at most, one point, the solution. It is therefore established that the equation $F(e)$ possesses a unique solution.

This equation was converted to the form $e = f(e)$ as follows:

$$e = e + kF(e) \triangleq f(e) \tag{37}$$

where k is positive and real.† Let the right-hand side, $e + kF(e)$, of this equation be denoted by $f(e)$. The graph of $f(e)$ versus e is shown in Fig. 5.3-11 for several values, k_i of k. First observe that Eq. (37) has a unique solution s which is the same as the solution of Eq. (36). Next observe that for positive values of k, which are "very close" to zero, $f(e)$ is very close to the line $y = e$, staying below that line for $e > s$ and above that line for $e < s$. Qualitatively we see that there must be a range of values $0 < k < k_m$ where the Lipschitz condition is met, and hence where convergence is guaranteed. In order to establish a value for k_m in our example, we assume that $f(.)$ is differentiable [recall that for $f(.)$ differentiable, boundness of $f'(.)$ by unity ensures that the Lipschitz condition is met], and state the expression for $f'(.)$:

$$f'(e) = 1 + k[-1 - 5(-1 - e)^4 - 3(4 - e)^2] \tag{38}$$

Inspection of Eq. (38) reveals that for any given k there exists a value e sufficiently far from the solution such that $|f'(e)| > 1$. Hence, we cannot choose k so that $|f'(e)| < 1$ for all e. We can, however, meet this condition for all values of e

†k is identical to α of Eq. (3).

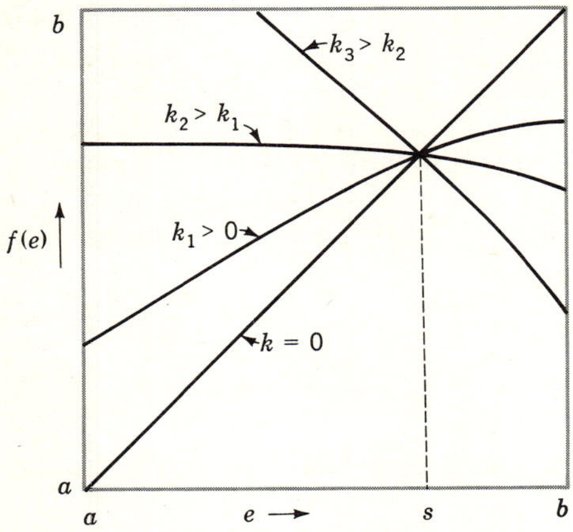

FIG. 5.3-11 The function f(e) of the example.

in an interval $[a,b]$ containing the solution; for if

$$\min_{a \le e \le b} [-1 - 5(-1 - e)^4 - 3(4 - e)^2] = -N \tag{39}$$

where N is positive, then we can guarantee convergence, from Eq. (38), for values of k such that

$$0 < k \le \frac{2}{N} \tag{40}$$

Usually, we will know approximately where the solution s is, so that finding a bound for k is possible. For example, in this problem, we may know (from a previous computation with "slightly" different voltage-source values) that the solution is in the range

$$0 < s < 1.5 \tag{41}$$

The slope of $[-1 - 5(-1 - e)^4 - 3(4 - e)^2]$ at these two values, 0 and 1.5, is -54 and -215, respectively. Hence, for values of k between 0 and 0.01, convergence is guaranteed.

Finally, let us consider the relationship of k to the convergence time. From our previous discussion we know that fastest convergence (minimum m) occurs when L approaches zero; in the case of our differentiable function $f(e)$ this means that we should choose k so that $f'(e) \simeq 0$ in the vicinity of the solution. From our previous calculations we recall that the term $[-1 - 5(-1 - e)^4 - 3(4 - e)^2]$ of Eq. (38) is -54 at $e = 0$ and -215 at $e = 1.5$. To make $f'(e) = 0$ in Eq. (38), k would therefore have to be $1/54 = 0.02$ if the solution were near $e = 0$, or $1/200 = 0.005$ if the solution were near $e = 1.5$. As we do not know where in the interval [0, 1.5] the solution is, it turns out that choice of a value for k about "halfway" between these two values is most reasonable; hence we choose $k = 0.01$, which indeed results in fairly rapid convergence, requiring only six iterations (Table 5.2-2).

Recapitulation

The major results of the section have been as follows:

Given the equation

$$x = f(x) \tag{42}$$

assume that

$$a \leq x \leq b \tag{43}$$

$$a \leq f(x) \leq b \tag{44}$$

$$f(.) \text{ continuous} \tag{45}$$

Furthermore, assume that the Lipschitz condition,

$$|f(x_1) - f(x_2)| \leq L|x_1 - x_2| \qquad L < 1 \tag{46}$$

holds for all x_1, x_2 in $[a,b]$. Alternately, if $f(x)$ is differentiable, assume that

$$\left|\frac{df(x)}{dx}\right| < 1 \qquad \text{for all } x \text{ in } [a,b] \tag{47}$$

Then

1. There exists a unique solution s such that

$$s = f(s) \qquad a \leq s \leq b \tag{48}$$

2. The Picard algorithm

$$x_{k+1} = f(x_k); \qquad k = 0, 1, 2, \ldots \tag{49}$$

converges to the solution, i.e.,

$$\lim_{k\to\infty} x_k = s \tag{50}$$

3. The time of convergence m is related to the Lipschitz constant L by

$$m \leq \frac{-1}{\ln(L)}$$

5.4 ITERATIVE SOLUTION OF $F(x) = 0$: NEWTON'S METHOD

In this section we shall examine another iterative procedure which, just as Picard's method, enjoys wide popularity in the solution of nonlinear algebraic equations. This technique is credited to Sir Isaac Newton; hence it is most often referred to as Newton's method.

When Newton's method converges, then it converges very fast as it is characterized by a *quadratic convergence rate*. By this we mean that in the vicinity of the solution, the error at the $(k + 1)$st iteration is proportional to the square of the error at the kth iteration. On the other hand, the requirements (sufficient conditions) that guarantee the convergence of Newton's method tend to be more stringent than those for Picard's method. In addition, Newton's method requires more computations per iteration than Picard's method. Nonetheless, Newton's method is quite attractive because of its fast convergence.

There are many ways of "deriving" Newton's method. Here we shall give two: one is based on Picard's method; the other is based on Taylor series expansion. Naturally both approaches yield the same method; yet they result in different and valuable insights on how this method works.

Newton's Method from Picard's Method

The basic problem under consideration is as follows: *Find the solution s of the equation*

$$F(x) = 0 \tag{1}$$

i.e., find a real number s such that

$$F(s) = 0 \tag{2}$$

For the time being we assume that Eq. (1) has a unique solution s.

Recall that in Sec. 5.2 [Eqs. (4) and (5)], we discussed the construction from function F, above, of a new function f, so that the equation $x = f(x)$ has the same unique solution as $F(x) = 0$, i.e., $x = s$. The objective of such a construction is, of course, the use of Picard's algorithm $x_{k+1} = f(x_k)$ to find the solution of Eq. (1). In that discussion, it was pointed out that there are generally many different such functions, f. In particular, we can start from Eq. (1), multiply both sides by a constant α, and add x to both sides, resulting in

$$x + \alpha F(x) = x$$

We can thus define $f(.)$ as

$$f(x) = x + \alpha F(x)$$

and we are guaranteed that this equation has the same solution as $F(x) = 0$, for any value of α.

Since there are many ways of constructing $f(.)$ from $F(.)$, we can become more demanding and require that *the function $f(.)$ be such that Picard's algorithm converges very rapidly in the vicinity of the solution.*

Let us now recall that if the function $f(.)$ in the equation $x = f(x)$ is differentiable, then the convergence of the Picard algorithm $x_{k+1} = f(x_k)$ is assured if $|f'(x)| < 1$ in the interval of interest. Thus, the magnitude of the derivative $f'(x)$ plays the role of the Lipschitz constant L. We also know that the smaller the Lipschitz constant L, the faster the convergence of Picard's algorithm. For example, if

$$f'(s) = 0$$

where s is the solution of $x = f(x)$, then we expect that Picard's algorithm will converge in *one step* when x_k is very near s.

On the basis of these observations we proceed with the following strategy:

Given the equation $F(x) = 0$, with unique solution s; determine a function $f(.)$ such that

The equation $x = f(x)$ has the *same solution* s (3*a*)

$f'(s) = 0$ (to exploit rapid convergence) (3*b*)

One way of accomplishing this objective for the class of functions $F(.)$ that are twice continuously differentiable on an interval $[a,b]$ with $F'(x) \neq 0$ for all x in $[a,b]$, is to define $f(.)$ as follows:

For all x,

$$f(x) \triangleq x - \frac{F(x)}{F'(x)} \tag{4}$$

Evidently, $f(.)$ exists since we assumed that $F'(.)$ exists and is nonzero. Moreover, observe that differentiation of $f(.)$ in Eq. (4) yields

$$\begin{aligned} f'(x) &= 1 - \frac{[F'(x)]^2 - F(x)F''(x)}{[F'(x)]^2} \\ &= \frac{F(x)F''(x)}{[F'(x)]^2} \end{aligned} \tag{5}$$

Observe that $f'(x)$ also exists since F is twice continuously differentiable, and $F'(x) \neq 0$. In addition $f(.)$ possesses the desired property for rapid convergence, i.e.,

$$f'(s) = 0 \tag{6}$$

since $F(s) = 0$ and $F'(s) \neq 0$. Also, it follows from Eq. (4) that

$$f(s) = s \tag{7}$$

Hence the function $f(.)$ constructed from $F(.)$ by Eq. (4) meets the two desired conditions (3). Now let us recall that the Picard algorithm $x_{k+1} = f(x_k)$ is guaranteed to converge, provided that $f(.)$ is continuous and continuously differentiable, and that $|f'(x)| < 1$ in the interval of interest. Since $F(.)$ is twice continuously differentiable and $F'(.)$ is nonzero, we need only require, for convergence purposes, that

$$-1 < \frac{F(x)F''(x)}{[F'(x)]^2} < 1 \tag{8}$$

We summarize next the above observations.

Newton's Method

OBJECTIVE: Find the solution s of the equation

$$F(x) = 0$$

ALGORITHM

Step 1: Choose x_0. Let $k = 0$.
Step 2: If $|F(x_k)/F'(x_k)| < \epsilon$ (where ϵ represents maximum tolerable error). Stop; the solution is then $s \simeq x_k$. Otherwise go to Step 3, below.
Step 3: Compute

$$x_{k+1} = x_k - \frac{F(x_k)}{F'(x_k)} \tag{9}$$

Increase k by 1, and go to Step 2, above.

The graphical interpretation of Eq. (9), i.e., the generation of the value x_{k+1} from x_k, is indicated in Fig. 5.4-1. Starting with the value of x_k, we draw the tangent line to the graph $y = F(x)$ at $x = x_k$; the next iterate x_{k+1} is at the intersection of this tangent line with the x axis. In Fig. 5.4-2 we illustrate a typical sequence x_0, x_1, x_2, x_3 generated by Newton's method; in this example $s \simeq x_4$.

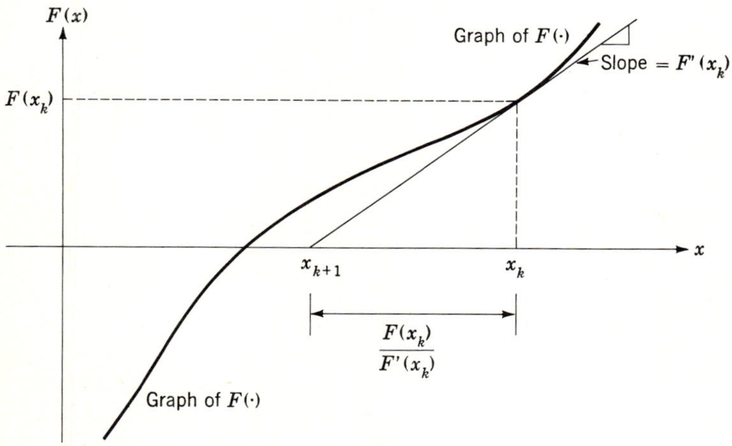

FIG. 5.4-1 Graphical interpretation of Newton's method.

Newton's Method from Taylor Series Expansion

We have seen above how Newton's method was obtained from a "fast" Picard algorithm. We shall indicate how Newton's method can be derived from a Taylor series expansion.

Again we are concerned with the determination of the solution s to the equation $F(x) = 0$. We are interested in generating a sequence x_0, x_1, x_2, . . . which converges to the solution s. Suppose that x_k is known and x_{k+1} is to be computed. If we expand $F(x_{k+1})$ about x_k in a Taylor series, we obtain

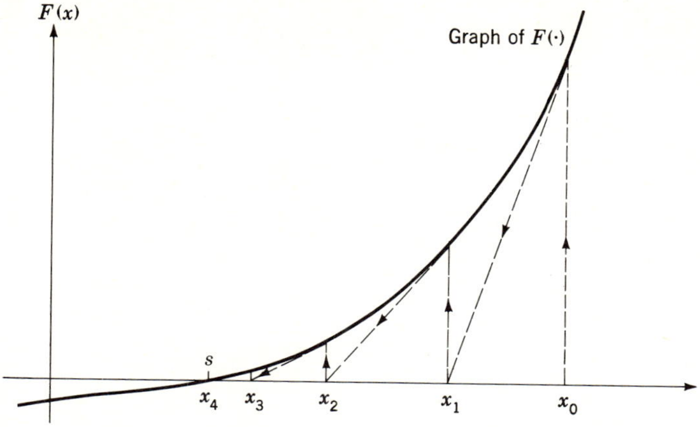

FIG. 5.4-2 Graphical trajectory of iteration by Newton's method.

$$F(x_{k+1}) = F(x_k) + F'(x_k)(x_{k+1} - x_k) + \tfrac{1}{2} F''(x_k)(x_{k+1} - x_k)^2 + \text{higher order terms} \quad (10)$$

Based on this expansion, the strategy behind Newton's method is as follows:

1. If the function $F(.)$ is linear, then establish the solution in one step.
2. If $F(.)$ is nonlinear, then pretend it is linear and keep trying.

Let us see how this approach works. If $F(.)$ is linear, then we can write

$$F(x) = ax + b \quad (11)$$

from where it follows that

$$F'(x) = a \qquad F''(x) = 0 \quad (12)$$

and the Taylor series (10) takes the form

$$F(x_{k+1}) = F(x_k) + F'(x_k)(x_{k+1} - x_k) \quad (13)$$

since the second and higher derivatives of $F(.)$ are zero. By *one* step in strategy 1 above we mean that $x_{k+1} = s$, which in turn implies that $F(x_{k+1}) = 0$. Thus, Eq. (13) reduces to

$$F(x_k) + F'(x_k)(x_{k+1} - x_k) = 0 \quad (14)$$

Solution of this equation for x_{k+1} yields Newton's algorithm,

$$x_{k+1} = x_k - \frac{F(x_k)}{F'(x_k)} \quad (15)$$

and, of course, in the linear case $x_{k+1} = s$. When $F(.)$ is nonlinear, then by pretending that $F(.)$ is linear, we ignore the quadratic and higher-order terms in Eq. (10).

Convergence of Newton's Method

We shall now state and prove a theorem which accounts for the popularity and widespread use of Newton's method.

THEOREM 5.4-1 Suppose that $F(x)$ is continuous in x and that it has a zero at s: i.e., $F(s) = 0$. Suppose further that $F'(x)$ and $F''(x)$ exist and are continuous, and that $F'(x) \neq 0$, for all x. Then the sequence of x_k's generated by

$$x_{k+1} = x_k - \frac{F(x_k)}{F'(x_k)} \tag{16}$$

will always converge to s, provided that x_0 is sufficiently close to s.

PROOF: The convergence of the Newton algorithm will be proved by showing that the conditions that guarantee convergence of the associated Picard algorithm hold. The proof relies on the definition of continuity.† Observe as we have already seen, that Newton's method is identical to Picard's method for the solution of

$$x = f(x) \tag{17}$$

where

$$f(x) = x - \frac{F(x)}{F'(x)} \tag{18}$$

The assumptions of the theorem guarantee that $f(x)$ is continuous and that

$$f'(x) = \frac{F(x)F''(x)}{[F'(x)]^2} \tag{19}$$

is continuous.

Since $f'(x)$ is continuous, then by definition for every $\epsilon > 0$ there is a $\delta > 0$ such that

$$|f'(x) - f'(s)| < \epsilon \qquad \text{for } |x - s| < \delta \tag{20}$$

Since $F(s) = 0$, then Eq. (19) yields $f'(s) = 0$, and Eq. (20) reduces to

$$|f'(x)| < \epsilon \qquad \text{for } |x - s| < \delta \tag{21}$$

Next we use the mean value theorem‡ to obtain

$$f(x) = f(s) + f'(\hat{x})(x - s) \qquad |\hat{x} - s| < |x - s| \tag{22}$$

†We recall that a function $g(x)$ is continuous at $x = x_0$ if for every $\epsilon > 0$, there exists $\delta > 0$ such that $|g(x) - g(x_0)| < \epsilon$ for all values of x for which $|x - x_0| < \delta$.

‡Introduced in Eq. (5.3-12).

But $f(s) = s$; hence we can rewrite Eq. (22) as follows:

$$f(x) - s = f'(\hat{x})(x - s) \tag{23}$$

Taking absolute values of both sides yields

$$|f(x) - s| = |f'(\hat{x})|\ |x - s| \tag{24}$$

But from Eqs. (21) and (24) we deduce that

$$|f(x) - s| \leq \epsilon\delta \tag{25}$$

or

$$s - \epsilon\delta \leq f(x) \leq s + \epsilon\delta \qquad \text{for } |x - s| < \delta \tag{26}$$

If we select ϵ so that

$$0 < \epsilon < 1 \tag{27}$$

then

$$s - \delta_1 \leq f(x) \leq s + \delta_1 \qquad \text{where } \delta_1 = \delta\epsilon \tag{28}$$

and

$$|f'(x)| < 1 \tag{29}$$

for

$$s - \delta_1 < x < s + \delta_1 \tag{30}$$

which are precisely the conditions which *guarantee* the convergence of Picard's algorithm to a unique solution.

Speed of Convergence

Earlier, we demonstrated that Picard's algorithm is characterized by *linear* convergence. We shall now show that Newton's method is characterized by *quadratic* convergence, i.e., the error at the $(k + 1)$st step is roughly proportional to the square of the error at the kth step. Thus, near the solution, Newton's method converges much more rapidly than Picard's method.

To establish analytically that convergence is quadratic, we proceed as follows: The Newton algorithm is

$$x_{k+1} \triangleq x_k - \frac{F(x_k)}{F'(x_k)} = f(x_k) \tag{31}$$

Taylor series expansion of $f(x_k)$ about $x = s$ yields

$$f(x_k) \simeq f(s) + f'(s)(x_k - s) + \tfrac{1}{2} f''(s)(x_k - s)^2 \tag{32}$$

But

$$f(s) = s \qquad f'(s) = 0 \qquad f''(s) = \frac{F''(s)}{F'(s)} \tag{33}$$

Hence, Eq. (31) reduces to

$$x_{k+1} \simeq s + \frac{1}{2}\frac{F''(s)}{F'(s)}(x_k - s)^2 \tag{34}$$

or

$$x_{k+1} - s \simeq \frac{1}{2}\frac{F''(s)}{F'(s)}(x_k - s)^2 \tag{35}$$

Therefore it follows that

$$|x_{k+1} - s| \lesssim K|x_k - s|^2 \qquad \text{where } K = \frac{1}{2}\left|\frac{F''(s)}{F'(s)}\right| \tag{36}$$

Example 5.4-1 *Our Old Example with Newton's Method* Applying Newton's method to the network example that we considered at the beginning of this chapter gives rise to the following algorithm:

$$e_{k+1} = f(e_k) = e_k - \frac{6 - e_k + (-1 - e_k)^5 + (4 - e_k)^3}{-1 - 5(-1 - e_k)^4 - 3(4 - e_k)^2}$$

Applying this algorithm with $e_0 = 0$ gives the results of Table 5.4-1, where the format is as in Sec. 5.2. Observe that convergence is achieved in four iterations without the benefit of adjusting α, while in the case of Picard's method, the fastest convergence takes six iterations and is very critically dependent on the value of α. The fastest Picard convergence and the results of Newton's method are compared in Fig. 5.4-3. Observe in Fig. 5.4-3 how the *logarithm* of the error decreases for the Picard and Newton method, indicating the linear and quadratic convergence, respectively.

TABLE 5.4-1 Newton's Iteration: Network of Fig. 5.2-1

K	*E*	*I1*	*I2*	*I3*	*SUM*
0	0.000000000*E*0	6.000000000*E*0	¯1.000000000*E*0	6.400000000*E*1	6.900000000*E*1
1	1.277777778*E*0	4.722222222*E*0	¯6.131359178*E*1	2.017301097*E*1	¯3.641835859*E*1
2	1.047022300*E*0	4.952977700*E*0	¯3.594287967*E*1	2.575019380*E*1	¯5.239708169*E*0
3	1.001441116*E*0	4.998558884*E*0	¯3.211545555*E*1	2.696110856*E*1	¯1.557881056*E*¯1
4	1.000001365*E*0	4.999998635*E*0	¯3.200010920*E*1	2.699996315*E*1	¯1.474198920*E*¯4

Example 5.4-2 *A Comparison of Picard's and Newton's Methods* In this example we seek the solution of the equation

$$x = e^{-x} \qquad \text{for } 0 \le x \le 1 \tag{37}$$

It is a straightforward matter to establish that the function $f(x) = e^{-x}$ satisfies all the sufficient conditions for convergence of Picard's method. Thus, we expect that the algorithm

$$x_{k+1} = e^{-x_k} \tag{38}$$

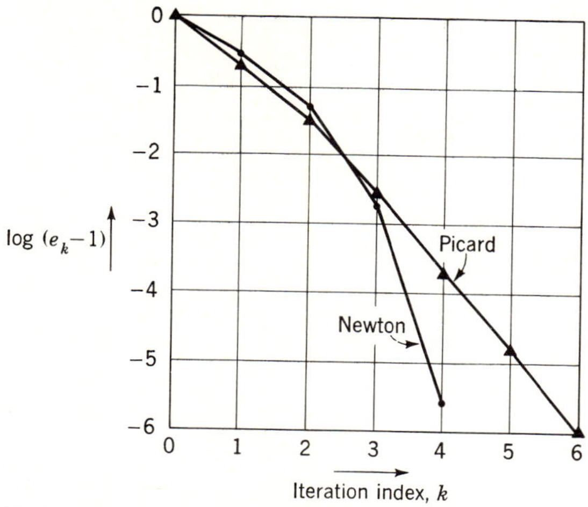

FIG. 5.4-3 Comparison of Newton and Picard convergence.

will converge on the solution s of Eq. (1). Starting with $x_0 = 0$ and a termination criterion of $|x_k - f(x_k)| < 0.005$, the Picard algorithm yields the results of Table 5.4-2, from where it can be seen that convergence is achieved after 11 iterations.

To apply Newton's method to Eq. (37), we must rewrite that equation in the form $F(x) = 0$ such that $F(s) = 0$. This is easily done as follows:

$$F(x) = x - e^{-x} \tag{39}$$

Since $F'(x) = 1 + e^{-x}$, Newton's algorithm is

$$x_{k+1} = x_k - \frac{x_k - e^{-x_k}}{1 + e^{-x_k}} \tag{40}$$

Starting with $x_0 = 0$ and the same terminating criterion, we obtain the results

TABLE 5.4-2 Picard and Newton Iteration for Example 5.4-2

Picard	Newton
ITERATE	*ITERATE*
X[0] = 0	X[0] = 0
X[1] = 1	X[1] = 0.5
X[2] = 0.3678794412	X[2] = 0.5663110032
X[3] = 0.6922006276	X[3] = 0.567143165
X[4] = 0.5004735006	*CONVERGENCE AFTER* 3 *ITERATIONS* S = 0.567143165
X[5] = 0.6062435351	
X[6] = 0.545395786	
X[7] = 0.5796123355	
X[8] = 0.5601154614	
X[9] = 0.5711431151	
X[10] = 0.5648793474	
X[11] = 0.568428725	
CONVERGENCE AFTER 11 *ITERATIONS* S = 0.568428725	

of Table 5.4-2, from where it can be seen that convergence is achieved after three iterations—a significant improvement over Picard's method. Figure 5.4-4 shows in graphical terms the results of Tables 5.4-1 and 5.4-2, and hence the way in which the error converges to zero for each method.

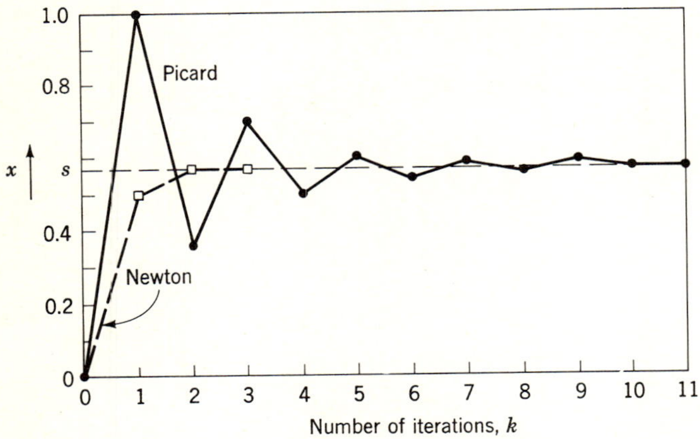

FIG. 5.4-4 Comparison of Newton and Picard convergence for Example 5.4-2.

5.5 SOLUTION OF $f(x) = x$: HIGHER-ORDER CONVERGENCE

Undoubtedly the reader has already discovered how to extend the foregoing Taylor series approach that led to Newton's method toward the development of higher-order algorithms. For example, given $F(.)$, we can envision methods for constructing $f(.)$ such that $f'(s)=f''(s)=0, f(s)=s$, and $F(s)=0$. The corresponding Taylor series expansion for the error would then have a nonzero third-degree term, and convergence would be *ternary;* i.e., the error at the $(k+1)$st iteration would be proportional to the cube of the error at the kth iteration.

Why then do we not dispense with lower-order convergence algorithms and use always the highest-order algorithm that we can construct? The answer, stated simply, is as follows: The faster the convergence speed of an algorithm, the less is its stability (loosely speaking, its chance of convergence) or, alternatively, the stricter are the conditions which must be imposed on $F(.)$, so that convergence is guaranteed. For example, the function shown in Fig. 5.5-1*a* satisfies the conditions for convergence of the Picard algorithm. The Newton method, however, could not apply to such a function "with corners," because the differentiability condition is not satisfied. Even if a function were differentiable, however, its derivative function could be such that Newton's method would still not converge, as shown in the example of Fig. 5.5-1*b*. It is therefore desirable to choose the order of iteration so that the relative penalty imposed by the constraints that must be applied on $f(.)$ is justified by the resultant benefits in convergence speed.

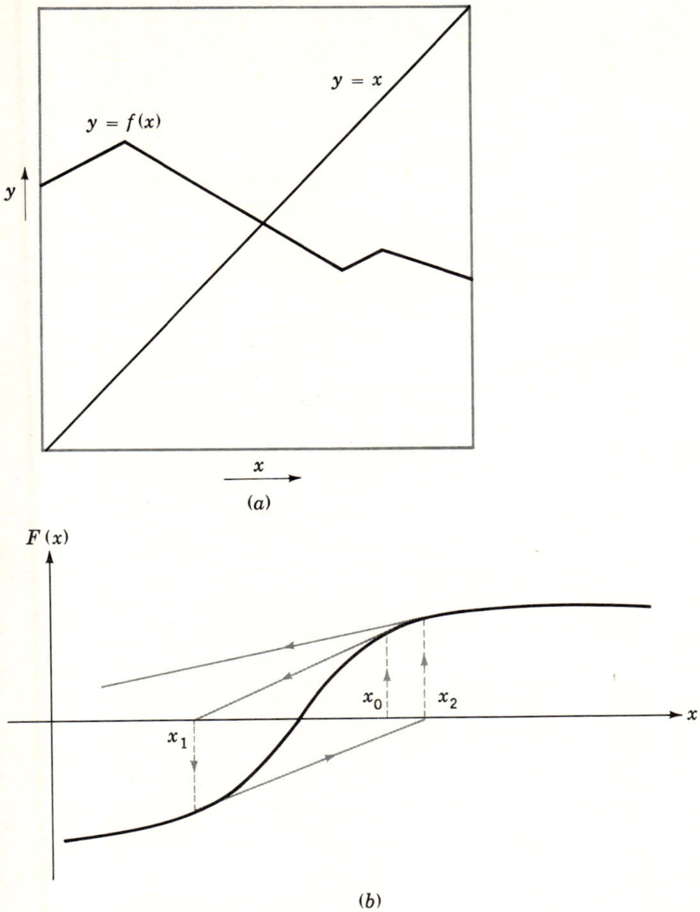

FIG. 5.5-1 *Divergence of Newton's method.*

5.6 CONCLUDING REMARKS

In this chapter we focused our attention on some special, yet popular, iterative methods for the solution of simple algebraic equations. Such equations arise when one wishes to analyze nonlinear memoryless networks and block diagram systems.

The basic notions behind Picard's and Newton's methods can readily be generalized for multivariable problems. In other words, these methods can be used to solve sets of nonlinear simultaneous equations. After the reader has been exposed to vectors, matrices, and linear algebra, he will appreciate the extension of these methods to "multidimensional" problems.

Needless to say, there are many other algorithms that can be used instead of the Picard or Newton algorithms. The interested reader can consult books on numerical analysis on this subject. We have not treated them here because we wished to concentrate on the fundamentals of iteration rather than to provide a library of algorithms. We hope that we have convinced the reader that algorithmic solutions are practical and that the associated notions of existence, uniqueness, convergence, and speed of convergence are important.

PROBLEMS

5.3-1 Each of the "pictorial" questions in the figure should be answered with an explanation of your reasoning. Notation is identical to that used in the text, and all problems refer to solution of the equation $x = f(x)$ by Picard's method.

(*a*) For which starting values of x_0 will Picard's algorithm converge on the solution?

(*b*) Explain why the algorithm will always converge. What is the convergence time?

(*c*) Discuss convergence for all possible starting values x_0 between a and b.

(*d*) Here $f(x) = e^{-x}$. Show that the algorithm will converge for all starting values $0 \leq x_0 \leq 1$.

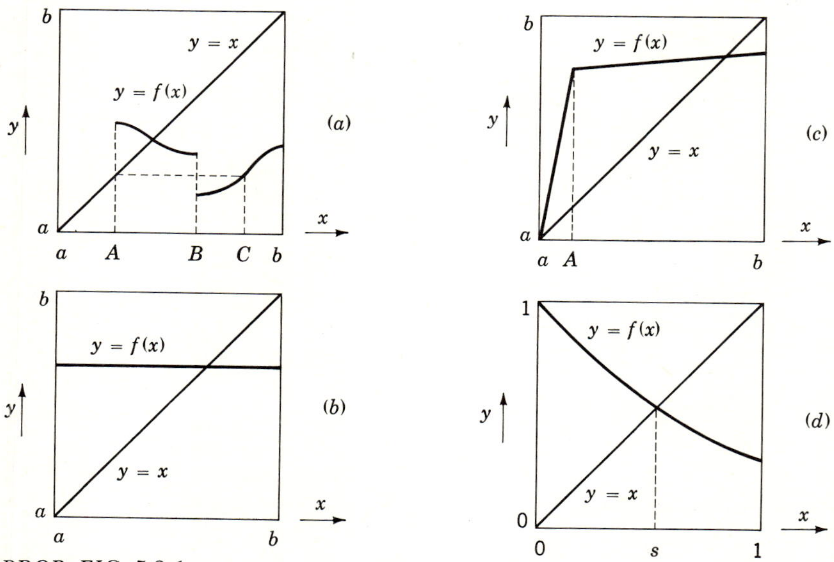

PROB. FIG. 5.3-1

5.3-2 In the memoryless system shown in the figure, $f'(.)$ denotes the derivative of $f(.)$, and the function $f(.)$ is specified by.

$$f(0) = 1 \tag{1}$$

$$f'(x) = 1/2 \qquad \text{for } 0 \leq x \leq 4 \tag{2}$$

$$f'(x) = 2 \qquad \text{for } 4 < x \leq 6 \tag{3}$$

$$f'(x) = 0 \qquad \text{for } x > 6 \qquad \text{or } x < 0 \tag{4}$$

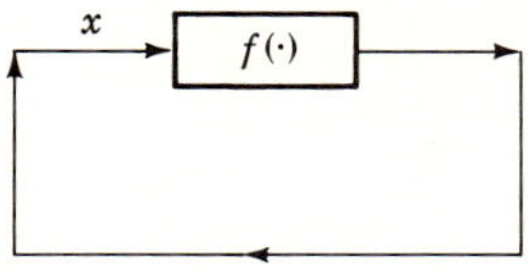

PROB. FIG. 5.3-2

(*a*) Discuss the existence and uniqueness of solutions(s) to the equation characterizing this system.

(*b*) Consider the algorithm

$$x_{1+1} = f(x_1)$$

Discuss the convergence of this algorithm as a function of the initial guess, x_0. Specify for which ranges of values of x_0 this algorithm converges (if at all) to the same answer.

5.3-3 Consider the equation

$$x = 0.9e^{-x} \tag{1}$$

(*a*) Is there a solution *s* in the interval $0 \leq x \leq 1$?
(*b*) Is this solution unique?
(*c*) Demonstrate that the Lipschitz condition holds.
(*d*) Consider the algorithm

$$x_{n+1} = f(x_n) \tag{2}$$

to solve Eq. (1). Let your first guess be $x_0 = 1$. Compute the sequence $x_0, x_1, x_2, \ldots$ until

$$|x_N - s| \leq 10^{-2}$$

5.3-4 Suppose that $f(.)$ is continuous and it *expands*, that is,

For all x_1, x_2,

$$|f(x_1) - f(x_2)| \geq L|x_1 - x_2| \qquad L > 1$$

(Note that this is the "opposite" of the Lipschitz condition, and that here $L > 1$.)

(*a*) Can the following condition hold?

For all $a \leq x \leq b$,

$$a \leq f(x) \leq b$$

Explain clearly why it can or cannot hold.

(*b*) Prove that if a solution to $x = f(x)$ exists, then it *must* be unique.

(c) Must a solution to $x = f(x)$ exist? If so, then prove it. If not, you must show a counterexample (a picture will do).

(d) Show that Picard's algorithm will always diverge from the unique solution (when that solution exists).

(e) Give the simplest algorithm you can think that will find the solution when the inverse function f^{-1} is known. [Note that for all x, y; $f^{-1}(x) = y$ if and only if $f(y) = x$.]

(f) Repeat (e) above when f^{-1} is unknown or/and you are not allowed to compute it. That is, in your algorithm you may compute $f(x)$, but you may *not*, given y, compute x so that $y = f(x)$.

5.3-5 Function f obeys the condition

For $x_2 > x_1$

$$-10(x_2 - x_1) < f(x_2) - f(x_1) < -(x_2 - x_1)$$

(a) If $f(0) = 22$, show that the equation $f(x) = x$ has a unique solution s. Give a least upper and a greatest lower bound for s.

(b) For what values of k will the following algorithm be guaranteed to converge on s, regardless of starting value? You may assume that $0 < k < 1$.

$$x_{i+1} = x_i + k(f(x_i) - x_i)$$

5.3-6 In the network shown in the figure, E is constant; and g_1, g_2 are linear conductances. A sequence of experiments is performed, and at the end of each jth experiment after i_j is measured, the value of V_j is changed to value V_{j+1} for the next experiment, according to the algorithm

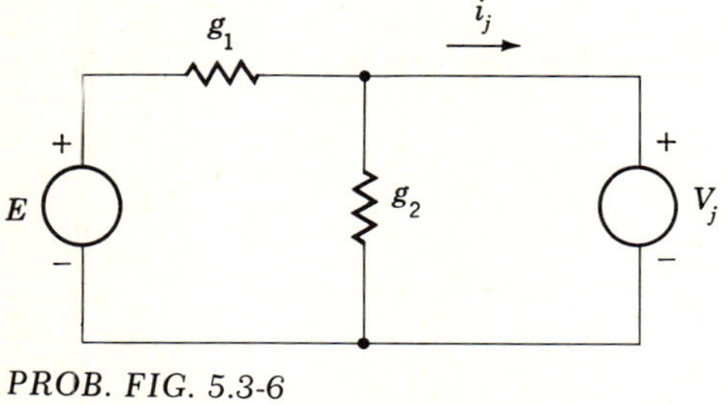

PROB. FIG. 5.3-6

$$V_{j+1} = V_j + Ki_j \qquad \text{with any initial } V_0$$

For what values of K will the sequence $V_0, V_1, V_2, \ldots$ converge on the voltage that appears across g_2 (when the voltage source V_j is removed from the circuit)?

5.3-7 The i-v characteristic of a nonlinear element is shown in part (a).

$$i = \begin{cases} e^v - 1 & v \geq 0 \\ 0 & v < 0 \end{cases}$$

(*a*) Explain why the intersection of the *line*

$$i = 1 - v$$

and the curve

$$i = \begin{cases} e^v - 1 & v \geq 0 \\ 0 & v < 0 \end{cases}$$

determines the voltage across and the current through the nonlinear element when it is connected to the source and resistance as shown in part (*b*). This line is commonly called the *load line* and the point of intersection the *quiescent operating point.* We shall propose an iterative scheme for locating this point. In the scheme, we guess the current at the operating point and then improve the guess by iteration. We continue the iteration until the voltage (at the terminals of the nonlinear element)

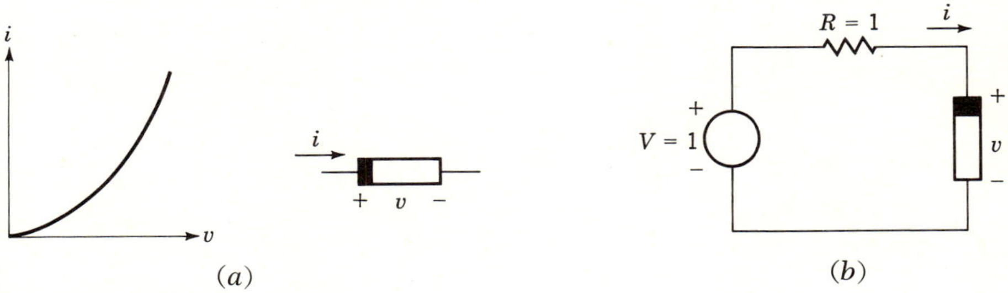

PROB. FIG. 5.3-7

$$v = \ln (i + 1)$$

is within ϵ ($\epsilon > 0$) of the voltage at the terminals of the series source and resistor,

$$v = 1 - i$$

Iteration Scheme

1. Choose any current i_0; let $n = 0$.
2. If $|1 - i_n - \ln (i_n + 1)| < \epsilon$, Stop ($i = i_n$); otherwise go to Step 3.
3. Compute $i_{n+1} = 1 - \ln (i_n + 1)$; increase n by 1 and go to Step 2, above.

(*b*) For what range of values of i_0, if any, does this rule converge, so that i_n approaches the current at the operating point to within ϵ?

(*c*) Select a value for i_0, and then find the current (within $\epsilon = 0.02$) and the corresponding voltage at the operating point by using the given iteration rule.

(*d*) Can you specify an iteration rule that converges if we initially guess the voltage at the operating point? This iteration should continue until the currents in the nonlinear element and in the series source and resistance balance to within ϵ.

5.3-8 Two resistors with characteristics $i_1 = f_1(v)$ and $i_2 = f_2(v)$ are connected in parallel as shown. The functions f_1 and f_2 obey the constraint

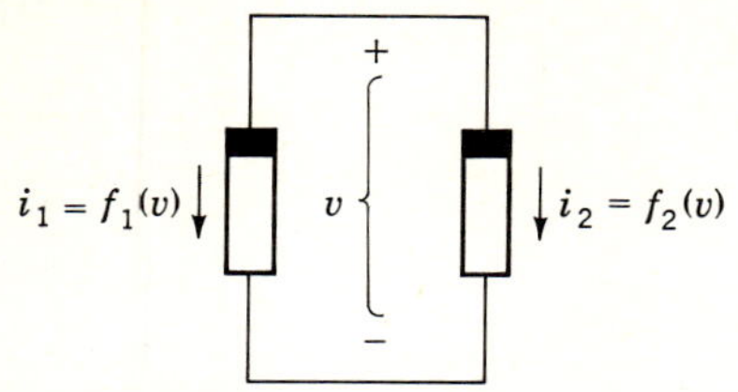

PROB. FIG. 5.3-8

$$0 < m_i(x_1 - x_2) < f_i(x_1) - f_i(x_2) < M_i(x_1 - x_2) < \infty$$

where m_i and M_i are constants. The following algorithm is proposed to solve the network:

Choose any v_0.

For $n = 0, 1, \ldots$

$$v_{n+1} = v_n - K(i_{1n} + i_{2n})$$

where $K > 0$.

(*a*) Show that there exists a range of values of K for which the sequence $v_0, v_1, \ldots, v_n, \ldots$ converges to the solution v of the network. [Hint: Put the network equation in the form $x = f(x)$.]

(*b*) Find a range of values of K for which the algorithm will be guaranteed to converge.

5.3-9 Consider the two functions $f(x)$ and $g(x)$. Assume that

(i) $f(x)$ and $g(x)$ are continuous in x.
(ii) $f(x)$ and $g(x)$ are real-valued for all real x.
(iii) For all real values x_1, x_2,

$$|f(x_1) - f(x_2)| \leq a|x_1 - x_2| \qquad a < 1$$

$$|g(x_1) - g(x_2)| \leq b|x_1 - x_2| \qquad b < 1$$

Now consider the equation

$$x = f(g(x)) \tag{1}$$

where f and g are the above functions, and assume that a solution exists.

(*a*) Prove that Eq. (1) has a unique real solution $x = s$.
(*b*) Prove that the algorithm

$$x_{k+1} = f(g(x_k)) \qquad k = 0, 1, 2, \ldots \tag{2}$$

converges to the solution s of Eq. (1).

(*c*) Is the convergence speed of the algorithm [Eq. (2)], in general, faster or slower than the convergence speed of each of the algorithms

$$x_{k+1} = f(x_K)$$

$$x_{k+1} = g(x_K)$$

5.3-10 Consider the network shown in the figure where $V_1, \ldots, V_m, x$ are volt-

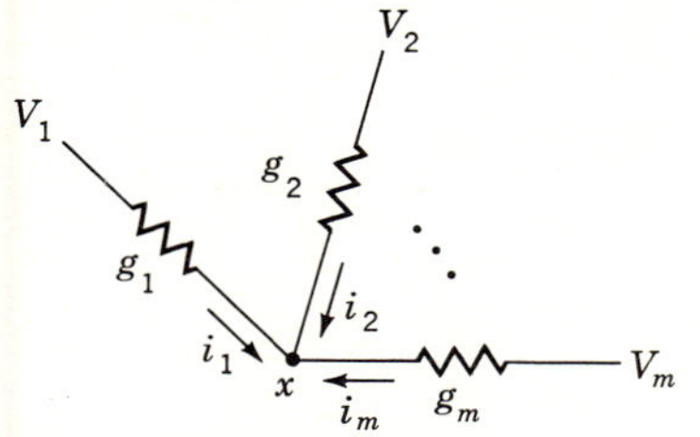

PROB. FIG. 5.3-10

ages with respect to some reference ground; and all resistors are linear with conductance g_m. Apply the following algorithm:

Choose any x_0.

$$x_{n+1} = x_n + K \sum_{j=1}^{m} i_{jn} \qquad n = 0, 1, 2, \ldots \tag{1}$$

That is, try x_n at the nth iteration; then perturb it by the sum of all the currents at that iteration (which should be zero when the solution $x = s$ is reached), scaled by a constant K.

(*a*) Find sufficient conditions on K, for convergence of the algorithm on the solution $x = s$.

(*b*) For a given $K = K_1$ subject to the conditions of (*a*) above, show that the error

$$\epsilon_n = |x_n - s|$$

converges *exponentially to zero;* and give the convergence time of the process.

(*c*) An exponential is uniquely defined by any *three* points through which it passes. Can you use this property to change algorithm (1) after x_0, x_1, x_2 are known, so that $x_3 = s$?

5.3-11 Let $y(x)$ be a function such that

For all x,

$$0 < m_1 \leq \frac{dy}{dx}(x) \leq m_2 < \infty$$

where m_1 and m_2 are constants. Suppose that you have such a function block which, when given x as its input, produces $y(x)$ at its output. You wish to accomplish *function inversion,* that is, given $\hat{y}$ as input to a new system, $\hat{x}$ must appear at its output, such that $\hat{y} = y(\hat{x})$. The task is to design this "new system" using the function block $y(x)$. Note that you *cannot* apply an input to the output of the function block and expect it to operate "backward."

The system shown in the figure is proposed for this task. Operation is as follows: Starting at some time t_0, the "sampler" closes the switch

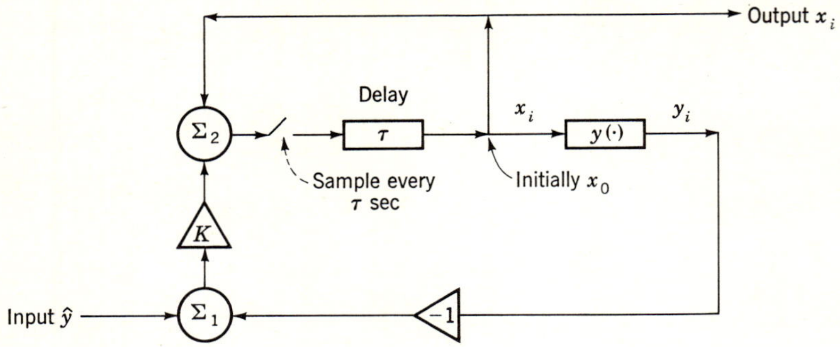

PROB. FIG. 5.3-11

for a very short time (compared to τ) and applies the Σ_2 output to the delay. After τ seconds the process is repeated, i.e., the sampler once again excites the delay input with the Σ_2 output. Let the sequences of input and output values to the function block be $x_0, x_1, x_2, \ldots, x_m$ and $y_0, y_1, \ldots, y_m$ at successive samples. The claim is that for sufficiently large m, x_m approaches $\hat{x}$ when $\hat{y}$ is applied as input to the system; that is, the output of this system *converges* on the desired answer.

(*a*) Draw a function $y(x)$ that meets the above condition. Choose any $\hat{x}$, and $\hat{y}$, and trace graphically the progress of this system by connecting with straight lines adjacent points $[x_i, y_i]$ and $[x_{i+1}, y_{i+1}]$: (1) for "small" K, and (2) for "large" K, $(K > 0)$.

(*b*) For exactly *what values of* K (as a function m_1 and m_2) is convergence guaranteed?

(*c*) Next, let the function be a straight line of slope M.

1. For what value of $K = K_0$ is convergence achieved in one sample, i.e., with x_1?
2. Choose a positive value of K smaller than K_0, and sketch on a graph x_j versus j, for $j = 0, 1, 2, \ldots$.
3. Note that KM is a good measure of the speed of convergence. What does this measure become when $K = K_0$?

(*d*) Coming back to the original function with slope bounded by m_1 and m_2, which value of K would you use, in the range of values determined by (*b*) above, so that convergence is fastest? What is the guaranteed slowest convergence under these circumstances?

5.4-1 In the text we show a way for constructing Newton's method based upon matching terms of the Taylor series. Follow the same approach to construct your own algorithm so that

(*a*) If $F(.)$ is a quadratic, i.e., if $F(x) = ax^2 + bx + c$, then your algorithm will find the solution in one step.

(*b*) If $F(.)$ is not quadratic, then pretend it is quadratic, and keep trying.

Show clearly all your steps, and include a brief discussion on what you feel are the merits and pitfalls of your algorithm. You may use a small example if you want to clarify your points.

5.4-2 Consider the equation $g(x) = x$ with a unique solution s. In this problem you are required to construct a new function f such that

1. The equation $f(x) = x$ has the same solution s.
2. $f'(s) = 0$. [Note that, in general, $g'(s) \neq 0$.]

(*a*) Give an expression for f in terms of g.

(*b*) Write Picard's algorithm for solving the equation $f(x) = x$; evidently this algorithm will find the solution of $g(s) = s$. Show on a square the graph of the lines $y = x$ (diagonal) and $y = g(x)$ for some $g(x)$, and show how one can obtain graphically x_{i+1} from x_i (*Hint:* tangent).

(*c*) Give as many conditions on g as you think are necessary to guarantee that the algorithm of (*b*) above will converge to the unique solution of $g(x) = x$.

(*d*) If these conditions are met, will the algorithm of (*b*) above be always, sometimes, or never faster than the algorithm $x_{i+1} = g(x_i)$? Fully explain your reasoning.

(*e*) Let $g(x) = e^{-x}$, and give the first five terms computed by (1) the algorithm $x_{i+1} = g(x_i)$ and (2) the algorithm of (*b*) above. Use $x_0 = 1$, and sketch the progress of both algorithms on the "square" with diagonal $y = x$ and the graph of function $y = e^{-x}$.

5.4-3 Consider the equation

$$x = 0.9e^{-x}$$

For $0 < x < \infty$.

(*a*) Use Newton's method to solve this equation, starting with the initial guess,

$$x_0 = 1$$

Compute the sequence x_n, until $|x_n - s| \leq 10^{-2}$.

(*b*) How does the convergence speed of this algorithm compare with that of Prob. 5.3-3?

5.4-4 Consider the equation $g(x) = 0$. Newton's method applied to this equation yields

$$x_{n+1} = x_n - \frac{g(x_n)}{g'(x_0)} \qquad n = 0, 1, 2, \ldots \tag{1}$$

Now consider a *modified Newton's method.* In this method the algorithm is

$$x_{n+1} = x_n - \frac{g(x_n)}{g'(x_0)} \qquad n = 0, 1, 2, \ldots \tag{2}$$

(*a*) Give a geometric interpretation of Eq. (2).

(*b*) Suppose that both Eqs. (1) and (2) converge to the unique solution s of $g(x) = 0$. Compare the rate of convergence of Eq. (2) with that of Eq. (1).

(*c*) Can you see any advantages in the use of the modified Newton's method?

(*d*) Apply the modified Newton's method to the equation $x = 0.9e^{-x}$ with $x_0 = 1$, and comment on its convergence as compared with the results found in Prob. 5.4-3.

5.4-5 The network shown in the figure is the familiar implementation (from Chap. 4) of an AND gate with n inputs. Each of the voltage sources

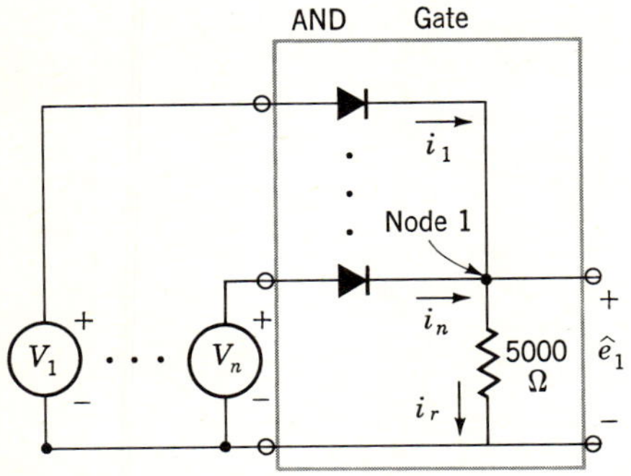

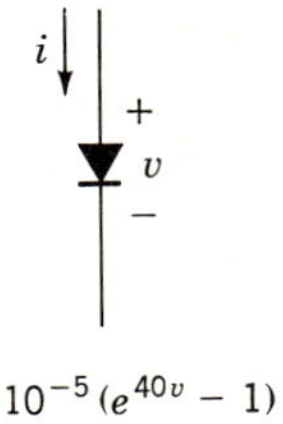

PROB. FIG. 5.4-5

(inputs) $V_1, \ldots V_n$ is either 0 or 5 V. If the diodes are ideal, $\hat{e} = 5$, if and only if all of the V_i are at 5 V, and at 0 V otherwise. The diodes, however, are not ideal, in fact, they have the *i-v* characteristic specified above.

(*a*) Determine $\hat{e}$ by the Picard algorithm,

$$e_{k+1} = e_k + h\left(\sum_{j=1}^{n} i_{jk} - i_{rk}\right) \qquad k = 0,1,2, \ldots$$

where i_{jk} denotes the current i_j when $\hat{e} = e_k$, and h is a constant which you will specify to obtain convergence. Let $n = 2, 4, 10, 50, 100$, and give a graph of $\hat{e}$ versus n.

(*b*) Repeat (*a*), using Newton's method on a KCL equation written about node 1.

(*c*) Compare briefly the convergence of the two algorithms, and discuss why 100-input AND gates are not practical.

5.5-1 Suppose function $f(x)$ is such that

$$M_1(x_2 - x_1) \le f(x_2) - f(x_1) \le M_2(x_2 - x_1)$$

for all values x_1 and x_2 such that

$$a < x_1 < x_2 < b$$

where M_1, M_2, a, and b are specified positive constants.

(*a*) Prove that if $f(a)$ is negative and $f(b)$ is positive, then $f(x)$ has a unique zero lying between a and b.

(b) Consider the following algorithm ($\epsilon > 0$ is a specified tolerance):

1. $x_0 = (b + a)/2 \qquad d_0 = (b - a)/4$

2. If $f(x_k) > \epsilon$ $\quad x_{k+1} = x_k - d_k$

If $f(x_k) < -\epsilon$ $\quad x_{k+1} = x_k + d_k$

If $|f(x_k)| \leq \epsilon$ $\quad$ Stop

3. $d_{k+1} = d_k/2$ $\quad$ Go to Step 2

Show that the algorithm reaches Stop for all positive ϵ.

(*c*) How is the speed of convergence related to ϵ?

(*d*) Suppose now that M_1 is negative. Show that $f(x)$ may have an odd number of zeros between (*a*) and (*b*). What does the algorithm of (*b*) do if f has, say, three zeros? Sketch an example.

chapter 6

Discrete Memoryless Systems

This chapter discusses a class of memoryless systems for which the inputs and outputs can assume only a finite number of values. In particular, we will be studying combinational logic, for which the number of values is restricted to two. Logic networks are essential components of digital computers and other digital systems. After looking briefly at some electric network realizations of logic gates, we will focus on Boolean algebra—a mathematical system used in the analysis and synthesis of combinational logic networks. Properties of the algebra will be investigated and its use in describing the behavior of combinational logic exploited.

6.1 INTRODUCTION

In this chapter, as in Chaps. 3 and 4, we will be discussing memoryless systems, i.e., systems for which the present behavior is determined solely by the present inputs. The systems of interest here, however, are characterized by the presence of discrete-valued signals.† These systems are treated separately because the associated analysis and synthesis techniques are quite different from those used for continuous systems. Furthermore, the concept of linearity is of reduced importance and does not influence the means of system characterization.

A major reason for the engineering use of systems employing discrete-valued signals is the reduction of the effects of unwanted signals (noise) on system performance. Signals are almost always used for the representation of information, and any unknown perturbation of these signals can result in the corruption of the information being represented. At an intuitive level, it is reasonable that the greater the amount of information a signal must represent, the greater the chance is of noise changing the represented value. As an example, consider a voltage that can range between 0 and 4 V and assume that we wish to use signals of this type to represent four different "pieces" of information. Since any physical system contains noise, we would like to pick a representation that offers as much noise "immunity" as possible. A possible scheme, using one variable v, is shown in Fig. 6.1-1, where the pieces of information have been called A, B, C, and D.

Voltage	*Information*
$0 \le v \le 1$	A
$1 < v \le 2$	B
$2 < v \le 3$	C
$3 < v \le 4$	D

FIG. 6.1-1 A representation of four pieces of information by one variable.

Now if B, for example, is ideally represented by 1.5 V, the representation can tolerate noise of approximately $\pm \frac{1}{2}$ V. Alternative ways of representing the information use more variables and have better noise immunity.

Voltages	*Information*
$0 \le v_1 \le 2,\ 0 \le v_2 \le 2$	A
$0 \le v_1 \le 2,\ 2 < v_2 \le 4$	B
$2 < v_1 \le 4,\ 0 \le v_2 \le 2$	C
$2 < v_1 \le 4,\ 2 < v_2 \le 4$	D

FIG. 6.1-2 Representation of four pieces of information by two variables.

†A signal is a physical way of representing information. It may be a voltage (or current) value, or a voltage versus time (current versus time) waveform.

Such a scheme, shown in Fig. 6.1-2, uses two voltages v_1 and v_2. If B is ideally represented by $v_1 = 1$ and $v_2 = 3$, the representation can tolerate noise of approximately ± 1 V on either voltage before any corruption occurs.

This example illustrates, in a simple fashion, the virtue of requiring a variable to represent as little information as possible. In the example of Fig. 6.1-2 we have used each signal in the simplest possible nontrivial fashion by distinguishing only between two ranges of values. The price paid for the resulting increased noise immunity is obvious. The representation of Fig. 6.1-2 requires twice as many variables as that of Fig. 6.1-1.

It might appear that, depending on the situation, various compromises might be made between the number of variables and the number of levels associated with each variable. This is rarely true. If we consider existing computers, for example, we find only analog computers which use continuous signals and digital computers which employ bivalued or binary signals exclusively. Analog computers are simply interconnections of block diagram components such as summers, multipliers, and integrators. We have already studied some of these elements and will discuss others, as well as some typical analog computer interconnections, in later chapters. There are certain applications such as simulation of ordinary and partial differential equations for which analog computers are well suited.† Digital computers are interconnections of two basic types of bivalued elements: those with memory and those without. The elements with memory can be broken down further into basic elements such as ferrite cores,‡ and elements formed from certain interconnections of memoryless and memory elements. These interconnections, as we will see in Chap. 10, have "feedback," i.e., output values that are influenced by themselves. The purpose of this chapter, however, is to study the basic memoryless building blocks of digital systems that operate on and produce bivalued signals. Memoryless systems of this type are referred to as *combinational logic* or *logic networks*, and have the input-output characterization of Fig. 6.1-3. Ideally, each input x_i can assume one of two values of a physical variable, which can be voltage, current, pressure, etc. Similarly, each output z_i can assume one of two values of usually the same physical variable. Furthermore, the values of the outputs at any time depend only on the values of the inputs at *that* time. It should be emphasized, however, that if one looks in detail at the nature of signals in a digital system, one sees a spectrum or range of actual values. In addition, even if the signals actually were very close to one value or the other, they must change if the system is "doing anything." When such a change occurs, it always requires a finite time, during which the signal assumes all

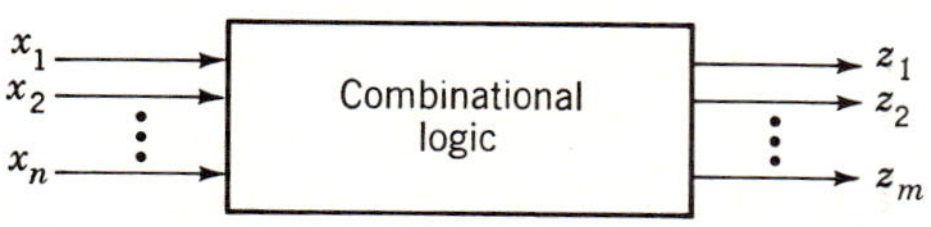

FIG. 6.1-3 Input-output characterization of combinational logic.

†Although their accuracy is of the order of 1 percent, they solve such equations very rapidly.

‡A doughnut-shaped piece of magnetic material with two states of magnetization. These form the basis of core memory.

intermediate values. For this reason, a detailed circuit-oriented study of the logic circuits to be introduced is a complex process not discussed in this chapter.

6.2 LOGIC CIRCUITS

The purpose of this section is to provide an introduction to some of the more common types of logic circuits. Interconnections of circuits of these types form the basis for almost all digital systems. In the example of Sec. 6.1 we assigned a range of values to represent a particular piece of information. In this section we will simply assign one of two "nominal" values to signals.

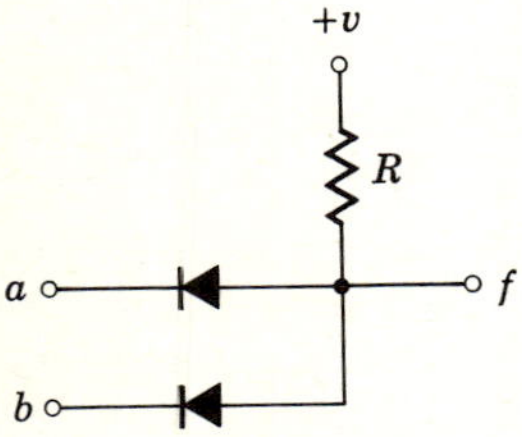

(a)

a	b	f
0	0	0
0	+v	0
+v	0	0
+v	+v	+v

(b)

FIG. 6.2-1 Example of diode-resistor logic.

The simplest logic circuits are of the diode-resistor variety. Two examples are shown in Fig. 6.2-1*a* and 6.2-2*a*. The inputs *a* and *b* are constrained to be 0 or $+v$ V. The four possible input combinations are listed in the left two columns of Figs. 6.2-1*b* and 6.2-2*b*, while the right column gives the output voltage for the input combinations, assuming ideal diodes. There is nothing special about these circuits (or any described in this section) that restricts them to processing bivalued signals. The circuits of Fig. 6.2-1*a* and 6.2-2*a* appear in Chap. 4 and were analyzed for a continuous domain of values *a* and *b*. We are simply restricting our analysis to two-signal values because of the way the circuits will be used.

The fundamental shortcoming of diode-resistor logic is that the

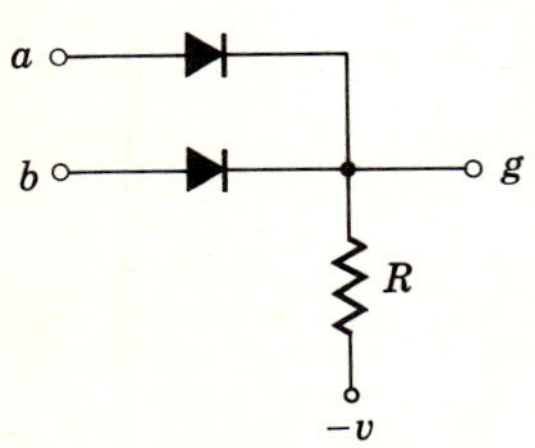

(a)

a	b	g
0	0	0
0	+v	+v
+v	0	+v
+v	+v	+v

(b)

FIG. 6.2-2 Another example of diode-resistor logic.

output voltage is not exactly 0 or $+v$ because of the characteristics of physical diodes (see Prob. 5.4-5). Thus, if a slightly degraded output voltage is used as an input to another diode-resistor logic circuit, the process becomes cumulative, ultimately making the original 0 and $+v$ values indistinguishable.

Resistor-transistor logic (RTL) is the simplest example of logic using an *active* element. The most commonly used circuit, along with its description, is shown in Fig. 6.2-3*a* and *b*. If either input is $+v$, the corresponding

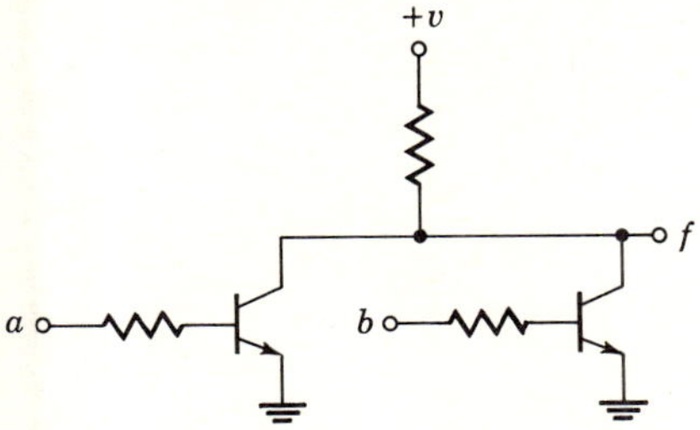

(a)

a	b	f
0	0	v
0	$+v$	0
$+v$	0	0
$+v$	$+v$	0

(b)

FIG. 6.2-3 Example of resistor-transistor logic (RTL).

transistor connects its three terminals together through an essential short circuit, and the output is essentially zero. More transistors can be paralleled to increase the number of inputs, although the total number rarely exceeds four.

A slightly more complicated logic form is known as diode-transistor logic (DTL). A typical circuit and its description are shown in Fig. 6.2-4*a* and *b*.

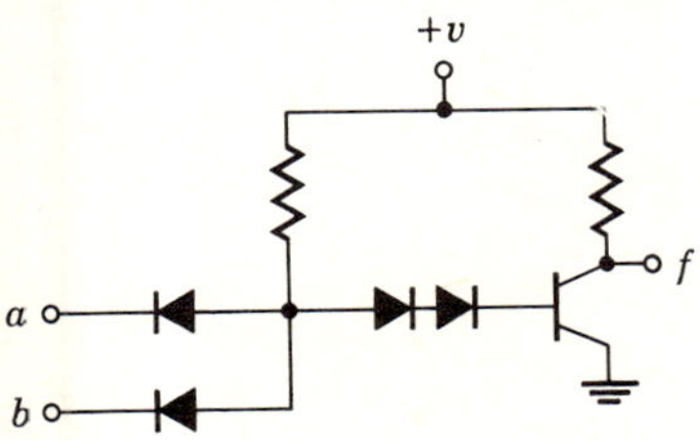

(a)

a	b	f
0	0	$+v$
0	$+v$	$+v$
$+v$	0	$+v$
$+v$	$+v$	0

(b)

FIG. 6.2-4 Example of diode-transistor logic (DTL).

These are but a few examples of logic circuits and are provided to establish a technological framework for the discussion to follow.

6.3 LOGIC GATES

The term "logic gate" is used to describe the broad class of electrical block diagram elements that are *memoryless* and are suited to the processing of bivalued signals. The circuits of Sec. 6.2 are examples of logic gates. Having seen these examples, we can now introduce representations for their behavior that do not refer to the physical nature of the input or output signals, i.e., whether they are voltages, currents, etc. This can be conveniently done by realizing that since all signals are assumed to be bivalued, they can be uniquely identified by the terms *high* (*H*) and *low* (*L*), understood to apply to the appropriate physical quantity. Since we are discussing memoryless systems, it is clear that any logic gate can be described in tabular form like the circuits in the previous section. Such a tabulation is considered a logic (or logical) description of the gate.

As an example, the circuit of Fig. 6.2-2*a* is a logic gate described by the table of Fig. 6.3-1*a*. We have assumed that $v > 0$; *L* has been assigned to 0 V, and *H* to $+v$. Since we are no longer interested in the actual realization, the

a	*b*	*g*
L	*L*	*L*
L	*H*	*H*
H	*L*	*H*
H	*H*	*H*

(*a*)

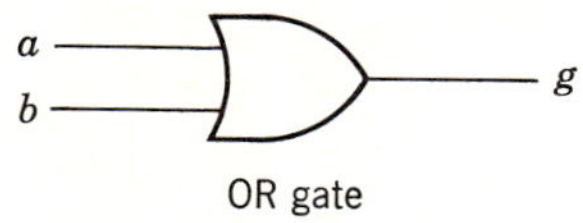

(*b*)

FIG. 6.3-1 Or gate and description.

circuit diagram can be replaced by a symbol. The gate we are discussing (remember this means the behavior described in Fig. 6.3-1*a*) is so common that it is assigned both a special name and symbol. It is called an OR gate and the associated symbol† is shown in Fig. 6.3-1*b*. The circuit of Fig. 6.2-2*a* is an OR gate in that when *H* and *L* are assigned to its inputs and output, the table of Fig. 6.3-1*a* results. The OR gate is defined, however, by the table of Fig. 6.3-1*a*, *not* by a realization. Another common gate is the AND gate, whose description and symbol are given in Fig. 6.3-2*a* and *b*. The reader should verify that the circuit of Fig. 6.2-1*a* is a realization of an AND gate.

a	*b*	*f*
L	*L*	*L*
L	*H*	*L*
H	*L*	*L*
H	*H*	*H*

(*a*)

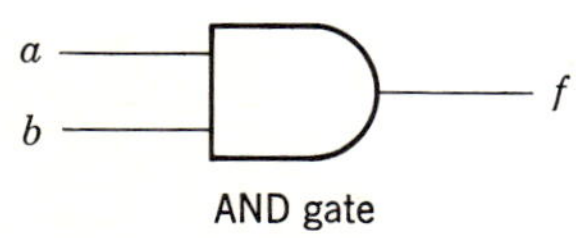

(*b*)

FIG. 6.3-2 AND gate and description.

The names OR and AND are quite descriptive of the behavior of the gates. The output of the OR gate is *H* if either *a or b* is *H*. (Or is used here in an inclusive sense.) The output of the AND gate is *H* if *a and b* are *H*. Both these definitions can be naturally extended to gates with more than two inputs. The output of an OR gate with inputs $x_1, \ldots, x_n$ is *H* if x_1 *or* x_2 *or* . . . *or* x_n is *H*. The output of an AND gate with inputs $x_1, \ldots, x_n$ is *H* if x_1 and x_2 and . . . and x_n are *H*. The corresponding symbols are shown in Fig. 6.3-3*a* and *b*.

The gate shown in Fig. 6.3-4 has a single input and output and is

†We should caution the reader that the symbology for logic gates is not standard. The symbols and interpretation we are using are fairly well accepted.

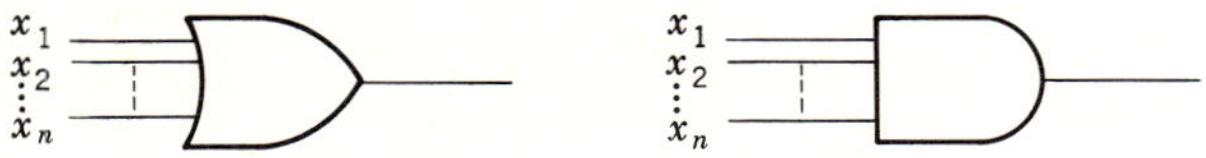

(a) (b)

FIG. 6.3-3 n-input OR gate and n-input AND gate.

called an INVERTER or NOT gate. An RTL gate with its two inputs connected is a realization of an INVERTER. Two more commonly used gates, the NAND and NOR gates, are described in Fig. 6.3-5*a* and *b* and Fig. 6.3-6*a* and *b*. NAND is a contraction of NOT-AND; NOR is a contraction of NOT-OR. Figure 6.2-4*a* is a realization of a NAND gate, and Fig. 6.2-3*a* is a realization of a NOR gate.

a	*b*
L	*H*
H	*L*

(a)

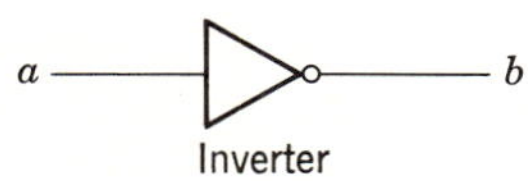

(b)

FIG. 6.3-4 INVERTER or NOT gate.

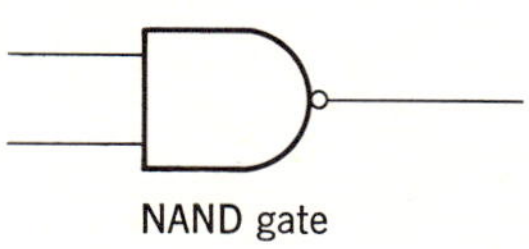

(a)

x_1	x_2	f
L	*L*	*H*
L	*H*	*H*
H	*L*	*H*
H	*H*	*L*

(b)

FIG. 6.3-5 NAND gate and description.

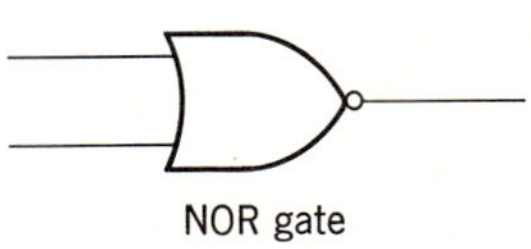

(a)

x_1	x_2	f
L	*L*	*H*
L	*H*	*L*
H	*L*	*L*
H	*H*	*L*

(b)

FIG. 6.3-6 NOR gate and description.

Larger combinational circuits are interconnections of circuits of the type described in Sec. 6.2. A relatively simple example is shown in Fig. 6.3-7. The analysis problem associated with this circuit, treated as a logic circuit, is to determine how *f* depends on the bivalued variables *a*, *b*, and *c*. We can identify each circuit enclosed by a box as a NOR gate and would be tempted to conclude that Fig. 6.3-8 is a correct logical representation of Fig. 6.3-7. This is in fact true,

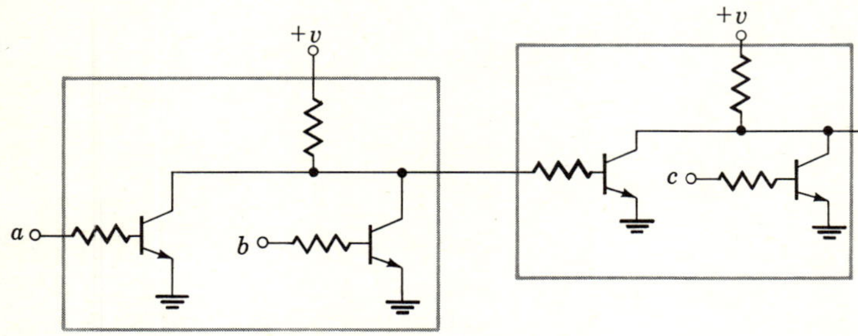

FIG. 6.3-7 A larger combinational circuit.

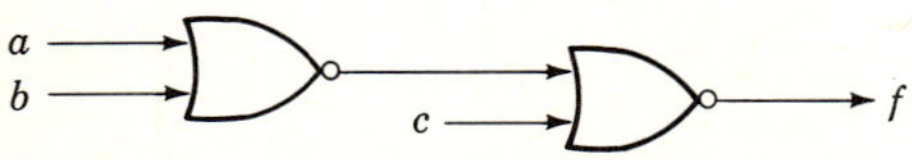

FIG. 6.3-8 The circuit of Fig. 6.3-7 as an interconnection of logic gates.

but it is important to understand why.

First observe that our logical description of gates are crude in the sense that they do not reflect any detailed electrical characteristics of the realization. Consequently, it is not valid to assume that the individual logical descriptions of the gates will be necessarily retained when they are interconnected. The important fact is that logical circuits *are designed so that they can be interconnected* without their individual logical descriptions being destroyed. To ensure that this occurs, however, logic manufacturers provide detailed rules, which if followed, allow a circuit such as that of Fig. 6.3-7 to be modeled by Fig. 6.3-8. The derivation of these rules, which are called *loading rules,* requires detailed circuit analysis of the circuit of Fig. 6.3-7. An informal description of the problems encountered is easy to understand, especially if we work with the simplest logical circuit, the INVERTER.

Assume that an inverter has the v_{in} versus v_{out} characteristic shown in Fig. 6.3-9. Hence if the input voltage is 0, the output is $+v$, and if the input

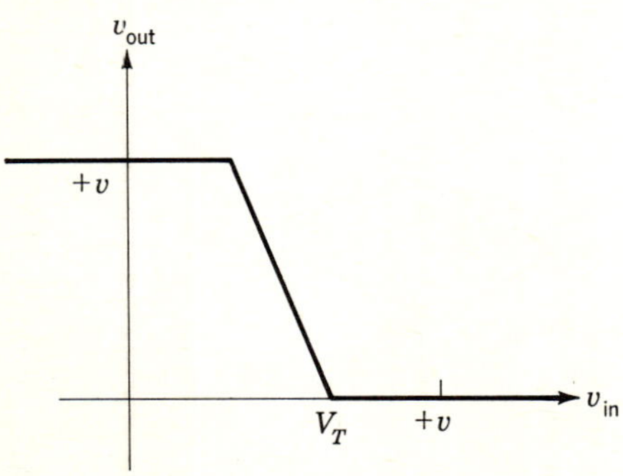

FIG. 6.3-9 Inverter input-output electrical characteristics.

voltage is $+v$, the output is 0. Now assume that we interconnect $n+1$ of these INVERTERS as shown in Fig. 6.3-10*a*. If we analyze this interconnection, we

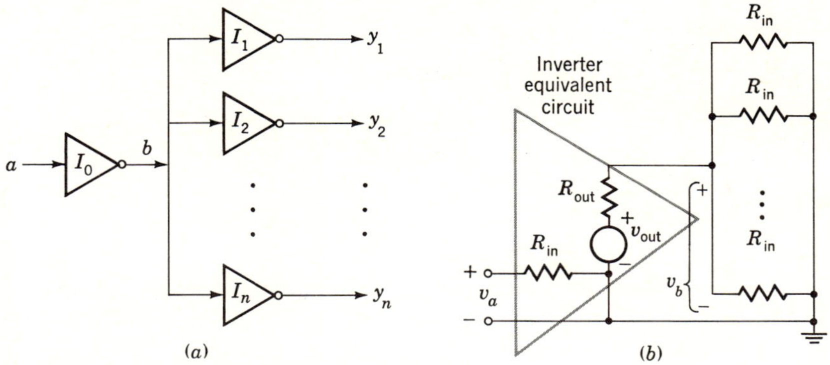

FIG. 6.3-10 Analysis of loading.

deduce that ideally, $y_1 = y_2 = \cdots = y_n = a$. As n increases beyond a certain value, the performance of this network becomes progressively less ideal. This degradation is easily explained by referring to Fig. 6.3-10*b*, where a voltage source v_{out} and a series resistance R_{out} represent the equivalent circuit seen at the output of inverter I_0, with $a = 0$. The input equivalent circuit of each inverter $I_1, \ldots, I_n$ is simply a resistor of value R_{in}. By the divider relationship, it follows that

$$v_b = \frac{R_{in}}{nR_{out} + R_{in}} v_{out}$$

Hence, as n increases, v_b (which is the input v_{in} to inverters $I_1, \ldots, I_n$) decreases. From the inverter characteristics of Fig. 6.3-9, as v_{in} decreases from value v, v_{out} stays constant until v_{in} equals V_T. Thus the value of n such that $v_b = V_T$ represents the maximum number of inverters that can be connected as in Fig. 6.3-10*a* and still ensure that $y_1 = y_2 = \cdots = y_n = 0$ when $a = 0$. A further increase of n results in a further decrease of v_b, hence a degradation of $y_1, y_2, \ldots, y_n$ ultimately to v instead of 0 V. This is a simple example of the loading problems that can be encountered in logic circuits. In what follows we shall assume that logic gates retain their characteristics when interconnected, and shall proceed with our development.

Our goal is to develop analysis and synthesis techniques for logic networks. Let us look at an example of the type of problem we wish to be able to solve before we examine the mechanisms we will use to actually solve the problem. Assume that we have two two-bit binary numbers that we wish to multiply together to obtain a binary representation of the product. That this is basically a combinational problem is clear, for each time the inputs assume a particular value, the output can be predicted. The next question concerns representation. Assume that we associate a low signal with the binary 0 and a high signal with the binary 1. Then the binary number 10, or decimal 2, is represented as *HL*. Finally, we must consider the number of output bits required. If we compute the "largest" product 3×3, the answer requires four bits to repre-

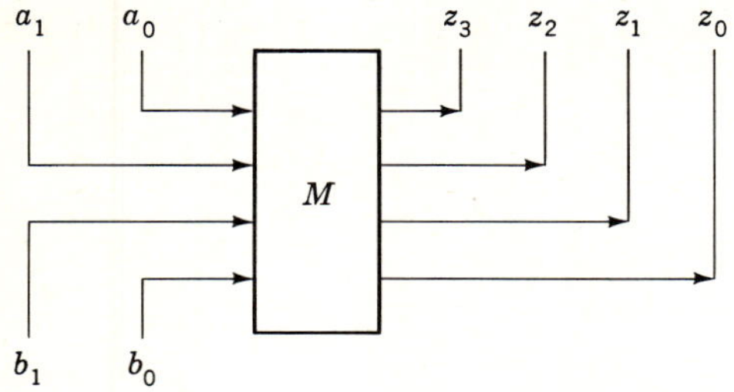

FIG. 6.3-11 Logic network M performing multiplication.

sent. Thus we wish to design a four-input four-output combinational circuit M, as shown in Fig. 6.3-11. The last piece of information required is a description of its behavior. By doing the multiplication we can arrive at the table given in Fig. 6.3-12. Now let us assume that we wish to construct, by interconnecting gates, a logical circuit that has this behavior. Furthermore, we would like some orderly way to arrive at this circuit and some assurance that it was a good realization by some measure, perhaps not having an excess of gates over those required by some other realization. Here the reader might wish to pause and contemplate the problem if the number of bits in the input words were 16 or 32 —common values for digital computer representations of numbers.

	Inputs				*Outputs*			
Product	a_1	a_0	b_1	b_0	z_3	z_2	z_1	z_0
0×0	L	L	L	L	L	L	L	L
1×0	L	H	L	L	L	L	L	L
2×0	H	L	L	L	L	L	L	L
3×0	H	H	L	L	L	L	L	L
0×1	L	L	L	H	L	L	L	L
1×1	L	H	L	H	L	L	L	H
2×1	H	L	L	H	L	L	H	L
3×1	H	H	L	H	L	L	H	H
0×2	L	L	H	L	L	L	L	L
1×2	L	H	H	L	L	L	H	L
2×2	H	L	H	L	L	H	L	L
3×2	H	H	H	L	L	H	H	L
0×3	L	L	H	H	L	L	L	L
1×3	L	H	H	H	L	L	H	H
2×3	H	L	H	H	L	H	H	L
3×3	H	H	H	H	H	L	L	H

FIG. 6.3-12 Desired input-output behavior of M.

We are now ready to consider certain ideas that will help us solve problems like that posed above. The first tool we require is a suitable mathematical framework in which to view the problems. Boolean algebra is such a framework.

6.4 BOOLEAN ALGEBRA

In this section we introduce a mathematical system called Boolean algebra. The particular algebra we are interested in has two elements, and we will study its use for the analysis and synthesis of combinational systems with bivalued signals. Boolean algebra provides us with a compact way of describing the behavior of combinational logic, and in some circumstances enables us to use algebraic manipulation to find "good realizations" of logic systems. We proceed with an axiomatic definition of the Boolean algebra having two elements.

DEFINITION 6.4-1 The two-element Boolean algebra consists of two elements called 0 and 1, together with two binary operations on the elements. These two operations, usually called addition (+) and multiplication (·), and the elements must satisfy the following axioms:

1. Each operation is associative and commutative, and each distributes over the other.
2. For every element b in the algebra, $b + 0 = b$ and $b \cdot 1 = b$,
3. For every element b in the algebra, there exists another element, denoted $\bar{b}$, such that $b + \bar{b} = 1$ and $b \cdot \bar{b} = 0$.

Let us look in detail at these three axioms. Axiom 1 states that if a, b, and c represent arbitrary elements of the algebra, then

$$a + b = b + a \qquad a \cdot b = b \cdot a \qquad \text{(commutativity)}$$

$$(a + b) + c = a + (b + c) \qquad (a \cdot b) \cdot c = a \cdot (b \cdot c) \qquad \text{(associativity)}$$

$$a + (b \cdot c) = (a + b) \cdot (a + c) \qquad a \cdot (b + c) = (a \cdot b) + (a \cdot c) \qquad \text{(distributivity)}$$

Although for convenience we use "+" and "·" as symbols for the two operations, a cursory examination reveals that the Boolean operations do not behave as the arithmetic operations associated with the same symbols; particularly, arithmetic addition does not distribute over arithmetic multiplication. As we do in arithmetic, we will write ab for $a \cdot b$, when no confusion will result. Other normal arithmetic conventions will also be adopted: $xy + z$ means $(x \cdot y) + z$, *not* $x \cdot (y + z)$.

Axiom 2 states two properties of the elements called 0 and 1. The reader is cautioned not to think of 0 and 1 as integers. They are simply names for elements, nothing more.

Finally, Axiom 3 states that for each element, namely, 0 and 1, there must be at least one other element, denoted $\bar{0}$ and $\bar{1}$, called the complement of the element, satisfying the stated equations. Using condition 2, we can deduce that 1 is *a* complement of 0 and that 0 is *a* complement of 1.

How are the operations + and · defined? Since + and · are binary operations, a satisfactory description is one that gives the result of the operation for all pairs of elements. Two such tabular descriptions are given in Fig. 6.4-1*a* and *b*. As an example of interpreting these tables, the upper right-hand entry in the first table is interpreted as $0 + 1 = 1$. Our first task is to verify that these

+	0	1
0	0	1
1	1	1

(*a*)

·	0	1
0	0	0
1	0	1

(*b*)

FIG. 6.4-1 The operations + and ·

tables satisfy all the conditions of Definition 6.4-1. The reader can verify immediately that both addition and multiplication as defined in Fig. 6.4-1 are commutative, as required. The associativity and distributivity can be verified by listing all eight ways of choosing a, b, and c as elements of the algebra, and using the tables of Fig. 6.4-1 to verify that, for example, for all eight cases, $(a + b) + c = a + (b + c)$ or $a \cdot (b + c) = (a \cdot b) + (a \cdot c)$. How were these tables developed, and are they unique? Axiom 2 of Definition 6.4-1 specifies uniquely the left column and top row of Fig. 6.4-1*a*, and right column and bottom row of Fig. 6.4-1*b*. The reader is encouraged to verify that any choices for the remaining entries, other than those shown, violate the conditions of Definition 6.4-1.

Before we can apply Boolean algebra to the analysis and synthesis of combinational logic, it is necessary to have available certain theorems. In proving these theorems we have attempted to show several of the common methods of proof.

THEOREM 6.4-1 The unique complement of 0 is 1 and the unique complement of 1 is 0.

PROOF (*by Contradiction*): We already know that 1 is *a* complement of 0. Assume that 0 is also a complement of 0; i.e., assume $\bar{0} = 0$. Then Axiom 3 requires that $0 + 0 = 1$, a contradiction. A similar proof can be used for the complement of 1.

Q.E.D.

THEOREM 6.4-2 Let b be an element of the algebra. Then $(\bar{\bar{b}}) = b$.

PROOF (*by Cases*): Let $b = 0$; then $\bar{b} = 1$ and $(\bar{\bar{b}}) = (\bar{1}) = 0$. Let $b = 1$; then $\bar{b} = 0$ and $(\bar{\bar{b}}) = (\bar{0}) = 1$. Q.E.D.

THEOREM 6.4-3 Let b be an element of the algebra. Then $b + b = b$ and $b \cdot b = b$.

PROOF (*Axiomatic*):

$$b = b + 0 \qquad\qquad b = b \cdot 1$$

$$= b + b\bar{b} \qquad\qquad = b \cdot (b + \bar{b})$$

(*by Distributivity*): $$= (b + b) \cdot (b + \bar{b}) \qquad\qquad = (b \cdot b) + (b \cdot \bar{b})$$

$$= (b + b) \cdot (1) \qquad\qquad = (b \cdot b) + (0)$$

$$= (b + b) \qquad\qquad = (b \cdot b)$$

Q.E.D.

An important characteristic of the algebra is revealed by the parallel proofs of Theorem 6.4-3. Any equality in the right-hand proof can be obtained by taking the corresponding equality in the left-hand proof, replacing all +'s by ·'s, and ·'s, by +'s, all 1s by 0s and all 0s by 1s. Is this simply a coincidence? Looking at Axioms 1 to 3 of Definition 6.4-1, we see that they remain *unchanged* when subjected to the same replacements. This must then be true of the validity of *any statement or equality deducible from the axioms*. Namely, an equality remains an equality if 0s and 1s and +'s and ·'s are interchanged. Each time an equality is derived, we get another equality by performing this interchange. Statements or theorems related in this manner are called *dual* statements or theorems. Thus the dual of $0 + 1 = 1$ is $1 \cdot 0 = 0$. For Theorem 6.4-4 below we prove both the theorem and its dual (Theorem 6.4-5). After this we shall simply state dual theorems.

THEOREM 6.4-4 If a and b are elements of the algebra, then $a + ab = a$.

PROOF:

$$\begin{aligned} a + ab &= a(1 + b) && \text{by distributivity of } \cdot \text{ over } + \\ &= a(1) && \text{See Fig. 6.4-1}b \\ &= a \end{aligned}$$

Q.E.D.

THEOREM 6.4-5 *Dual of Theorem 6.4-4*

$$a \cdot (a + b) = a$$

PROOF:

$$\begin{aligned} a \cdot (a + b) &= a + (0 \cdot b) && \text{by distributivity of } + \text{ over } \cdot \\ &= a + 0 && \text{See Fig. 6.4-1}a \\ &= a \end{aligned}$$

Q.E.D.

THEOREM 6.4-6 If a and b are elements of the algebra, $ab + a\bar{b} = a$.

PROOF:

$$\begin{aligned} ab + a\bar{b} &= a(b + \bar{b}) \\ &= a \cdot 1 \\ &= a \end{aligned}$$

Q.E.D.

THEOREM 6.4-7 Dual of Theorem 6.4-6

$$(a+b)(a+\bar{b}) = a$$

THEOREM 6.4-8 If a and b are elements of the algebra, then $(\overline{a+b}) = (\bar{a}) \cdot (\bar{b})$.

PROOF (by Exhaustion or Perfect Induction): A satisfactory and convenient way to prove theorems in Boolean algebra is to use perfect induction. To do this, simply try the "theorem" for all combinations of elements in the algebra. If the equality is satisfied for all cases, then it is a theorem. This method of proof is illustrated below for Theorem 6.4-8. We see that the columns labeled $(\overline{a+b})$ and $\bar{a} \cdot \bar{b}$ are identical, and the theorem is proved.

a	b	$a+b$	$(\overline{a+b})$	$\bar{a}$	$\bar{b}$	$\bar{a} \cdot \bar{b}$
0	0	0	1	1	1	1
0	1	1	0	1	0	0
1	0	1	0	0	1	0
1	1	1	0	0	0	0

Note that in a proof by exhaustion we let each variable in the theorem assume the value 0 and 1. Since in the above theorem, there are only two distinct variables a and b, there are only four possible cases. If the theorem had 16 distinct variables, then to prove it we would have to try $2^{16} = 65{,}536$ cases, truly an "exhaustive" proof.

THEOREM 6.4-9 Dual of Theorem 6.4-8

$$(\overline{a \cdot b}) = \bar{a} + \bar{b}$$

Theorems 6.4-8 and 6.4-9 are commonly called *DeMorgan's laws or theorems.*

THEOREM 6.4-10

$$a + \bar{a}b = a + b$$

PROOF (Axiomatic):

$$a + \bar{a}b = (a+\bar{a}) \cdot (a+b)$$

$$= a + b \qquad \text{Q.E.D.}$$

These theorems are sufficient for our purposes. Before we discuss Boolean functions and their application to combinational logic, it is worthwhile to mention two pitfalls to be avoided in proving theorems in Boolean algebra. If the equality $ab = ac$ is obtained, it is very tempting to do a little "division" and conclude that $b = c$. Likewise it is tempting to deduce from $a + b = a + c$ that $b = c$. The reader should have no trouble finding counterexamples for both cases.

6.5 BOOLEAN FUNCTIONS AND EXPRESSIONS

In this section Boolean functions and an algebraic means for describing them are introduced. As we know, a function (or mapping) is a directed relation from one set to another. The function *assigns* to every member of one set (the *domain* of the function) a member of the other set (the *range*). The domain of a Boolean function of k variables is the set of all combinations of k 0s and 1s—the so-called *k-tuples* of 0s and 1s. The range of all Boolean functions will be the set of the two elements 0 and 1.

DEFINITION 6.5-1 A *Boolean function* of k variables is a mapping from the domain of k-tuples of 0s and 1s to the range of 1-tuples of 0s and 1s.

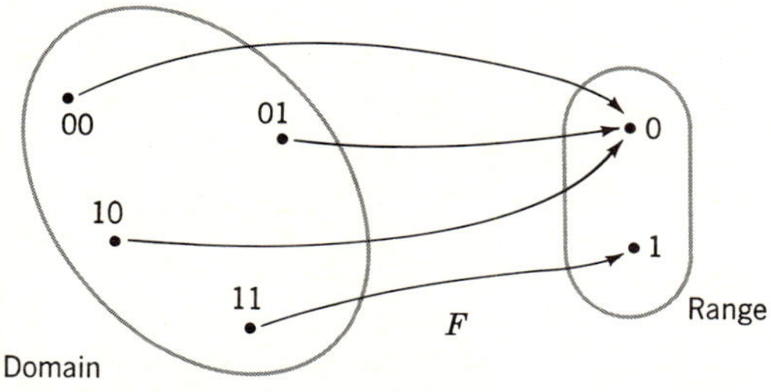

FIG. 6.5-1 A Boolean function of two variables.

Figure 6.5-1 describes a function F of two variables. Although this method of description is quite graphic, it is neither compact nor convenient. A preferable manner of description is a tabular form known as the *truth table* for the function. We have already used similar tables to describe logic gates. Figure 6.5-2 gives the truth table for the function F of Fig. 6.5-1. The left columns of a truth table for a function of k variables list all members of the domain;

x_1	x_2	F
0	0	0
0	1	0
1	0	0
1	1	1

FIG. 6.5-2 Truth table for the Boolean function of Fig. 6.5-1.

the entries in the right column are the corresponding elements of the range.

Boolean expressions, defined below, provide a convenient algebraic description of Boolean functions, one that is more suited to our purposes than a truth table. A *Boolean constant* is a specific element of the algebra, i.e., either 0 or 1. A *Boolean variable* is a symbol that can represent either element of the algebra. We have already used variables in some theorems in Sec. 6.4.

DEFINITION 6.5-2 E is a Boolean expression if and only if it is derivable from a finite number of applications of the following three rules. (We write "expression" for Boolean expression)

1. Any variable or constant is an expression.
2. If G is an expression, then so is $\overline{G}$.
3. If G and H are expressions, then so are $G + H$ and $G \cdot H$.

The following are all expressions: 0, 1, x, $\overline{x}$, $\overline{x} + y$, and xy. If only constants are used to form an expression, the expression can be interpreted as denoting a specific element of the algebra. This is done by evaluating the expression using the operations of the algebra. For example, $0 + 1 \cdot (0 + 1 + \overline{0}) = 1$, and is hence said to denote or evaluate to 1. If, however, variables are used in the formation of an expression, the element denoted by the expression depends on the elements assigned to the variables. The expression $x + \overline{y}$, when evaluated for $x = 0$ and $y = 1$, denotes the element 0, and when evaluated for $x = 1$ and $y = 1$, it denotes the element 1. If an expression contains both constants and variables, the rules of the algebra can be used to eliminate the constants. The expression $x + 1 \cdot (y + \overline{z} + 0) + 0 \cdot w$ always denotes the same element as $x + (y + \overline{z})$. We will be primarily concerned with expressions formed from variables.

We can now employ expressions to describe functions. Consider an expression containing k variables. For every combination of assignments or values for the variables, the expression denotes a specific element, either 0 or 1. The expression can be considered to describe the function that maps each k-tuple of variable values into the element that the expression denotes. As an example, consider the expression $x\overline{y} + z$. The truth table of Fig. 6.5-3 gives the evaluation of the expression for all combinations of values for x, y, and z. The

x	y	z	$x\overline{y} + z$
0	0	0	0
0	0	1	1
0	1	0	0
0	1	1	1
1	0	0	1
1	0	1	1
1	1	0	0
1	1	1	1

FIG. 6.5-3 Truth table for Boolean function denoted by the expression $x\overline{y} + z$.

formation of this table can be illustrated using the entry $x = 0$, $y = z = 1$. Substituting these values into the expression gives $[0 \cdot (\overline{1})] + 1 = 0 + 1 = 1$, the entry appearing in the column marked $x\overline{y} + z$.

While any Boolean expression describes, with our interpretation, a Boolean function, we have not yet shown that any Boolean function can be described by an expression. In the process of showing this to be true, we will derive some standard-form expressions for describing Boolean functions. Consider the Boolean function F of x, y, and z given by the truth table of Fig. 6.5-4a. To find an expression describing this function, we will decompose the function into "simpler" functions, find expressions for the simpler functions, and then construct the desired expression. Figure 6.5-4*b* shows four *functions*, F_1, F_2, F_3, and F_4, each of which maps only one combination of the variables x, y, and z into 1. If we had *expressions* f_1, f_2, f_3, f_4 describing these four functions, the expression $f_1 + f_2 + f_3 + f_4$ would describe the function F.

x	y	z	F
0	0	0	0
0	0	1	1
0	1	0	1
0	1	1	0
1	0	0	0
1	0	1	1
1	1	0	0
1	1	1	1

(*a*)

x	y	z	F_1	F_2	F_3	F_4
0	0	0	0	0	0	0
0	0	1	1	0	0	0
0	1	0	0	1	0	0
0	1	1	0	0	0	0
1	0	0	0	0	0	0
1	0	1	0	0	1	0
1	1	0	0	0	0	0
1	1	1	0	0	0	1

(*b*)

FIG. 6.5-4 A Boolean function and its decomposition into four functions such that $F = F_1 + F_2 + F_3 + F_4$.

If we examine the function described by xyz, we see that it has value 1 if and only if x *and* y *and* z have value 1. Likewise, the "product" $x\bar{y}z$ has value 1 if and only if x and $\bar{y}$ and z have value 1, implying that $x=1$, $y=0$, and $z=1$. Thus the expression xyz describes the function F_4, and $x\bar{y}z$ describes the function F_3. To describe F_2, we need an expression that is 1 if and only if $x = 0$, $y = 1$, and $z = 0$; $\bar{x}y\bar{z}$ is such an expression. It is also easy to verify that there is no other three fold product chosen from x, $\bar{x}$, y, $\bar{y}$, z, and $\bar{z}$ that has this property. Finally $\bar{x}\bar{y}z$ describes F_1. Then the expression below describes the function F of Fig. 6.5-4*a*:

$$\bar{x}\bar{y}z + \bar{x}y\bar{z} + x\bar{y}z + xyz = F(x,y,z)$$

As another example, the reader should verify that the function G given in Fig. 6.5-5 is described by the expression shown in that figure.

x	y	z	G
0	0	0	0
0	0	1	1
0	1	0	0
0	1	1	1
1	0	0	0
1	0	1	1
1	1	0	1
1	1	1	1

$$G(x,y,z) = \bar{x}\bar{y}z + \bar{x}yz + x\bar{y}z + xy\bar{z} + xyz$$

FIG. 6.5-5 A Boolean function G and an expression describing G.

Without formalizing the procedure, it should be clear that this method will yield an expression describing any function. We will call this form the *standard sum of products* form or *normal disjunctive* form. That the expression is a sum of products is obvious. The word standard applies to the fact that each product term contains *exactly* one appearance of each variable, either complemented or uncomplemented. We can now state:

THEOREM 6.5-1 Every Boolean function can be described by a standard sum-of-products expression, and every Boolean expression can be transformed into a standard sum-of-products expression that describes the same function.

PROOF: We will accept the procedure described above as proof of the first statement. To put any expression in standard sum-of-products form, find the function described by the expression, and then apply the above procedure.

Q.E.D.

An alternate but symmetric approach to the derivation of expressions gives the *standard product of sums*. This approach focuses on the domain elements for which the function is *zero*. The given function is decomposed into a product of functions, each of which is zero for exactly one domain element. The standard product-of-sums expression for the function of Fig. 6.5-3 is

$$(x + y + z)(x + \bar{y} + z)(\bar{x} + \bar{y} + z)$$

This form is also called the *normal conjunctive form*.

It is expedient to define the results of certain operations on functions in terms of operations on expressions for the functions. We will define the sum $F_1 + F_2$ of two functions to be the function described by the sum of an expression for F_1 and an expression for F_2. Likewise the product F_1F_2 of two functions is a function described by the product of an expression for F_1 and an expression for F_2. Finally, the complement $\bar{F}$ of a function is the function described by the complement of an expression for the function F. For example, if $F = x_1\bar{x}_2(x_3 + x_4)$, then $\bar{F} = \overline{x_1\bar{x}_2(x_3 + x_4)} = \overline{x_1\bar{x}_2} + \overline{(x_3 + x_4)} = \bar{x}_1 + x_2 + \bar{x}_3\bar{x}_4$. Here we have made repeated use of Theorems 6.4-8 and 6.4-9 to obtain the complement of the expression for F.

The next section introduces the use of Boolean functions and expressions as descriptive tools for combinational logic circuits.

6.6 COMBINATIONAL LOGIC AND BOOLEAN FUNCTIONS

The material in this section is most appropriately introduced by an example.

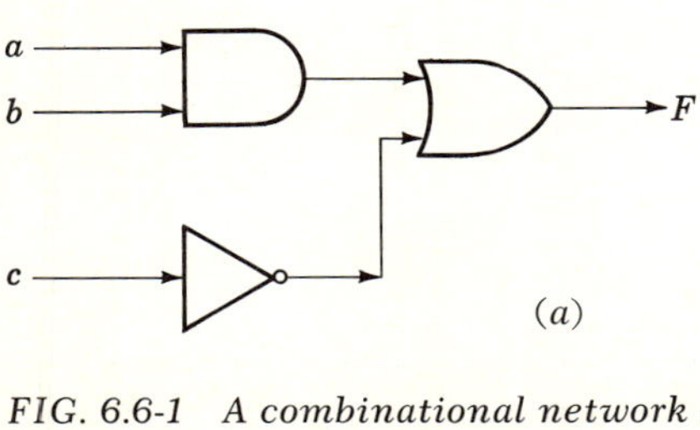

(*a*)

a	b	c	F
L	L	L	H
L	L	H	L
L	H	L	H
L	H	H	L
H	L	L	H
H	L	H	L
H	H	L	H
H	H	H	H

(*b*)

FIG. 6.6-1 A combinational network and its input-output behavior.

Example 6.6-1 *A Combinational Network* Consider the combinational network shown in Fig. 6.6-1*a*. The input-output description of this network is given in Fig. 6.6-1*b*. Our goal is to find a Boolean expression that "describes" this terminal behavior. Let us transform the table of Fig. 6.6-1*b* by assigning the Boolean element 0 to L and the element 1 to H. Doing so gives the table of Fig.

a	b	c	F
0	0	0	1
0	0	1	0
0	1	0	1
0	1	1	0
1	0	0	1
1	0	1	0
1	1	0	1
1	1	1	1

(*a*)

$$F = (a + b + \bar{c})(a + \bar{b} + \bar{c})(\bar{a} + b + \bar{c})$$

(*b*)

FIG. 6.6-2 A corresponding function and a describing expression.

6.6-2*a*. The motivation for this transformation should be obvious. Figure 6.6-2*b* gives an expression for this function. Observe that this Boolean function describes the logic in the sense that if we reverse the transformation, we get back the table of Fig. 6.6-1*b*.

In the above example, we chose a simple way of assigning Boolean constants to L and H. There is no particular reason for assigning 0 to L; it could just as well have been assigned to H, with 1 assigned to L. Furthermore, the choice of assigning 0 to L or to H could have been made independently for each of the columns a, b, c, and F. There are 2^4 such choices, but we will be concerned with only two. If the function is derived from the combinational logic by assigning 0 to L and 1 to H, the resulting function is called a *positive logic interpretation* of the logic. If the function is derived from the combinational logic by assigning 0 to H and 1 to L, the resulting function is called a *negative logic* interpretation. The function† given in Fig. 6.6-2*b* is a positive logic interpretation of the logic of Fig. 6.6-1*a*. A negative logic truth table and the corresponding

a	b	c	G
1	1	1	0
1	1	0	1
1	0	1	0
1	0	0	1
0	1	1	0
0	1	0	1
0	0	1	0
0	0	0	0

(*a*)

$$G = ab\bar{c} + a\bar{b}\bar{c} + \bar{a}b\bar{c}$$

(*b*)

FIG. 6.6-3 Negative logic interpretation of the table of Fig. 6.6-1b and a corresponding expression.

†We will sometimes follow the common practice of calling an expression a function. What we mean, of course, is the function described or denoted by the expression.

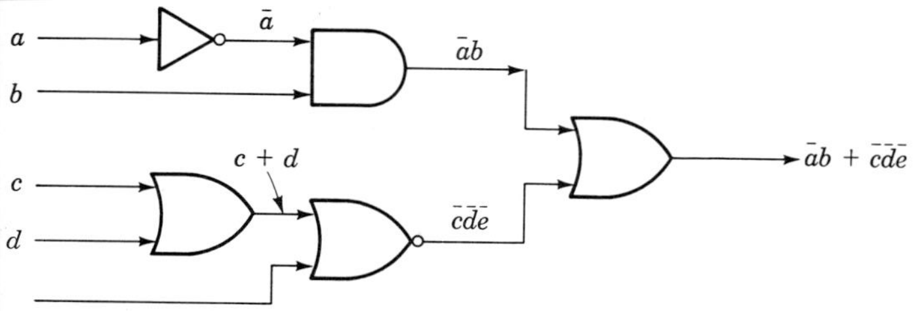

FIG. 6.6-5 Finding a Boolean expression.

†Throughout this chapter we use arrows to facilitate "reading" the diagrams. In engineering practice such arrows are usually omitted.

function are given in Fig. 6.6-3*a* and *b*. Observe that the functions of Figs. 6.6-2*b* and 6.6-3*b* are not the same.

The above observations can be generalized, permitting us (1) to give a Boolean expression for any single-output explicit combinational network and (2) to design a combinational network corresponding to any Boolean expression. We proceed with the formal definition of combinational logic.

The element set E along with the positive and negative logic Boolean expressions for each element or gate is shown in Fig. 6.6-4.

Implicit combinational networks behave like the implicit block diagram systems of Chap. 3, in the sense that their input-output behavior is characterized by relations and not by functions. Accordingly, for a given combination of input values either (1) no output value will be possible, or (2) there will be exactly one output value, or (3) there will be more than one possible output values. Two examples illustrating these three cases are shown in Fig. 6.6-6*a* and *b*.

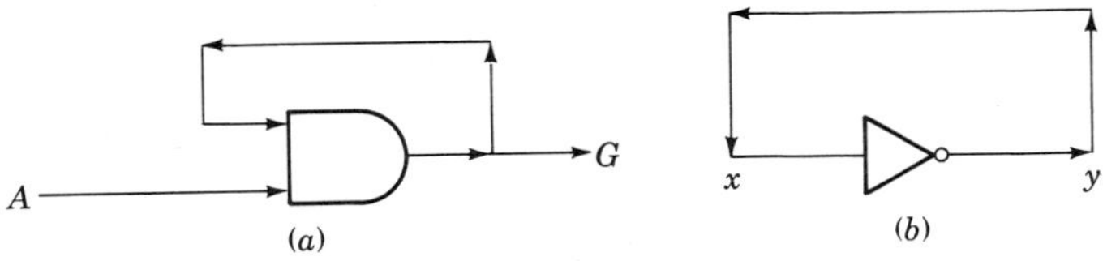

FIG. 6.6-6 Examples of implicit combinational networks.

Observe that under a positive logic interpretation, if $A = 0$ in the network of Fig. 6.6-6*a*, then G can only be 0; hence there is *exactly one output* value. If, however, $A = 1$, then either $G = 0$ or $G = 1$ are the *two possible output values*. Finally, observe that in the network of Fig. 6.6-6*b*, regardless of whether x is 0 or 1, there is *no possible value* for y. More formally, and in terms of our discussion of Chap. 3, the relation characterizing the network of Fig. 6.6-6*a* is $\{(0,0),(1,0),(1,1)\}$; i.e., the only possible pairs of values (A,G) are the three indicated in braces, while the relation characterizing the network of Fig. 6.6-6*b* is $\{\ \}$, i.e., the null relation which has no possible pairs of (x,y) values. The reader is undoubtedly concerned at this point with the relationship of those idealized models to reality. What would happen, for example, if the two networks of Fig. 6.6-6 were wired out of commercially available gates? In Fig. 6.6-6*a*, setting A to 1 would "latch" G to either 1 or 0, depending on (1) the way the input $A = 1$ was impressed in time, and (2) the relationship of time delays present in the AND gate and in the rest of the network. This implies the presence of delays in the real network which were not taken into account in the model of Fig. 6.6-6*a*. Hence the network must be modeled as a digital system *with memory*, which is the topic of Chap. 10. As we shall see in Chap. 10, basic memory elements such as flip-flops are intentionally constructed from looped interconnections of logic gates and (inherent) delays. To complete the argument, if a commercially available inverter were connected as in Fig. 6.6-6*b*, it would either burn up, or exhibit an output which is neither 0 nor 1 but somewhere in between.

Now assume that we are given the simple function $F = a + b$ and asked to find an interconnection of gates that is described by this function. Two problems arise. First, since no logic interpretation is given, we would feel free to say that either an OR gate with a positive logic interpretation or an AND gate with a negative logic interpretation is a suitable answer. Second, although the circuit of Fig. 6.6-7 is a positive logic realization, it is probably less desirable than a single OR gate. The first of these problems can be easily handled; the second is the topic of the next section.

DEFINITION 6.6-1 A combinational circuit is said to be a positive (negative) logic realization of a function $F(x_1, \ldots, x_n)$ if the positive (negative) logic interpretation of the circuit is equal to $F(x_1, \ldots, x_n)$.

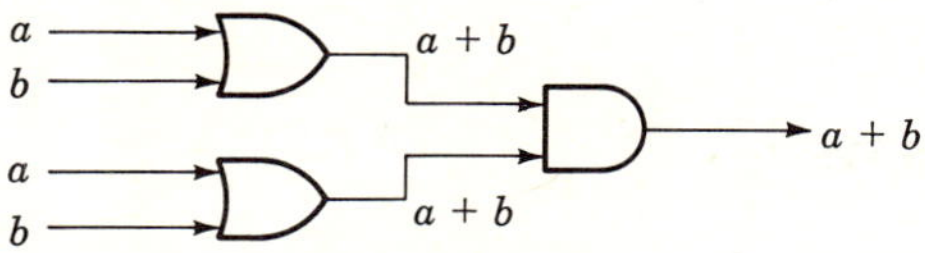

Fig. 6.6-7 A realization of $f = a + b$.

In the remainder of this book we will be concerned only with positive logic realizations.

It is easy to show, by example, how to find a realization of any function. Consider the function $F = \bar{x}_1 x_2 \bar{x}_3 + x_1 x_2 \bar{x}_3 + x_1 \bar{x}_2 \bar{x}_3 + \bar{x}_1 x_2 x_3 + x_1 x_2 x_3$, which is written in *standard sum of products* form. A realization of each product term can be achieved with inverters and a single AND gate, as shown for the term $\bar{x}_1 x_2 \bar{x}_3$ in Fig. 6.6-8. Realizations of this form can be used with a single OR gate, as shown in Fig. 6.6-9, to create a realization of F. This realization has two levels (excluding input inverters), and all the gates in the first level

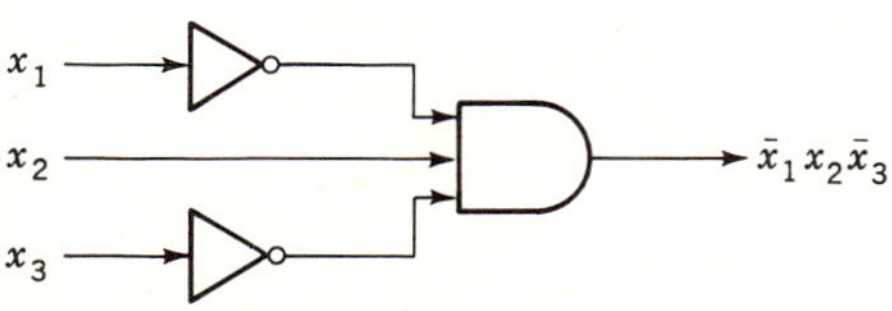

FIG. 6.6-8 Realization of a product term.

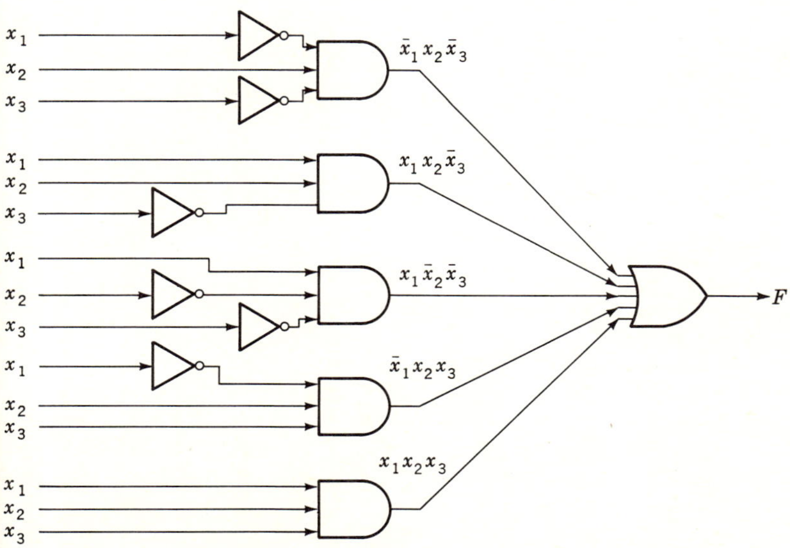

FIG. 6.6-9 AND-OR realization of F.

are of the same variety (AND gates). Realizations of this form are called *two-level AND-OR realizations.* The procedure used in this example is completely general, and we can state that *any Boolean function has a two-level AND-OR realization* derivable from the standard sum-of-products expression for the function.

An analogous realization can be derived from the standard product of sums and is called a *two-level OR-AND realization.* The standard product-of-sums expression for the function F of Fig. 6.6-9 is $(x_1 + x_2 + x_3) \cdot (x_1 + x_2 + \bar{x}_3) \cdot (\bar{x}_1 + x_2 + \bar{x}_3)$; and any term, $(x_1 + x_2 + \bar{x}_3)$ for example, can be realized as shown in Fig. 6.6-10. Realizations of this form can be combined, using an AND gate to

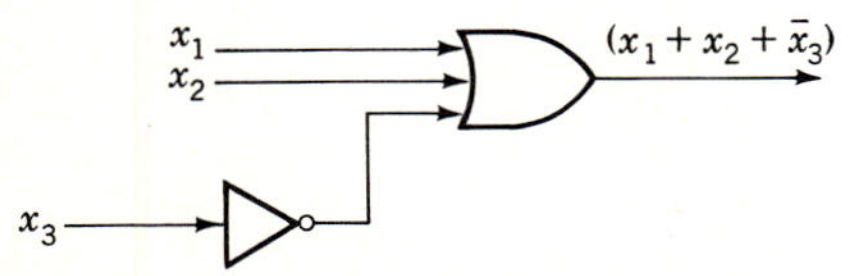

FIG. 6.6-10 Realization of a sum term.

realize F in an alternate form as shown in Fig. 6.6-11. This procedure can be applied to any standard product of sums and gives a two level OR-AND realization.

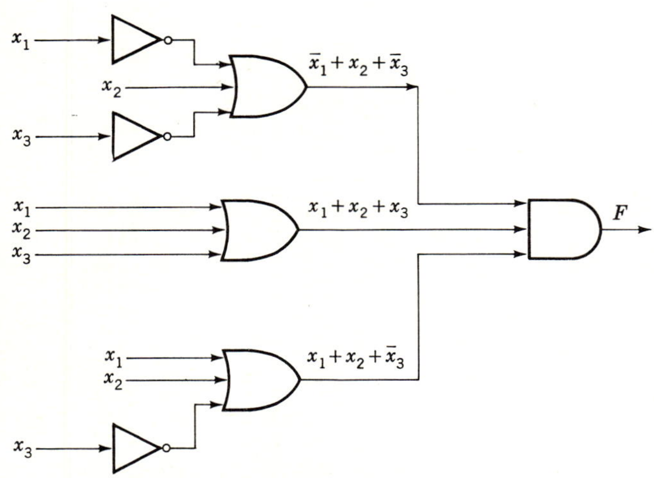

FIG. 6.6-11 OR-AND realization of F.

The two-level AND-OR and OR-AND realizations are not the only ways to realize Boolean functions. In particular, given *any Boolean expression,* we can construct a network from two-input AND-OR gates and from inverters. This follows from Definition 6.5-2 of Boolean expressions. That is, any Boolean expression is, by definition, of the form $\overline{E}_1$, $E_1 + E_2$, $E_1 \cdot E_2$, a constant (0 or 1) or a variable. Thus any Boolean expression can be realized at the output of an inverter, an OR gate, an AND gate, as a constant input or as an input variable. If such an expression is realized by a gate output, then the gate inputs will be expressions; hence they will be realizable by yet other gates or constants or input variables. This process is guaranteed to result in an *explicit* combinational network that realizes any given Boolean expression. For example, suppose we wish to realize the expression

$$F=((\overline{\overline{A}\cdot B})\cdot(C\overline{D})+A)+1$$

Straightforward application of the above procedure yields the network of Fig.

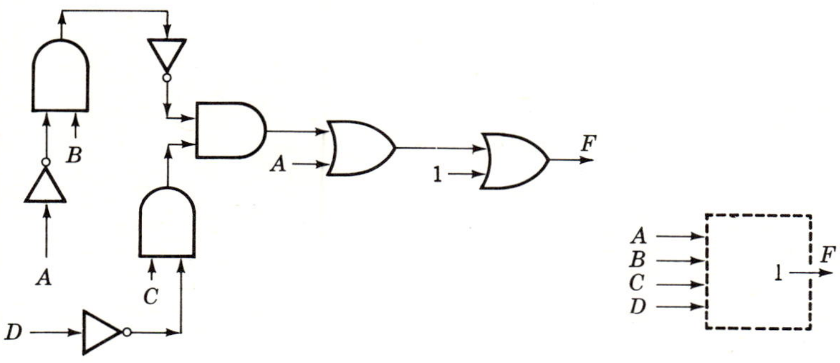

(a) (b)

FIG. 6.6-12 Realization of F.

6.6-12*a*. The reader should have no difficulty deducing that the expression 1 (constant) denotes the same Boolean function as *F*. That is, the systems of Fig. 6.6-12*a* and *b* are *equivalent* (as defined in Chap. 3).

Since a given Boolean function can be characterized by many (in fact, an infinite number of) expressions and *equivalent* combinational networks, we are confronted with the challenging task of designing a network that is economical in some sense and that realizes a given Boolean function. For instance, in designing a hardware multiplier for a computer, we would like to achieve multiplication with a combinational network which is as fast as possible and as economical as possible. Although we have some procedures, discussed in the next section, for conducting such synthesis, the problem is basically very complex and unsolved in its general form.

Certainly a given function has a large number of different realizations. Theoretically, this number is infinite, since an arbitrary number of redundant gates may be added to any realization. The purpose of the next section is to introduce a reasonable way for finding realizations that are "good" in a sense to be shortly defined.

6.7 MINIMIZATION OF BOOLEAN FUNCTIONS

Since gates cost money, dissipate heat, take up space, and have a finite delay time, it is desirable to find realizations that are minimal in some sense. The circuit shown in Fig. 6.7-1 is another realization of the function *F* realized by the gate network of Fig. 6.6-9. The circuit of Fig. 6.7-1 is intuitively a "better" realization than that of Fig. 6.6-9. For example, the realization in Fig. 6.7-1 requires only three gates while that of Fig. 6.6-9 requires 12 gates.

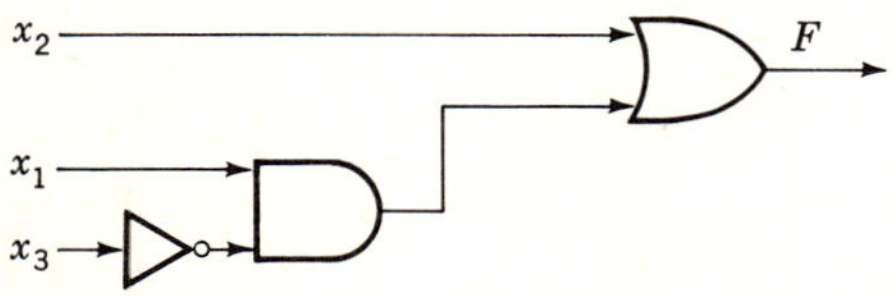

FIG. 6.7-1 A more economical system equivalent to that of Fig. 6.6-9.

The aim of this section is to demonstrate how certain algebraic manipulations can lead to realizations such as that in Fig. 6.7-1. We can illustrate how these manipulations occur, starting with the standard sum of products for F, given following Definition 6.6-1:

$$\bar{x}_1x_2\bar{x}_3 + \bar{x}_1x_2x_3 + x_1x_2x_3 + x_1x_2\bar{x}_3 + x_1\bar{x}_2\bar{x}_3$$

This expression can be rewritten and simplified as shown below, where the term $x_1x_2\bar{x}_3$ has been repeated.

$$\begin{aligned}\bar{x}_1x_2(\bar{x}_3 + x_3) + x_1x_2(x_3 + \bar{x}_3) + x_1\bar{x}_3(x_2 + \bar{x}_2)\\ &= \bar{x}_1x_2 + x_1x_2 + x_1\bar{x}_3\\ &= x_2(\bar{x}_1 + x_1) + x_1\bar{x}_3\\ &= x_2 + x_1\bar{x}_3\end{aligned}$$

Note that the simple realization of Fig. 6.7-1 is an obvious realization of the expression $x_2 + x_1\bar{x}_3$. The rule used for simplification is: Two product terms that are identical, except for the appearance of a variable in one and the complement of the variable in the other, are factored, and the relation $x + \bar{x} = 1$ is repeatedly used. While these manipulations can be carried out directly on the expression,† a much more convenient representation called the *Karnaugh map* can be used for expressions of few (up to six) variables.

Karnaugh maps are a convenient vehicle for the *graphical* representation of Boolean functions. Figure 6.7-2 shows some Karnaugh maps. Each square of a Karnaugh map is labeled with coordinates corresponding to variable assignments. The map of Fig. 6.7-2*a* is used to represent functions of two variables, while those of Fig. 6.7-2*b* and *c* are for functions of three and four variables, respectively. The entries in Fig. 6.7-2*c* are for later reference. To represent a function, the value of the function for a given combination of variable values is recorded in the appropriate square. The Karnaugh map for the function of Fig. 6.7-3*a* is given in Fig. 6.7-3*b*. Quite often just the 1s of the function are entered, with the blank entries read as 0s.

†And are done this way in minimizing expressions of many variables by computer.

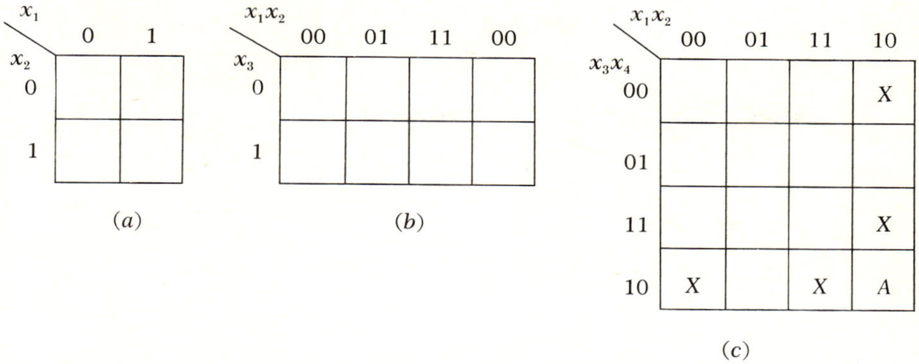

FIG. 6.7-2 Karnaugh maps of two, three, and four variables.

x	y	z	F
0	0	0	0
0	0	1	1
0	1	0	0
0	1	1	1
1	0	0	1
1	0	1	1
1	1	0	0
1	1	1	1

(a)

$F(x \cdot y \cdot z) = x\bar{y} + z$

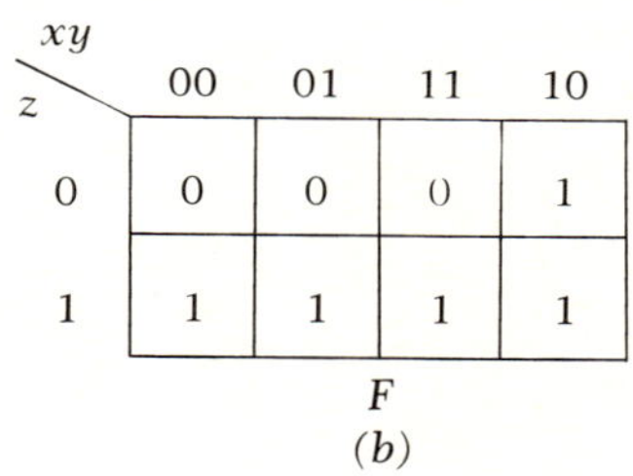

FIG. 6.7-3 A Boolean function and its Karnaugh map display.

The utility of Karnaugh maps for minimization hinges on the particular manner in which the coordinates are arranged. The squares marked *A* and X in Fig. 6.7-2*c* will be used to describe this arrangement. Observe that the coordinates of each square marked X differ in the value of *only* one variable from the coordinates of square *A*. Also notice that each X square is physically adjacent to square *A* if the map is thought of as closed on itself both horizontally and vertically. Alternatively, one can think of the map of Fig. 6.7-2*c* as being on a torus or donut. Two squares that are horizontally or vertically adjacent have coordinates that differ in only one variable value. We encourage the reader to investigate these adjacencies and especially to think of the *end-around* nature of the labeling.

To illustrate the use of the Karnaugh map, consider the function *G* whose map appears in Fig. 6.7-4. Observe that the product terms associated with the 1s in squares *C* and *D* are $\bar{x}_1\bar{x}_2\bar{x}_3x_4$ and $\bar{x}_1x_2\bar{x}_3x_4$, respectively.

Notice, however, that the sum of the above two terms can be written as

$$\bar{x}_1\bar{x}_2\bar{x}_3x_4 + \bar{x}_1x_2\bar{x}_3x_4 = \bar{x}_1\bar{x}_3x_4(\bar{x}_2 + x_2) = \bar{x}_1\bar{x}_3x_4$$

x_3x_4 \ x_1x_2	00	01	11	10
00		1_A	1_B	
01	1_C	1_D	1_E	
11		1_F	1_G	1_H
10		1_J	1_K	1_L

G

FIG. 6.7-4 Karnaugh map of G.

using distributivity and the fact that $x + \bar{x} = 1$ as we did in the introductory example. This reduction makes "graphical sense" too, since the ones of squares C and D are adjacent. Hence if $x_1 = 0$, $x_3 = 0$ and $x_4 = 1$, regardless of whether x_2 is 1 or 0, G is 1. Reasoning in a similar fashion, we can examine the product terms corresponding to the squares G, H, K, and L. The sum of these four terms can be reduced as follows:

$$x_1x_2x_3x_4 + x_1\bar{x}_2x_3x_4 + x_1x_2x_3\bar{x}_4 + x_1\bar{x}_2x_3\bar{x}_4$$

$$= x_1x_3[x_2x_4 + x_2\bar{x}_4 + \bar{x}_2x_4 + \bar{x}_2\bar{x}_4]$$

$$= x_1x_3[x_2(x_4 + \bar{x}_4) + \bar{x}_2(x_4 + \bar{x}_4]$$

$$= x_1x_3(x_2 + \bar{x}_2) = x_1x_3$$

This reduction can be interpreted by observing that G and H are adjacent, and hence x_2 can be eliminated from an expression for these two 1s; while K and L are also adjacent, and x_2 can again be eliminated. Then, because of the relative placements of these groups, x_4 can be eliminated from an expression representing all four 1s.

The reader should verify that, by similar reasoning, the sum corresponding to the eight 1s in squares A, B, D, E, F, G, J, and K can be represented simply by x_2. Finally, since we have taken care of *every* 1 of the functions in *at least one* product, we can write the function G of Fig. 6.7-4 as $\bar{x}_1\bar{x}_3x_4 + x_1x_3 + x_2$.

The 1s from which the three terms originated are shown as graphical "patches" in Fig. 6.7-5. These patches have two properties: (1) the sum of the standard product terms corresponding to the 1s in each patch can be represented by a simple product of some variables (either complemented or uncomplemented), and (2) patches contain only 1s, and every 1 of the function is included in at least one patch. For every set of patches that satisfies these two conditions for a function G, it is the case that G can be represented as a sum of the product terms associated with the patches. The resulting expression is known as a *sum of products* expression for G. It is to expressions of this nature that we wish to assign a measure of "goodness" or minimality.

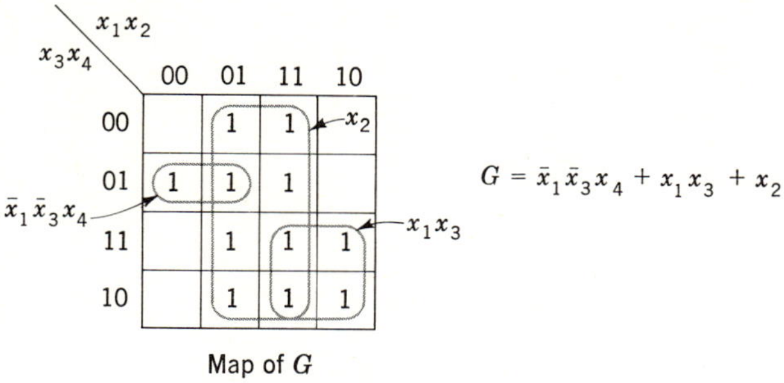

FIG. 6.7-5 Patches corresponding to the three terms of G.

The measure of minimality or cost we will use can best be explained by example. Consider two different sum-of-products expressions for the same function. If they have different numbers of product terms, the expression with the fewer product terms is of lower cost. If they have the same number of product terms, then the expression with the fewer literals is of lower cost. A *literal* is an occurrence of a variable, either complemented or uncomplemented; e.g., the expression $\bar{x}_1\bar{x}_3x_4 + x_1x_3 + x_2$ contains six literals. A *minimal sum of products* expression for a function is any expression such that there exist no expressions of lower cost according to the above criteria. It is necessary to introduce some additional concepts before we describe how to find minimal-cost sum-of-product expressions.

DEFINITION 6.7-1 Let $F(x_1, \ldots, x_n)$ and $G(x_1, \ldots, x_n)$ be Boolean functions. F includes G, and we write $F \geq G$, if and only if F can be expressed as $F = G + H$, where H is another function. If $F \geq G$ and $F \neq G$, then we write $F > G$. If it is not the case that $F \geq G$, then we write $F \ngeq G$.

Informally, $F \geq G$ means that $G = 1$ implies $F = 1$. In Fig. 6.7-5, the function represented includes the products associated with each patch. If any function is written as a sum of products, the function must include each product.

DEFINITION 6.7-2 Let $F(x_1, \ldots, x_n)$ be a Boolean function. A *prime implicant* p of F is a product of literals (with each variable appearing at most once) such that (1) $F \geq p$, and (2) if *any* literal is removed from p to give a product p', then $F \ngeq p'$.

To illustrate, let us determine whether $\bar{x}_1x_2x_4$ is a prime implicant of the function H of Fig. 6.7-6. This product clearly satisfies the first condition since it is 1 only for squares B and D where H is 1. It does not satisfy the second condition, however, because if the literal x_2 is removed, the product $\bar{x}_1x_4$ is still included by H, i.e., $H > \bar{x}_1x_4$. The reader can easily verify that $\bar{x}_1x_4$ is a prime implicant of G. Informally then, *a prime implicant of H is a patch of ones of H that is not contained in yet a larger patch of ones of H*. All prime implicants of H are indicated by solid patches in Fig. 6.7-6.

We can now give a practical procedure for finding all prime impli-

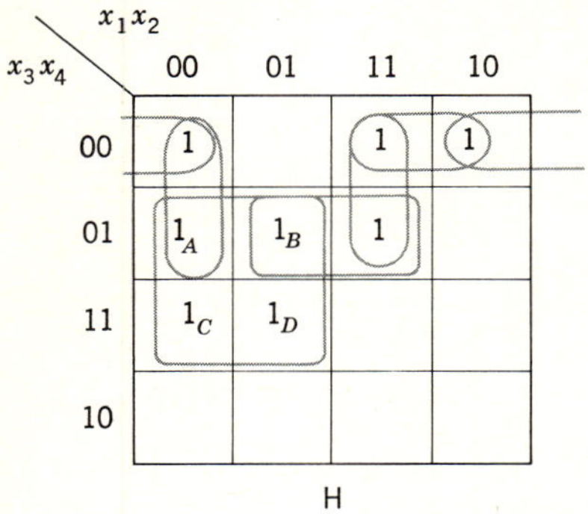

FIG. 6.7-6 Prime implicants of H. Term $\bar{x}_1x_2x_4$ is not a prime implicant.

cants of a function. Take each 1 of the function, and find the largest patch or patches of 1s associated with a product of literals that includes the particular 1. Note that every patch corresponding to a product of literals contains 2^k ones, where $k=0, 1, 2, \ldots$. Thus, for $k=0$, a patch contains a single 1 if there are no adjacent 1s. The 1 in the upper left-hand corner of the map of Fig. 6.7-6 belongs to two prime implicants, each represented by a threefold product. Figure 6.7-7 shows some possible patches that may represent prime implicants.

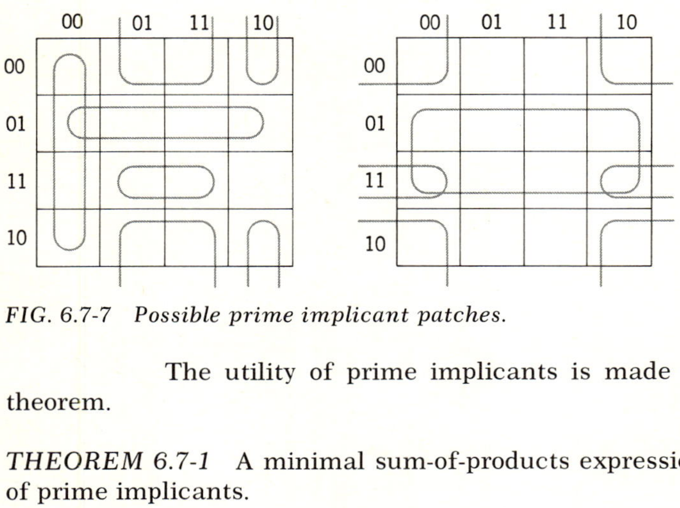

FIG. 6.7-7 Possible prime implicant patches.

The utility of prime implicants is made clear by the following theorem.

THEOREM 6.7-1 A minimal sum-of-products expression is necessarily a sum of prime implicants.

PROOF: Assume that we have a sum-of-products expression for a function F and that at least one product term s is not a prime implicant of F. It must be the case that $F \geq s$, or the expression would be incorrect. But since s is not a prime implicant, at least one literal can be eliminated to produce s'. Then $F \geq s' > s$, and s' can replace s in the original expression to produce an expression that is of lower cost. This contradicts the assumption that the original expression was minimal. The process can be repeated until all products are reduced to prime implicants.

Q.E.D.

Theorem 6.7-1 does not say that any sum of prime implicants is a minimal sum of products. It does say, however, that at least one sum of prime implicants is a minimal sum of products. Certainly to find this minimal expression we need consider only sums in which each prime implicant appears at most once. Since all functions have a finite number of prime implicants, we can formulate an effective† procedure for finding all minimal sums of products for a function F as follows: (1) Find all prime implicants p_i of F; (2) examine all sums of prime implicants that represent the function and for which each prime implicant appears at most once, and select those sums that have minimal cost. Those sums are all minimal sums of products.

Since we have all the prime implicants of the function H of Fig. 6.7-6, we can illustrate the second step of the above procedure. Certainly, the prime implicant $\bar{x}_1x_4$ must be included in every minimal sum as it is the only one that includes the 1s in squares C and D. This leaves four uncovered 1s, and since all the remaining prime implicants cover only two 1s, the best we can do is to use two more prime implicants. We will choose $x_1x_2\bar{x}_3$ and $\bar{x}_2\bar{x}_3\bar{x}_4$. The prime implicants chosen represent the function and give the expression and patches shown in Fig. 6.7-8. For this example the choice of prime implicants giving the minimal sum of products is unique. This is not always true. In particular, Figure 6.7-9*a* gives all prime implicants of a function G for which the choice is

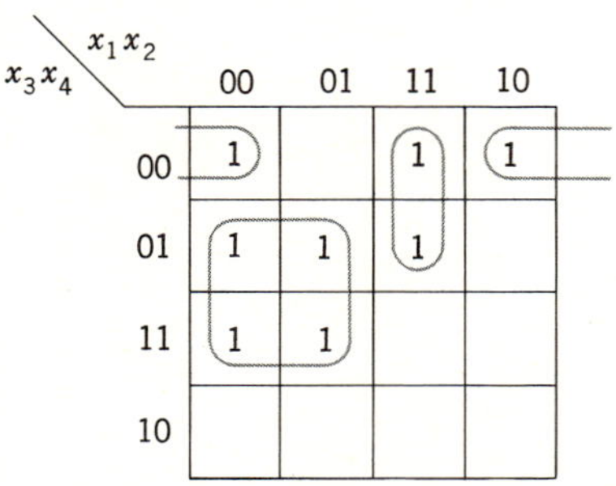

FIG. 6.7-8 Minimization of H: $H = \bar{x}_1x_4 + x_1x_2\bar{x}_3 + \bar{x}_2\bar{x}_3\bar{x}_4$.

not unique. The *two* minimal sums of products are shown in Fig. 6.7-9*b* and *c* and clearly have equal cost.

The examples used have had four variables for two reasons. Two- and three-variable functions do not offer enough variety, and with more than four variables Karnaugh maps get cumbersome. Karnaugh maps are rarely used with more than six variables, and other techniques must then be employed. Procedures for producing either one or all minimal sums of products are ideally

†A procedure that can be carried out by computer and is guaranteed to terminate in finite time.

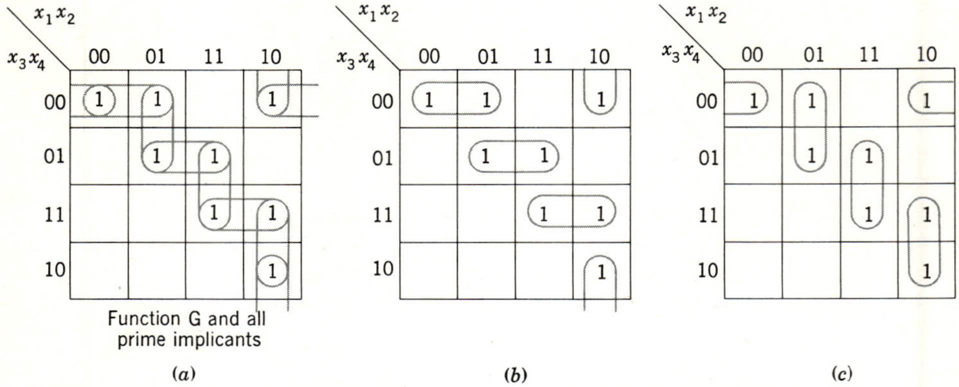

FIG. 6.7-9 Function G and two minimal sum-of-product realizations.

suited for digital computers, and programs exist to perform the computation for a large number of variables. One widely used technique is known as the Quine-McCluskey procedure.†

A procedure analogous to the one presented above can be used to derive *minimal products of sums.* This approach is based on the fact that $(x_1 + x_2 + \bar{x}_3) \cdot (\bar{x}_1 + x_2 + \bar{x}_3) = (x_2 + \bar{x}_3)$. The sum terms that correspond to prime implicants are traditionally called *dual prime implicants* (even though they are not the duals of the prime implicants). The reader should be able to construct the entire procedure with the help of the following example. Figure 6.7-10 gives the dual prime implicants for the function H of Fig. 6.7-8. In Fig. 6.7-10 no dual prime implicants can be omitted from any expression, and $H = (x_1 + \bar{x}_2 + x_4)(\bar{x}_1 + x_2 + \bar{x}_4)(\bar{x}_1 + \bar{x}_3)(\bar{x}_3 + x_4)$ is a minimal product of sums.

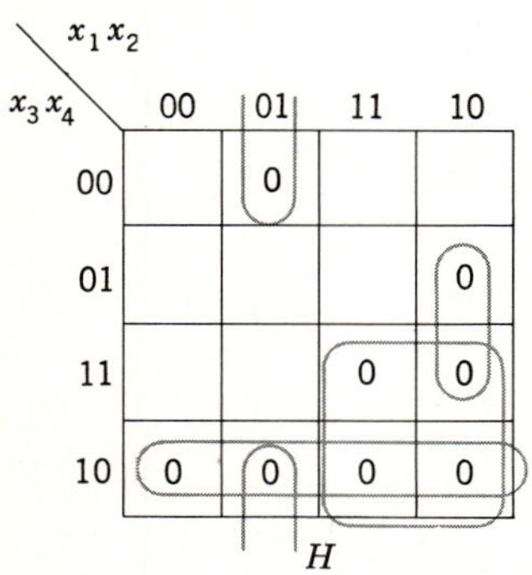

FIG. 6.7-10 "Dual" prime implicants of H.

The use of minimal sum-of-products or product-of-sums expressions for deriving economical realizations of Boolean functions is straightforward. Consider the two-level AND-OR realization derived in the obvious manner

†See Prob. 6.7-8.

from a minimal sum-of-products expression for a function. Of all such realizations (two-level AND-OR) no realization has fewer AND and OR gates than one derived from a minimal sum-of-products expression. Figure 6.7-11 shows such a realization for the function H of Fig. 6.7-8, using the minimal sum-of-products expression $\bar{x}_1x_4 + x_1x_2\bar{x}_3 + \bar{x}_2\bar{x}_3\bar{x}_4$. Figure 6.7-12 shows a two-level OR-AND realization for H derived from the minimal product-of-sums expression associated with Fig. 6.7-10.

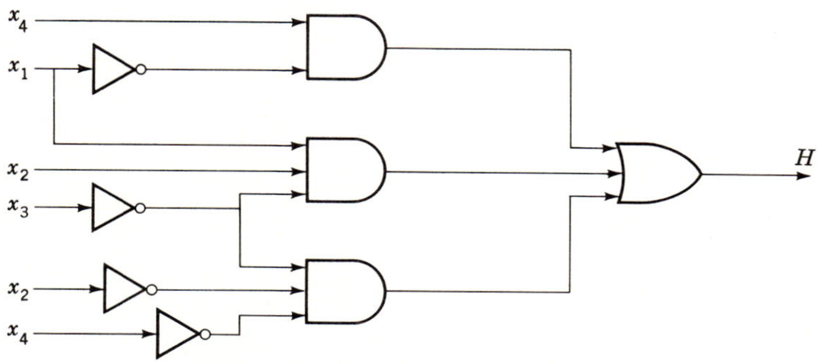

FIG. 6.7-11 Minimal AND-OR realization of H: $H = \bar{x}_1x_4 + x_1x_2\bar{x}_3 + \bar{x}_2\bar{x}_3\bar{x}_4$.

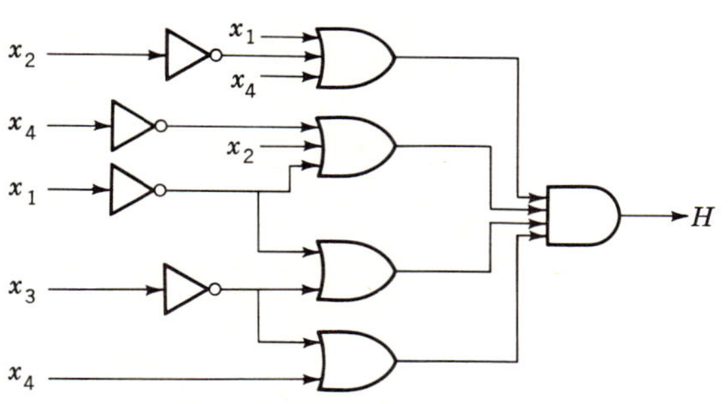

FIG. 6.7-12 Minimal OR-AND realization of H.

The expression associated with Fig. 6.7-11 can be modified in appearance by factoring to give $H = \bar{x}_1x_4 + \bar{x}_3(x_1x_2 + \bar{x}_2\bar{x}_4)$. If we use this expression as a guide for a realization, we arrive at the logic of Fig. 6.7-13. This realization is intuitively poorer than that of Fig. 6.7-11 for two reasons. First, it requires ten gates compared to the eight for the realization of Fig. 6.7-11. Second, since it is not two-level, signal changes at the input have more gates to pass through to get to the output, and hence more delay is encountered and the circuit is inherently slower. It does, however, have a redeeming characteristic in that it requires only two-input AND and OR gates. In certain design situations, only two-input gates are available, and factoring is a useful aid to finding realizations. Similar comments can be made about the minimal product-of-sums form. For example, the expression associated with Fig. 6.7-12 can be factored to yield $[x_4 + (x_1 + \bar{x}_2)\bar{x}_3]\,[\bar{x}_1 + (x_2 + \bar{x}_4)\bar{x}_3]$.

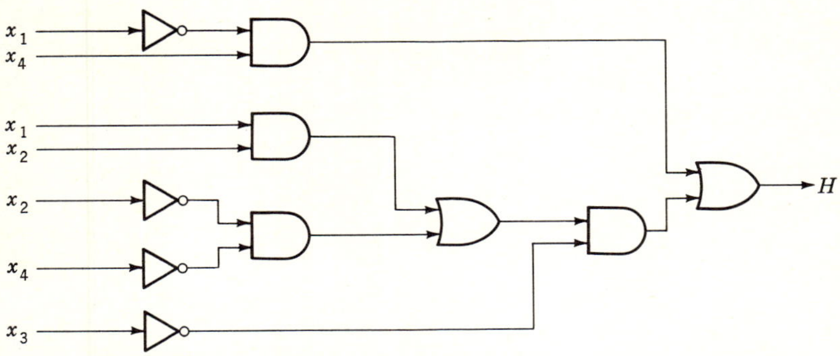

FIG. 6.7-13 An alternate realization of H using two-input AND and OR gates.

In some sense, the cost measure used in this section is artifically chosen to go with the concept of a sum of prime implicants representing the function. If a more realistic measure is chosen, such as the total number of gates, *with no topological constraint,*† then we know of no practical methods for finding a minimal realization. Thus, experience and ingenuity play an important part in the design of actual logic circuits. Furthermore, there are other practical issues that influence the choice of one realization over another. The introductory nature of this book does not permit a discussion of these issues.

At this point we are ready to apply the techniques we have developed to some complete design examples.

6.8 EXAMPLES, FUNCTIONAL COMPLETENESS, AND DON'T CARES

Two very common logic functions are the parity functions shown in Fig. 6.8-1*a* and *b*. Function *F* is called *odd parity,* and function *G* is called

x_1x_2	F
0 0	0
0 1	1
1 0	1
1 1	0

(*a*)

x_1x_2	G
0 0	1
0 1	0
1 0	0
1 1	1

(*b*)

FIG. 6.8-1 Odd and even parity functions.

even parity. Function G also detects equality of x_1 and x_2 by producing a 1 output. Figure 6.8-2 gives two-level realizations for each of the functions.

The realization of Fig. 6.8-2*a* will be now used to illustrate a powerful property of the NOR gate. (The positive logic interpretation of the NOR gate is given in Fig. 6.6-4.) First, observe that if two inverters are "inserted"

†That is, without a restriction such as two levels of gates.

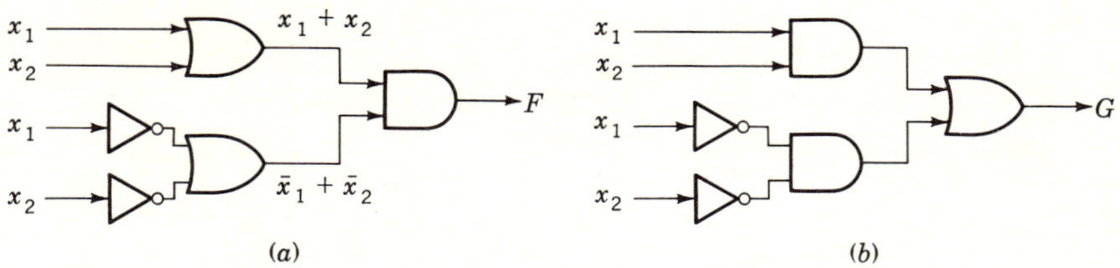

FIG. 6.8-2 *Two-level realization of* (a) *odd and* (b) *even parity.*

in series at every input of the AND gate of Fig. 6.8-2*a*, as shown in Fig. 6.8-3*a*, then the overall input-output behavior of the network remains unchanged.

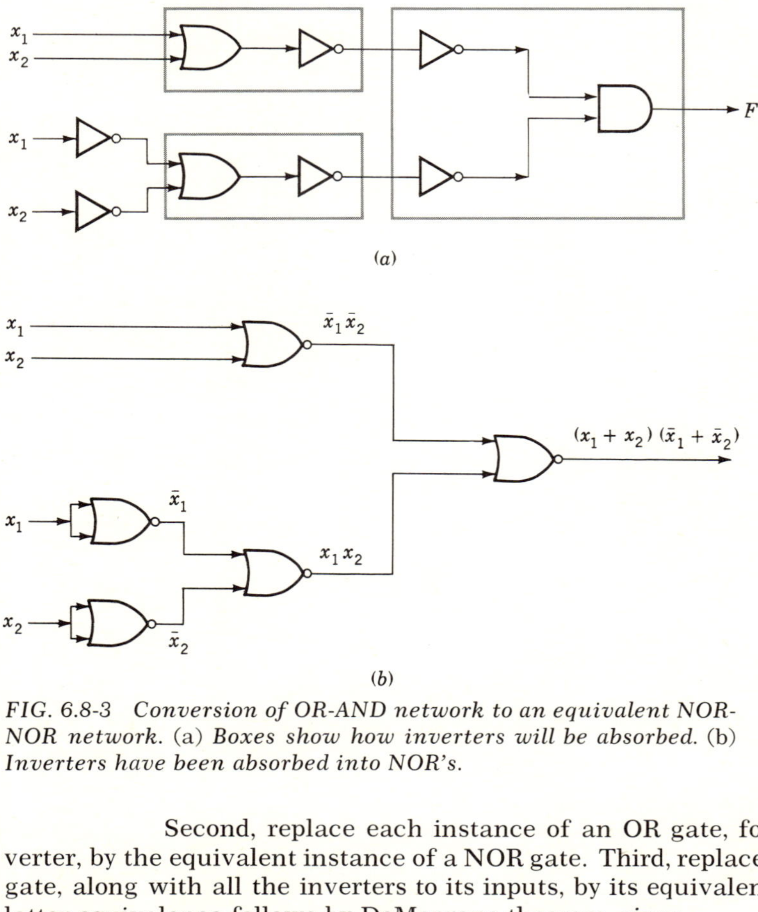

FIG. 6.8-3 *Conversion of OR-AND network to an equivalent NOR-NOR network.* (a) *Boxes show how inverters will be absorbed.* (b) *Inverters have been absorbed into NOR's.*

Second, replace each instance of an OR gate, followed by an inverter, by the equivalent instance of a NOR gate. Third, replace the output AND gate, along with all the inverters to its inputs, by its equivalent NOR gate; the latter equivalence follows by DeMorgans theorem, since

$$\bar{x}_1 \cdot \bar{x}_2 \cdot \bar{x}_3 \cdot \cdots \cdot \bar{x}_n = \overline{x_1 + x_2 + x_3 + \cdots + x_n}$$

Finally, replace any inverters that have not yet been "absorbed," by a NOR with both inputs tied together. Clearly, the resulting network, shown in Fig. 6.8-3*b*, consists entirely of NOR gates and is equivalent to the network of Fig. 6.8-3*a*. Certainly, this procedure can be applied to *any* two-level OR-AND network, resulting in a two-level NOR-NOR network with the same total number of gates. An alternative approach is shown in Fig. 6.8-4, where NOR-gate realizations of $\bar{a}$, ab, and $a + b$ are given. These realizations can be substituted for the INVERTER, AND, and OR gates in any realization of a function, to show

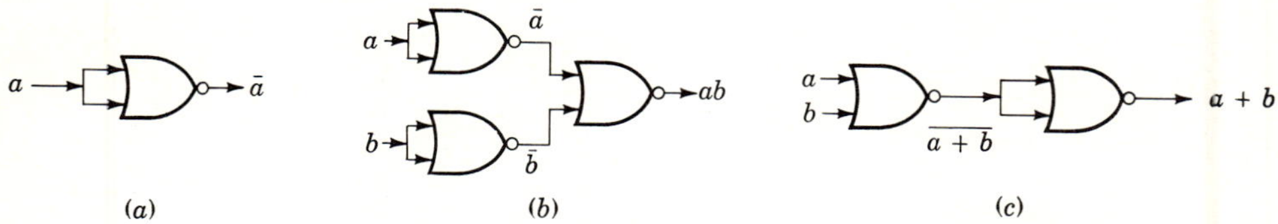

FIG. 6.8-4 Realization of (a) INVERTER, (b) AND, and (c) OR gates with NOR gates.

that NOR gates are sufficient to realize any Boolean function. We say that a set of gates is *functionally complete* if any Boolean function can be realized using gates from that set. Thus the set {AND,OR,NOT} is functionally complete, as is the set {NOR}. When a functionally complete set has only one gate we say that the gate is *universal*. The NAND gate is also universal, and the sets {AND, NOT}, {OR,NOT} are also functionally complete. The reader is invited to convince himself of these facts by synthesizing a known universal gate out of the gates in the above three sets. Knowing that a set of gates is functionally complete has important practical implications, since keeping these gates in stock will guarantee realization of *any* desired Boolean function.

The two-level AND-OR realization of odd parity F given in Fig. 6.8-5*a* can be converted to the NAND only realization shown in Fig. 6.8-5*b*. The

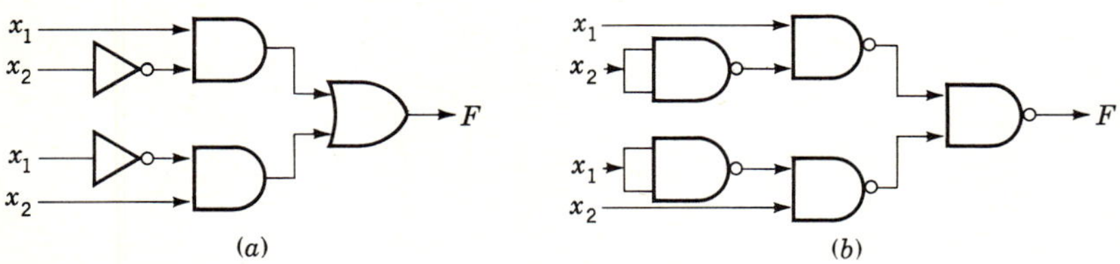

FIG. 6.8-5 AND-OR and NAND-NAND realization of F.

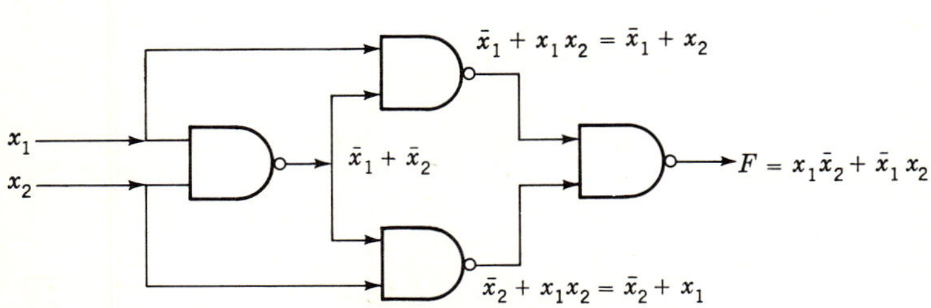

FIG. 6.8-6 An alternate multilevel NAND gate realization of F.

reader is encouraged to find the equivalent of Fig. 6.8-4 using only NAND gates. An alternate realization of F that requires only four NAND gates is shown in Fig. 6.8-6. This is an example of a realization that is not easily derived by algebraic manipulation.

We can now return to the digital multiplier specified in Fig. 6.3-12. Figure 6.8-7 shows the maps for outputs z_0, z_1, z_2, and z_3 of this example (assuming, as usual, a positive logic interpretation). We can write minimal sums of

$b_1b_0 \backslash a_1a_0$	00	01	11	10
00	0	0	0	0
01	0	0	0	0
11	0	0	1	0
10	0	0	0	0

z_3

$b_1b_0 \backslash a_1a_0$	00	01	11	10
00	0	0	0	0
01	0	0	0	0
11	0	0	0	1
10	0	0	1	1

z_2

$b_1b_0 \backslash a_1a_0$	00	01	11	10
00	0	0	0	0
01	0	0	1	1
11	0	1	0	1
10	0	1	1	0

z_1

$b_1b_0 \backslash a_1a_0$	00	01	11	10
00	0	0	0	0
01	0	1	1	0
11	0	1	1	0
10	0	0	0	0

z_0

FIG. 6.8-7 Input-output behavior of multiplier M.

products for these four functions, by inspection, as follows:

$$z_3 = a_1a_0b_1b_0$$

$$z_2 = a_1b_1(\bar{a}_0 + \bar{b}_0)$$

$$z_1 = a_1\bar{b}_1b_0 + a_1\bar{a}_0b_0 + \bar{a}_1a_0b_1 + a_0b_1\bar{b}_0$$

$$z_2 = a_0b_0$$

To gain additional experience with manipulation, we can factor some of these expressions to give those below.

$$z_3 = a_1a_0b_1b_0$$

$$z_2 = a_1b_1(\bar{a}_0 + \bar{b}_0)$$

$$z_1 = a_1b_0(\bar{a}_0 + \bar{b}_1) + a_0b_1(\bar{a}_1 + \bar{b}_0)$$

$$z_0 = a_0b_0$$

At this point we could proceed and give a realization for each function. We can, however, make some additional observations that will be useful. Looking at the expression for z_1 and defining $x = a_1b_0$ and $y = a_0b_1$, we can write

$$z_1 = x\bar{y} + \bar{x}y$$

which we recognize as the odd-parity function of Figs. 6.8-2a and 6.8-5 to 6.8-6. Thus we could use the circuit of Fig. 6.8-6, for example, as a component of our realization. If we proceed along this line, we will realize z_1 as shown in Fig. 6.8-8. Having done this, we observe that an additional INVERTER will generate

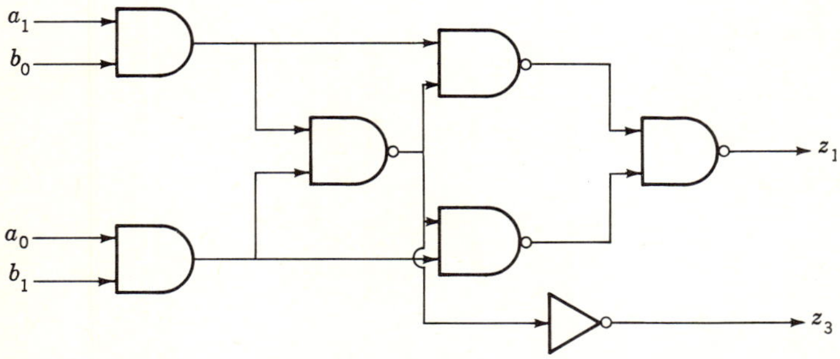

FIG. 6.8-8 Realization of digital multiplier: z_1 and z_3.

z_3 (also shown in Fig. 6.8-8). Finally, we note that z_2 can be written as

$$z_2 = a_1 b_1 \bar{z}_0$$

which can be realized along with z_0 as shown in Fig. 6.8-9. This example is

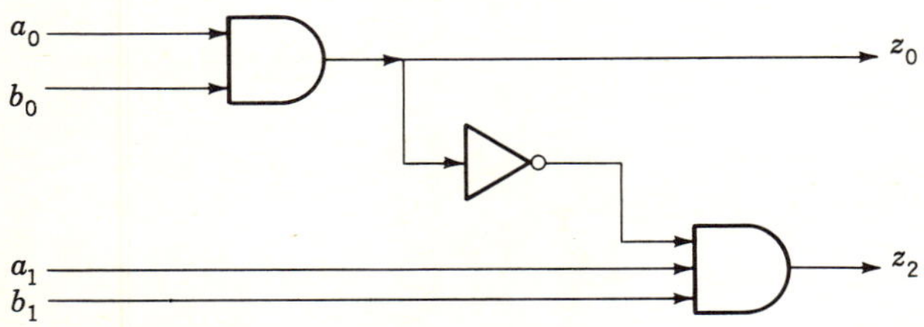

FIG. 6.8-9 The digital multiplier continued: z_0 and z_2.

intended to give a feeling for that portion of logic design that comes between theory and final realization.

Finally, to illustrate the so-called *don't care synthesis,* we will design a multiple-output logic circuit that takes a four-bit binary-coded decimal† number x as an input, and produces as an output a binary-coded decimal number equal to $12 - x$. The code to be used on the input and output is shown in Fig. 6.8-10. It is also known that *the decimal value of the input* will never exceed 12. The problem statement does not specify any behavior for certain input combinations, namely, those of decimal value 13, 14, and 15. In such cases, caused by external constraints on the input variables, the designer is free to specify the missing input-output behavior to reduce the complexity of the realization. Figure 6.8-11 gives separate Karnaugh maps for each output function; the Ø

†Binary coded decimal (BCD) numbers are four-bit representations of the decimals.

	$x_3x_2x_1x_0$	$z_3z_2z_1z_0$
0	0 0 0 0	1 1 0 0
1	0 0 0 1	1 0 1 1
2	0 0 1 0	1 0 1 0
3	0 0 1 1	1 0 0 1
4	0 1 0 0	1 0 0 0
5	0 1 0 1	0 1 1 1
6	0 1 1 0	0 1 1 0
7	0 1 1 1	0 1 0 1
8	1 0 0 0	0 1 0 0
9	1 0 0 1	0 0 1 1
10	1 0 1 0	0 0 1 0
11	1 0 1 1	0 0 0 1
12	1 1 0 0	0 0 0 0

FIG. 6.8-10 Input-output specification.

entries are those outputs which are not specified by the problem statement. Such entries are known as *don't cares*. It is generally, although not always, desirable to specify don't cares so as to obtain the best minimal sum of products or product of sums. The don't cares have been specified in the maps of Fig. 6.8-12 to satisfy this condition. The reasoning for assigning the don't cares of z_0 is most transparent. The choice made results in $z_0 = x_0$; hence no logic is required to realize z_0. The reader should convince himself that the other assign-

z_3:

x_1x_0 \ x_3x_2	00	01	11	10
00	1	1	0	0
01	1	0	Ø	0
11	1	0	Ø	0
10	1	0	Ø	0

z_2:

x_1x_0 \ x_3x_2	00	01	11	10
00	1	0	0	1
01	0	1	Ø	0
11	0	1	Ø	0
10	0	1	Ø	0

z_1:

x_1x_0 \ x_3x_2	00	01	11	10
00	0	0	0	0
01	1	1	Ø	1
11	0	0	Ø	0
10	1	1	Ø	1

z_0:

x_1x_0 \ x_3x_2	00	01	11	10
00	0	0	0	0
01	1	1	Ø	1
11	1	1	Ø	1
10	0	0	Ø	0

FIG. 6.8-11 The specification of Fig. 6.8-10 showing don't cares.

ments shown in Fig. 6.8-12 result in minimal sum-of-products expressions for z_1, z_2, and z_3.

In conclusion, observe that the most straightforward approach to deriving a realization for a multiple-output network is to realize each output

x_1x_0 \ x_3x_2	00	01	11	10
00	1	1	0	0
01	1	0	0	0
11	1	0	0	0
10	1	0	0	0

z_3

x_1x_0 \ x_3x_2	00	01	11	10
00	1	0	0	1
01	0	1	1	0
11	0	1	1	0
10	0	1	1	0

z_2

x_1x_0 \ x_3x_2	00	01	11	10
00	0	0	0	0
01	1	1	1	1
11	0	0	0	0
10	1	1	1	1

z_1

x_1x_0 \ x_3x_2	00	01	11	10
00	0	0	0	0
01	1	1	1	1
11	1	1	1	1
10	0	0	0	0

z_0

$$z_3 = \bar{x}_3\bar{x}_2 + \bar{x}_3\bar{x}_1\bar{x}_0$$

$$z_2 = \bar{x}_2\bar{x}_1\bar{x}_0 + x_2x_0 + x_2x_1$$

$$z_1 = \bar{x}_1x_0 + x_1\bar{x}_0$$

$$z_0 = x_0$$

FIG. 6.8-12 Assignment of don't cares.

function separately. This approach, regardless of how clever one is with each realization, generally does not produce as good a realization as is possible through sharing. As another example of the type of sharing that can be exploited in a multiple-output circuit, consider the network of Fig. 6.8-13 which realizes all four of the above functions with only nine NOR gates. This circuit was designed assuming that complemented as well as uncomplemented input variables were available and that only two-input NOR gates could be used. The reader should analyze this realization as an exercise. It is also instructive to realize each function separately, using the same design constraints, and to compare the realizations.

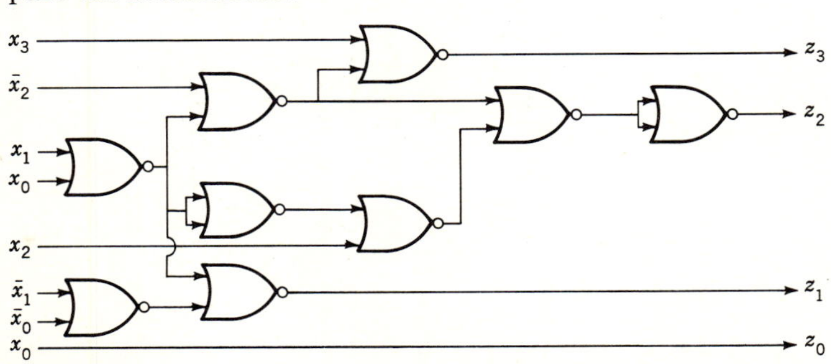

FIG. 6.8-13 An alternate realization of z_3, z_2, z_1 and z_0.

PROBLEMS

6.2-1 *An OR Gate; Fan-out Limitations and Noise Thresholds.* Consider the circuit shown in the figure where all diodes are ideal and the input voltages are nominally either at 0 or 5 V with respect to ground.

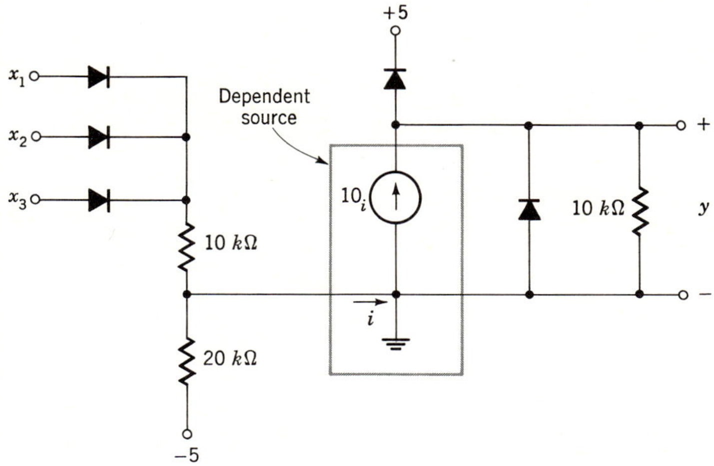

PROB. FIG. 6.2-1

(*a*) Analyze the network so that you can give an input-output table for the overall gate, denoting 5 V by *H* and 0 V by *L*. This is the *nominal* behavior.

(*b*) Assume that the output of this gate will be connected to n inputs of similar gates. What is the minimum value of n (known as the *fan-out*), for which y *will not* attain nominal value (i.e., exactly 5 or exactly 0 V) under any nominal combination of input values?

(*c*) With the above gate unloaded ($n=0$), calculate the minimum V_H and maximum V_L ($0 < V_L < V_H < 5$) so that if each x_i is at either V_H or V_L, the output y is at either 0 or 5 V, as given by the table of (*a*) above. What happens (qualitatively) to V_H and V_L as n increases from 0?

6.4-1 The elements of a *four-element* Boolean algebra are 0, 1, a, and b. The + table is given here. The axioms which characterize a four-element Boolean algebra are identical to those of the two-element algebra, except of course, for the number of elements. Find tables for multiplication and complementation so that a Boolean algebra results.

+	0	a	1	b
0	0	a	1	b
a	a	a	1	1
1	1	1	1	1
b	b	1	1	b

6.4-2 Write the complements of

$$(a + \overline{b}\overline{c}d)[\overline{a}\overline{d} + f(b\overline{c} + e)] \qquad (1)$$

$$a\overline{b}c + (\overline{a} + b + d)(ab\overline{d} + \overline{c}) \qquad (2)$$

6.4-3 Show that, if expressions f and g obey the relation

$$f(x_1, \ldots, x_n; +\ \cdot) = g(x_1, \ldots, x_n; +\ \cdot) \qquad (1)$$

then

$$f(x_1, \ldots, x_n; \cdot\ +) = g(x_1, \ldots, x_n; \cdot\ +) \qquad (2)$$

where in Eq. (2), $\cdot$ and $+$ have been interchanged with $+$ and $\cdot$ of Eq. (1), respectively.
(*Hint:* By induction, or/and by duality.)

6.5-1 State whether each of the following objects is or is not a Boolean expression, and briefly justify your answer wherever your answer is no.

$$(\overline{A + B} + C)D + BC \qquad (1)$$

$$A + AB + AB + AB + AB + AB \qquad (2)$$

$$A + AB + AB + AB + \cdots \qquad \text{(infinite number of terms)} \qquad (3)$$

$$AB = C + D \qquad (4)$$

6.5-2 Simplify the following Boolean expressions:

$$F = A + AB + BCD + AG + B \qquad (1)$$

$$F = (\overline{A + B})C + (\overline{A + C})B + \overline{B} \qquad (2)$$

$$F = (A + B + C)(A + D)(B) \qquad (3)$$

6.5-3 Demonstrate, without using perfect induction, that the following equations are valid.

$$\overline{a}\overline{b} + ab + bc = ab + \overline{a}\overline{b} + \overline{a}c \qquad (1)$$

$$ab + \overline{c}\overline{d} + \overline{a}bc\overline{d} + a\overline{b}\overline{c}d = (a + \overline{d})(b + \overline{c}) \qquad (2)$$

$$bc + abd + a\overline{c} = bc + a\overline{c} \qquad (3)$$

6.5-4 Prove that there are $(2)^{2^n}$ distinct Boolean functions of n variables.

6.5-5 Solve the following three systems of simultaneous Boolean equations:

$$\left.\begin{aligned} E &= \overline{CD} + CD \\ E &= C = D \end{aligned}\right\} (1) \qquad \left.\begin{aligned} F &= \overline{G} \\ G &= \overline{F} \end{aligned}\right\} (2) \qquad \left.\begin{aligned} G &= \overline{A}B + A\overline{B} \\ B &= G \end{aligned}\right\} (3)$$

What conclusions can you draw about the similarities and differences between such equation systems and the more common systems of algebraic equations?

6.5-6 Two functions are said to be *dual* if they can be denoted by dual expressions.

(*a*) Find a function of three variables that is its own dual. (Such a function is called *self-dual.*)

(*b*) Let f_d denote the dual of a function $f(x_1, \ldots, x_n)$. Prove that

$$g(x_0,x_1, \ldots, x_n) = x_0 f + \bar{x}_0 f_d$$

is a self-dual function, i.e., $g_d = g$.

(*c*) Can your example for (*a*) be written as $x_0 f + \bar{x}_0 f_d$?

6.6-1 Construct a table and give a standard sum-of-products expression relating the output y to the inputs x_1, x_2, and x_3 in the figure for a positive logic interpretation.

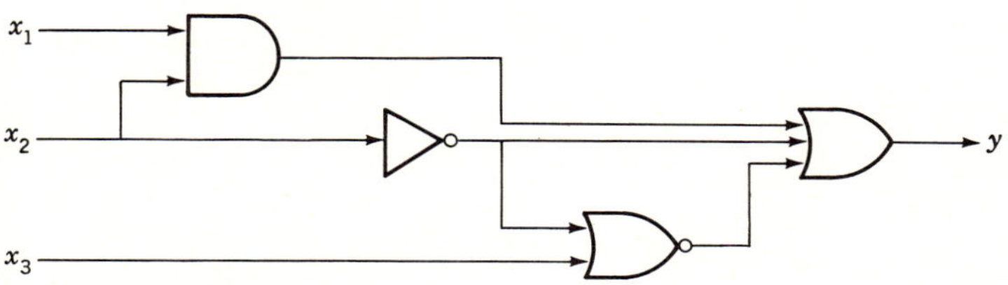

PROB. FIG. 6.6-1

6.6-2 Identify whether each of the systems shown in the figure is or is not a combinational network. If it is, specify whether it is explicit or implicit, and give a table of a function or relation which characterizes the input-output behavior (positive logic interpretation).

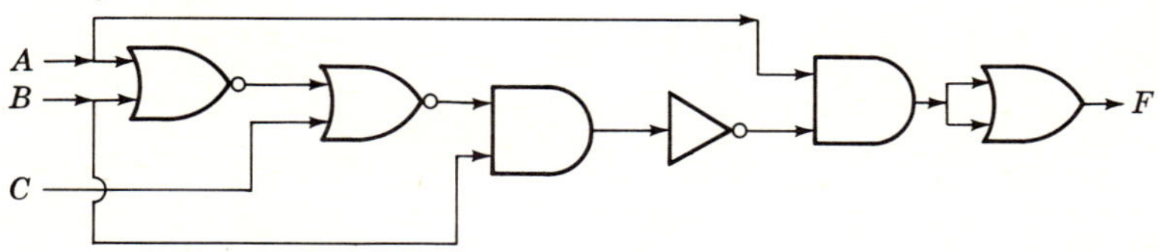

(*a*)

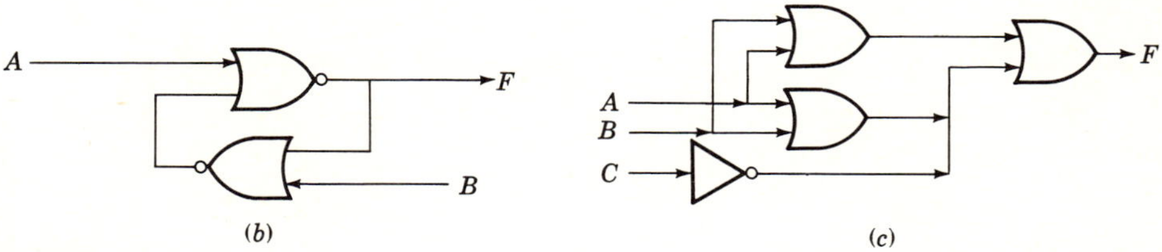

(*b*) (*c*)

PROB. FIG. 6.6-2

6.6-3 *Minimum-time Networks.* Given a very large supply of *two-input* OR gates, you are required to design an explicit logic network of m inputs and one output that is equivalent to an m-input OR gate.

(*a*) Show that for any $m \geq 2$, exactly $m-1$ OR gates must be used.

(*Hint:* Let k be the number of OR gates, and write a balance equation that ensures "everything is connected to something").

(*b*) If $m = 2^n$ ($n = 1, 2, \ldots$), and if each OR gate has unit delay, give the two network *forms* that exhibit longest and shortest overall time delay, respectively. What is this delay in terms of n in either case?

6.7-1 A logical gate, ⊛, has been developed. Its operation is as specified in part *a* of the figure. Give the Karnaugh map for the function $f(a,b,c,d)$ realized by the network in part *b*.

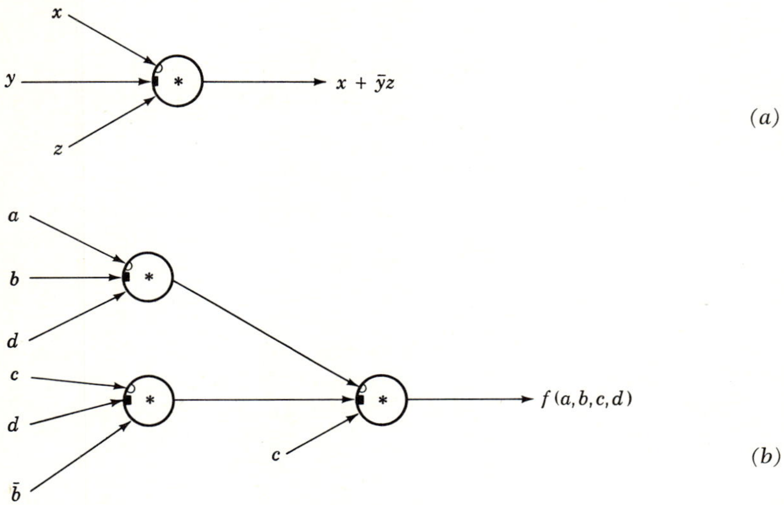

PROB. FIG. 6.7-1

6.7-2 (*a*) Write a minimum sum-of-products expression for F.

(*b*) Synthesize the Boolean function using only *NOR* gates with two inputs.

(*c*) With variables *a, b, c, d, and their complements* available as inputs, synthesize the function using *one* OR and any desired number of AND gates.

(*d*) Repeat (*c*) above with OR and AND interchanged in the problem statement.

ab \ cd	00	01	11	10
00	0	1	1	0
01	1	0	0	1
11	1	0	0	1
10	0	1	1	0

Map of F

PROB. FIG. 6.7-2

6.7-3 A corporation has four voting members, A, B, C, and D, with 30, 25, 40, and 5 shares, respectively. When they vote, a "yes" decision can be reached if and only if more than 50 shares vote in favor of that decision. Let each member have a push button which produces a ONE if A, B, C, and D, respectively, vote in favor. Design a "secret voting" network of AND, OR, NOT gates with inputs A, B, C, D, and output F which is ONE (a light turns on) only if a "yes" decision is reached. Use minimum sum-of-products and minimum product-of-sums forms, as defined in the text, to see which gives rise to a more economical network.

6.7-4 A new gate G has been developed. Its symbol and input-output function are shown in part I, where $f(w,x,y,z) = wyz + xyz$.

(*a*) Develop and explain a procedure for realizing Boolean functions of four variables, using two-level G-OR form, i.e., one output OR gate and a minimum number of input G gates.

(*b*) Using complemented and/or uncomplemented variables A, B, C, and D as inputs, realize the function shown in part II in minimum two-level G-OR form (no ANDs allowed).

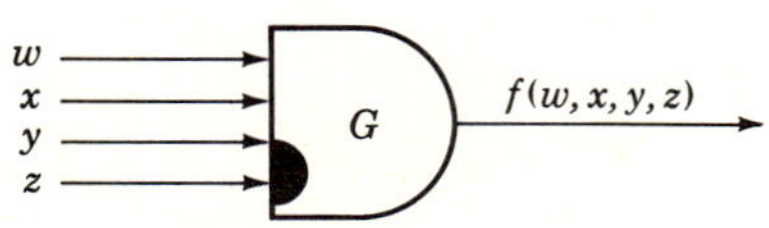

I

CD \ AB	00	01	11	10
00	0	0	0	1
01	0	0	1	1
11	0	1	1	0
10	1	1	0	0

II

6.7-5 Consider the function f given in the Karnaugh map in the figure.

cd \ ab	00	01	11	10
00	1	1	0	1
01	0	1	1	0
11	0	1	1	1
10	1	1	0	1

PROB. FIG. 6.7-5

(*a*) Find *all* the prime implicants of f.
(*b*) Find all minimum sum-of-products expressions for f.
(*c*) Find all the *dual* prime implicants of f.
(*d*) Find all minimum product-of-sums expressions for f.

6.7-6 *Binary Addition.* Let $x_n, x_{n-1}, \ldots, x_1$ and $y_n, y_{n-1}, \ldots, y_1$ be two binary numbers that must be added to give $z_{n+1}, z_n, \ldots, z_1$. Addition will be performed by n identical logic networks, each of which is known as a *full adder*. Each full adder is connected as shown in the figure, where $j \neq 1$, $j \neq n+1$, c_{j-1} is the "carry in," and c_j is the "carry out" necessary for conventional binary addition.

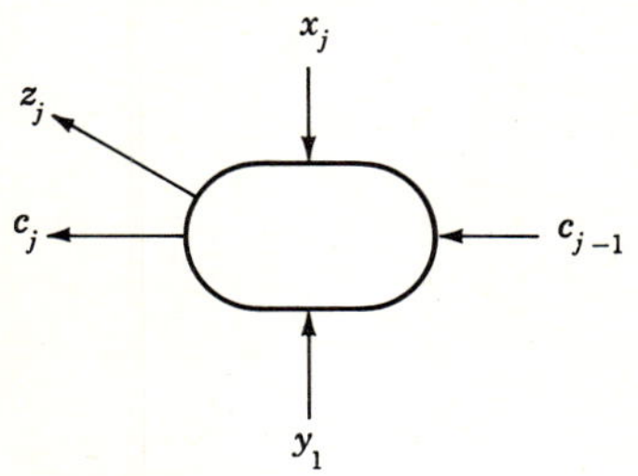

PROB. FIG. 6.7-6

(*a*) Design the full adder shown in the figure in two-level NAND-NAND form for $j = 2, 3, \ldots, n$.

(*b*) Design the full adders at the two extremes ($j = 1$ and $j = n+1$) in NAND-NAND form, exploiting the special situations that hold in either case.

6.7-7 Using *perfect* DACs (digital-to-analog converters) and ADCs (analog-to-digital converters) discussed in Chap. 4, consider the method shown in the figure of adding the binary integers x and y of Prob. 6.7-6. Discuss briefly

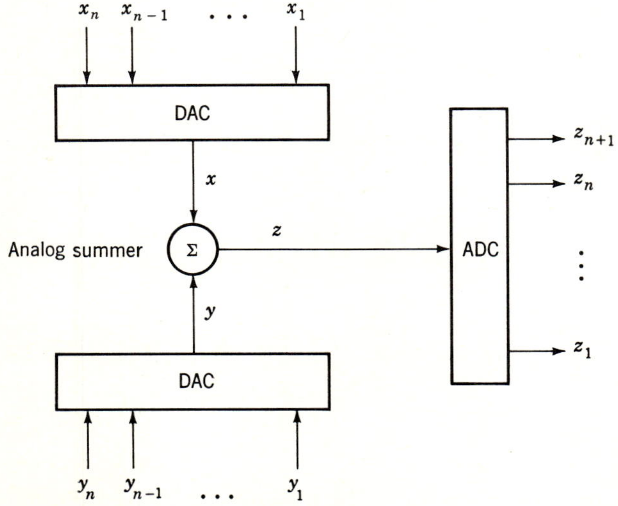

PROB. FIG. 6.7-7

why in digital computers addition is *not* performed in this way. Try in particular to relate the inaccuracies ϵ_1 and ϵ_2 of the analog summer due to temperature, noise, aging, etc., expressed as $z = (x + y)(1 + \epsilon_1) + \epsilon_2$, to the number of bits that can be successfully handled by the above scheme. How big can n be if $\epsilon_1 \simeq \epsilon_2 \simeq 0.01$?

6.7-8 *Quine-McCluskey Procedure for AND-OR Synthesis by Computer.* Recall that Karnaugh map minimization is based on identifying adjacent 1s and combining them in patches. Forget the maps, and consider a *list* of all the combinations of input values (*input words*) where a given function F is 1. For example, if $F(x_1,x_2,x_3)$ is 1 whenever two or more of the x_i are 1, then the list would be: 011, 110, 101, 111. Note that if two words are "adjacent," they will differ in exactly one value; e.g., 101 and 111 are adjacent. If two adjacent words are combined in a "patch," then use a dash in place of the differing values to denote that patch. For the above example, 101 and 111 give the patch denoted by 1–1.

Given any function F, construct the above list, and *order* it in *ascending number of ones.*

(*a*) Give a simple rule for detecting from the given list *all* pairs of adjacent 1s. You may express your rule in terms of a very brief flow chart, and you must end up with a new list of all these patches of two 1s, i.e., a list where every entry has one dash.

(*b*) Using only the new list, give again a simple rule for detecting *all* groups of *four* 1s that would have formed a patch on a Karnaugh map. Your result must be a new list, every entry of which has two dashes and denotes a patch of four 1s of F.

(*c*) Extend the process of (*a*) and (*b*) above so that your rule keeps generating new lists, each time with one more dash per entry, until the largest patches of F have been discovered.

(*d*) Give a simple rule that detects from your lists *all the prime implicants of F.*

(*e*) Give a final rule which determines by an exhaustive search a minimum sum of products of F.

6.8-1 Let f_A, f_0 be the functions corresponding to the positive logic interpretation of n-input AND and OR gates, respectively. Let f_x be some function from $\{0,1\}^n$ to $\{0,1\}$. Show that if $\{f_x,f_A\}$ is functionally complete, so is $\{f_x,f_0\}$. In different words, show that if a supply of f_x and AND gates can realize any Boolean function of n arguments, then a supply of f_x and OR gates has the same logical power.

6.8-2 A defective three-input gate behaves as shown below:

A	B	C	F
0	0	0	1
0	0	1	1
0	1	0	x
0	1	1	x
1	0	0	0
1	0	1	x
1	1	0	1
1	1	1	0

That is, its output F is 1 or 0 for the combinations shown in the table. The value x means that under these combinations of inputs, the gate *burns out.* Since you were given a few thousand of these worthless gates, you are asked to determine a "cure" or "rule," so that gates can be used in combinational networks in such a way that you can (i) realize any given Boolean function of any number of arguments, and (ii) guarantee that no gate will ever burn out (owing to the dangerous input combinations ever occurring).

6.8-3 Show that the set {AND,OR} is *not* functionally complete, provided, of course, that complemented input variables are not available. Note that it is much more difficult to prove noncompleteness than completeness, since you must assure yourself that regardless of how clever, how complex, and how large a network of OR and AND gates is (there is an infinite number of such networks), it cannot realize some function, e.g., $f(x) = \bar{x}$. (*Hint:* Give a careful *inductive* argument, showing that if an input "increases" from 0 to 1, it is impossible for *any* gate output of *any* network of AND and OR gates to "decrease" from 1 to 0.)

6.8-4 The variables A, B, C, D are outputs of a system which constrains them as follows:

$$D = (A + B + C) \cdot D$$

$$B = (A + C + D) \cdot B$$

You are to design a minimum sum-of-products network whose output F is 1 if and only if two or more of A, B, C, D have the value 1. (*Hint:* Indicate the above constraints as don't cares on the map of F.)

6.8-5 *An Optimal Algorithm for Don't Care Synthesis.* Consider the following algorithm: Given a function F of n variables with some don't cares:

1. Assign to all don't cares the value 1, resulting in function F_1, and identify *all* prime implicants of F_1.
2. Assign to all don't cares of F the value 0, resulting in function F_0.
3. Choose a minimal number of prime implicants of F_1 that realize F_0 in minimum sum-of-products form.

The claim is that this algorithm results in the minimal AND-OR network realizing F; i.e., it is impossible by assigning 0 or 1 values to the don't cares of F to obtain a "more minimal" AND-OR network.

(*a*) Apply the procedure to the function F which has 1s at input ($ABCD$) combinations 1100, 1101, 1001, 0111, 1011 and don't cares at 0101, 1111, 0110.

(*b*) Argue briefly but carefully why this algorithm results in a *minimal* AND-OR network.

chapter 7

Dynamic Systems and Networks: State Analysis

This chapter is concerned with the formulation and solution of equations that characterize simple dynamic systems (continuous block diagram systems with memory), and resistor-inductor-capacitor networks. Formulation and analysis involve the use of certain special internal variables—the state variables—which act as the system memory and which lead to the familiar state model. The resulting equations, which are ordinary differential equations, are examined, solved, and related to the systems and networks that they characterize. Considerable emphasis is placed on linear systems, and hence on linear differential equations. Also discussed is the construction of dynamic system basic blocks out of networks with memory.

7.1 INTRODUCTION

In the preceding chapters we have discussed systems and networks having a common characteristic—the absence of memory. In this chapter we start our discussion of systems with memory, with the so-called continuous dynamic systems and resistor-inductor-capacitor (RLC) networks.

These continuous dynamic systems and networks are quite useful in the processing of information (e.g., analog computers and electronic filters), in communications (e.g., telephone, radio, television, and radar) and in control (e.g., servomechanisms and associated electronics). They are generally faster, less expensive, and less accurate than their counterpart digital systems. Besides the immediate and widespread need and use of such electric and electronic systems in everyday applications, the associated mathematical tools and characterizations are *directly* applicable to a much wider range of physical phenomena, such as vibrating strings, sound generation and processing, automobile suspension systems, mechanisms, and even to simple models of economic and social dynamics.

As we have done with memoryless systems, we shall first establish means for formulating appropriate equations that describe dynamic systems and networks. Our formulation, which leads to the state model discussed in Chap. 2, involves the identification of certain variables typically internal to the system. These *state variables*, as they are called, have the property of "remembering" at any time through their *value* all that is necessary for determining the input-output behavior of the system beyond that time. In a sense, they retain for the purposes of future input-output behavior all the significant aspects of the input's past history. The equations resulting from this formulation are ordinary differential equations, but since they involve state variables, they are also called *state differential equations.* This state formulation and state analysis of dynamic systems and networks distinguishes this chapter from the next, where the characterizing equations relate input functions of time to output functions of time without regard to any internal variables. State analysis is characterized by relative ease in formulating and solving (by digital computer) the system equations. Input-output analysis is particularly useful in the study of linear systems and in their characterization by the so-called system functions in the frequency domain. Both approaches lead to useful and often different insights into system and network behavior, which are important in the analysis and synthesis of these systems.

Following the formulation of state differential equations, we discuss their solution, especially when they are linear, and derive several general results. Particular attention is devoted to the solution of such equations in the case of one state variable and to the relationship between that solution and the network or system characterized by the equation. Helpful terms such as *time constants,* new concepts such as the *impulse response,* and familiar ideas such as superposition are introduced in order to facilitate understanding and analysis of what might otherwise appear to be complicated situations. Finally, simple systems and differential equations in two variables are briefly discussed, in order to illustrate the multivariable extension of the foregoing ideas.

As in the case of memoryless continuous systems and resistive networks, this chapter as well as the next one concentrates on basic concepts and explicit analytical solutions. When such solutions are not possible, we resort to numerical integration, discussed in Chap. 9, in a manner analogous to our use of iteration for the memoryless case. Moreover, networks with memory bear the same relationship to continuous dynamic systems as do resistive networks to memoryless block diagram systems. That is, both the networks and the systems are modeled by the same kinds of equations, and the networks can be used, as we discuss in this chapter, to construct the basic elements of dynamic systems.

7.2 FIRST-ORDER BLOCK DIAGRAM SYSTEMS

This section discusses continuous dynamic systems with one memory element—the *integrator*. A system made up of a single integrator and one or more function blocks (discussed in Chap. 3) is a *first-order system*. Such a system *must* have at least one output and may have one or more inputs. If it has no input,† then the system is called *autonomous* (from the Greek self-governed). In terms of the system-defining convention of Chap. 2, the class of first-order systems contains the element set E shown in Table 7.2-1.

TABLE 7.2-1 Element Set E for First-order Systems

E	
Integrator 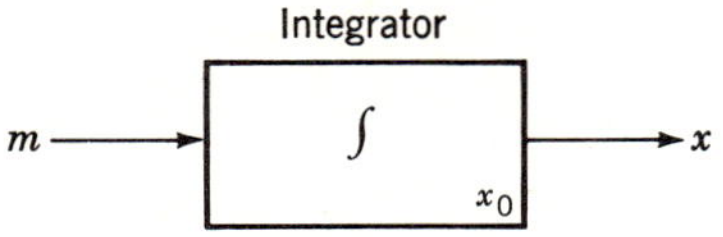	Function block 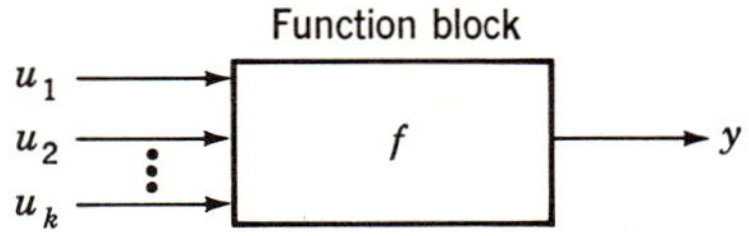
$x(t) = \int_{t_0}^{t} m(\tau)d\tau + x_0$	$y = f(u_1, u_2, \cdots, u_k)$

The interconnection rules R are the same as those for function-block systems and combinational networks. That is, for interconnection purposes, the integrator adheres to the same interconnection rules as a one-input one-output function block, and it may be used in explicit or implicit systems with or without inputs.

The integrator has the following input-output behavior:

$$\frac{dx}{dt}(t) = m(t)$$

or in integral form,

†A system without inputs is not necessarily "dull." A useful or interesting output may very well be the result of what is stored within the system's memory.

$$x(t) = \int_{t_0}^{t} m(\tau)\, d\tau + x_0$$

where t is time, $m(t)$ is the input, $x(t)$ the output, t_0 is some time origin, and x_0 is the *initial value* $x(t_0)$, or the constant held by the integrator at time t_0. An example of the input-output behavior of such an integrator is shown in Fig. 7.2-1. Note, that as with memoryless block diagram systems, we are not concerned with the physical quantities that correspond to x and m, but rather with the values that these variables take, and the way in which these values are related

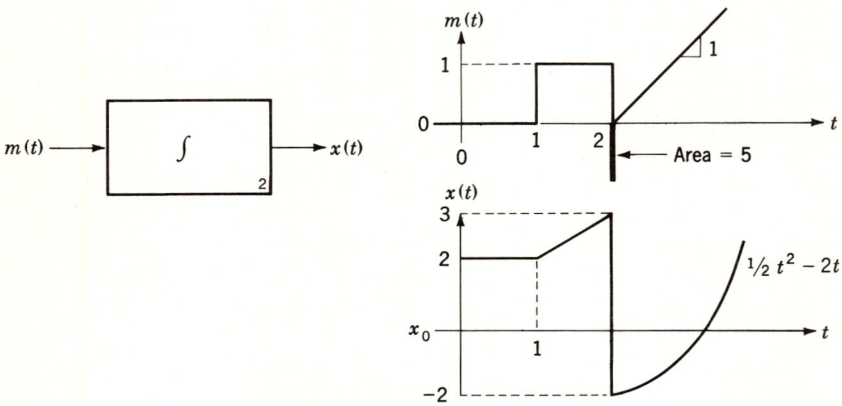

FIG. 7.2-1 Example of integrator input-output behavior. Heavy line represents a tall negative "pulse" of area 5.

to each other and to time. Observe further that the integrator is *not memoryless,* since knowing its input value at time t is not sufficient for determining the output value at that time. Instead, what does determine the output value at time t is the past history of the input, retained by the integrator as the time integral of the input $m(t)$.

One can think of infinitely many ways in which an integrator and several function blocks can be interconnected to form a first-order system. Such systems, however, can be classified in one of two categories according to whether the integrator is or is not contained in a loop of elements. For example, in Fig. 7.2-2a the integrator is not, while in Fig. 7.2-2b and c it is contained in at least one such loop. Note that these two categories are *not* the same as implicit and explicit systems since, for example, the integrator of such a system may not be in a loop while several function blocks are.

Systems belonging to the first category, such as the one of Fig. 7.2-2a, can be analyzed in a rather straightforward manner through function composition and integration; that is, given the inputs of such a system, we can always construct a forward-going, explicit process, at any stage of which we either integrate a function or compose it with another function. Thus, for the system of Fig. 7.2-2a, the output can be *explicitly* determined from the inputs by

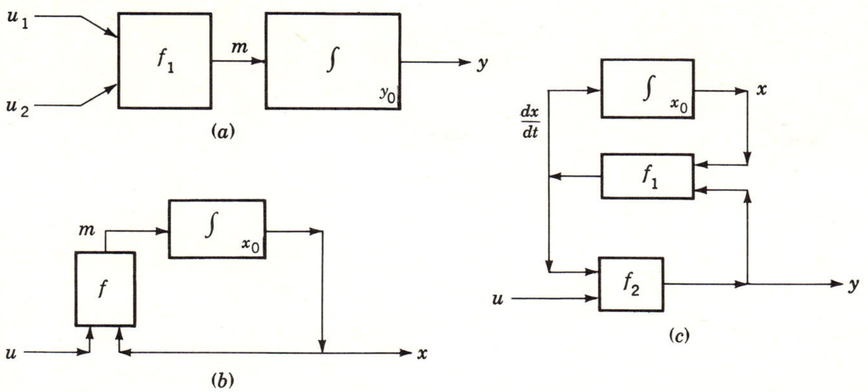

FIG. 7.2-2 Examples of first-order systems with the integrator (a) *not participating and* (b, c) *participating in a loop.*

$$y(t) = y_0 + \int_{t_0}^{t} f_1[u_1(\tau),u_2(\tau)]\, d\tau$$

If, for example, f_1 were the sum function, and if $t_0 = 0$, $y_0 = 0$, $u_1(t) = t$, and $u_2(t) = \cos(t)$; then we could easily determine $y(t)$ as

$$y(t) = \tfrac{1}{2}t^2 + \sin(t)$$

The second category of systems, however, does not lead to such straight-forward results. Consider for example the system of Fig. 7.2-2*b* which can be characterized as follows:

$$x(t) = x_0 + \int_{t_0}^{t} f[u(\tau),x(\tau)]\, d\tau$$

Observe that the form of this equation is implicit; i.e., to find $x(t)$ on the left side of the equation, we must know $x(t)$ for use on the right side of the equation. The system of Fig. 7.2-2*b* can also be characterized by the following equation, which can be obtained either by differentiation of the above equation or by looking at Fig. 7.2-2*b* in a different way.

$$\frac{dx}{dt} = f(u,x) \qquad \text{initially, } x(t_0) = x_0 \tag{1}$$

That is, at any time, the rate dx/dt is determined by the *values* u and x of $u(t)$ and $x(t)$, respectively, at *that very same time*. Equation (1) is called a *first-order ordinary differential equation*, the solution of which we will undertake in subsequent sections.

If a first-order system contains loops of memoryless elements, then in the corresponding differential equation, the function f of Eq. (1) will not be given explicitly, but will be in implicit form.† Specifically, the system of Fig.

†As we have noted in Chap. 3, the use of function blocks in system loops does not necessarily lead to equivalent functions but to relations.

7.2-2c is characterized by

$$\frac{dx}{dt} = f_1(x,y) \tag{2}$$

where

$$y = f_2[f_1(x,y),u] \tag{3}$$

Observe here that at any time, given the values for x and u, y can be determined by solving the implicit nonlinear Eq. (3) by iteration (Chap. 5). That value can then be used in Eq. (2) to give dx/dt at that time.

The above observations lead us to a systematic way of formulating the equations that characterize a first-order block diagram system. Our approach will be to "pull" the integrator out of the system, as shown in Fig. 7.2-3. The rest of the system is then memoryless, since by definition, there is only one

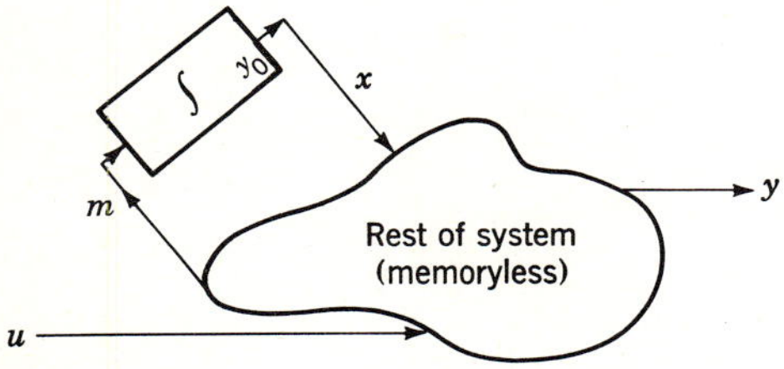

FIG. 7.2-3 Pulling the integrator out of the system.

memory element in the system—the integrator. The memoryless system of Fig. 7.2-3 has inputs u and x and outputs m and y. We proceed then by expressing these outputs in terms of the two inputs,

$$m = f(x,u) \tag{4}$$

$$y = g(x,u) \tag{5}$$

If the memoryless system has no loops, then functions f and g can be easily obtained. If that system has loops, then we will be concerned only with the cases in which f and g exist, although we will be generally unable to state these functions explicitly, and instead will have to solve one or more implicit equations, usually by iteration. The integrator of Fig. 7.2-3 can be characterized by its defining equation,

$$\frac{dx}{dt} = m \qquad x(t_0) = x_0 \tag{6}$$

Equations (4) to (6) yield the general characterizing equations for a first-order system:

$$\frac{dx}{dt} = f(x,u) \qquad x(t_0) = x_0 \tag{7}$$

$$y = g(x,u) \tag{8}$$

Equation (7) is what we have already identified as an ordinary first-order differential equation. After we solve it, we will know $x(t)$, which we can then substitute in Eq. (8) to establish $y(t)$.

To summarize, the formulation of equations that characterize a first-order block diagram system entails:

1. Writing a differential equation which gives the value of the integrator input dx/dt, given the value of the integrator output $x(t)$ and the value of the system input $u(t)$, and identifying the initial value x_0 of $x(t)$.
2, Writing an equation which at any time t gives the value of the system output $y(t)$, given the values of $u(t)$ and $x(t)$.

The above characterization is precisely the state model characterization discussed in Chap. 2 and shown in part c of Table 7.2-2. Here $x(t)$ is the state variable, dx/dt is the state update, and x_0 is the initial state.

Example 7.2-1 *Formulation of Equations for a Linear System* Consider the system of Fig. 7.2-4, where a, b, c, and d are scalers. We can relate by inspec-

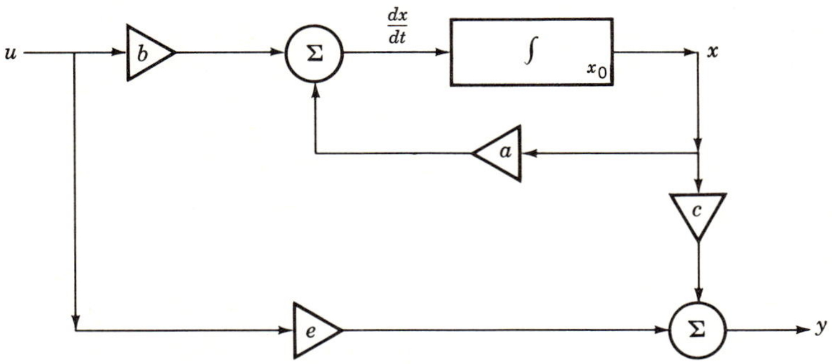

FIG. 7.2-4 Example of a linear first-order dynamic system.

tion, the integrator input dx/dt to x and u as follows:

$$\frac{dx}{dt}(t) = ax(t) + bu(t); \qquad \text{initially } x(t_0) = x_0 \tag{9}$$

Likewise, the system output is related to x and u as follows:

$$y(t) = cx(t) + eu(t) \tag{10}$$

Equations (9) and (10) fully characterize the system of Fig. 7.2-4.

Example 7.2-2 *Formulation of Equations for a Nonlinear System*

From Fig. 7.2-5, the characterizing equations are

$$\frac{dx}{dt} = u - f(x) \qquad x(t_0) = x_0 \tag{11}$$

$$y = x \tag{12}$$

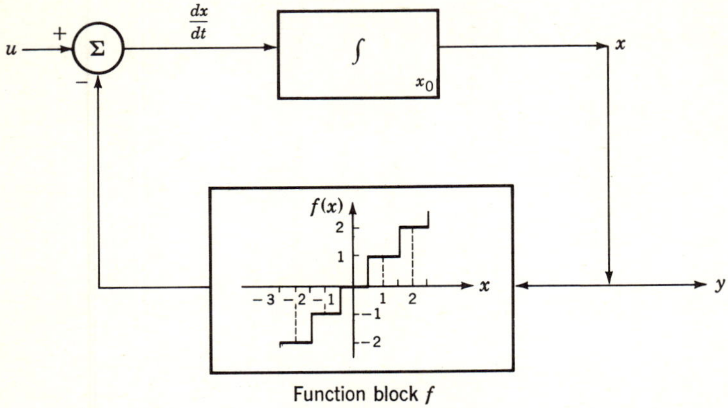

FIG. 7.2-5 Example of a nonlinear first-order dynamic system.

where the nonlinear function f is given by

For $k = \ldots, -2, -1, 0, 1, 2, \ldots,$

$$f(x) = k \qquad \text{iff } k - 0.5 \le x < k + 0.5$$

TABLE 7.2-2 Some First-order Dynamic Systems and Their State Characterizations

System	*Equations*
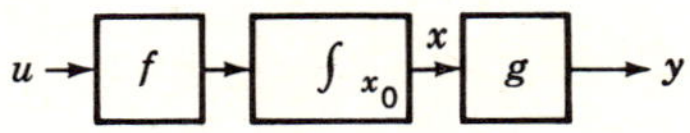 (*a*)	$\dfrac{dx}{dt} = f(u) \qquad x(t_0) = x_0$ $y = g(x)$
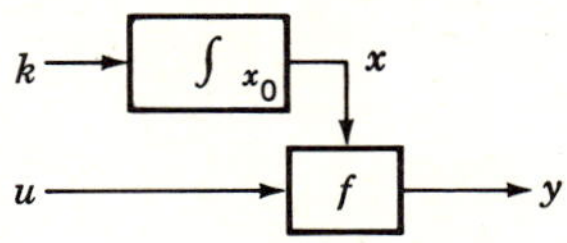 (*b*)	$\dfrac{dx}{dt} = k \qquad x(t_0) = x_0$ $y = f(x,u)$
General state model form 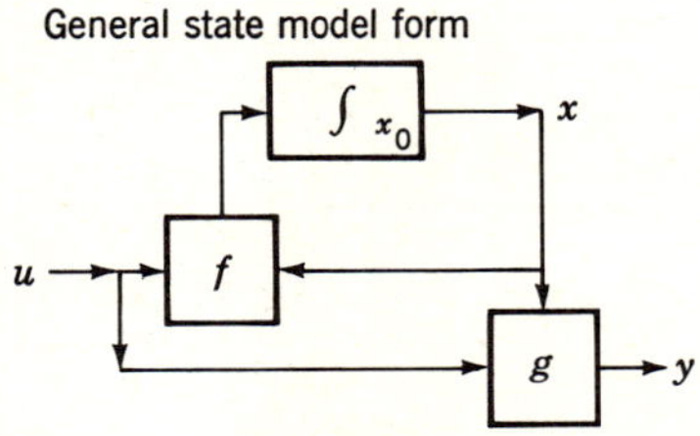 (*c*)	$\dfrac{dx}{dt} = f(x,u) \qquad x(t_0) = x_0$ $y = g(x,u)$ *Note:* This is the general form of a first-order system, and is in fact the state model form discussed in Chap. 2.

Note, that in Eqs. (11) and (12) for convenience we have dropped the variable t; i.e., we write $f(x)$ instead of $f(x(t))$, although we fully understand that dx/dt, u, y, and x represent the values $\frac{dx}{dt}(t)$, $u(t)$, $y(t)$, and $x(t)$, at any time t.

Example 7.2-3 *Miscellaneous System Forms and Their Characterizing Equations* Table 7.2-2 gives several first-order systems and their characterizing equations.

We will discuss the solution of these equations after we see how first-order electric networks give rise to the same kind of differential equations.

7.3 CAPACITORS AND INDUCTORS: THE BUILDING BLOCKS OF DYNAMIC NETWORKS

First-order electric networks are generally made up of one inductor and one or more resistors or of one capacitor and one or more resistors, as well as dependent and independent sources. We start our discussion, accordingly, with these two types of basic elements.

Mathematical Characterization

Capacitors and inductors are two-terminal network elements which, like resistors, exhibit a voltage across and a current through their terminals, when connected in an electric network. We have seen in Chap. 4 that resistors constrain current to voltage through a function. Capacitors and inductors differ from resistors in that they constrain *the time integral* of one of these variables to the other variable, respectively. That is, a capacitor constrains the time integral of current (the charge) to voltage, whereas an inductor constrains the time integral of voltage (the flux) to current.

The element set E of resistor-inductor-capacitor networks (RLC nets) along with the corresponding constraints and linear cases, but with the exception of sources, is shown in Table 7.3-1. Study this table carefully, since it contains important relationships among fundamental quantities that will be used throughout the book.

This table indicates that, in general, for a capacitor we will have either

$$q = f_c(v) \tag{1}$$

or

$$v = f_s(q) \tag{2}$$

where v is voltage and q is charge.

The former is called a *voltage-controlled* capacitor and the latter a *charge-controlled* capacitor. When f_c and f_s are differentiable, f_c' is called the *incremental capacitance* and f_s' the *incremental elastance*. Similiarly, for an inductor

$$\phi = f_L(i) \tag{3}$$

$$i = f_\Gamma(\phi) \tag{4}$$

where ϕ is flux linkage.

As in the above cases, we distinguish between *current-controlled* [Eq. (3)] and *flux-controlled* [Eq. (4)] inductors; and when f_L and f_Γ are differentiable, we speak of the *incremental inductance* f_L' and the *incremental inverse inductance* f_Γ'. In both inductors and capacitors, the functions $f_c, f_s, f_L,$ and f_Γ will be in general continuous.†

TABLE 7.3-1 Resistors, Inductors, and Capacitors

	Resistors	*Capacitors*	*Inductors*
Symbol	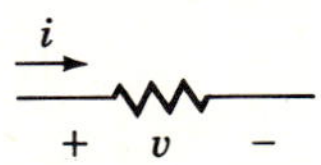	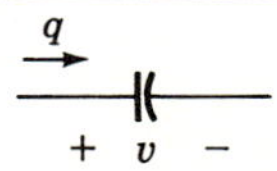	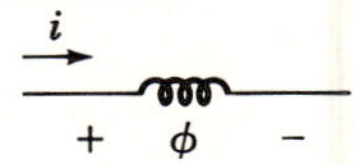
Graph of constraint	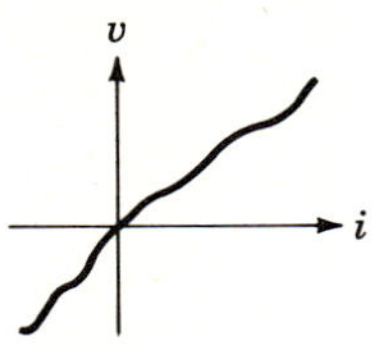	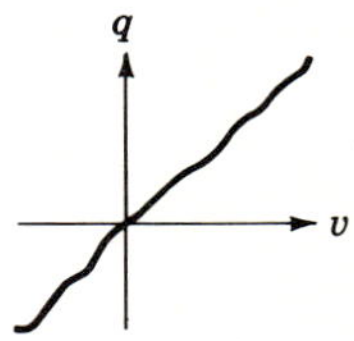	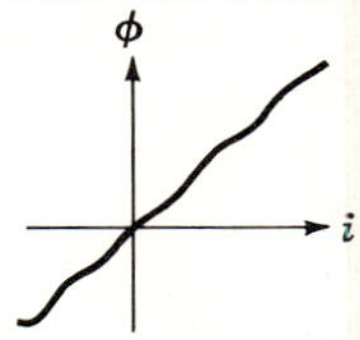
Linear elements: Graph	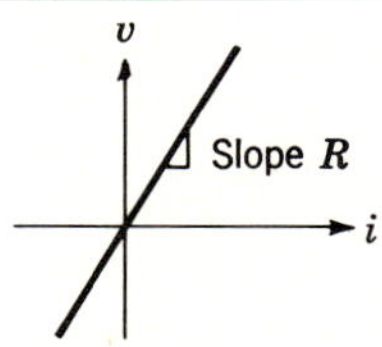	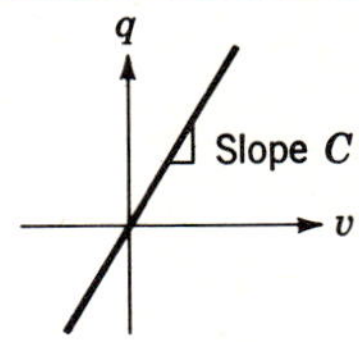	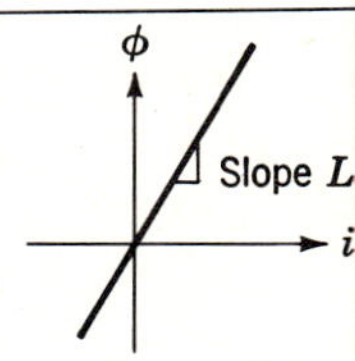
Linear elements: Fundamental relationship	R is the *resistance* $G = 1/R$ is the *conductance* $v = Ri$ $i = Gv$	C is the *capacitance* $S = 1/C$ is the *elastance* $q = Cv$ $v = Sq$	L is the *inductance* $\Gamma = 1/L$ is the *inverse inductance* $\phi = Li$ $i = \Gamma\phi$
Linear elements: *i-v* relationship	$v = Ri$ $i = Gv$ No initial conditiorr on v or i	$i = C\dfrac{dv}{dt}$ $\dfrac{dv}{dt} = Si$ Initially, $v(t_0) = v_0$	$v = L\dfrac{di}{dt}$ $\dfrac{di}{dt} = \Gamma v$ Initially, $i(t_0) = i_0$

†Although from a mathematical point of view these functions could be nonmonotonic, all the physical capacitors and inductors that we know can be adequately modeled by monotonic functions.

In the special but important case of *linear capacitors,* Eqs. (1) and (2) assume the special form

$$q = Cv \tag{5}$$

or

$$v = Sq \qquad \text{where } S = \frac{1}{C} \tag{6}$$

Since current is defined as the time rate of change of charge, $i = dq/dt$, we can also write for Eq. (5) its corresponding differential or integral form:

$$i = C\frac{dv}{dt} \tag{7}$$

or

$$v(t) = \frac{1}{C}\int_{t_0}^{t} i\,(\tau)\,d\tau + v_0$$

where $v(t_0) = v_0$, and C is called the *capacitance* and S the *elastance.* Similarly, in the case of *linear inductors,* Eqs. (3) and (4) become

$$\phi = Li \tag{8}$$

or

$$i = \Gamma\phi \qquad \text{where } \Gamma = \frac{1}{L} \tag{9}$$

Since voltage is defined as the time rate of change of flux, $v = d\phi/dt$, we can also write for Eq.(8) its differential or integral form,

$$v = L\frac{di}{dt} \tag{10}$$

or

$$i(t) = \frac{1}{L}\int_{t_0}^{t} v(\tau)\,d\tau + i_0$$

where $i(t_0) = i_0$. Here L is called the *inductance,* and Γ the *inverse inductance.*

Occasionally, a distinction of *time-varying* or *time-invariant* inductors or capacitors is made, according to whether f_L or f_C vary or not with time. Throughout the book we will be concerned with time-invariant capacitors or inductors, so we will merely use the adjectives linear or nonlinear, and it will be understood that we mean time-invariant.

Finally, the units of the quantities, R, G, C, S, L, Γ in the mks system are as follows:

R ohms
G mhos
C farads
S darafs
L henrys
Γ inverse henrys

We proceed now with several examples that illustrate the above ideas:

Example 7.3-1 Consider the q-v constraint of a linear capacitor shown in Fig. 7.3-1a. This capacitor is connected to a voltage source as shown in Fig.

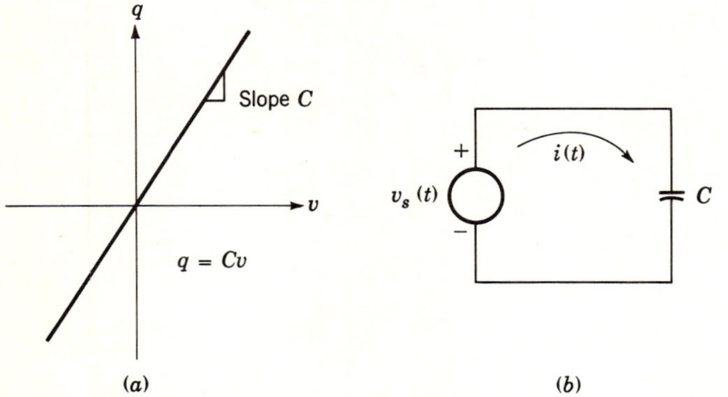

FIG. 7.3-1 Voltage-charge characteristic and network of Example 7.3-1.

7.3-1b. It is also given that $v_s(t) = 10 \sin 2t$, and $C = 10^{-4}$.

From the relationship $q = Cv$ we have

$$q(t) = 10^{-3} \sin 2t$$

Through differentiation, $i = dq/dt$, we obtain the final result:

$$i(t) = (2 \times 10^{-3}) \cos 2t$$

Example 7.3-2 Figure 7.3-2a shows the characteristics of a voltage-controlled capacitor, and Fig. 7.3-2b the characteristics of a charge-controlled capacitor.

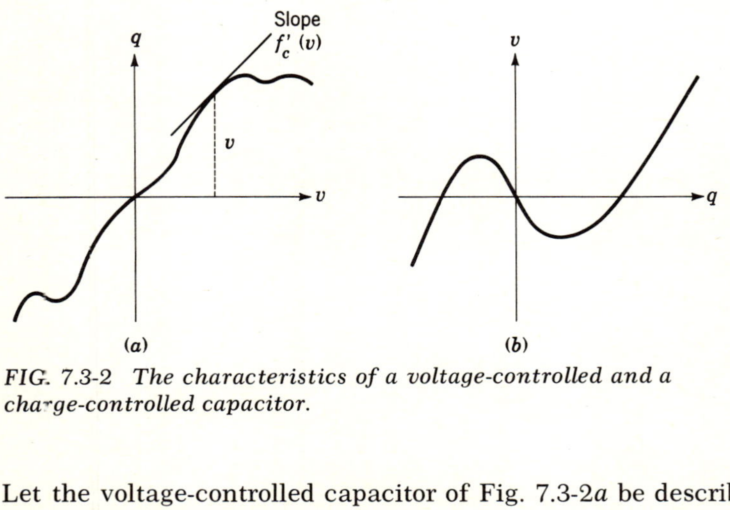

FIG. 7.3-2 The characteristics of a voltage-controlled and a charge-controlled capacitor.

Let the voltage-controlled capacitor of Fig. 7.3-2a be described by the relation

$$q = f_C(v) \tag{11}$$

Assume that $f_C(v)$ is differentiable. Since the current i and the charge q are related by

$$i = \frac{dq}{dt}$$

we can differentiate Eq. (11) to find

$$i = \frac{dq}{dt} = \frac{d}{dt} f_C(v) = f_C'(v) \frac{dv}{dt} \tag{12}$$

where we have used the chain rule and the fact that the function f_C depends only upon the voltage v and is independent of any other parameter, including time. Note that the function $\frac{df_C}{dv}(v)$ is what we have already defined as *incremental capacitance,* and is graphically depicted in Fig. 7.3-2a as the slope of the graph at any given value of v. Thus, if, for example,

$$q = C_0 v + C_1 v^3$$

where C_0 and C_1 are positive constants, we can use Eq. (12) to establish that

$$i = (C_0 + 3C_1 v^2) \frac{dv}{dt}$$

If, moreover, the voltage v across the capacitor is given by

$$v(t) = 1 - e^{-t}$$

then the current, as a function of time, will be

$$i(t) = [C_0 + 3C_1(1 - e^{-t})^2]e^{-t}$$

We proceed now with a few remarks about the physical nature of capacitors and inductors for the reader who would like to have some idea of the physical structure of these elements.

Physical Aspects of Capacitors and Inductors

A simple physical capacitor is a device which is usually constructed by separating, with a dielectric material, two surfaces (plates) made from a conducting material, as illustrated in Fig. 7.3-3. Such a capacitor can "store" electric energy by retaining electric charge on its plates. The role of the dielectric material is to prevent the flow of the electric charge, i.e., electrons, between the conducting plates. A perfect dielectric "material" is vacuum. Other dielectric materials are air, paper, plastic, and mica.

Whenever two conducting materials, insulated from each other, are at different potentials, they store a certain amount of charge. Experimental results as well as electromagnetic theory indicate that the amount of stored charge q at any time t depends upon the voltage v which exists across the capacitor at that time. We can thus view the capacitor as a device which relates or constrains the charge q stored in it to the voltage across it. When the dielectric is air or mica, this dependence between charge and voltage is very nearly linear, and the capacitor can be adequately modeled by Eqs. (5), (6), or (7).

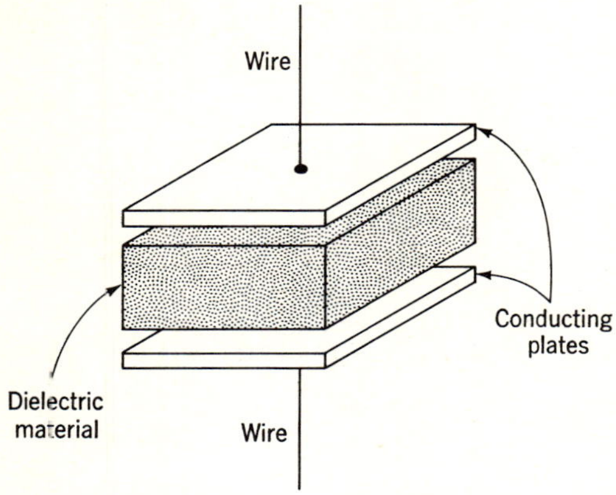

FIG. 7.3-3 Physical elements of a capacitor.

In general, the mathematical relations between charge $q(t)$ and voltage $v(t)$ can be derived from physical considerations. For example, in a linear capacitor, the capacitance can be computed with good accuracy from the dimensions of the conducting surfaces, the distance between these surfaces, and the dielectric constant of the material which is introduced between the conducting surfaces. The higher the dielectric constant, the higher the value of the capacitance. The dielectric constant in turn may vary with physical variables such as temperature and pressure (in gases). In particular, there are materials such as titanium oxides and barium titanate which are broadly classified as *ferroelectric* and whose dielectric constant depends upon the strength of the applied electric field. Such materials can therefore be used to construct *nonlinear capacitors*, modeled by Eq. (1) or (2).

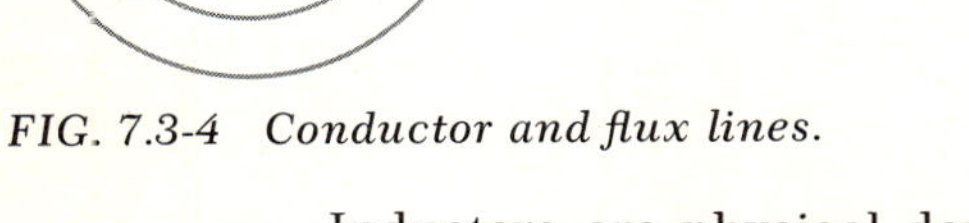

FIG. 7.3-4 Conductor and flux lines.

Inductors are physical devices which, like capacitors, can store energy. However, unlike capacitors which store electric energy, inductors store magnetic energy. A wire which carries a current is said to produce a magnetic field in the surrounding space. Faraday, who was one of the first experimenters in this area (ca. 1830), visualized the magnetic field as a collection of *flux lines*. In a long wire, these flux lines form a pattern of concentric circles as illustrated in Fig. 7.3-4. If the wire is wound into a coil, as shown in Fig. 7.3-5, the magnetic flux lines assume an oval-shaped form.

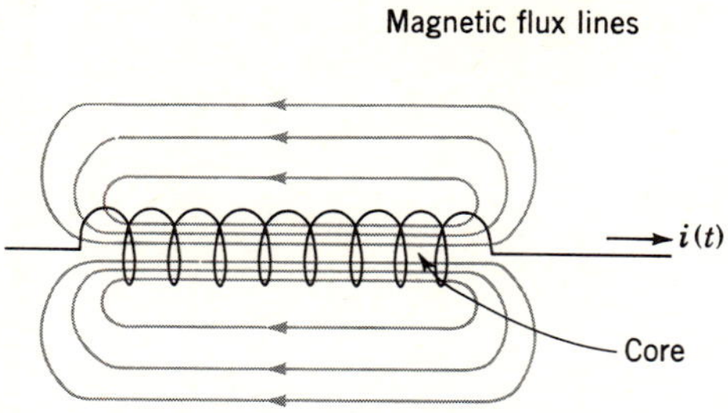

FIG. 7.3-5 Physical elements of an inductor.

In such a coil, the flux (or (flux linkage) is defined as the product of the number of flux lines and the number of turns in the coil "linked" by the magnetic flux lines. Faraday in 1831 showed experimentally that a current flow is generated in a closed circuit whenever there is a variation in the magnetic flux lines linking the circuit. Since current flow usually implies the existence of a voltage in the circuit, it was deduced that there is an (induced) voltage arising from a time variation of the magnetic flux. In 1833, Lenz stated the following law: *The direction of the induced voltage is such that any produced current tends to oppose the change of flux that produces it.* In 1845 Neumann stated the law: *When the magnetic flux linking a circuit is changed, then a voltage is induced whose magnitude is proportional to the time rate of change of the flux linkage.* The result of all these experimental laws (or the relevant consequences of electromagnetic theory) can be abstracted by saying that an inductor provides a relation or constraint between the current through it and the flux linking it, as indicated by Eqs. (3) and (4). The material through which the flux lines pass (the core of the coil, as it is often called) affects the relationship between flux and current. For example, if the material is air, then the relationship is adequately modeled by linear functions, as in Eqs. (8), (9), and (10). If, on the other hand, the material is iron, it exhibits a nonlinear relationship called saturation.†

Power and Energy in Network Elements

Recall from Sec. 4.3 that the electric power $p(t)$ entering any two-terminal network element at any time t is defined as the product of the voltage $v(t)$ across that element and the current $i(t)$ through the element.

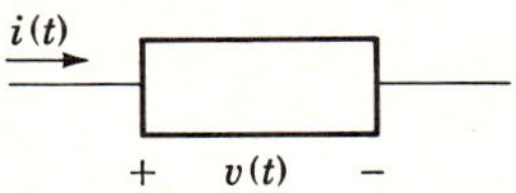

$$p(t) = i(t)v(t) \tag{13}$$

This definition holds as long as the current reference direction is from the + to the − signs of the voltage reference direction. Otherwise, Eq. (13) gives the power leaving the element.

†The incremental inductance decreases with increasing current.

Recall further that the electric energy $w(t)$ entering such an element in a time interval $[t_0,t_1]$ is, from elementary physics, the time integral of power. That is,

$$w(t_1,t_0) = \int_{t_0}^{t_1} p(t)\,dt = \int_{t_0}^{t_1} v(t)i(t)\,dt \tag{14}$$

In physical *resistors*, whose characteristics are in the first and third quadrants of the i-v plane, the product $v(t)i(t)$ is *always* positive or zero. Hence, power may only enter a resistor, and the process is irreversible. This power is generally dissipated as heat and is radiated or conducted away from the resistor. Recall that in linear resistors $v = Ri$ and $R = 1/G$. Hence, we have from Eq. (13)

$$p(t) = i^2(t)R \tag{15a}$$

or

$$p(t) = v^2(t)G \tag{15b}$$

Furthermore, the energy dissipated in a linear resistor is given from Eqs. (14) and (15) by

$$w(t_1,t_0) = R\int_{t_0}^{t_1} i^2(t)\,dt \tag{16a}$$

or

$$w(t_1,t_0) = G\int_{t_0}^{t_1} v^2(t)\,dt \tag{16b}$$

In *capacitors* it is possible for the voltage and current to have opposite as well as equal signs, since capacitors constrain q to v and not i to v. Consider a charge-controlled capacitor whose constraint is $v = f_s(q)$, as shown in Fig. 7.3-6. From this relation and the fact that $i = dq/dt$ we obtain, via Eq. (14), the result

$$w(t_1,t_0) = \int_{t_0}^{t_1} v(t)i(t)\,dt = \int_{t_0}^{t_1} v(t)\frac{dq(t)}{dt}\,dt = \int_{q_0}^{q_1} f_s(q)\,dq \tag{17}$$

where

$$q_0 \triangleq q(t_0) \qquad q_1 \triangleq q(t_1)$$

We can therefore interpret the energy $w(t_1,t_0)$ entering the capacitor as the area under the f_s graph of Fig. 7.3-6, from the initial charge q_0 to the final charge q_1.

Evidently, if we force the current in the other direction (say, through a current source connected across the capacitor) so that in a time interval $[t_2,t_3]$ we go from charge q_1 to charge q_0, the energy entering the capacitor will be, by the same argument, the negative of the shaded area of Fig. 7.3-6; that is, the energy that entered the capacitor in the time interval $[t_0,t_1]$, leaves the capacitor in the time interval $[t_2,t_3]$. *We conclude, then, that a capacitor does not burn up or dissipate energy but rather stores energy.* For a linear capacitor, where $v = Sq$, it follows that

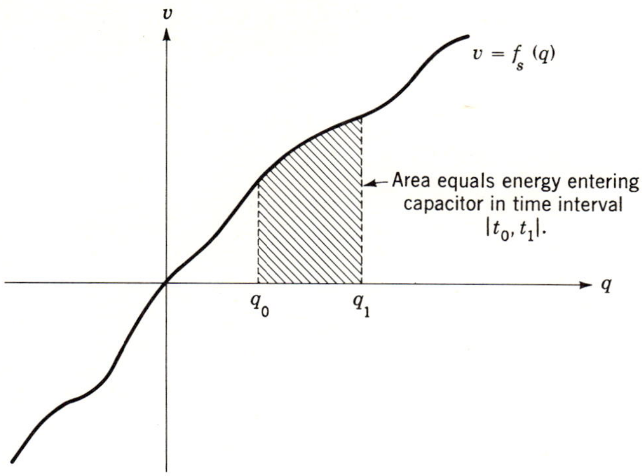

FIG. 7.3-6 Energy entering nonlinear capacitor.

$$w(t_1,t_0) = \frac{Sq_1^2}{2} - \frac{Sq_0^2}{2} \tag{18}$$

Specifically, if we start with $q_0 = 0$ at t_0, i.e., with the capacitor initially uncharged, we see that the energy stored within the capacitor at any subsequent time is *only a function of the charge at that time*. Hence, it makes good sense to express the *stored energy* in a linear capacitor as follows:

$$w(q) = \frac{Sq^2}{2} \tag{19a}$$

or

$$w(v) = \frac{Cv^2}{2} \tag{19b}$$

where we have used the equation $q = Cv$ characterizing a linear capacitor.

An entirely analogous situation holds for inductors, where the energy stored within an inductor equals the area under the f_Γ graph from ϕ_0 to ϕ_1. For linear inductors, it follows that the stored energy is given by

$$w(\phi) = \frac{\Gamma\phi^2}{2} \tag{20a}$$

$$w(i) = \frac{Li^2}{2} \tag{20b}$$

Electrical Quantities and Their Interrelationship

The relationships among the electrical quantities of current, voltage, charge, flux, power, and energy can be represented diagrammatically in terms of the three types of network elements and certain natural laws. These relationships are shown in Fig. 7.3-7*b*. The appropriate reference directions are shown in Fig. 7.3-7*a*.

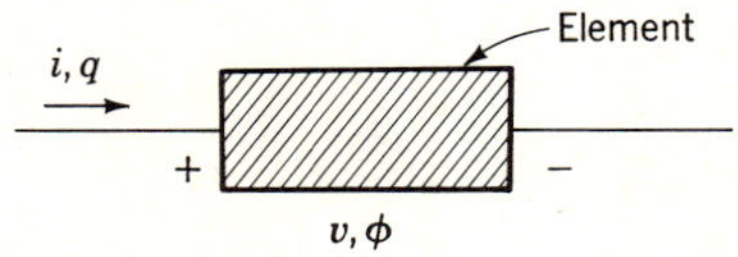

(*a*)

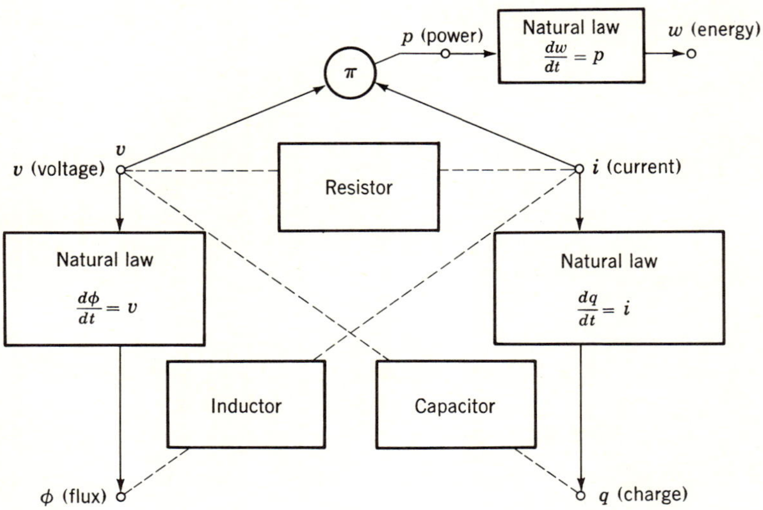

(*b*)

FIG. 7.3-7 Relationship among electrical quantities.

Figure 7.3-7 says that a resistor constrains the voltage across it to the current through it; an inductor constrains the flux (which is the integral of the voltage across it) to the current through it, and a capacitor constrains the charge stored in it to the voltage across it. The natural laws are as follows: The rate of change of charge passing through an element equals the current through that element; the rate of change of flux across an element equals the voltage across it; power entering the element at any time equals the voltage across times the current through that element at that time, and the time rate of change of energy entering an element equals the power entering that element.

From an engineering point of view, it is often important to characterize the response of a network, using the variables which are most easily measured. It is difficult to measure directly charge, flux linkage, power, or energy in an electric network. On the other hand, it is possible to measure, with good accuracy, current and voltage. It is for this reason that we shall often, but not always, characterize the response of a network, using currents and voltages.

Another reason is that the basic axioms of network theory, such as Kirchhoff's laws, are based on experimental evidence. Since every experiment involves measurements, it is important that the choice of the variables that describe a network should conform with our ability to measure these variables.

7.4 FIRST-ORDER NETWORKS

First-order electric networks are those networks that can be modeled by first-order block diagram systems. That is, given (1) an input $u(t)$, which may be either an independent voltage or current source, (2) a network structure and element characteristics, (3) one or more required voltages or currents [for simplicity, we shall think of one unknown $y(t)$], we must be able to find *exactly one* variable $x(t)$, the *state variable* as we call it, and functions f and g so that

At any time $t \geq t_0$

$$\frac{dx}{dt} = f(x,u) \qquad \text{initially } x(t_0) = x_0 \tag{1a}$$

and

$$y = g(x,u) \tag{1b}$$

where dx/dt, x, u, and y are values of the variables at any time t. Note that if several inputs $u_1, u_2, \ldots$ and several outputs $y_1, y_2, \ldots$ are involved, we still have a first-order network, as long as we have only *one equation* of the form (1a), e.g.,

$$\begin{aligned} \frac{dx}{dt} &= f(x,u_1,u_2, \ldots) \qquad \text{initially } x(t_0) = x_0 \\ y_1 &= g_1(x,u_1u_2, \ldots) \\ y_2 &= g_2(x,u_1,u_2 \ldots) \\ &\cdots\cdots\cdots\cdots \end{aligned} \tag{1}$$

The reason that we insist on *one* first-order differential equation of the form (1a) is that, as we have already remarked and as we shall see in the next section, we can solve this equation *first* to establish

$$x(t) \text{ for all } t \geq t_0$$

and we can thereafter use these known values of $x(t)$ along with the known values of the input $u(t)$ or inputs $u_1, u_2 \ldots$ to establish $y(t)$ or the outputs y_1, $y_2, \ldots$.

Usually, first-order electric networks will consist of only one capacitor or of only one inductor and possibly several resistors. It is possible, however, to interconnect two or more capacitors and/or two or more inductors in such a way that the resulting network is still of first order. For example, two linear inductors connected in series, with no other elements attached to their common node, could be replaced by one inductor with value equal to the sum of the above two inductor values, and network performance would be unaffected. We leave such "special cases," for later, however, and proceed here with the formulation of Eq. (1) for typical first-order networks.

Formulation of the State and Output Equations

Consider the network of Fig. 7.4-1*a* where $q_c(t_0) = q_0$, and where the nonlinear resistor is voltage-controlled and the nonlinear capacitor is current-

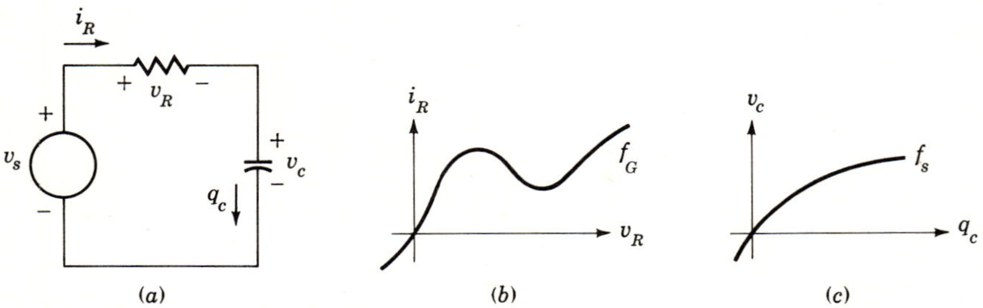

FIG. 7.4-1 Example of a first-order capacitive network.

controlled. That is, in connection with the reference directions of Fig. 7.4-1*a*,

For all v_R,

$$i_R = f_G(v_R) \tag{2}$$

and for all q_C,

$$v_C = f_s(q_C) \tag{3}$$

where f_G and f_s might be as shown in Fig. 7.4-1*b* and *c*, respectively. The independent quantity in this network is the input $v_s(t)$. Required outputs will be all the branch currents and voltages; by KCL, however, there is only one current i_R common to all three elements; hence our unknowns are

$$v_C(t),\ v_R(t),\ i_R(t) \qquad \text{for all } t \geq t_0 \tag{4}$$

To formulate the equations in the form of a first-order block diagram system we might think as follows: Our system must have an integrator whose output is the *state variable* $x(t)$. Can the state variable be i_R? The answer is no, since dx/dt would then be di_R/dt, and we are far from knowing di_R/dt in terms of i_R and the input v_s, as Eq. (1*a*) necessitates. Let us try $v_R(t)$ as the state variable. This is again not possible, for we cannot determine dv_R/dt as a function of $v_s(t)$ and $v_R(t)$. How about $v_C(t)$ for a state variable? This is again not possible, since $dv_C/dt = f_s'(q)\, dq/dt$, and we know neither f_s' [though we know $f(s)$] nor $q_C(t)$ (without using a second integrator to integrate dq_C/dt). We have left for last the variable q_C, which meets all our requirement for a state variable, as summa-

TABLE 7.4-1 Tests that a State Variable Must Meet

1. Its derivative is a function of the input and of the state variable. Specifically, by inspection of Fig. 7.4-1,

   For all $t \geq t_0$,

   $$\frac{dq_C}{dt} = i_R = f_G(v_s - v_C) = f_G(v_s - f_s(q_C)) \tag{5}$$

   and

   $$q_C(t_0) = q_0$$

2. All the outputs can be expressed as functions of the state variable and of the input. Specifically, our three desired outputs of Eq. (4) are related to q_C and v_s as follows:

   For all $t \geq t_0$,

   $$v_C = f_s(q_C) \qquad \text{by the capacitor constraint} \tag{6}$$

   $$v_R = v_s - v_C \qquad \text{by KVL} \tag{7}$$

   $$i_R = f_G(v_R) \qquad \text{by the resistor constraint} \tag{8}$$

rized in Table 7.4-1, where v_C, q_C, v_R, v_s, and i_R denote values of these variables at time $t \geq t_0$. Equation (5) is the desired first-order differential equation (D.E.), and Eqs. (6) to (8) are the output equations. Rearranging these equations we have the desired final formulation which is in the form of Eq. (1).

State D.E.	$\frac{dq_C}{dt} = f_G(v_s - f_s(q_C)) \qquad q_C(t_0) = q_0$	(9*a*)
Output Equations	$v_C = f_s(q_C)$	(9*b*)
	$v_R = v_s - f_s(q_C)$	(9*a*)
	$i_R = f_G(v_s - f_s(q_C))$	(9*d*)

Observe that although these equations were formulated from the network of Fig. 7.4-1*a*, they can be thought of as equations formulated from a first-order dynamic system.

The corresponding dynamic system, characterized by Eqs. (9) is shown in Fig. 7.4-2. Observe that the system has only one integrator, and all three outputs as well as the state update are functions of q_C and v_s; we have emphasized this in Fig. 7.4-2 by enclosing in a larger box all these functions which make up the memoryless portion of the state model.

Example 7.4-1 *A Linear RL Network: Formulation of Equations* Consider the network of Fig. 7.4-3, where the resistor is *linear* with resistance R, and the inductor is *linear* with inductance L. Let $i(t)$ be the input.

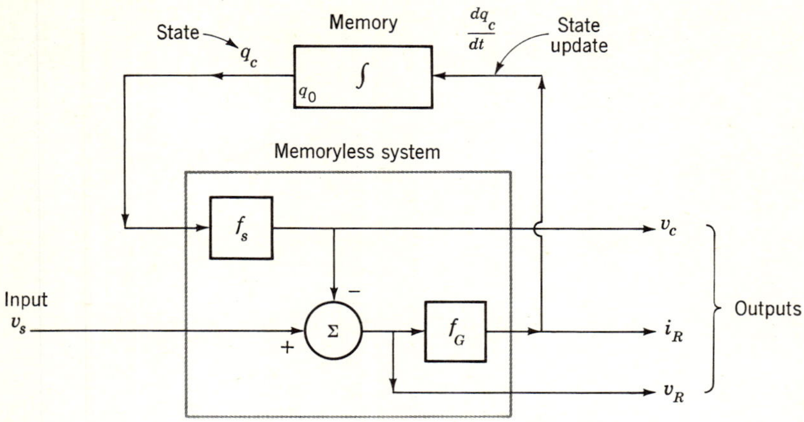

FIG. 7.4-2 State model for the network of Fig. 7.4-1.

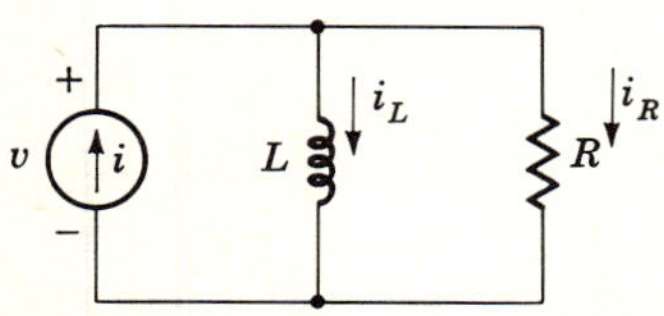

FIG. 7.4-3 A current-driven linear RL network.

Since in a capacitor we let the charge be the state variable, a good guess for a state variable in this example is the flux ϕ_L which meets all the tests of a state variable, as the reader is invited to verify. The resulting equations are:

State† D.E.	$\frac{d\phi_L}{dt} = R\left(i - \frac{\phi_L}{L}\right) \qquad \phi(t_0) = \phi_0$	(10
Output Equations	$i_L = \frac{\phi_L}{L}$ $i_R = i - i_L$ $v = Ri_R$	

Note that since in a linear inductor ϕ_L and i_L are linearly related ($\phi_L = Li_L$), we could equally well have used i_L as the state variable. The resulting equation would then be

†The derivation is straightforward: $d\phi_L/dt = v = Ri_R = R(i - i_L) = R(i - \phi_L/L)$.

State D.E.	$\frac{di_L}{dt} = \frac{R}{L}(i - i_L)$	$i_L(t_0) = i_{L_0}$	(11)
Output Equations	i_L	(the state variable)	
	$i_R = i - i_L$		
	$v = R(i - i_L)$		

Equations (11) are linearly related to Eqs. (10). They vividly demonstrate the fact that, in general, there will be more than one network variable that can be used as a state variable.

The state model corresponding to Eqs. (11) is shown in Fig. 7.4-4.

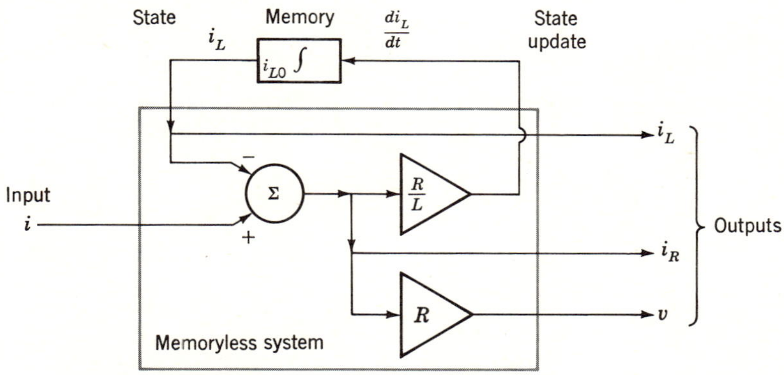

FIG. 7.4-4 State model for the network of Fig. 7.4-3.

Observe that this is a linear system; i.e., the functions f and g of Eq. (1) are *linear,* consisting of the composition of summers and scalers.

What we have observed here will in general be the "rule." That is, *for networks with one linear inductor we will use the inductor current as a state variable, and for networks with one linear capacitor we will use the capacitor voltage as a state variable.* For nonlinear networks we shall try to use the charge or flux. There will be, of course, networks with a single energy-storage element which we cannot put in this form, as demonstrated by the following example of a linear capacitor across a voltage source.

Example 7.4-2 *A Network that Cannot Be Put in State Model Form* See Fig. 7.4-5. Here, we cannot use v as a state variable, since its derivative dv/dt (and hence i_C) is not a function of the input value v_s and of the state-variable value v, but rather of the value dv_s/dt. Neither can we use i or v_s as state variables, since we do not know their derivatives. Yet, we can express the behavior of the network by the following equation:

$$i = C\frac{dv_s}{dt}$$

which tells us that $i(t)$ will be the time derivative of $v_s(t)$, weighted by C.

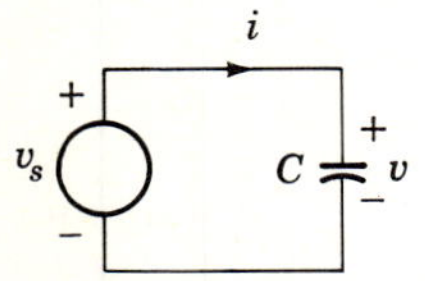

FIG. 7.4-5 Example of a network that cannot be put in state model form with an integrator as the memory element.

Alternatively we might be confronted with the *second-order* network of Fig. 7.4-6 which can only be adequately described in terms of *two* state variables, i and v_C.

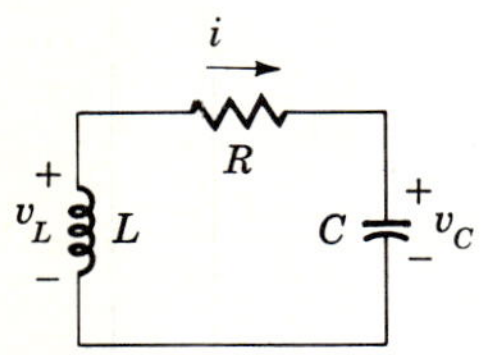

FIG. 7.4-6 Example of a second-order network.

Making an Integrator out of Network Elements

In Sec. 4.7 we saw how resistors and sources can be combined to construct memoryless block diagram elements such as summers, scalers, and function blocks. The same approach is possible with block diagram elements with memory. That is, resistors, *capacitors,* and sources can be (and are) used to construct intergrators for use in analog computers.

Perhaps the simplest integrator that we can conceive could be made out of a single capacitor, of unit capacitance, as shown in Fig. 7.4-7*a*.

Here, if we let $i(t)$ be considered the input and $v(t)$ the output, we have by the defining relation of a linear capacitor the equation

$$v(t) = \int_{t_0}^{t} i(\tau)\, d\tau + v(t_0) \tag{12}$$

which is precisely the equation characterizing an integrator. Unfortunately,

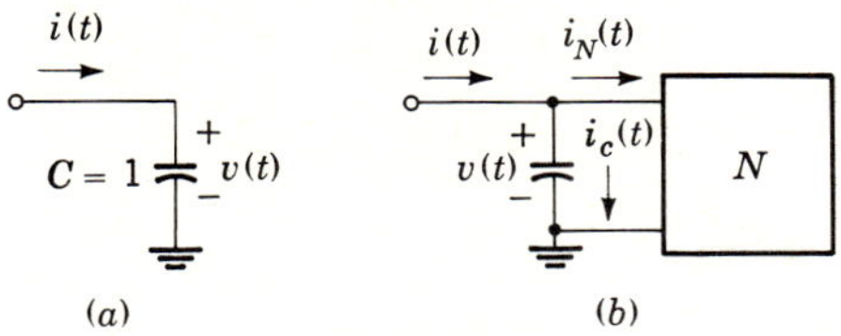

FIG. 7.4-7 A simple but unsatisfactory integrator.

however, our "integrator" ceases to behave like an integrator as soon as it is put to some use, i.e., as soon as it is connected to some other network N, which draws a current $i_N(t)$, as shown in Fig. 7.4-7*b*. For in this case the output $v(t)$ is no longer the time integral of the input $i(t)$, but of the capacitor current $i_C(t)$, which is by KCL different from $i(t)$.

A possible "cure" to this problem is the introduction of a dependent voltage source which ensures that $i_N = 0$, as shown in Fig. 7.4-8.

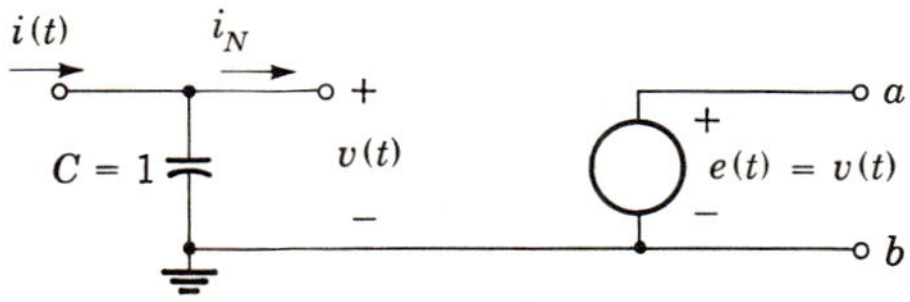

FIG. 7.4-8 A better, but still inadequate, integrator.

Here, the dependent source produces an output $e(t)$ which is equal to $v(t)$ of Eq. (12); moreover, the current through the capacitor is always $i(t)$, regardless of the current drawn at output terminals a and b. Unfortunately, this approach is unsatisfactory because of technological limitations which make dependent sources with a *constant* multiplying factor impossible or very costly. That is, in the dependent-source relationship $e(t) = Kv(t)$, it is impossible or very costly to keep K equal to 1 (or any other constant value); temperature, aging, and even the current drawn at terminals a and b will normally affect the value of K, typically by a factor of 2, if the dependent source is made out of transistors.

To avoid this dependence on K, we resort to the same approach that we used in Sec. 4.7 for the realization of summers. That is, we use a dependent source with a very large K (an operational amplifier or *opamp* as we have called it) in a closed loop, so that variations of K affect the output by a negligible amount. A network with these properties is shown in Fig. 7.4-9.

Observe that, by definition of the linear capacitor,

$$\frac{dv_C}{dt} = \frac{i_C}{C} \tag{13}$$

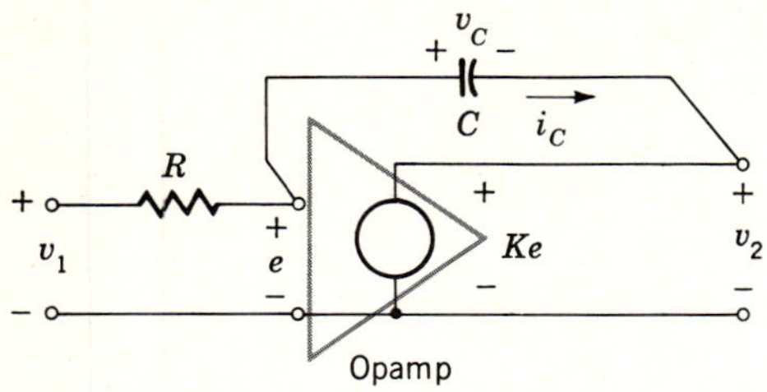

FIG. 7.4-9 Use of an opamp to construct an integrator.

But since i_C flows through R, it is also true that

$$i_C = \frac{v_1 - e}{R} \tag{14}$$

Observe further that by KVL

$$Ke + v_C = e$$

Hence

$$e = \frac{v_C}{1 - K} \tag{15}$$

Combination of Eqs. (13) to (15) yields

$$\frac{dv_C}{dt} = \frac{1}{RC}\left(v_1 - \frac{v_C}{1 - K}\right) \tag{16}$$

which is a first-order differential equation. Since, however, for an opamp K is very large in magnitude (10^6), we can rewrite Eq. (16) as

$$\frac{dv_C}{dt} \simeq \frac{1}{RC} v_1 \tag{17}$$

where the approximation is good to about one part per million, or one part per 500,000 if K varies by a factor of 2. Equation (17) can be written in integral form as follows:

$$v_c(t) \simeq \frac{1}{RC}\int_{t_0}^{t} v_1(\tau)\, d\tau + v_c(t_0) \tag{18}$$

Since, $v_2 = Ke$ and $e = v_C/(1 - K)$, it follows that

$$v_2 = \frac{K}{1 - K} v_C$$

Hence, again within one or two parts per million,

$$v_2 \simeq v_C \tag{19}$$

Setting† $RC = 1$ and combining Eqs. (18) and (19), we have the final result,

$$v_2(t) \simeq -\int_{t_0}^{t} v_1(\tau)\, d\tau + v_C(t_0) \tag{20}$$

Equation (20) is the desired input-output relationship of an integrator (except for a minus sign), where v_1 is the input, v_2 the output, and $v_2(t_0)$ the initial value of the output [equal to the voltage $v_C(t_0)$ stored initially across the capacitor].

7.5 SOLUTION OF A SIMPLE FIRST-ORDER ORDINARY DIFFERENTIAL EQUATION

In the foregoing, we saw that first-order systems and networks give rise to two equations of the form

$$\frac{dx}{dt} = f(x,u) \qquad x(t_0) = x_0 \tag{1a}$$

$$y = g(x,u) \tag{1b}$$

where u is the value $u(t)$ of the input at any time t, x is the value $x(t)$ of the state, dx/dt is the value $dx(t)/dt$ of the state update, y is the value $y(t)$ of the output, and f and g are two functions. We have remarked that Eq. (1a) can be solved first to establish the desired function $x(.)$. Thereafter, values of this function can be used in Eq. (1b) to establish the output. The purpose of this section is to introduce first-order differential equations and to illustrate in an informal way the solution of such a simple equation. We proceed now with a more precise statement of what is a first-order differential equation, and what is the meaning of a solution of such an equation.

DEFINITION 7.5-1 A first-order ordinary differential equation is defined as follows‡:

$$\frac{dx}{dt} = f(x,u) \qquad \text{and} \qquad x(t_0) = x_0 \tag{2}$$

Here variable t, which usually denotes time, takes on real values; the *given* quantities are

1. A *forcing function* $u(.)$ from real numbers to real numbers,

†Practically, the value of a resistor can be held constant to about one part per thousand, and the value of a capacitor to about one part per 200, so that the constant $1/RC$ affects the accuracy of integration in a more serious way than K.

‡The more conventional mathematical definition is $dx/dt = g(x,t)$. Since we know the input function $u(t)$, we can relate, without loss of any generality, the two definitions: i.e., $f(x,u(t)) = g(x,t)$, for all x and t.

2. An *initial value* x_0 (a real number), and
3. A function $f(.\,,.)$ from pairs of real numbers to real numbers.

The *required* quantity, which when found constitutes a solution, is a function $x(.)$, such that for the initial time t_0, $x(t_0) = x_0$ and for every value of $t \geq t_0$, the real numbers $\frac{dx}{dt}(t)$, $x(t)$, and $u(t)$ satisfy the following equation:

$$\frac{dx}{dt}(t) = f(x(t), u(t))$$

Several questions arise immediately: Is there, in the first place, a solution $x(.)$, and if there is one, could there be others $x_1(.)$, $x_2(.)$ . . . , as well? These questions of existence and uniqueness, respectively, arise in the solution of equations, as we have already seen in Chap. 5 for $x = f(x)$. For the present case, we state without proof a general theorem that concerns existence and uniqueness of a solution to Eq. (2).

THEOREM 7.5-1 If (1) $u(t)$ is bounded and has a finite number of jumps in a finite time interval; and if (2) $f(x,u)$ is finite for all finite x and u, and continuous for all x and u; and (3) if for all u, x_1, x_2 there is a finite constant L such that

$$|f(x_1,u) - f(x_2,u)| \leq L\, |x_1 - x_2|$$

then Eq. (2) has a unique solution.

In a different language, the above condition (1) on $u(t)$ states that the input waveform $u(t)$ cannot "blow up" (to an infinite value), but is allowed to jump (for example, its graph in time could look like a staircase), as long as it does not jump an infinite number of times. The second condition on f states that it too must not "blow up" for any x and u, and that it must be continuous in both x and u; that is, if y, x, and u are visualized as three orthogonal coordinates, then the surface $y = f(x,u)$ cannot have "steps" or jumps, though it may very well have "corners." Finally, the Lipschitz condition† (3) says that any two points lying on the plane (u = constant) are connected by a line whose slope magnitude is less than some finite L. A *stricter* condition on $f(x,u)$ which implies this Lipschitz condition is that $\partial f/\partial x$ be continuous (that is, $f(.\,,.)$ can have no corners along the x direction).

The above conditions are not very strict from an engineering point of view since to our knowledge they are obeyed by all waveforms $u(t)$ generated by physical devices and all functions $f(.\,,.)$ representing first-order networks and systems. Why, then, do we worry about existence and uniqueness? Although the general answer to this question is "in order to increase our knowledge and understanding," we can cite some very practical cases where difficulty arises, because in our *modeling* of a physical system we have violated one or more of these conditions. Specifically, in solving first-order D.E.'s by digital computer we very often encounter the existence of solutions other than the one we seek.

†Note that here L can be any finite number, whereas in Chap. 5, in order to ensure convergence of Picard's algorithm L had to be less than 1.

These parasitic solutions, as they are called, are "excited" by roundoff errors and can blow up to overflow values, thereby hiding the solution of interest. The main reason for this trouble is that the digital computer is *not* a continuous device. But more on this subject later. Let us return now to the way in which a simple D.E. "works."

Walking through a Nonlinear First-order Differential Equation

Consider the differential equation characterizing the system of Fig. 7.5-1*a*,

$$\frac{dx}{dt} = u - f(x) \qquad x(0) = 2 \tag{3}$$

where $u = -\frac{1}{2}$, and $f(x)$ is shown in Fig. 7.5-1*b*.

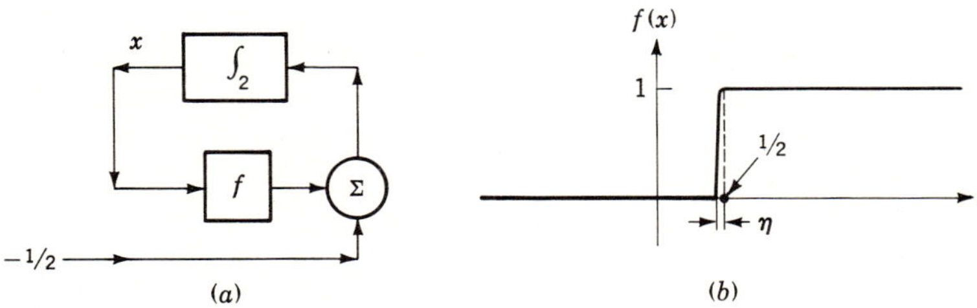

FIG. 7.5-1 A nonlinear first-order system, and the function f.

Note that $f(.)$, which is continuous, undergoes at $x = \frac{1}{2}$ a very rapid transition, from value 0 to value 1. The precise nature of this transition is not very important for our discussion; what is important, however, is that it is a sharp transition. Let us now "walk through" the differential equation in order to find the solution.

At $t = 0$, the starting time, we have the following situation:

$$x(0) = 2 \qquad f(x(0)) = f(2) = 1$$

Hence, by Eq. (3) we can evaluate dx/dt at $t = 0$. This yields

$$\frac{dx}{dt}(0) = -\frac{1}{2} - 1 = -\frac{3}{2} \tag{4}$$

A very small amount of time, Δt, later we expect that we will have

$$x(\Delta t) \cong x(0) + \frac{dx}{dt}\Delta t = 2 - \frac{3}{2}\Delta t$$

Since $\Delta t < < 2$, we will have at time Δt.

$$x(\Delta t) \cong 2 - \frac{3}{2}\Delta t \qquad f(x(\Delta t)) = f(2 - \frac{3}{2}\Delta t) = 1$$

and from Eq. (3) we have

$$\frac{dx}{dt}(\Delta t) = -\,{}^3\!/_2 \tag{5}$$

From Eqs. (4) and (5) we conclude that the condition

$$\frac{dx}{dt} = -\,{}^3\!/_2 \tag{6}$$

will last as long as x is greater than ½. To find up to what time Eq. (6) will hold, we assume that the equation holds and we obtain an expression for $x(t)$ by straight-forward integration of Eq. (6). Thus, we have

$$x(t) = x_0 - \int_0^t \frac{dx}{dt}\,dt = 2 - {}^3\!/_2\,t$$

Hence at $t = 1$,

$$x(1) = {}^1\!/_2$$

The solution in the interval $0 \leq t \leq 1$ is shown in Fig. 7.5-2.

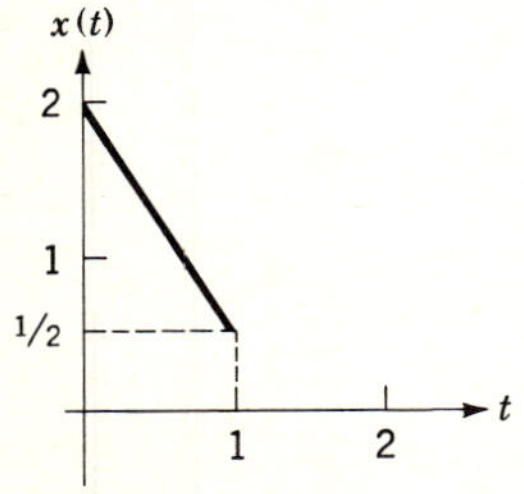

FIG. 7.5-2 Solution in the interval $0 \leq t \leq 1$.

For $t > 1$ the following will happen: As t exceeds the value 1, $x(t)$ will decrease below ½, and hence $f(x(t))$ will decrease very rapidly [because of the sharp transition of $f(\cdot)$]. As $f(x(t))$ decreases, dx/dt will change from $-{}^3\!/_2$ toward $-{}^1\!/_2$, and we expect that in a very short time, say, at $t = 1 + \epsilon$ where ϵ is positive and small, we will have "crossed" the transition of $f(.)$. That is, we will have the case

$$x(1+\epsilon) = {}^1\!/_2 - \eta \qquad \text{and} \qquad f(x(1+\epsilon)) = 0$$

where η is the closest distance, from $x = {}^1\!/_2$, to the point where the value of $f({}^1\!/_2 - \eta)$ is zero (see Fig. 7.5-1*b*). Hence by Eq. (3) we will have

For $t > 1 + \epsilon$

$$\frac{dx}{dt} = -{}^1\!/_2$$

which will last forever, and x will keep decreasing as t increases. In graphical terms, the entire solution is shown in Fig. 7.5-3, along with a plot of dx/dt versus t.

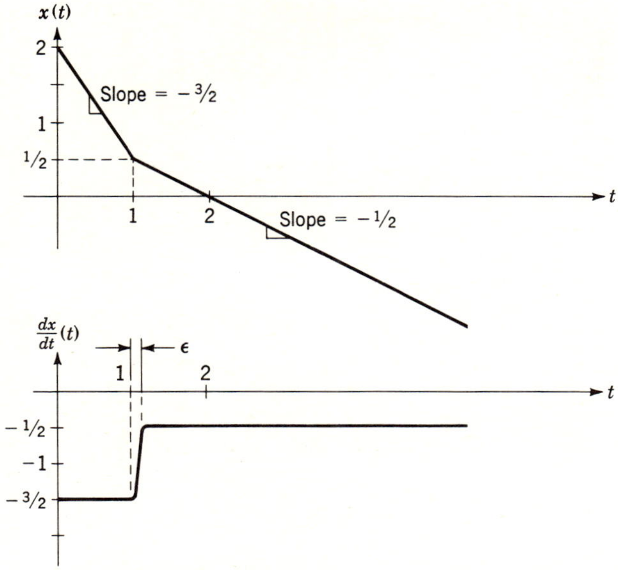

FIG. 7.5-3 The entire solution.

The reader is invited to check that $x(t)$ of Fig. 7.5-3 is indeed the solution: i.e., $x(0) = 2$; and for all $t \geq 0$, $dx/dt = u - f(x)$. We can now visualize the entire solution in the following intuitive way, which can be sometimes helpful when we are dealing with nonlinear differential equations.

Initially, $x(0) = 2$; we can visualize this starting "point" on the graph of $f(x)$ shown in Fig. 7.5-4 as point a. As time increases, we assume that this point "slides" to the left at a rate of $dx/dt = -3/2$. As it slides, $f(x(t))$ remains

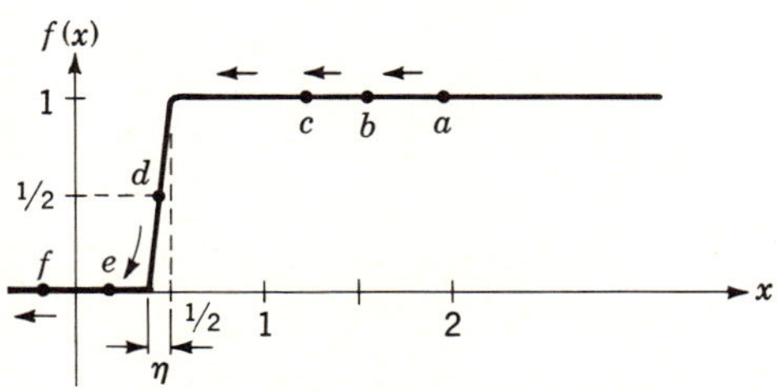

FIG. 7.5-4 The "sliding" solution point.

constant (for $t < 1$) at a value of 1, thus confirming from Eq. (3) that $dx/dt = -3/2$; i.e., our assumption of the point sliding at a rate of $-3/2$ is correct. When the point reaches $x = 1/2$, it will slide down the hill with a rate that varies from $-3/2$ at the beginning of the slide to $-1/2$ at the end of the slide. This "crossing" of the transition of f will last ϵ seconds, or approximately

$$\epsilon \cong \frac{\eta}{q}$$

where η is the "width" of the transition in Fig. 7.5-4 and q is the average rate halfway between the starting and final rate. For example, when our point is at

d (Fig. 7.5-4), the value of $f(x)$ will be ½, and hence dx/dt will be, from Eq. (3),

$$\frac{dx}{dt} = -\tfrac{1}{2} - \tfrac{1}{2} = -1$$

7.6 SOLUTION OF FIRST-ORDER DIFFERENTIAL EQUATIONS

In this section we shall be primarily concerned with the linear differential equation (D.E.):

$$\frac{dx}{dt} = ax + bu \qquad x(t_0) = \xi \tag{1}$$

Here a and b are given constant real numbers, the *coefficients;* $u(.)$ is a given function of time called the *forcing function* or *excitation;* and ξ is a given *initial* or *boundary condition* of x at time t_0. By a solution of this D.E., we mean a function of time $x(.)$ which satisfies (1) the initial condition for $t = t_0$ and (2) the equation $\frac{dx}{dt}(t) = ax(t) + bu(t)$ for every value of time $t \geq t_0$.

A graphical interpretation of the above statements should further clarify what we want to do. The given D.E. (1) will usually arise from a fixed network or system, such as the one shown in Fig. 7.6-1*a*, which can be thought of

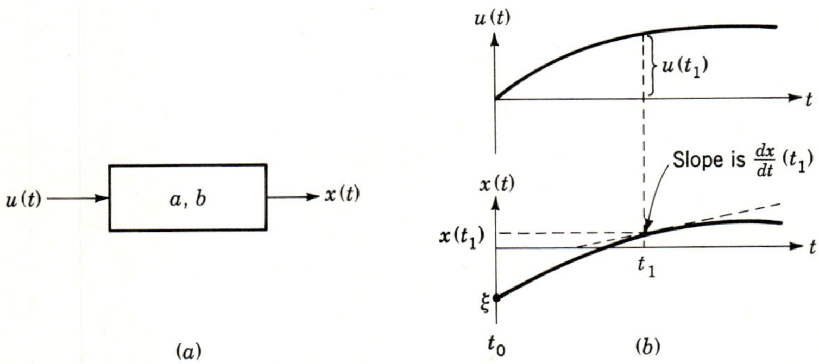

FIG. 7.6-1 A linear dynamic system and a graphical interpretation of its D. E. solution.

as being driven or excited by an input function of time $u(t)$ for $t \geq t_0$. Such an excitation is shown on the top part of Fig. 7.6-1*b*. The fixed nature of the system is reflected in the two constants a and b and in the equation $dx/dt = ax + bu$ which constrains a, b, $x(t)$, and $u(t)$. The output or solution $x(t)$ is only initially specified. That is, the function we seek is constrained to "start" from value ξ when $t = t_0$, as marked in the bottom part of Fig. 7.6-1*b*. Solving the given D.E. means finding a function $x(t)$, shown at the bottom part of Fig. 7.6-1*b*, which starting from the value ξ at t_0, satisfies the given D.E. for every value of $t \geq 0$.

In other words, if in Fig. 7.6-1*b* we chose a time t anywhere to the right of t_0, then the three quantities (shown in that figure) $u(t_1)$, $x(t_1)$, and $(dx/dt)(t_1)$ must satisfy the equation

$$\frac{dx}{dt}(t_1) = ax(t_1) + bu(t_1)$$

The uniqueness theorem (7.5-1) that we have already stated guarantees that there is *exactly* one and only one possible function $x(t)$ that satisfies these requirements, when the conditions of the theorem are met.

The Unforced Case

An important special case of Eq. (1) occurs when the input $u(t)$ is zero for all t, in other words when there is no excitation. Accordingly, in this case, Eq. (1) becomes

$$\frac{dx_h}{dt} = ax_h \qquad x_h(t_0) = \xi \tag{2}$$

and is called the *unforced* or *homogeneous* first-order linear differential equation, corresponding to Eq. (1). The solution of Eq. (2), called the zero-input unforced, homogeneous solution can be easily obtained by direct integration of Eq. (2) after it has been rewritten as follows:

$$\frac{dx_h}{x_h} = a\,dt \qquad x_h(t_0) = \xi \tag{3}$$

Integration of both sides from t_0 to t yields

$$\ln x_h(t) - \ln x_h(t_0) = a(t - t_0)$$

which is rewritten as

$$\boxed{\text{Unforced solution} \qquad x_h(t) = \xi e^{a(t-t_0)}} \tag{4}$$

For positive a, $x_h(t)$ grows in magnitude with increasing time, as shown in Fig. 7.6-2.

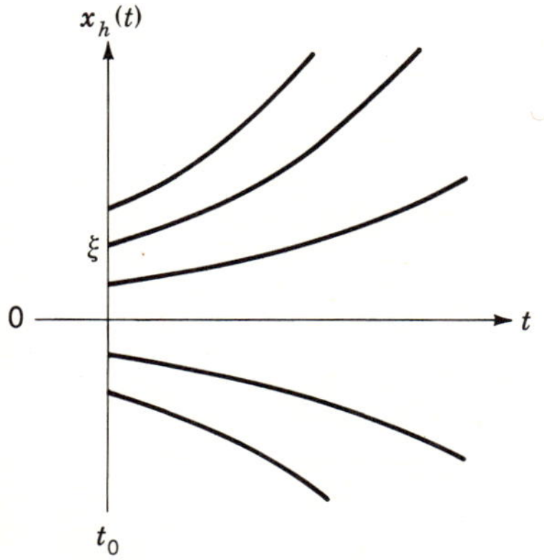

FIG. 7.6-2 Homogeneous solutions for positive a and various initial values ξ.

For D.E.'s with negative a, $x_h(t)$ decays in magnitude with increasing time, as shown in Fig. 7.6-3.

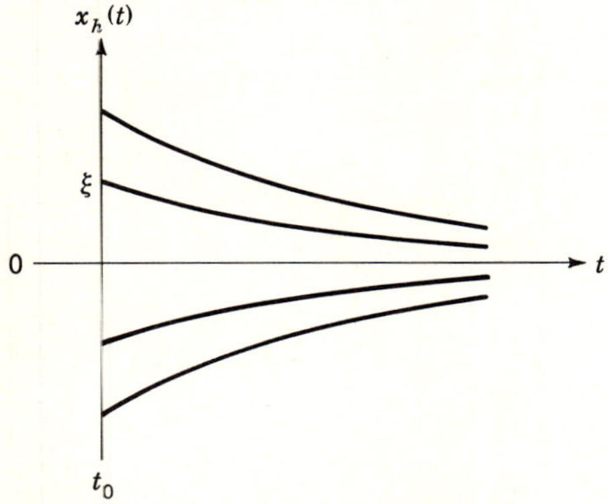

FIG. 7.6-3 Homogeneous solutions for negative a, and various initial values ξ.

When $a=0$, then $x_h(t)=\xi$, a constat for all $t \geq t_0$. These waveforms are significant in that they demonstrate how *every* linear first-order dynamic system will behave or, as we say, "relax" when the excitation is removed.

The Forced Case

Returning to Eq. (1), we now seek the solution $x(t)$ when there is an excitation present. To simplify matters, we combine the known constant b and known function $u(.)$ into a known function $v(.)$. Thus, we wish to solve the D.E.,

$$\frac{dx}{dt} = ax + v \qquad x(t_0) = \xi \tag{5}$$

where $v(t)=bu(t)$, for all $t \geq t_0$. Since this equation is the homogeneous equation with the "drive" $v(t)$ superposed, we ask what function $q(t)$ must we add to the homogeneous solution $x_h(t)$ of Eq. (4) to obtain the solution of Eq. (5). That is, we seek $q(t)$ such that

$$x(t) = x_h(t) + q(t) \tag{6}$$

Substitution of Eq. (6) into Eq. (5) yields

$$\frac{d(x_h + q)}{dt} = a(x_h + q) + v \qquad x_h(t_0) + q(t_0) = \xi \tag{7}$$

Rearranging Eqs. (7) yields

$$\frac{dx_h}{dt} + \frac{dq}{dt} = ax_h + aq + v \qquad x_h(t_0) + q(t_0) = \xi \tag{8}$$

Equations (2) and (8), in turn, yield

$$\frac{dq}{dt} = aq + v \qquad q(t_0) = 0 \tag{9}$$

Hence, the function q (that we must add to x_h to get x) is itself a solution of exactly the same D.E., with one important distinction—*the initial condition on q is zero,* not ξ. Because Eq. (5), hence Eq. (9), will be typically a differential equation in a state variable, we call the solution to Eq. (9) the *zero-state* solution or response of the D.E.

The above discussion has merely established the fact that

Desired solution = zero-input solution + zero-state solution

The zero-state solution, i.e., the solution to Eq. (9), is given by the following theorem, where we have set $t_0 = 0$ to simplify matters.

THEOREM 7.6-1 If the excitation $v(t)$ is bounded and piecewise-continuous, with a finite number of jumps in a finite time interval, then the unique zero-state solution $q(t)$ to Eq. (9) with $t_0 = 0$ is given by

For all $t \leq 0$,

$$q(t) = e^{at} \int_0^t e^{-a\tau} v(\tau)\, d\tau \tag{10}$$

PROOF: $q(t)$ as given by Eq. (10),

1. Satisfies the initial condition $q(0) = 0$, since

$$\int_0^0 e^{-a\tau} v(\tau)\, d\tau = 0$$

   because the integrand is bounded.
2. Satisfies Eq. (9) for all $t \geq 0$: i.e., substituting Eq. (10) into Eq. (9) yields

$$\frac{dq}{dt} = ae^{at} \int_0^t e^{-a\tau} v(\tau)\, d\tau + e^{at} \frac{d}{dt} \int_0^t e^{-a\tau} v(\tau)\, d\tau$$

$$= ae^{at} \int_0^t e^{-a\tau} v(\tau)\, d\tau + e^{at}\, e^{-at}\, v(t)$$

$$= aq(t) + v(t)$$

Hence, $q(t)$ as given by Eq. (10) is a solution of Eq. (9). However Eq. (9) satisfies all the conditions of Theorem 7.5-1, as the reader is invited to determine. Therefore $q(t)$ is the unique solution to Eq. (9). Q.E.D.

Since, by Eq. (6), the overall solution is the sum of the zero-state (10) and zero-input (4) solutions, it follows that

THEOREM 7.6-2 The unique solution to the D.E.,

$$\frac{dx}{dt} = ax + v \qquad x(0) = \xi$$

is

$$x(t) = \xi e^{at} + e^{at} \int_0^t e^{-a\tau} v(\tau) \, d\tau$$

which may be rewritten in the desired final form,

$$\boxed{x(t) = e^{at}[\xi + \int_0^t e^{-a\tau} v(\tau) \, d\tau]} \tag{11}$$

First, we wish to point out that although what we have done is technically in order, it is less than satisfying to "pull a solution out of a hat," as we did in Theorem 7.6-1. We intend to remedy this in the following section where we will show how the above solution can be derived.

Second, the solution as given by Eq. (11) above is still not in explicit form. If, for the given excitation $v(.)$, the product $e^{-a\tau}v(\tau)$ can be integrated analytically, then we can give explicit solutions, as we do in the examples that follow. If, on the other hand, the above product is not in a recognizable form, we will generally resort to numerical integration, discussed in Chap. 9.

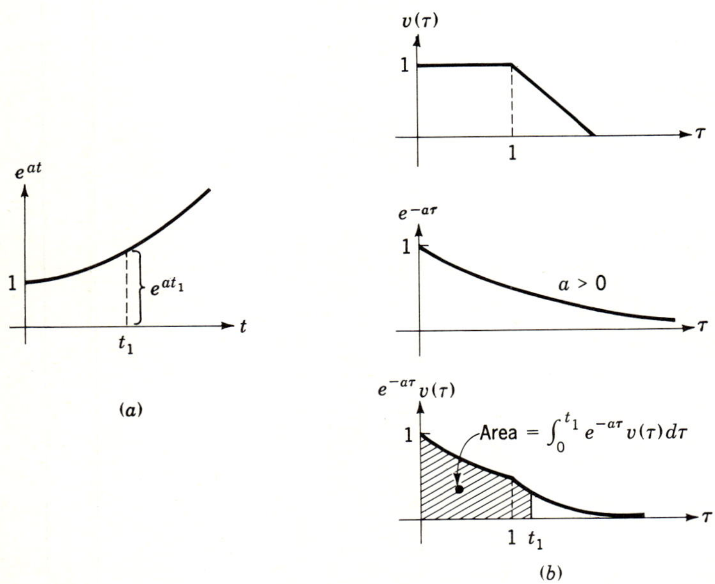

FIG. 7.6-4 Graphical illustration of Eq. (10).

Finally, before we proceed with some typical cases of specific excitations, we should like to clarify in graphical terms the meaning of Eq. (10). As shown in Fig. 7.6-4, the zero-state solution is the product of two functions. One is e^{at}. The other is the area under the excitation, weighted by $e^{-a\tau}$. Thus, for any value $t_1 > 0$ of t, the solution is the product of e^{at} (as shown in Fig. 7.6-4*a*) and

$$\int_0^{t_1} e^{-a\tau}v(\tau)\,d\tau \qquad \text{(shaded area in the bottom of Fig. 7.6-4}b\text{)}$$

We are now ready to consider the solutions to the linear first-order D.E. (1) for the three special but important cases of constant, exponential, and sinusoidal excitations.

Constant Excitation

We seek the solution of the D.E.

$$\frac{dx}{dt} = ax + V \qquad x(0) = \xi \tag{12}$$

where V is a constant. Substitution of V for $v(\tau)$ in Eq. (11) yields for the solution

$$x(t) = e^{at}\left(\xi + V\int_0^t e^{-a\tau}\,d\tau\right)$$

$$= e^{at}\xi - \frac{1}{a}(1 - e^{at})V \tag{13}$$

Figure 7.6-5 shows the form of this solution for several choices of a and ξ. Note that (1) when $a < 0$, the solution starts from value ξ and goes exponentially

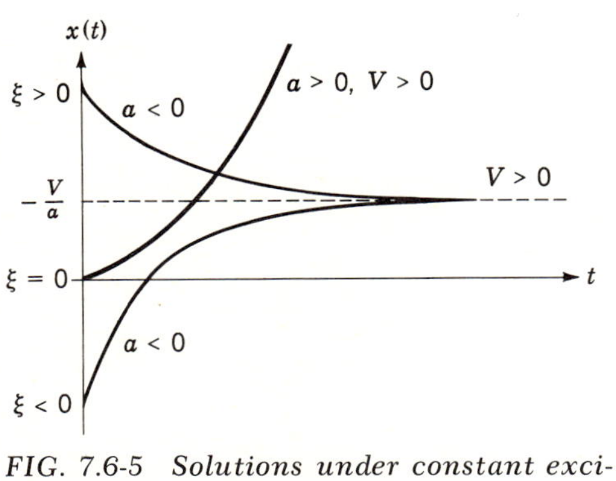

FIG. 7.6-5 Solutions under constant excitation for several a, and ξ.

to value $-V/a$; (2) when $a > 0$, the solution starts from ξ and grows in magnitude exponentially; (3) when $a = 0$, the solution is $\xi + Vt$ for all t. We will return to this case in considerable detail in the following section.

Exponential Excitation

Here the D.E. is again as before, but the excitation is exponential. That is, we seek the solution to

$$\frac{dx}{dt} = ax + be^{st} \qquad x(0) = \xi \tag{14}$$

where a, b, and s are constants. Use of Eq. (11) with this excitation, and straightforward integration result in the solution

$$\begin{aligned} x(t) &= \frac{b}{s-a}(e^{st} - e^{at}) + e^{at}\,\xi \qquad && \textit{for } s \neq a \\ x(t) &= bte^{at} + e^{at}\xi && \textit{for } s = a \end{aligned} \tag{15}$$

Figure 7.6-6 illustrates the nature of the solution for specific values of the

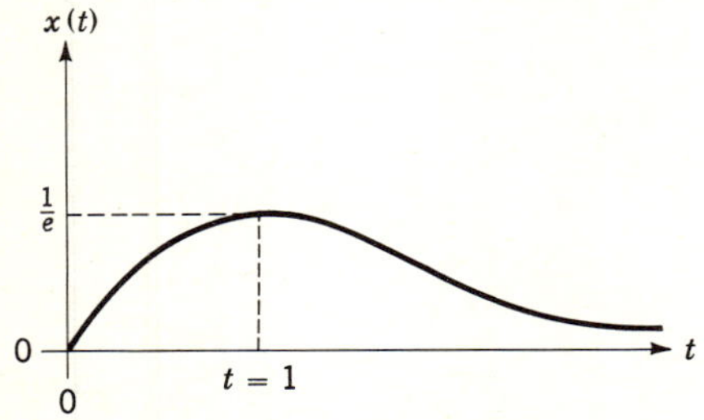

FIG. 7.6-6 Solution under exponential excitation: $s = a = -1$; $\xi = 0$; $b = 1$.

parameters. As we have done in the proof of Theorem 7.6-1, it is easy to verify by substitution in the original D.E. that a certain function is indeed the solution. For example, if $s = a$, i.e., if the D.E. is

$$\frac{dx}{dt} = ax + be^{at} \qquad x(0) = \xi$$

then we can ascertain that the bottom equation (15) gives the solution, since

1. For $t = 0$, $x(0) = b0e^{a0} + e^{a0}\xi = \xi$, i.e., the function $x(.)$ satisfies the initial condition.
2. For every $t > 0$, using $x(t) = bte^{at} + e^{at}\xi$, we have

$$\frac{dx}{dt} = abte^{at} + be^{at} + ae^{at}\xi$$

$$ax = abte^{at} + ae^{at}\xi$$

and indeed

$$\frac{dx}{dt} = ax + be^{at}$$

i.e., function $x(.)$ satisfies the D.E. for all $t > 0$.

Sinusoidal Excitation

When the excitation is a sine wave, then the solution of

$$\frac{dx}{dt} = ax + b \sin \omega t \qquad x(0) = \xi$$

is again obtained through use of Eq. (11), and is

$$x(t) = e^{at}\xi + \frac{\omega b}{a^2 + \omega^2} e^{at} - \frac{\omega b}{a^2 + \omega^2}(a \sin \omega t + \omega \cos \omega t) \tag{16}$$

A typical solution waveform is shown in Fig. 7.6-7, for $a < 0$, from where it can be seen that $x(t)$ starts (at $t = 0$) with value ξ and decays "exponentially" to a sinusoid of zero average value (since the first two terms of Eq. (16) vanish as $t \to \infty$). We will reexamine this type of solution in greater detail, in later sections.

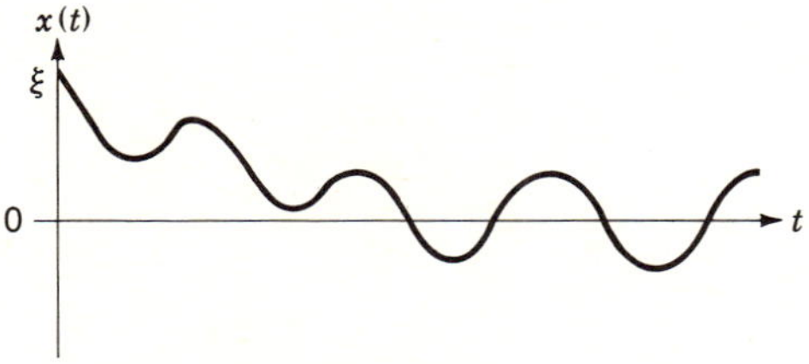

FIG. 7.6-7 Solution under sinusoidal excitation.

Nonlinear First-Order Differential Equations

Recall that a nonlinear first-order D.E. is of the form

$$\frac{dx}{dt} = f(x,u) \qquad x(t_0) = x_0 \tag{17}$$

where f is a nonlinear function of two arguments, $u(.)$ is, a function of time—the *excitation*, and $x(.)$ is another function of time—the desired solution, with initial value x_0.

Unlike the case of linear D.E.'s, there are no general explicit analytical ways for obtaining the solution to Eq. (17). If functions f and u are given in the form of expressions, then *we may be able to guess an expression* for the solution function x, as shown in the following example.

Example 7.6-1 *Analytical Solution of a Nonlinear First-order D.E..* Suppose we wish to solve the D.E.

$$\frac{dx}{dt} = 2\sqrt{x} \qquad x(0) = 1 \tag{18}$$

Here, the function f of Eq. (17) is $2\sqrt{(.)}$. In trying to solve this D.E., we seek a function $x(t)$, the time derivative of which equals twice its square root. Such a function, we guess, is of the form $x(t) = t^2$, since $\frac{d}{dt}(t^2) = 2\sqrt{t^2}$. However, the function t^2 does not satisfy the initial condition $x(0) = 1$. So we proceed to refine our guess, and we try

$$x(t) = (t+1)^2 \tag{19}$$

Indeed, $x(0) = 1$ and $dx/dt = 2(t+1) = 2\sqrt{(t+1)^2}$. Hence, the function given by Eq. (19) is a solution to the given D.E.

Guessing a solution is not a very satisfactory approach, especially since we cannot formalize our guessing procedures—they seem to consist of a search over the pairs of expressions and associated derivatives known to us, until

$$\text{Derivative} = f(\text{expression})$$

The more common approach used to find the solution of nonlinear differential equations involves the use of a computer program implementing an appropriate *numerical integration* algorithm. This is the main topic of Chap. 9. We give a preview here, using an inefficient but simple algorithm, due to Euler.

Example 7.6-2 *Numerical Solution of a Nonlinear D.E.* Consider the D.E. in Example 7.6-1, i.e.,

$$\frac{dx}{dt} = 2\sqrt{x} \qquad x(0) = 1$$

Integrating this D.E. yields, for any choice of t_0 and t ($t > t_0$).

$$x(t) = x(t_0) + 2\int_{t_0}^{t} \sqrt{x(\tau)}\, d\tau \tag{20}$$

Euler's algorithm, motivated from Eq. (20), *estimates* a sequence of equidistant-in-time values $x(0)$, $x(h)$, $x(2h)$, . . . , $x(nh)$, . . . as follows:

$$x((n+1)h) = x(nh) + 2h\sqrt{x(nh)} \tag{21}$$

Note that h is chosen small enough so that the integral $2\int_{nh}^{(n+1)h} \sqrt{x(\tau)}\, d\tau$ is approximated by the "rectangle area" $2hx(nh)$.

Using Euler's algorithm (21) with the given initial value $x(0) = 1$ and $h = 0.1$ yields the sequence shown in the leftmost column of Table 7.6-1. In the next column, the computation is repeated with $h = 0.01$, but is printed only at increments of 0.1. The third column shows the values obtained by evaluating the exact solution (19). The reader is invited to compare these results and to try to understand why the use of smaller h yields a sequence of values which are closer to the exact solution.

TABLE 7.6-1 Comparison of Numerical Solutions to Exact Solutions

t	x: $h = 0.1$	$h = 0.01$	*Exact*
0	1	1	1
0.1	1.2	1.208956566	1.21
0.2	1.419089023	1.437822062	1.44
0.3	1.657340069	1.686604069	1.69
0.4	1.91481551	1.955309001	1.96
0.5	2.191569228	2.24394236	2.25
0.6	2.487648219	2.552508914	2.56
0.7	2.803093823	2.881012837	2.89
0.8	3.137942674	3.229457812	3.24
0.9	3.492227456	3.597847113	3.61
1	3.865977504	3.98618367	4
1.1	4.259219287	4,39447012	4.41
1.2	4.671976809	4.82270885	4.84
1.3	5.10427193	5.270902028	5.29
1.4	5.556124647	5.739051636	5.76
1.5	6.027553311	6.227159489	6.25
1.6	6.518574831	6.735227259	6.76
1.7	7.029204827	7.263256487	7.29
1.8	7.559457779	7.811248602	7.84
1.9	8.109347142	8.379204929	8.41
2	8.678885451	8.967126704	9

7.7 STEP AND PULSE RESPONSE OF LINEAR FIRST-ORDER NETWORKS AND SYSTEMS

In this section we shall combine the formulation tasks of Sec. 7.2 to 7.4 with the solution of D.E.'s discussed in Secs. 7.5 and 7.6 in order to relate directly the solution of a first-order network or system to network and system parameters. In the process, we shall establish several helpful relations which are quite fundamental and which will help in the treatment of high-order systems.

The *RC* Network; Step Response

The basic network which we shall consider is shown in Fig. 7.7-1. We shall denote the capacitor voltage by $x(t)$; recall that $x(t)$ is a state variable

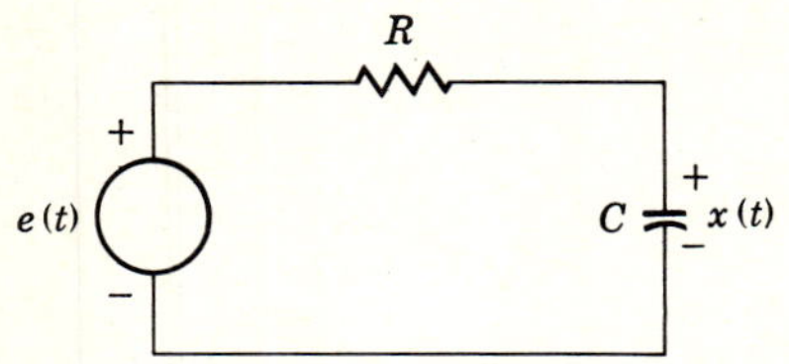

FIG. 7.7-1 The linear RC network.

for this network.

We shall first analyze the network response under the assumption that the source voltage $e(t)$ is a *step function* of amplitude V. More precisely, we have

$$e(t) = \begin{cases} 0 & \text{for all } t < 0 \\ V & \text{for all } t \geq 0 \end{cases} \tag{1}$$

which is also illustrated in Figure 7.7-2.

The differential equation linking $\frac{dx}{dt}(t)$ to $x(t)$ and to $e(t)$ is easily derived, as we have indicated in Sec. 7.4. It is as follows:

$$\frac{dx}{dt} = -\frac{1}{RC}x + \frac{1}{RC}e \tag{2}$$

Next, we must specify a boundary condition, i.e., a value for the capacitor voltage (state variable) at some instant of time. Let us first assume

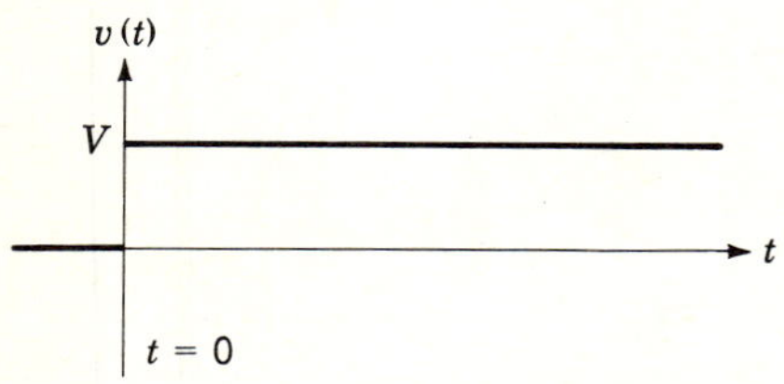

FIG. 7.7-2 Source voltage for the network of Fig. 7.7-1.

that the capacitor voltage is 0 at time 0.

$$x(0) = 0 \tag{3}$$

We can now analyze the entire network by finding $x(t)$ for all $t \geq 0$. To do this we write the general solution (11) of Sec. 7.6 to the differential equation (5) of Sec. 7.6 with the boundary condition (3).

$$x(t) = e^{at}\int_0^t e^{-a\tau}v(\tau)\,d\tau \tag{4}$$

We can relate Eq. (5) of Sec. 7.6 to Eq. (2) as follows:

$$a = -\frac{1}{RC}$$

$$v(t) = \frac{1}{RC} e(t)$$

Substitution of these two terms and Eq. (1) in Eq. (4) yields

$$x(t) = e^{-(1/RC)t} \int_0^t e^{(1/RC)\tau} \frac{V}{RC} \, d\tau \tag{5}$$

which yields

$$x(t) = V(1 - e^{-t/RC}) \qquad \text{for } t \geq 0 \tag{6}$$

This is, as expected, the same solution that we obtained for constant excitation, Eq. (13) of Sec. 7.6. This solution $x(t)$ is shown in Fig. 7.7-3 (where both R and C are greater than 0).

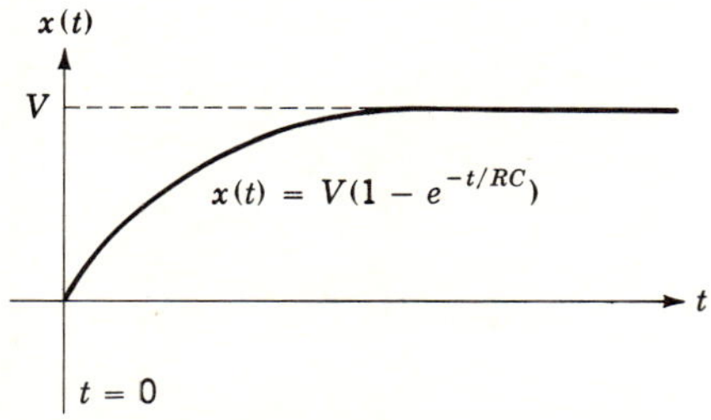

FIG. 7.7-3 The step response of the RC network.

Observe from Eq. (6) or from Fig. 7.7-3 that the capacitor charges up exponentially to the applied constant voltage. The rate of change of $x(t)$ depends on V, R, and C. In particular, as illustrated in Fig. 7.7-4, the smaller the value of RC, the faster the charging of the capacitor, in the sense that $x(t)$ approaches V faster. As $t \to \infty$, then

$$\lim_{t \to \infty} x(t) = V \tag{7}$$

which means that eventually, or as we say when *steady state* is reached (when $dx/dt = 0$) the voltage across C is the same as the applied voltage.

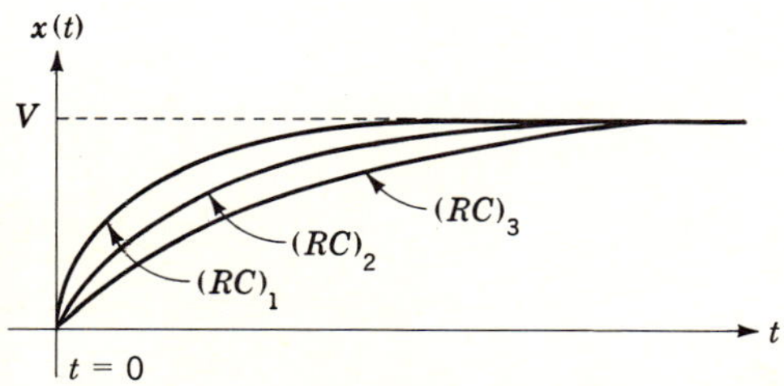

FIG. 7.7-4 Function $x(t)$ for three values of RC: $(RC)_1 < (RC)_2 < (RC)_3$.

Up to now, we have not been concerned with the fact that $e(t)$ of Eq. (1) is a step function, since, for values of time of interest to us ($t \geq 0$), $e(t)$ is a constant excitation V, and $x(0)$ is specified to be 0. Observe, however, that since the excitation is also constant $[e(t) = 0]$ for the (infinitely) long time $t < 0$, it is reasonable to expect from Eq. (7) that $x(t) = 0$ for $t < 0$. Hence, the initial value $x(0)$ of Eq. (3) can be thought of as being equal to the "final" value of x in the period of time $t < 0$. We shall briefly justify this view later in this section, and we shall say a lot more about it in Chap. 8, in connection with suddenly applied excitation. For the time being, however, we shall accept that when the excitation undergoes a step change, we are able to equate the final value of x before the step change to the initial value of x after the change.

Consider next the same RC network with the same constant input V, for $t \geq 0$, but let the initial condition be V_0. That is, the capacitor is initially charged to V_0, and we are therefore solving the equation

$$\frac{dx}{dt} = -\frac{1}{RC}x + \frac{1}{RC}V \qquad x(0) = V_0 \tag{8}$$

Use of Eq. (11) from Sec. 7.6 yields for the solution

$$x(t) = V_0 + (V - V_0)(1 - e^{-t/RC}) \qquad t \geq 0 \tag{9}$$

Graphically, we see from Fig. 7.7-5 that the capacitor voltage starts from value V_0 and ultimately ends at value V in an exponential way. To visualize this exponential better, call the quantity $RC = \tau$, the *time constant* of the network

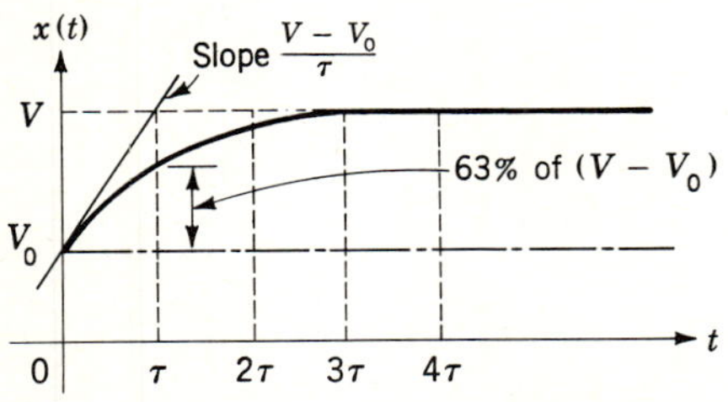

FIG. 7.7-5 Step response of RC network; time-constant relationships.

(no relation to the variable τ used in Eq. (13) of Sec. 7.6). Observe that the initial time rate of change of $x(t)$ is, from differentiation of Eq. (9),

$$\frac{dx}{dt}(0) = \frac{V - V_0}{\tau} \tag{10}$$

Hence, a tangent to the graph of $x(t)$ at the point $[0, V_0]$ will intersect the line of constant V at τ seconds later as shown in Fig. 7.7-5. The value of $x(t)$ at that time ($t = \tau$) is, by Eq. (9),

$$x(\tau) = V_0 + (V - V_0)(1 - e^{-1})$$

or, in numerical terms, $x(t)$ has traversed 63 percent of its overall change $V - V_0$. At $t = 4\tau$, $x(t)$ is, from Eq. (9),

$$x(4\tau) = V_0 + (V - V_0)(1 - e^{-4})$$

Again in numerical terms at that time ($t = 4\tau$), $x(t)$ has traversed 98 percent of its net change, and so it is practically at V.

Knowledge of the above numerical information is helpful in deducing rapidly the response of an RC network to a more complex input. Suppose, for example, that the input $e(t)$ is the following "square wave," and initially

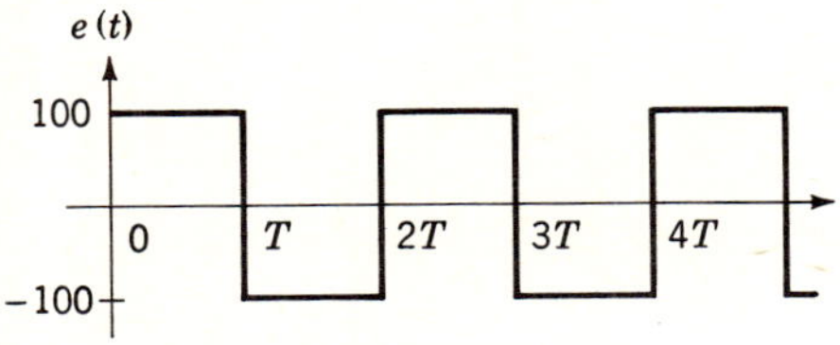

FIG. 7.7-6 *A square-wave drive.*

$x(0) = 0$; note here that the period T is equal to the time constant $\tau = RC$ of the network. From our previous discussion, we can solve for the response in the interval $0 \le t < T$; then we can use, the final value of $x(t)$ in that interval as an initial value for the next interval, $T \le t < 2T$, and so on. Moreover, since the period is τ, we know that in each such interval $x(t)$ will tend toward the constant value of $e(t)$ in that interval and will in fact traverse 63 percent of its final change (if it is allowed to reach that constant value). Thus, the response $x(t)$ for the forcing function of Fig. 7.7-6 and for $x(0) = 0$ is shown in Fig. 7.7-7.

Observe here that in the first interval, $0 \le t < T$, $x(t)$ tends from an initial value of 0 to a value of 100 V; yet by time $t = T$ it reaches 63 percent of its ultimate objective, i.e., 63 V. Note that at $t = 0$ the slope of $x(t)$ is indicated by

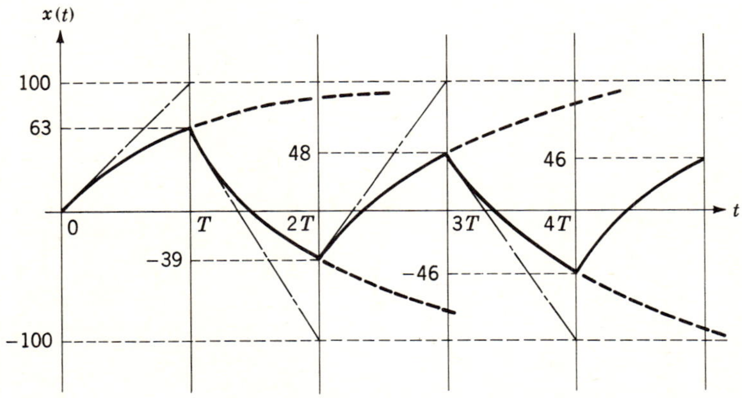

FIG. 7.7-7 *Constructing the response to the square-wave drive.*

the dashed line intersecting the point $t = T$, $x = 100$. When the second interval starts ($t = T$) the capacitor has a voltage of 63, which, with respect to that interval, is now considered the initial voltage. Here, $x(t)$ starting from 63 tends toward -100 [because $e(t)$ is now -100]. By time $t = 2T$ it has covered only 63 percent of its ultimate change (of $100 + 63 = 163$); i.e., it has covered $(0.63) \times (163) = 102$ in the negative direction, starting from its initial value (in the second time interval) of 63; hence $x(2T)$ is at $63 - 102 = -39$ V. Evidently, in the third interval, the starting value is -39 and the whole idea is repeated. Ultimately, we see that the response goes as much above as it does below the horizontal axis. Suppose that it goes b units above and b units below that axis. We can find b, by using our previous reasoning; i.e., starting from $-b$, $x(t)$ will tend toward 100 and in T seconds will cover 63 percent of $100 + b$, which should bring it $b + b$ units above $-b$. That is,

$$(0.63)(100 + b) = 2b$$

the solution of which is $b = 46$ V. Hence after several Ts the response will vary between a maximum value of 46 and a minimum value of -46.

Finally, note that when we change from interval to interval, i.e., when the input jumps, the differential equation *is* satisfied, since the slope dx/dt also jumps. In fact, if we were to calculate dx/dt slightly before and slightly after $t = T$, we would find it to be $37/T$ and $-163/T$, or a net jump of $-200/T$. The drive also jumps from $+100$ to -100, a net jump of -200. Note, however, that the response $x(t)$, *does not* jump at $t + T$. Thus we, see that

$$\text{Jump in slope} = \text{jump in forcing function} \tag{11}$$

which agrees with our D.E., (2). Specifically, if we write that equation for two times, one slightly before (at T^-) and one slightly after (at T^+) the time $t = T$, we have

$$\frac{dx}{dt}(T^-) = -\frac{1}{\tau}x(T^-) + \frac{1}{\tau}e(T^-)$$

$$\frac{dx}{dt}(T^+) = -\frac{1}{\tau}x(T^+) + \frac{1}{\tau}e(T^+)$$

Substracting the first of these two equations from the second yields the same result as Eq. (11), since $x(T^-) = x(T^+)$.

The *RC* Network: Narrow-pulse Response

The step input of voltage is a very common form of input excitation. Another common form is the narrow pulse, which is called in its "limit" an impulse. We shall now examine the response of the *RC* network to a narrow pulse. By a narrow-pulse forcing function, we mean that the voltage $e(t)$ in Eq. (2) is equal to $v_p(t)$, where

$$v_p(t) \triangleq \begin{cases} 0 & \text{for } t < 0 \\ \dfrac{k}{\Delta} & \text{for } 0 \leq t < \Delta \\ 0 & \text{for } t \geq \Delta \end{cases} \tag{12}$$

where $\triangle$ is very very small, and k is a real constant. Two pulses with different values of $\triangle$ are illustrated in Fig. 7.7-8.

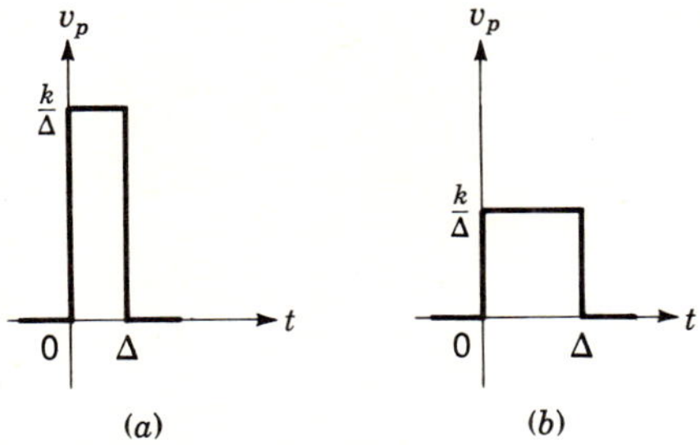

FIG. 7.7-8 Two examples of narrow pulses for different values of Δ.

From our previous discussion, the capacitor voltage will rise toward value $K/\triangle$ as follows:

$$x(t) = \frac{k}{\triangle}(1 - e^{-t/\tau}) \qquad \text{for } 0 \leq t \leq \triangle \tag{13}$$

At $t = \triangle$ the capacitor reaches $x(\triangle) = (k/\triangle)(1 - e^{-\triangle/\tau})$. As $\triangle \to 0, x(\triangle) \to k/\tau$, since

$$\lim_{\triangle \to 0} e^{-\triangle/\tau} = 1 - \frac{\triangle}{\tau} + \frac{1}{2!}\left(\frac{\triangle}{\tau}\right)^2 - \cdots$$

Thereafter, (for $t \geq \triangle$), $x(t)$ will tend from value k/τ toward zero (since the drive is zero); hence the response for $t \geq \triangle$ will be

For $t \geq \triangle$,

$$x(t) = x(\triangle)\, e^{-(t-\triangle)/\tau} \tag{14}$$

Thus, the narrow-pulse response will be as shown in Fig. 7.7-9*a*, *b*, and *c* for values $\triangle = 1/2$, $\triangle = 1/4$, and $\triangle \to 0$, respectively. The limiting form of Fig. 7.7-9*c* is called the *impulse response* of the network and is given by

For $t < 0$,

$$x(t) = 0 \tag{15}$$

For $t \geq 0$,

$$x(t) = \frac{k}{\tau} e^{-t/\tau}$$

Observe that $\triangle$ does not appear in Eqs. (15), and the significant parameter of the applied "impulse" is the *area* under its graph, i.e., the time integral of $v_p(t)$, whose value is k. The only other quantity, τ, appearing in Eqs. (15) is independent of the applied impulse, and is dependent strictly on the time constant of the network or system to which the impulse is applied. Observe further that for

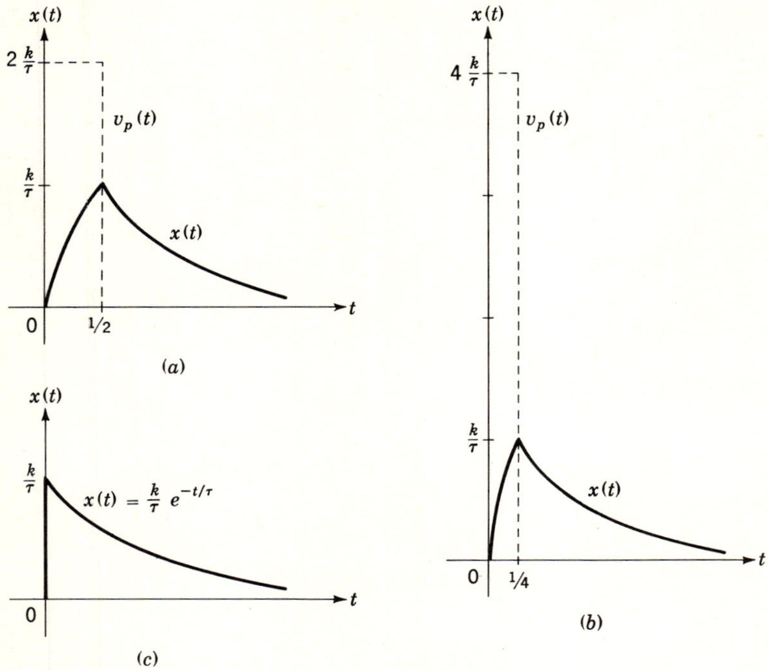

FIG. 7.7-9 Narrow-pulse response of the RC network ($RC = 1$): (a) $\Delta = 1/2$; (b) $\Delta = 1/4$; (c) $\Delta \to 0$.

negative t, $x(0) = 0$; yet for $t \geq 0$, $x(t) = (k/\tau)e^{-t/\tau}$. That is, $x(t)$ jumps at $t = 0$ from 0 to k/τ, as it should, in order for its derivative to match in Eq. (2) the infinitely high and infinitely thin pulse of area k.

Linearity and Superposition

Consider the linear D.E.,

$$\frac{dx}{dt}(t) = ax(t) + v_1(t) \qquad x(0) = x_0 \tag{16}$$

and let $x(t)$ denote its solution. Consider next the D.E.

$$\frac{dy}{dt}(t) = ay(t) + v_2(t) \qquad y(0) = y_0 \tag{17}$$

and let $y(t)$ denote its solution. Evidently, Eqs. (16) and (17) have the same unforced equations; they differ only in the forcing functions and in the initial conditions. In physical terms, Eqs. (16) and (17) represent identical linear networks or block diagram systems, with the exception that the drives or excitations applied to these systems, as well as their initial state, are different.

A term-by-term summation of Eqs. (16) and (17) results after some rearrangement in

$$\frac{d}{dt}[x(t) + y(t)] = a[x(t) + y(t)] + [v_1(t) + v_2(t)]$$

$$x(0) + y(0) = x_0 + y_0 \qquad (18)$$

If we let

$$\begin{aligned} z(t) &= x(t) + y(t) \\ z(0) &= z_0 = x_0 + y_0 \\ v(t) &= v_1(t) + v_2(t) \end{aligned} \qquad (19)$$

then Eqs. (18) become

$$\frac{dz}{dt}(t) = az(t) + v(t) \qquad z(0) = z_0 \qquad (20)$$

Observe that according to Eq. (20), $z(t)$ is the solution of the *same* system or network with the exception that the forcing function is the *sum of the forcing functions* of Eqs. (16) and (17); and the initial condition is the *sum of the initial conditions* of Eqs. (16) and (17). Since the solution $z(t)$ is the sum of the solutions of Eqs. (16) and (17) we conclude that:

THEOREM 7.7-1 The solution of a linear D.E. with a forcing function which is the sum of two forcing functions is the sum of the two solutions obtained by applying each forcing function alone, as long as the initial value of the combined solution is the sum of the initial values of the constituent solutions.

The above statement is another manifestation of the *principle of superposition* in linear systems. Whereas in linear resistive networks and memoryless systems we superposed real numbers, in linear networks and systems with memory *we superpose entire functions.* Observe finally that an obvious corollary of the above statement is the principle of homogeneity. That is, if we multiply the forcing function *and the initial condition* by a constant, the solution is also multiplied by that constant.

A special, yet frequently occurring case is that of *zero initial condition.* That is, the system or network to which an input is applied is initially "at rest." In such a case we do not need to worry about explicitly summing or scaling initial conditions since

$$0 + 0 = 0 \text{ and } k \times 0 = 0 \qquad \text{for real } k$$

Hence, we need only superpose or scale solutions in order to find the response of a system to two simultaneously applied inputs or to a scaled input, respectively.

Example 7.7-1 *Superposition of Impulse Responses* Consider the RC network of Fig. 7.7-1 with $\tau = 2$,† and $x(0) = 0$, and let the voltage-source waveform consist of two impulses as shown in Fig. 7.7-10a. Note that we represent each impulse as an arrow with the area of that impulse listed next to the arrow.

From our discussion of impulse responses we know, by inspection,

†Recall that $\tau = RC$.

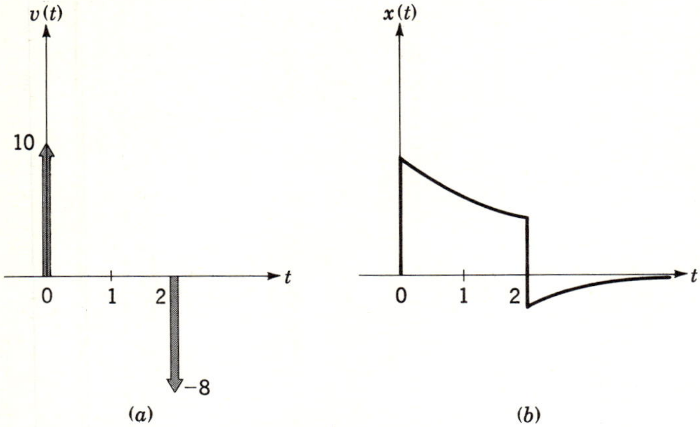

FIG. 7.7-10 Superposition of impulse responses; (a) *the excitation, and* (b) *the response.*

that the response to the impulse at the origin, when the second impulse is absent, is

$$x_1(t) = {}^{10}\!/_2\, e^{-t/2} \qquad \text{for } t \geq 0$$

If the first impulse were not present, the response to the second impulse would have been

$$x_2(t) = 0 \qquad \text{for } 0 \leq t < 2$$

$$x_2(t) = -\,{}^{8}\!/_2\, e^{-(t-2)/2} \qquad \text{for } t \geq 2$$

Since $x(0^-) = 0$ in either of the above two hypothetical experiments, we conclude that the response to both impulses applied together is

$$x(t) = x_1(t) + x_2(t)$$

or

$$x(0^-) = 0$$

$$x(t) = 5e^{-t/2} \qquad \text{for } 0 \leq t < 2$$

$$x(t) = 5e^{-t/2} - 4e^{-(t-2)/2} \qquad \text{for } t \geq 2$$

The third of these equations can be rearranged as follows:

$$x(t) = 5(e^{-1})(e^{-(t-2)/2}) - 4e^{-(t-2)/2} \qquad \text{for } t \geq 2$$

$$x(t) = \left(\frac{5}{e} - 4\right)e^{-(t-2)/2} \qquad \text{for } t \geq 2$$

We could have obtained this result without manipulating the above equations, since at $t = 2$ (one time constant away from $t = 0$) we know that $x(t)$ will be at $1/e$ (or about 37 percent) of its initial value 5.

Justification of Theorem 7.6-1

We are now in a position to justify Theorem 7.6-1 and Eq. (10) of Sec. 7.6 which were "pulled out of a hat." First, let us summarize what we have done in terms of impulse responses:

We have shown that for the first-order network characterized by

$$\frac{dx}{dt} = -\frac{1}{\tau}x + \frac{1}{\tau}e \qquad x(0) = 0 \tag{21}$$

the response $x(t)$ to an impulse of area k, applied at $t = 0$ is

$$x(t) = \frac{k}{\tau}e^{-t/\tau} \qquad \text{for } t \geq 0 \tag{22}$$

We have also shown in Example 7.7-1 that the response of the same network to an impulse of area k applied at $t = p$ and acting alone, is

$$x(t) = \frac{k}{\tau}e^{\,(t-p)/\tau} \qquad \text{for } t \geq p \tag{23}$$

If we rewrite Eq. (21) in the more general form,

$$\frac{dx}{dt} = ax + v(t)$$

i.e., if we let

$$a = -\frac{1}{\tau} \qquad v = \frac{1}{\tau}e \tag{24}$$

then Eq. (22) becomes

$$x(t) = -ake^{at}$$

and the forcing function $v(t)$ becomes an impulse of area $-ak$. Hence, if the forcing function is an impulse at the origin of area m, then the response is

$$x(t) = me^{at} \qquad \text{for } t \geq 0 \tag{25}$$

Similarly the response to an impulse of area q occurring at $t=p$ is, from Eq. (23) and from the above

$$\begin{aligned} x(t) &= 0 && \text{for } 0 \leq t < p \\ x(t) &= qe^{a(t-p)} && \text{for } t \geq p \end{aligned} \tag{26}$$

Our strategy will be to approximate an arbitrary forcing function by a sum of impulses and then to obtain by superposition the response to this sum. Consider an arbitrary forcing function $v(t)$, as shown in Fig. 7.7-11. We shall assume that $x(0) = 0$, and we shall divide the time interval $[0,p]$ into n equal intervals, i.e., $t_i = (i/n)p$ $(i = 0, 1, \ldots, n)$. We define next an approximation $v_n(t)$ to $v(t)$ as follows:

$$v_n(t) = v(t_i) \qquad \text{for } t_i \le t < t_{i+1} \tag{27}$$

Figure 7.7-11 illustrates $v_8(t)$.

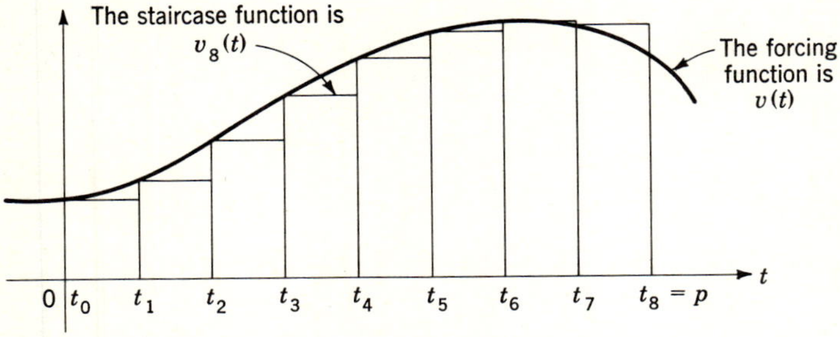

FIG. 7.7-11 The function $v_8(t)$ which approximates $v(t)$ with eight rectangular pulses.

We can express $v_n(t)$ as the sum of n rectangular pulses $q_i(t)$ defined as

For $i = 0, 1, \ldots, n-1$,

$$q_i(t) = \begin{cases} 0 & \text{for } t < t_i \\ v(t_i) & \text{for } t_i \le t < t_{i+1} \\ 0 & \text{for } t_{i+1} \le t \end{cases} \tag{28}$$

That is,

$$v_n(t) = \sum_{i=0}^{n-1} q_i(t) \tag{29}$$

The response to this forcing function is, by superposition, the sum of the responses to each $q_i(t)$ acting alone; note that $x(0^-) = 0$. Let $z_i(t)$ be the response to a single pulse $q_i(t)$. Then the response $x_n(t)$ to the forcing function $v_n(t)$ is, from Eq. (29) and by superposition, as follows:

$$x_n(t) = \sum_{i=0}^{n-1} z_i(t) \tag{30}$$

As n becomes very large, $q_i(t)$ approaches an impulse occurring at time t_i and having area $(p/n)v(t_i)$. Hence the response z_i to one such impulse becomes the impulse response which, by Eq. (26) and the above, is

$$z_i(t) = 0 \qquad \text{for } 0 \le t < t_i$$

$$z_i(t) = \left(\frac{p}{n}\right) v(t_i) e^{a(t-t_i)} \qquad \text{for } t \ge t_i$$

If we let $t_{n-1} \le t < t_n$, so as to ensure that n impulses have already acted, all n exponentials will be present, and Eq. (30) becomes

$$x_n(t) = \sum_{i=0}^{n-1} \left(\frac{p}{n}\right) v(t_i) e^{a(t-t_i)} \qquad \text{for } t_{n-1} \le t < t_n$$

$$= e^{at} \sum_{i=0}^{n-1} \left(\frac{p}{n}\right) v(t_i) e^{-at_i} \tag{31}$$

As n approaches infinity, let t_i approach s. Then, (1) the term p/n approaches the infinitesimal quantity ds; (2) $v(t_i)$ approaches $v(s)$; (3) e^{-at_i} approaches e^{-as}, and (4) the sum approaches the integral of the product of these three quantities, i.e.,

For $t_{n-1} < t \le t_n$

$$\lim_{n\to\infty} x_n(t) = e^{at} \int_0^p v(s) e^{-as}\, ds \tag{32}$$

Since this equation holds for $t_{n-1} < t \le t_n$, it holds for $t = t_n = p$. Hence, substituting $t = p$ in Eq. (32) and letting $\lim_{n\to\infty} x_n(t) = x(t)$ yields

$$x(p) = e^{ap} \int_0^p v(s) e^{-as}\, ds \tag{33}$$

which is the desired result [(10) of Sec. 7.6] with the variables renamed.

At this point, the reader may have some misgivings about the apparent cyclical nature of what we have done. That is, we used Eq. (10) of Sec. 7.6 to determine the step response of an *RC* network; this result was in turn used to derive the impulse response of the *RC* network; finally the impulse response was used to justify this equation. Actually, there is nothing wrong with this approach. In particular, the fact that the equation was not proved initially is immaterial to its use in deriving the step response of the *RC* network, since the step response does indeed satisfy all the requirements of a solution, and Theorem 7.5-1 guarantees the existence of only one solution—hence the one that we did obtain—regardless of the general validity of Eq. (10) of Sec. 7.6. Besides, this equation regardless of how it was obtained, does satisfy all the requirements of a solution, as we demonstrated upon its introduction.

In Chap. 11 we shall return in considerable detail to impulse responses and the convolution integral, as Eq. (33) is called. We proceed now with an example of an opamp system which is modeled by a first-order linear D.E..

A First-order Opamp System

Consider the dynamic system shown in Fig. 7.7-12, where, as usual, K is very large ($K \approx 10^6$); R_1, R_2, and C are linear network elements; and $u(t)$ is a constant of value U. The capacitor is initially charged to V_0 as shown in the figure. We seek the waveform $y(t)$ for $t \ge 0$.

The characterizing equations for this system are, by inspection of

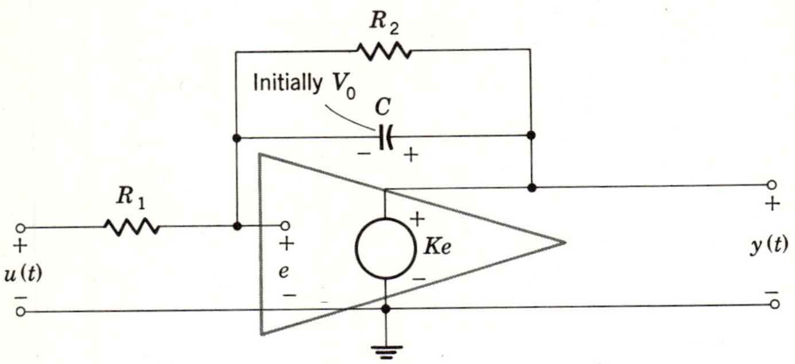

FIG. 7.7-12 First-order opamp system.

Fig. 7.7-12, as follows:

$$\frac{u-e}{R_1}=\frac{e-y}{R_2}+C\frac{d(e-y)}{dt}$$

$$y = Ke \qquad y(0)-e(0)=V_0$$

Elimination of e results in the following equation:

$$\frac{u}{R_1}-\frac{y}{KR_1}=\left(\frac{1}{KR_2}-\frac{1}{R_2}\right)y+C\left(\frac{1}{K}-1\right)\frac{dy}{dt} \tag{34}$$

Since K is very large, the above equation may be simplified as follows:

$$\frac{dy}{dt}\simeq-\frac{1}{R_2C}y-\frac{1}{R_1C}u \qquad y(0)=V_0 \tag{35}$$

This is a first-order linear D.E., and since the drive is constant of value U/R_1C, we are tempted to rearrange Eq. (35) so that it is of the form of Eq. (2). The result is

$$\frac{dy}{dt}=-\frac{1}{R_2C}y+\frac{1}{R_2C}\left(-\frac{R_2}{R_1}U\right) \qquad y(0)=V_0 \tag{36}$$

The solution of this equation is, as we know from our earlier work, an exponential starting at value V_0 at $t=0$ and going toward the value $-(R_2/R_1)\,U$ as $t\to\infty$. This result is shown in Fig. 7.7-13, along with the associated expression. Observe that the time constant is determined by R_2 and C, whereas the final value is determined by R_2, R_1, and U.

At this point, the reader may wonder why the above opamp system should ever be used in place of the simpler RC network of Fig. 7.7-1 which has the same input-output behavior. The reason is that the opamp system can have a "gain" greater than 1. That is, the overall change of $y(t)$ can be greater than the difference $U-V_0$ (which is not the case with a simple RC network.) The gain is in fact adjusted by choosing the ratio R_2/R_1 at the desired value. Thus, for example, if $R_1 = 1$ kΩ, $R_2 = 50$ kΩ, and $C = 1$ μF, and if $V_0 = 0$ V and $U = 0.01$ V, the output will be given by

$$y(t)=-0.5(1-e^{-t/0.05}) \tag{37}$$

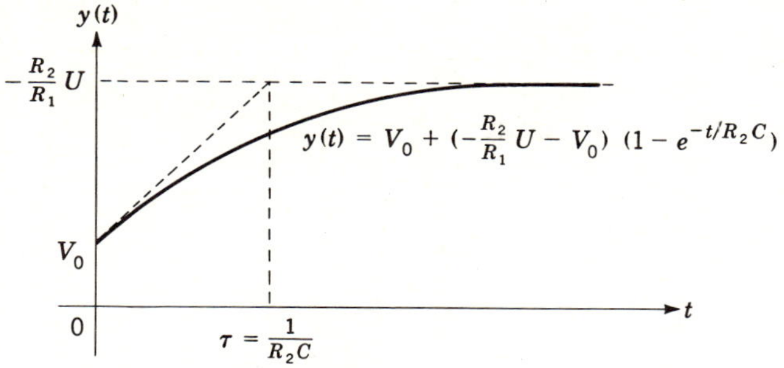

FIG. 7.7-13 *Output of opamp system.*

That is, while the input is 0.01 V, the output undergoes a 0.5 V change with a time constant of 50 ms.

We proceed now with a special and useful nonlinear first-order circuit made up of diodes and capacitors.

A Diode-Capacitor Network

Consider the network of Fig. 7.7-14a, where the diode has the ideal

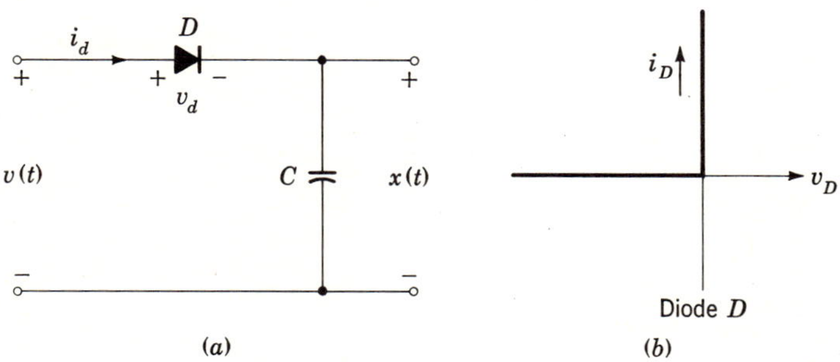

FIG. 7.7-14 *Diode-capacitor network.*

characteristics of Fig. 7.7-14*b*, i.e., whenever $i_D > 0$, $v_D = 0$, and whenever $v_D < 0$, $i_D = 0$. Another way of looking at this diode is by assuming a very large "forward" conductance and a very large "reverse" resistance, i.e.,

For $i_D > 0$,

$$\frac{1}{R_f} = \frac{i_d}{v_d} \to \infty \tag{38a}$$

For $v_D < 0$,

$$R_r = \frac{v_d}{i_d} \to \infty \tag{38b}$$

We wish to establish what $x(t)$ will do, given $x(0)$ and an arbitrary $v(t)$.

Observe that the network can be characterized by two differential equations according to whether the diode is conducting or not. Thus, when it conducts, we have

$$\frac{dx}{dt} = \frac{v(t) - x(t)}{R_f C} \tag{39}$$

and when it is in "reverse," we have

$$\frac{dx}{dt} = \frac{v(t) - x(t)}{R_r C}$$

Observe that what we shall call the *forward time constant*, $\tau_f = R_f C$, approaches 0 because of Eq. (38*a*), while the *reverse time constant*, $\tau_R = R_R C$, approaches infinity because of Eq. (38*b*). Hence when $x(t)$ tends toward a higher voltage, i.e., when $v(t) > x(t)$, we are dealing with the forward time constant and $x(t)$ rises rapidly, indeed instantly, as $\tau_f \to 0$, to the value of $v(t)$. Conversely, when $x(t)$ tends toward a lower voltage, i.e., when $v(t) < x(t)$, we are dealing with the reverse time constant and $x(t)$ drops very slowly; indeed it does not drop at all, as $\tau_R \to \infty$. This behavior is shown in Fig. 7.7-15 where $x(0) = 0$.

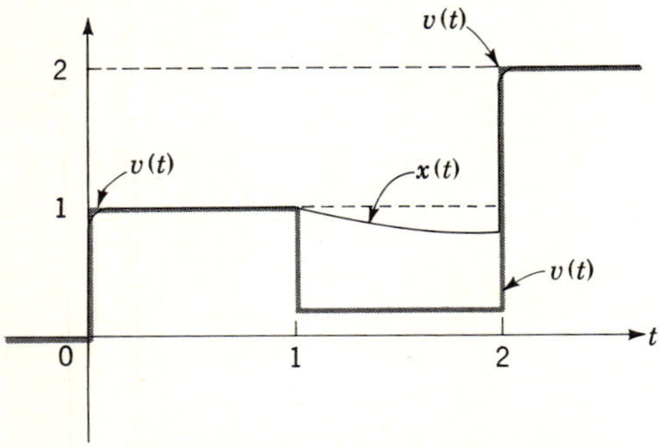

FIG. 7.7-15 Response of the diode-capacitor network with finite nonzero forward and reverse resistances.

Observe here that we have sketched the response assuming very large *but finite* τ_R and very small *but nonzero* τ_f. As these two quantities approach their respective limits, we have the response shown in Fig. 7.7-16.

We can now generalize the idealized behavior of this network to an arbitrary drive by stating

$$x(t) = \max_{0 \leq \tau \leq t} [v(\tau), x(0)] \tag{40}$$

That is, the response at any time t is the maximum of the initial state and of all the values of the input from time zero to time t. This simple network has many interesting applications which are discussed at the end of this chapter in terms of problems. One especially interesting application is rectification. That is, if

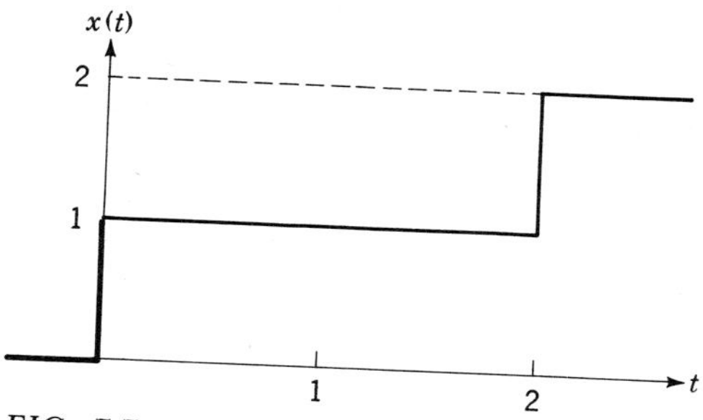

FIG. 7.7-16 *Response of the diode-capacitor network in the limiting case:* $R_f \to 0$, $R_r \to \infty$.

$v(t) = V \sin \omega t$ then it follows from Eq. (40) that $x(t) = V$, and we have converted a sinusoidal or alternating voltage $V \sin \omega t$ into a constant or direct voltage V.

7.8 SECOND-ORDER SYSTEMS AND NETWORKS

The words "second order" characterize, for the purposes of this section, *the class of systems and networks that can be modeled by two simultaneous first-order ordinary differential equations* (the state equations) and one or more "output" equations. As we shall see in the following chapter, a large subclass of such systems can also be characterized by a single second-order differential equation in one variable. The solution of two (or for that matter, of several) simultaneous first-order D.E.'s is a straightforward proposition for a digital computer.† Unfortunately, the *manual* solution of two such simultaneous equations involves, in the general case, too much detail, with the result that it is of little educational value. For this reason, we shall solve manually in the following chapter those systems of two first-order D.E.'s that can be converted to *one* second-order D.E. in order to gain insight about second-order system behavior.

Formulating Two Simultaneous Differential Equations

The formulation of equations for second-order systems and networks is identical to the formulation of first-order system and network equations, except that we now try to identify *two* state variables, which lead to two coupled (or simultaneous) first-order ordinary differential equations, and two initial conditions. As before, the number of remaining "output" equations is not restricted and can be none or more. These ideas are made more precise in the following definition, where to keep our equations simple we have used one input and one output variable. Example 7.8-1 further clarifies the formulation process.

DEFINITION 7.8-1 A system or network with input u and output y is of second order if two state variables x_1, x_2 can be identified so that

†This is a topic of extensive discussion in the second volume where emphasis is placed on multivariable systems.

In second-order networks, as in first-order networks, we will use as state variables the capacitor charges and the inductor fluxes. We will therefore

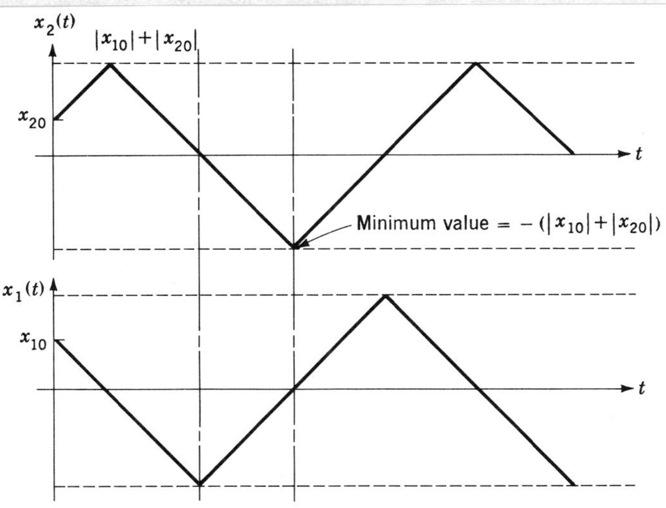

FIG. 7.8-6 *Behavior in time of the state variables.*

established. For example, if $x_{10} = 2$ and $x_{20} = 1$, then the state variables and outputs will be as shown in Fig. 7.8-7. If $x_{10} = -3$ and $x_{20} = 2$, the response will be as shown in Fig. 7.8-8.

Note in both these figures our various observations, which are summarized as follows:

FIG. 7.8-9 *The LC tank circuit.*

For $t \geq t_0$,	$\frac{dx_1}{dt} = f_1(x_1,x_2,u)$	$x_1(t_0) = x_{10}$	(1a)
Two			(1b)

Here, the function sgn means "the sign of," i.e.,

$$\text{For } x \geq 0, \text{ sgn } x = 1$$

$$\text{For } x < 0, \text{ sgn } x = -1$$

and initial time is $t_0 = 0$.

We shall solve this equation in a geometrically meaningful way, so that we may visualize the solution, given any pair of initial states x_{10}, x_{20}.

We start with the portrayal of the x_1-x_2 plane, the *state space*, as it is called; every point of this plane corresponds to a state x_1, x_2 of our system, and conversely every state of our system corresponds to a point on the plane. Figure 7.8-4*a* shows dx_1/dt, and Fig. 7.8-4*b* shows dx_2/dt at every point of this plane.

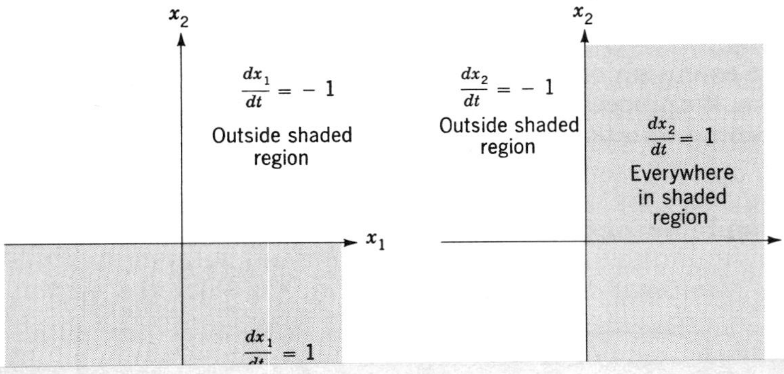

where x_1 is the inductor current; x_2 is the capacitor voltage; and ξ_1, ξ_2 are the initial current stored in the inductor and initial voltage stored on the capacitor, respectively.

If we let $\xi_1 = 0$ and $\xi_2 = 1$, we can "guess" by inspection of the derivative relationships in Eqs. (3) that

$$x_1(t) = \sin t$$
$$x_2(t) = \cos t \qquad (4)$$

Indeed $x_1(0) = 0$, $x_2(0) = 1$, and

$$\text{For } t \geq 0 \begin{cases} \frac{dx_1}{dt}(t) = \cos t = x_2(t) \\ \frac{dx_2}{dt}(t) = -\sin t = -x_1(t) \end{cases}$$

Hence, x_1 and x_2 of Eqs. (4) are a solution, indeed the only solution to Eqs. (3).

The last two examples illustrate how simple coupled first-order D. E.'s "work." The *periodicity* (in time) of solutions to second-order systems is a useful attribute not possessed by first-order systems. Thus, linear and nonlinear second- (or higher-) order systems must be used whenever we desire to generate periodic waveforms. Of course, periodicity is not the only phenomenon exhibited by second-order systems. At least two other interesting phenomena are *resonance* and *filtering*, as we shall see in the following chapter.

PROBLEMS

7.2-1 Write state differential and output equations, if they can be written, for each system shown in the figure.

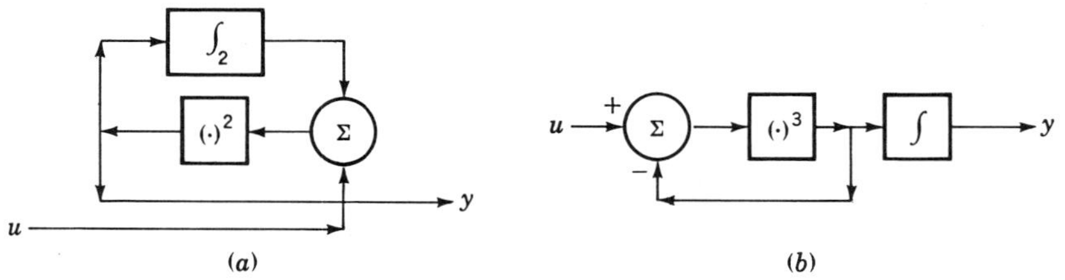

PROB. FIG. 7.2-1

7.3-1 An ideal constant-voltage source V, a nonlinear resistance, and a constant capacitance C are connected in series at time $t = 0$. The capacitance is initially uncharged. For the resistance, $i_R = f(v_R)$, f is continuous, $f(0) = 0$, and $i_R v_R$ is positive for all nonzero v_R.

(*a*) Show that the final capacitor voltage, i.e., the capacitor voltage at a very large time, is V.

(*b*) Prove that the final value of capacitance energy is equal to the energy lost as heat in the resistance.

7.4-1 Formulate, if possible, state equations for each network *a* through *e* shown in the figure. (*R*, *L*, and *C* are linear elements.)

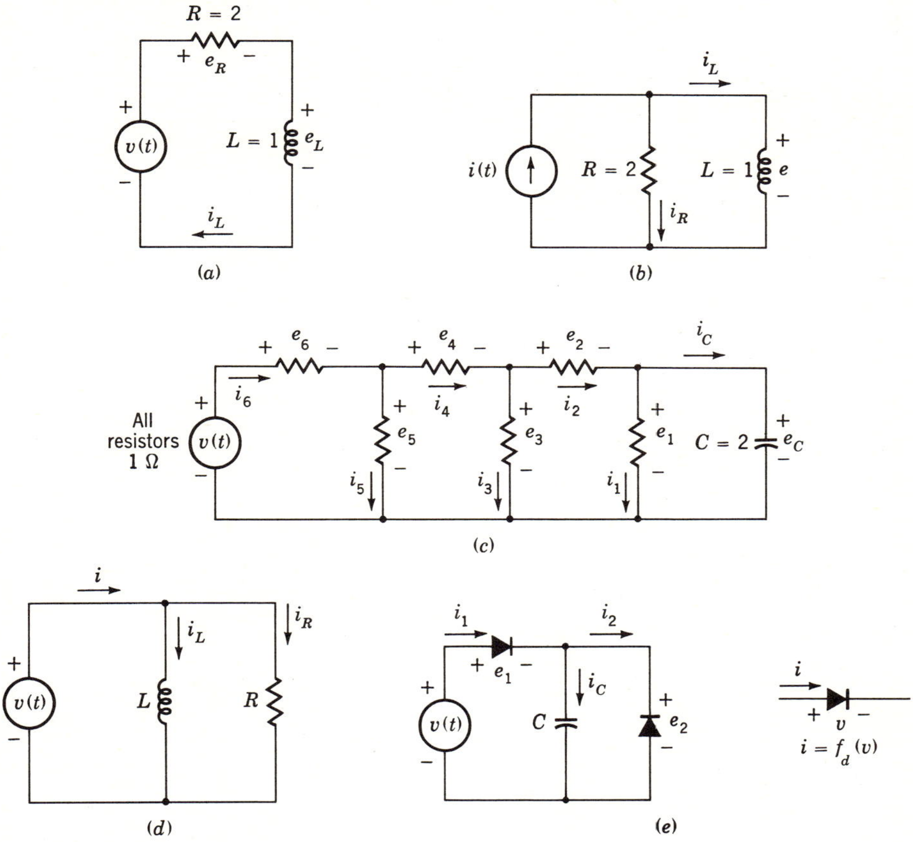

PROB. FIG. 7.4-1

7.4-2 (*a*) Write a differential equation relating $v(t)$ to the two sources $i_s(t)$ and $v_s(t)$, shown in the figure. $v(0) = 0$.

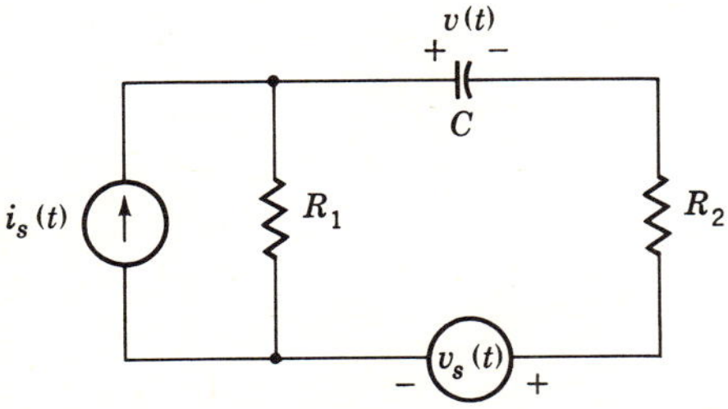

PROB. FIG. 7.4-2

(*b*) Note that the resulting differential equation contains the different sources as separate terms on the right-hand side. Since $v(0) = 0$, superposition can be used directly. This amounts to finding the response to each source separately with all other sources set to zero and then summing the responses. From (*a*) draw the appropriate networks and write the differential equations relating $v(t)$ to each of the two sources, with the other source set to zero.

7.5-1 Consider the network shown in the figure containing a linear inductor and an ideal diode. Suppose that the excitation is

$$v_s(t) = +1 \qquad \text{for } 0 \le t < 1$$

$$v_s(t) = -1 \qquad \text{for } 1 \le t < 3$$

$$v_s(t) = +1 \qquad \text{for } 3 \le t \le 4$$

Find the current $i(t)$, given that $i(0) = 0$, for $0 \le t \le 4$.

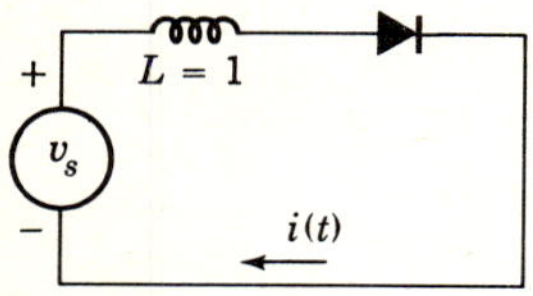

PROB. FIG. 7.5-1

7.5-2 Consider the parallel connection of the linear inductor ($L = 1$) and of the ideal diode across the independent current source $I(t)$, as shown in part I. The diode characteristics are shown in part II. Assume that at $t = 0$ there is no current flowing through the inductor.

(*a*) Show that the current $i_L(t)$ at any time $t > 0$ is equal to the past minimum value of $i(\tau)$ in the time interval $0 \le \tau \le t$ where $i(0) = 0$.

Next, suppose that $I(t)$ is the waveform shown in part III.

(*b*) Sketch and dimension $i_L(t)$ and $v(t)$ versus t for $t > 0$.

(*c*) Compute the energy delivered by the current source from $t = 0$ to $t = 10$ sec.

(*d*) Sketch and dimension the energy stored in the inductor versus $t(0 \le t \le 10)$. What is this stored energy at $t = 10$?

(*e*) From (*c*) and (*d*) show that no energy has been dissipated by the ideal diode. Can the ideal diode ever dissipate any energy? Justify your answer.

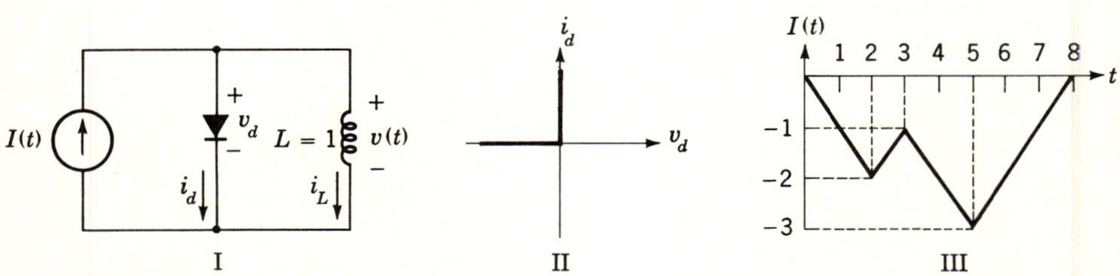

PROB. FIG. 7.5-2

7.6-1 In the network shown in part *a* of the figure we want $x(t)$ to be as shown in part b. Specify $x(0)$ and $v(t)$, for $t > 0$, by an expression and a sketch, labeling accurately all the important parameters, so that $x(t)$ is as shown above.

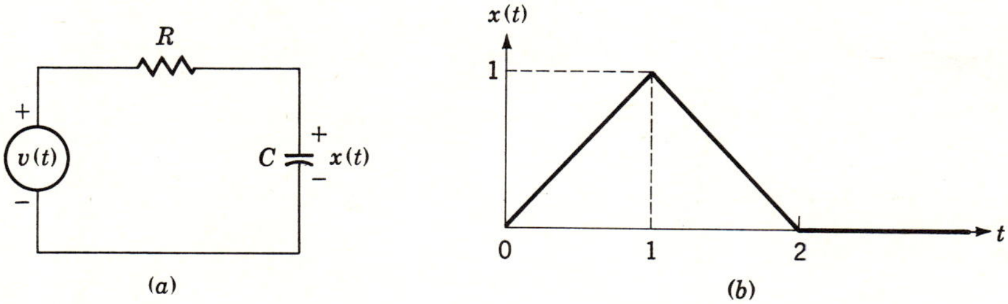

PROB. FIG. 7.6-1

7.6-2 A nonlinear resistance is such that its voltage is the cube of its current. This resistance is suddenly connected to a charged capacitor as shown in the figure. Find the capacitor voltage $v(t)$ as a function of time. (*Hint:* Separate variables to integrate.)

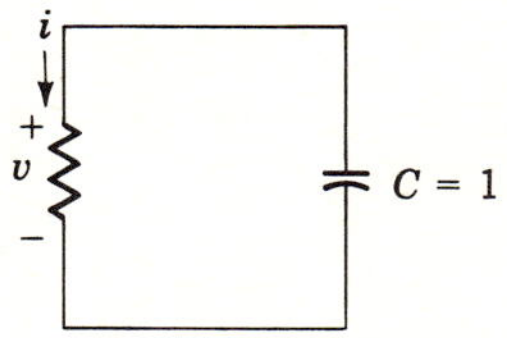

PROB. FIG. 7.6-2

7.6-3 Given the following differential equation:

$$\frac{dx}{dt} = -2x + 3u(t)$$

(a) Let $u(t)$ be the constant 5 for $t \geq 0$, and let $x(0) = 10$. Give the zero-input, zero-state, and overall solutions to the above D.E.

(b) Let $u(t)$ be as shown in the figure, and let $x(0) = 0$. Find $x(t)$.

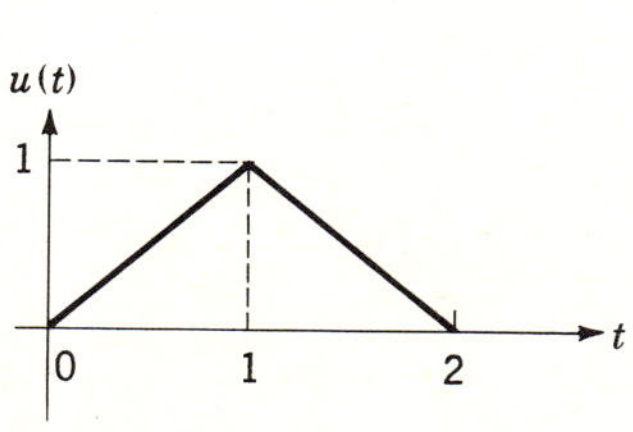

PROB. FIG. 7.6-3

(c) Sketch and carefully label an excitation waveform $u(t)$, such that $x(t)$ is identical to the waveform of $u(t)$ in (b) above. Again $x(0) = 0$.

7.6-4 A time function is periodic with period T if $f(t) = f(t + T)$, for all time t. Suppose a first-order linear system is driven by a periodic excitation; i.e., its state variable $x(t)$ satisfies the following D.E.:

$$\frac{dx}{dt}(t) = ax(t) + f(t) \qquad f(t) = f(t + T)$$

Find a value of $x(0)$ such that $x(t)$ is also periodic with period T.

7.7-1 (*a*) Determine the differential equation relating the response $i(t)$ to the excitation $V_s(t)$ in the network shown in the figure.

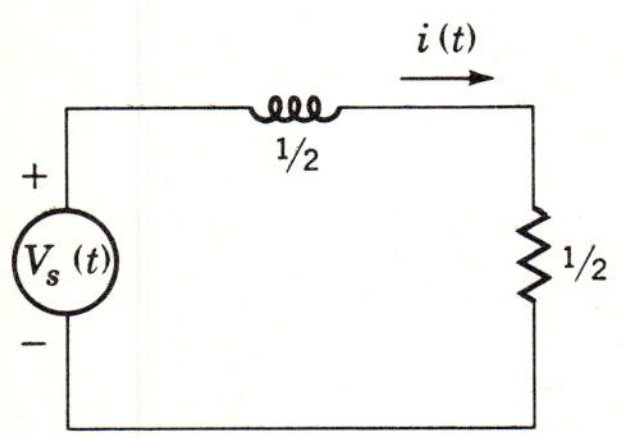

PROB. FIG. 7.7-1

(*b*) Given that $i(0) = 0$, solve the differential equation found in (*a*) for each of the following excitations, each of which is defined for $t \geq 0$.

$$V_s(t) = 3 \tag{1}$$
$$V_s(t) = \sin t \tag{2}$$
$$V_s(t) = \sin t + 3 \cos t \tag{3}$$

7.7-2 Consider the network in the figure which contains two switches. For $t < 0$,

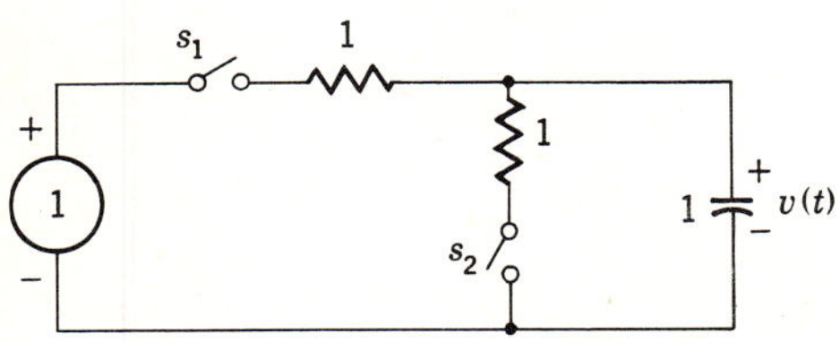

PROB. FIG. 7.7-2

both switches s_1 and s_2 are open, and $v(t) = 0$. Switch s_1 closes at $t = 0$, and switch s_2 closes at $t = \ln 2$. Express and sketch $v(t)$ for $t \geq 0$.

7.7-3 Consider the network in the figure which contains an ideal diode. Given that $v(0) = 0$ and that

$$v_s(t) = \begin{cases} 0 \text{ V} & \text{for all } t < 0 \\ 3 \text{ V} & \text{for all } t \geq 0 \end{cases}$$

Determine and plot the capacitor voltage $v(t)$ for $t \geq 0$.

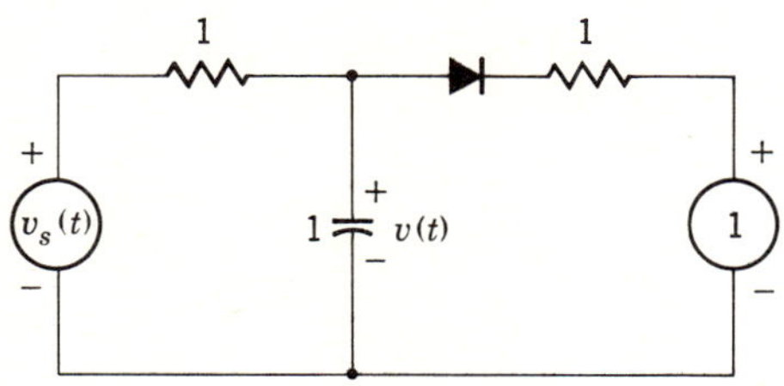

PROB. FIG. 7.7-3

7.7-4 In the network of part *a* of the figure a linear inductor ($L = 1$) and a nonlinear time-invariant resistor with the *i-v* characteristics of part *b* are connected in parallel across the independent current source $i(t)$. Suppose that at $t = 0$ the initial inductor current is

$$i_L(0) = 3 \text{ A}$$

and that the source current is

$$i(t) = 7 \text{ A} \qquad \text{for all } t \geq 0$$

Sketch the inductor current $i_L(t)$ (for $t \geq 0$) versus time, and dimension important coordinates. Give analytical expressions for $i_L(t)$ on your sketch.

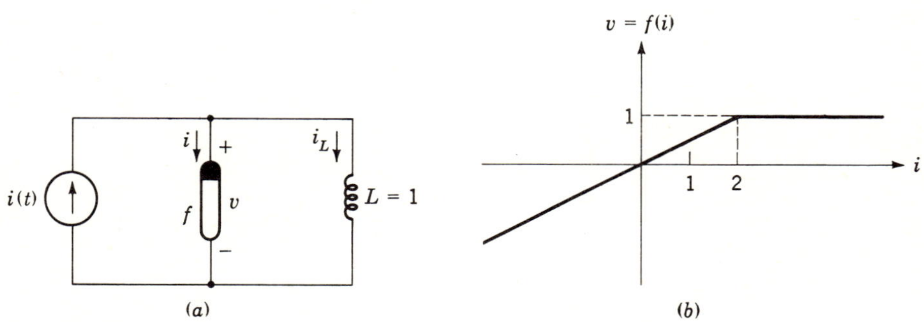

PROB. FIG. 7.7-4

7.7-5 Consider the network shown in the figure. Assume that the capacitor is initially uncharged. Assume further that the input voltage $u(t)$ is

$$u(t) = t \qquad \text{for } t \geq 0$$

It is desired that the capacitor voltage behave as follows:

$$v(t) = \tfrac{1}{2}t^2 \qquad \text{for } t \geq 0$$

Add a dependent source somewhere in the network to obtain this capacitor voltage. Show your reasoning, and specify the type and location of the source as well as its dependence relation.

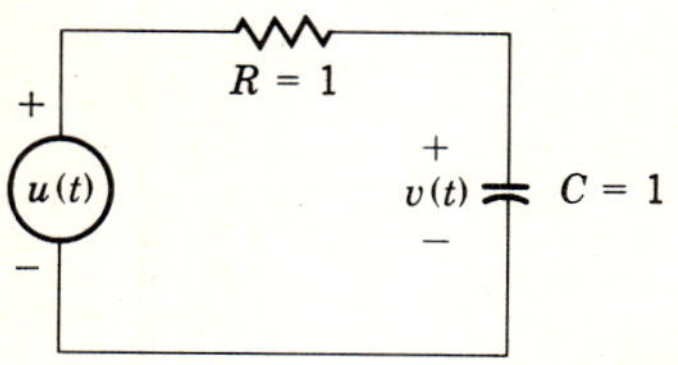

PROB. FIG. 7.7-5

7.7-6 Network N is linear and purely resistive. A constant unit voltage is applied to the terminals A-A'. When a 2-F capacitor (initially uncharged)

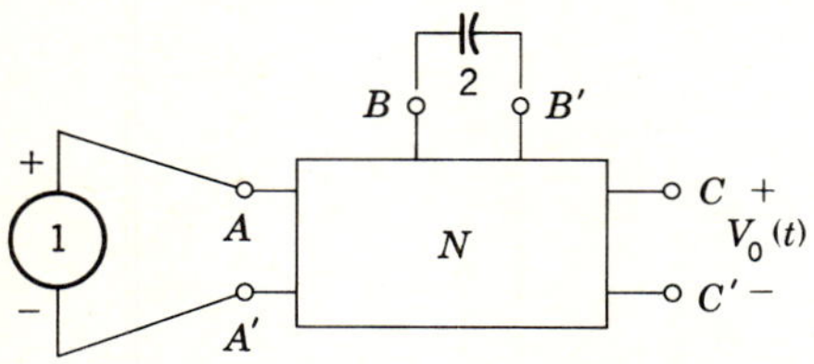

PROB. FIG. 7.7-6

is connected to terminals B-B' during this experiment, the output voltage V_0 is given by

$$V_0 = \tfrac{1}{2} + \tfrac{1}{8}e^{-1/4t} \qquad \text{for all } t \geq 0$$

Determine the output voltage $V_0(t)$ if a 2-H inductor with initially zero flux is connected to B-B' during the experiment, instead of the capacitor.

7.7-7 In part I of the figure, both capacitors are initially uncharged. The source voltage $v_1(t)$ is shown in part II, where

$$v_1(t) = \begin{cases} 0 & t < 0 \\ \dfrac{V}{\Delta}t & 0 \leq t \leq \Delta \\ V & t > \Delta \end{cases}$$

Our objective is to determine $v_2(t)$ when Δ approaches zero [i.e., when $v_1(t)$ is a step].

(*a*) Write the state equation for the network. Solve this equation for $v_2(t)$ as $\Delta \to 0$. You should specify the limiting $v_2(t)$ for all time.

(*b*) What should be the relation among R_1, R_2, C_1, and C_2 so that $v_2(t)$ is also a step, as $\Delta \to 0$? What is the height of this step in terms of the given parameters?

(*c*) When we measure voltage with a measuring device, we usually want the device to draw as little current as possible. The equivalent input circuit of an ordinary voltage-measuring device, such as an

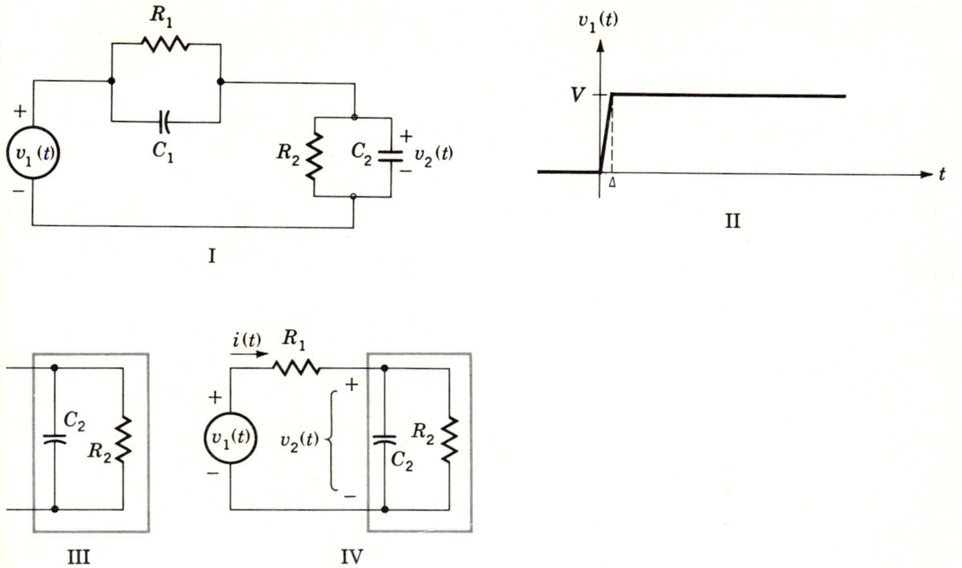

PROB. FIG. 7.7-7

oscilloscope, is a parallel *RC* network (part III). R_2 and C_2 are called the input resistance and capacitance of the device. One way to reduce the current drawn by the measuring device is to connect a resistance R_1 in series, as shown in part IV. However, with R_1 inserted, the voltage $v_2(t)$ is not a faithful reproduction of $v_1(t)$. Since the oscilloscope displays a waveform proportional to $v_2(t)$, it is important that $v_2(t)$ be proportional to $v_1(t)$. (The overall proportionality constant can be accounted for with the amplitude scales of the oscilloscope.)

From (*b*) we know that $v_2(t)$ can be made proportional to $v_1(t)$ by the addition of a capacitance C_1 across R_1. In oscilloscopes, R_1 and C_1 are usually placed in a probe at the end of a length of coaxial cable. For a typical oscilloscope, $R_2 = 10^6\ \Omega$ and $C_2 = 18 \times 10^{-12}$ F. What should be the value of R_1 and C_1 so that $v_2(t) = \frac{1}{10} v_1(t)$?

(*d*) In measuring a voltage $v_2(t)$, compare the current that would be drawn by the oscilloscope using the probe of (*c*) with the current that would be drawn by the oscilloscope if the probe were not used.

(*e*) Call the value of C_1 found in (*c*) $C_{1\ crit}$. Sketch $v_2(t)$ as $\Delta \to 0$, when R_1, R_2, and C_2 are held fixed and

(i) $C_1 > C_{1\ crit.}$
(ii) $C_1 = C_{1\ crit.}$
(iii) $C_1 < C_{1\ crit.}$

7.7-8 *Voltage Doubler.* The network in the figure is known as a voltage doubler. It simultaneously rectifies alternating to direct current and doubles its value. The source waveform is $v_s(t) = 100 \sin 400t$ for $t \geq 0$. The capacitors are initially uncharged.

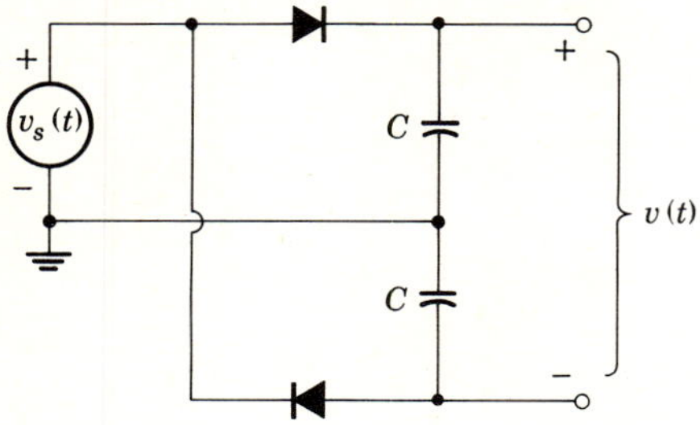

PROB. FIG. 7.7-8

(*a*) Assuming ideal diodes, give an expression for $v(t)$.

(*b*) Assuming that the diodes have finite nonzero forward and reverse resistance, sketch $v(t)$.

7.8-1 Choose state variables, and formulate state and output equations for the systems and networks *a* through *d* in the figure. Consider all shown currents and voltages as outputs in part *d*.

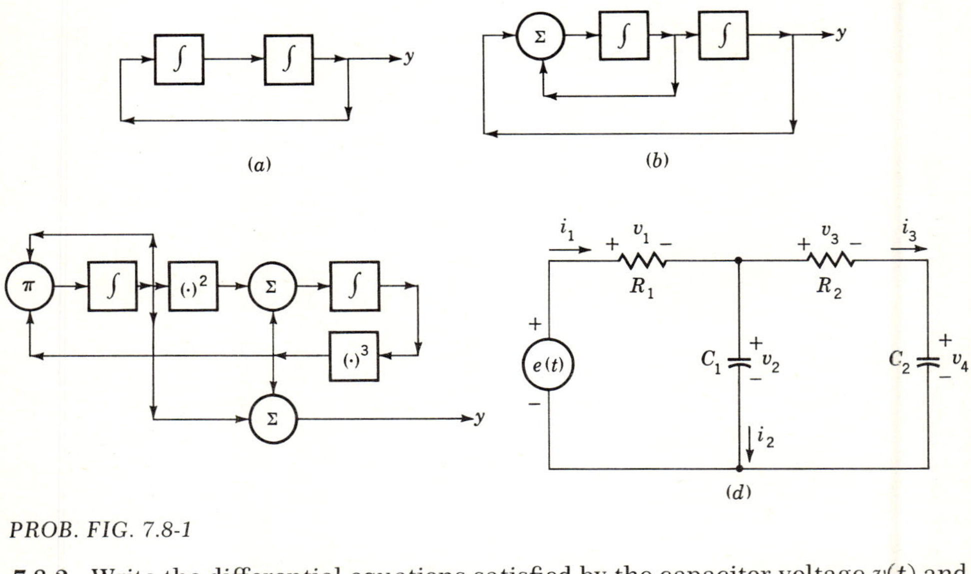

PROB. FIG. 7.8-1

7.8-2 Write the differential equations satisfied by the capacitor voltage $v(t)$ and the inductor current $i(t)$ for the network shown in the figure. Construct

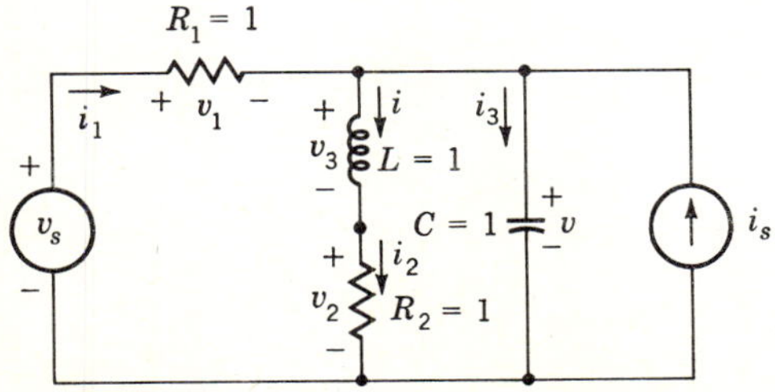

PROB. FIG. 7.8-2

an equivalent block diagram for the network, using $v(t)$ and $i(t)$ as the state variables, and considering all other network variables as outputs.

7.8-3 Write the state differential equations on state variables $v(t)$ and $i(t)$. The values of all elements are in ohms, farads, and henrys. Construct an

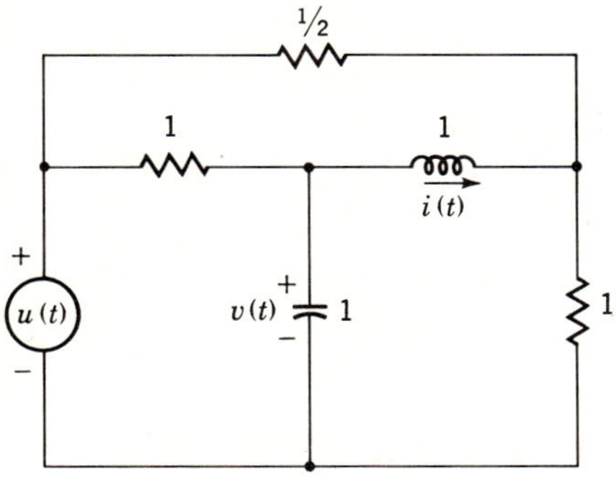

PROB. FIG. 7.8-3

equivalent block diagram system containing only integrators, adders, and scalers which obey the same equations as the network.

7.8-4 With $i(0) = e(0) = 0$ and $v(t) = 1$ for all $t \geq 0$, find $i(t)$.

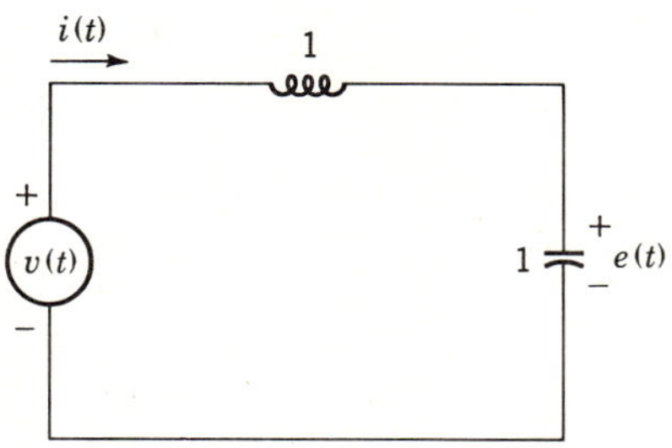

PROB. FIG. 7.8-4

7.8-5 Observe that in the network shown in the figure a *dependent* current source is connected across the capacitor. Determine $i(0)$, $v(0)$, and the value of K so that

$$i(t) = 5 \sin 3t \qquad \text{for all } t > 0$$

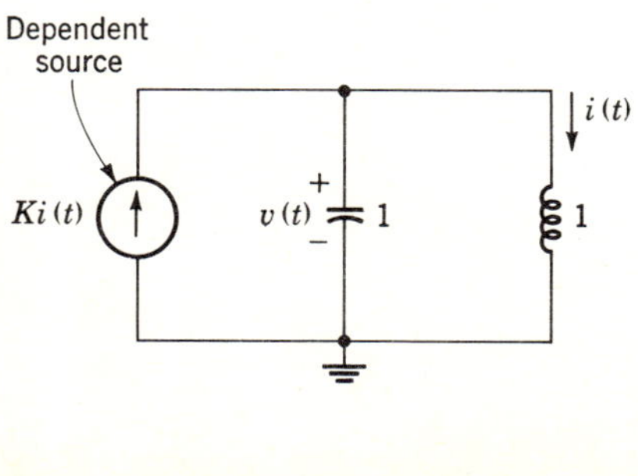

PROB. FIG. 7.8-5

7.8-6 You admittedly have no formal procedures for handling this problem; yet you should be able to have some fun and arrive at an answer by an ordered trial-and-error method. That is, you should propose a probable solution, and then analyze it to see whether it is indeed a solution. In the process, you will "learn" something about the independence and formulation issues. You are given the state form,

$$\left.\begin{aligned} \frac{dx_1}{dt} &= -a_{11}x_1 + a_{12}x_2 + u_1(t) \\ \frac{dx_2}{dt} &= a_{21}x_1 - a_{22}x_2 + u_2(t) \end{aligned}\right\} \quad (1)$$

where the a_{ij} are all positive, obeying the constraints $a_{11} > a_{12}$, and $a_{22} > a_{21}$; $u_1(t)$ and $u_2(t)$ are continuous time functions. Using three conductances, g_1, g_2, g_3; two current sources, $I_1(t)$ and $I_2(t)$; a unit capacitor and a capacitor of value C; construct a network whose state differential equation is (1) above. Be sure to specify unambiguously (1) your circuit and (2) the quantities

$$g_1,\ g_2,\ g_3,\ I_1(t),\ I_2(t),\ C$$

in terms of the given quantities

$$a_{11},\ a_{12},\ a_{21},\ a_{22},\ u_1(t),\ u_2(t)$$

As usual, the capacitor voltages will be the state variables.

chapter 8

Dynamic Systems and Networks; Input-Output Analysis

This chapter discusses continuous linear dynamic systems and networks from an input-output point of view. In contrast to the state-analysis approach of Chap. 7, which leads to simultaneous first-order differential equations in terms of appropriate internal variables, the input-output approach of this chapter leads to a single differential equation which relates first- and higher-order derivatives of the input to first- and higher-order derivatives of the output. Such linear high-order differential equations will be solved for the case of suddenly applied complex exponential inputs, and the solutions will be related to the so-called transient and steady-state behavior of the corresponding systems and networks. Alternate input-output descriptions called system functions will be introduced for characterizing steady-state behavior in terms of frequency. A special type of system function called impedance, which is applicable to dynamic networks that contain linear resistors, capacitors, and inductors, will be examined in some detail.

8.1 INTRODUCTION

In the previous chapter, the analysis of dynamic systems and networks was conducted from a state-variable point of view. For simple networks it was demonstrated that capacitor voltages and inductor currents can normally act as state variables, whereas for dynamic systems, integrator outputs are the typical state variables. In either case formulation of equations resulted in:

1. A set of first-order differential equations relating the state variables and the inputs, and
2. A set of algebraic equations specifying the output as a function of the state variables and the inputs.

From a philosophical point of view, the state-variable approach corresponds to a "microscopic" analysis, focused on the internal structure and interconnections of the system.

This chapter presents another approach to the description and analysis of systems, called the *input-output* approach. This method is designed to readily provide information about the manner in which input signals are transformed into output signals by the system. Particularly, the concept of state variables is not explicit in the analysis. It should be stressed that neither the state variable nor the input-output approach is deficient. Each has its place, and it is important to be aware of why and when one is to be preferred over the other.

The starting point for input-output analysis is the view of a system as a "black box" with inputs and outputs,† as shown in Fig. 8.1-1. One of the

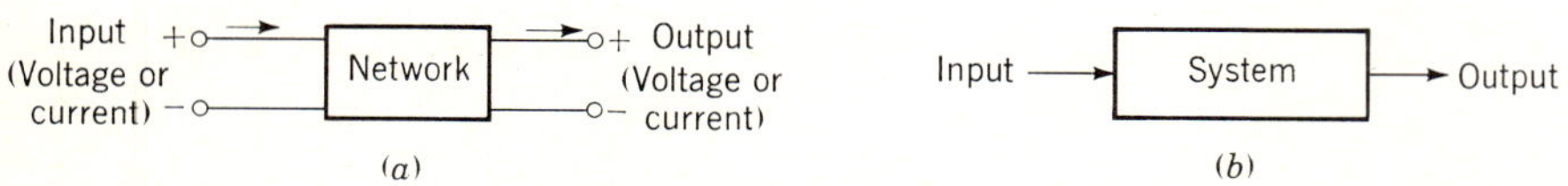

FIG. 8.1-1 A network and a block diagram system viewed as "black boxes."

motivations for this approach is to permit the simple analysis of systems which are constructed by connecting single-input single-output systems. If, for example, one is interested in the input-output behavior of the overall system of Fig. 8.1-2, then similiar information is clearly desired for the subsystems S_1, S_2, and S_3. If equations are available that explicitly describe how the input and output of the subsystems are related, then the analysis of the overall system is relatively straightforward.‡

In this chapter we shall examine networks and systems from the input-output point of view. We shall focus our attention on systems whose input and output are related by a single high-order linear differential equation with constant coefficients. Methods for solving these differential equations

†In this chapter only systems with a single input and output will be considered.

‡It is assumed that the behavior of a system is independent of what its output is connected to. This is not true for most networks, and is only true for certain block diagram systems.

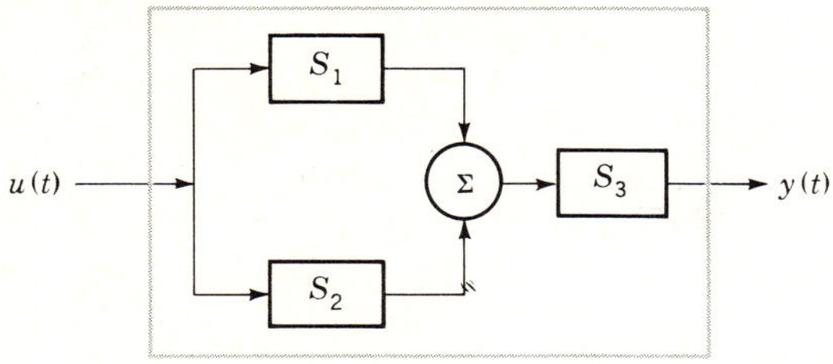

FIG. 8.1-2 A composition of single-input single-output systems.

when the input is a complex exponential will be presented. These methods lead in a natural way to the concepts of *system function, poles, zeros, frequency response,* and in the case of networks, to the concept of *impedance.*

8.2 EXAMPLES OF INPUT-OUTPUT REPRESENTATIONS

In this section, we shall derive differential equations linking the input and the output of both a block diagram system and a network. To obtain these equations, we shall start from the state-variable equations, and we shall relate input to output by eliminating the state variables.

Example 8.2-1 *A Block Diagram System* Consider the system illustrated in Fig. 8.2-1, which is constructed from integrators, adders, and scalers. Let $x_1(t)$,

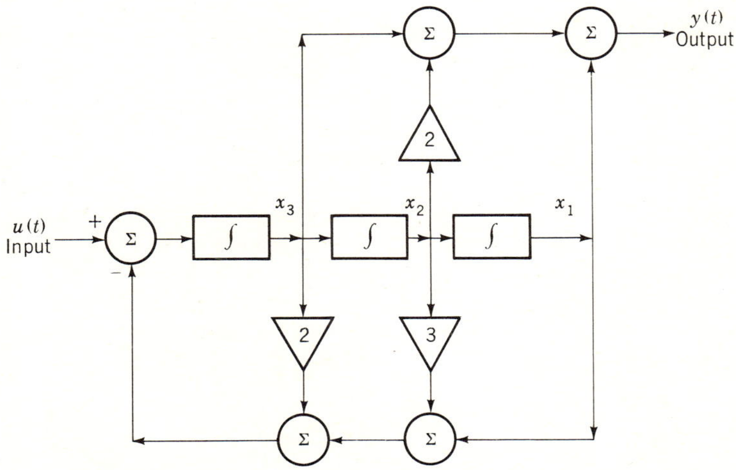

FIG. 8.2-1 Block diagram system of Example 8.2-1.

$x_2(t)$, and $x_3(t)$ denote the state variables; $u(t)$ the input; and $y(t)$ the output. Assume that at the initial time $t = 0$ *the system is at rest,* i.e., that

$$x_1(0) = x_2(0) = x_3(0) = 0 \tag{1}$$

As seen from Fig. 8.2-1, the state variables $x_1(t)$, $x_2(t)$, $x_3(t)$ are related by the differential equation,†

$$\begin{aligned}\dot{x}_1(t) &= x_2(t)\\ \dot{x}_2(t) &= x_3(t)\\ \dot{x}_3(t) &= -x_1(t) - 3x_2(t) - 2x_3(t) + u(t)\end{aligned} \tag{2}$$

while the output $y(t)$ is related to the state variables by

$$y(t) = x_1(t) + 2x_2(t) + x_3(t) \tag{3}$$

To obtain a single equation relating input and output, we must eliminate x_1, x_2, x_3 in Eqs. (2) and (3). This can be done by differentiating Eq. (3) and substituting from Eqs. (2), as shown in Eq. (4).

$$\frac{dy}{dt} = \dot{x}_1(t) + 2\dot{x}_2(t) + \dot{x}_3(t) = -x_1(t) - 2x_2(t) + u(t) \tag{4}$$

Differentiating Eq. (4), we obtain

$$\frac{d^2y}{dt^2} = -\dot{x}_1(t) - 2\dot{x}_2(t) + \frac{du}{dt} = -x_2(t) - 2x_3(t) + \frac{du}{dt} \tag{5}$$

Finally, differentiation of Eq. (5) yields

$$\frac{d^3y}{dt^3} = -\dot{x}_2(t) - 2\dot{x}_3(t) + \frac{d^2u}{dt^2} = 2x_1(t) + 6x_2(t) + 3x_3(t) - 2u(t) + \frac{d^2u}{dt^2} \tag{6}$$

Next, by eliminating x_1, x_2, x_3 in Eqs. (3) to (6), we arrive at the relation

$$\frac{d^3y}{dt^3} + 2\frac{d^2y}{dt^2} + 3\frac{dy}{dt} + y = \frac{d^2u}{dt^2} + 2\frac{du}{dt} + u \tag{7}$$

Equation (7) is a *third-order linear differential equation with constant coefficients* relating a linear combination of the input and its time derivatives to a linear combination of the output and its derivatives. The method used above to go from Eqs. (2) to Eq. (7) can be used for any linear state representation of a system, i.e., a representation similar to Eqs. (2) and (3). Thus any block diagram system consisting of integrators, summers and, scalers can be characterized by a single high-order differential equation relating the input and the output. Observe, however, from Eqs. (1), (3), (4), and (5) that

$$y(0) = 0 \qquad \frac{dy}{dt}(0) = u(0) \qquad \text{and} \qquad \frac{d^2y}{dt^2} = \frac{du}{dt}(0)$$

In general, in order to establish the initial values of $y(t)$ and its first $m-1$ derivatives, the initial values of $u(t)$ and its first $m-1$ derivatives must be known.

†The abbreviation $\dot{x}(t)$ stands for $\frac{dx}{dt}(t)$.

Example 8.2-2 *An RLC Network* Consider the parallel *RLC* network driven by an independent current source, as illustrated in Fig. 8.2-2.

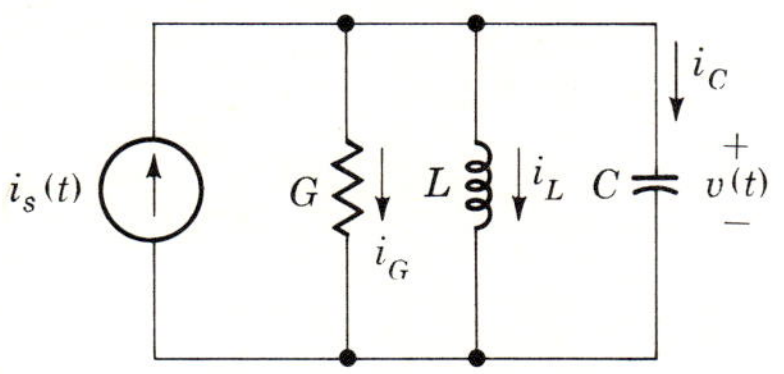

FIG. 8.2-2 *RLC network.*

Suppose that we wish to find a relation between the source current $i_S(t)$ (input) and the capacitor current $i_C(t)$ (output). From KCL,

$$i_C(t) + i_L(t) + i_G(t) = i_S(t) \tag{8}$$

Using the *v-i* relations for the branches, Eq. (8) can be written as

$$i_C(t) + \frac{1}{L} i_L(t_0) + \frac{1}{L}\int_{t_0}^{t} v(\tau)\, d\tau + Gv(t) = i_S(t) \tag{9}$$

Therefore,

$$\frac{di_C(t)}{dt} + \frac{1}{L} v(t) + G\frac{dv(t)}{dt} = \frac{di_S(t)}{dt} \tag{10}$$

But, since $i_C(t) = C\, dv(t)/dt$, we obtain

$$\frac{di_C(t)}{dt} + \frac{1}{L} v(t_0) + \frac{1}{LC}\int_{t_0}^{t} i_C(\tau)\, d\tau + \frac{G}{C} i_C(t) = \frac{di_S(t)}{dt} \tag{11}$$

and an additional time differentiation yields the desired result

$$\frac{d^2 i_C(t)}{dt^2} + \frac{G}{C}\frac{di_C(t)}{dt} + \frac{1}{LC} i_C(t) = \frac{d^2 i_S(t)}{dt^2} \tag{12}$$

Equation (12) is a second-order differential equation with constant coefficients and is similar in form to Eq. (7). This equation can be solved if the initial conditions $i_C(0)$ and $(di_C/dt)(0)$ are specified. Although it is not obvious from this example, we will see that a single high-order linear differential equation can be generally derived relating any two network variables of a linear network.

Generalization

The above examples illustrate systems whose input $u(t)$ and output $y(t)$ can be related by an equation that requires *a linear combination of the output and its time derivatives to equal a linear combination of the input and its derivatives.* The general form of such a relation is given by Eq. (13),†

$$\frac{d^N y}{dt^N} + a_{N-1}\frac{d^{N-1}y}{dt^{N-1}} + \cdots + a_1\frac{dy}{dt} + a_0 y = b_M\frac{d^M u}{dt^M} + b_{M-1}\frac{d^{M-1}u}{dt^{M-1}} + \cdots + b_1\frac{du}{dt} + b_0 u \tag{13}$$

where $y(0)$, $(dy/dt)(0)$, . . . , $(d^{N-1}y/dt)(0)$ are the *initial conditions.* We will be concerned with such equations only for the case $M \leq N$. Equation (13) is called an N th-order linear differential equation with constant coefficients. The variable t (time) is called the *independent variable,* the variable y (output) is called the *dependent variable,* and u (input) is called the *forcing function.*

The remainder of this chapter is devoted to techniques for deriving Eq. (13) for various systems and solving it for a certain class of forcing functions.

8.3 SINGULARITY FUNCTIONS

In this chapter we will often be driving or exciting systems with inputs that are usually called *suddenly applied excitations.* These are functions that are zero prior to some time and then assume a given character. An example of such a function is given in Eqs. (1).

$$\begin{aligned} x(t) &= 0 \qquad t < 0 \\ x(t) &= 1 + t^2 \qquad t > 0 \end{aligned} \tag{1}$$

The purpose of this section is to describe a family of functions that make the description and manipulation of functions like that of Eq. (1) compact and convenient. These functions are called *singularity functions.* Our approach in this section will be relatively informal. Our intent is to provide the reader with the ability to use two common singularity functions, the *step* and the *impulse,* the most common of these being the *unit step.*

DEFINITION 8.3-1 The unit step at $t = t_1$ is denoted by $u_{-1}(t - t_1)$ and is defined by

$$u_{-1}(t - t_1) = \begin{cases} 0 & \text{for all } t < t_1 \\ 1 & \text{for all } t > t_1 \\ \text{undefined but finite at } t = t_1 \end{cases} \tag{2}$$

The unit step occurring at $t = 0$ is denoted by $u_{-1}(t)$.

The unit step, as it is normally represented, is shown in Fig. 8.3-1*a*. It is an idealization of a continuous and differentiable time function such as the one shown in Fig. 8.3-1*b*. The "infinite" slope of the step is a convenience. Any use of the step could be replaced by a function like that of Fig. 8.3-1*b* with a steep but finite slope, without significant effect. The only problem is determining how steep is steep enough for a given situation. Infinite steepness is an easy solution.

†Here $d^N y/dt^N$ means the Nth time derivative of y.

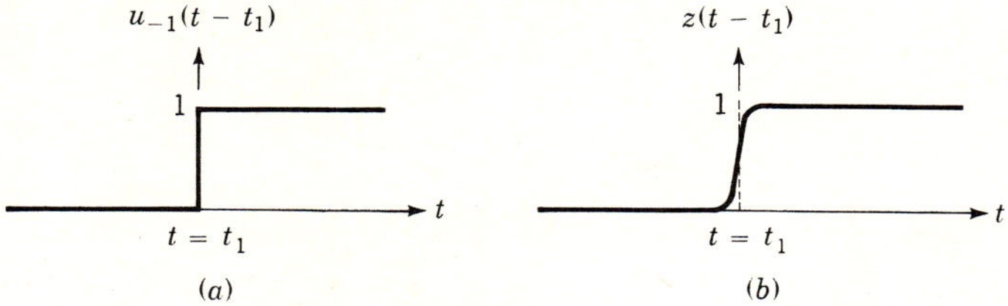

FIG. 8.3-1 (a) *The ideal unit step $u_{-1}(t - t_1)$ which is the idealization of* (b) *the waveform $z(t - t_1)$.*

The unit step can be used to indicate the "start" of any waveform, and hence provides a notational convenience. If $f(t)$ is a function defined for all time, then

$$f(t)\, u_{-1}(t - t_1) = \begin{cases} f(t) & \text{for all } t > t_1 \\ 0 & \text{for all } t < t_1 \end{cases} \tag{3}$$

This is illustrated in Fig. 8.3-2.

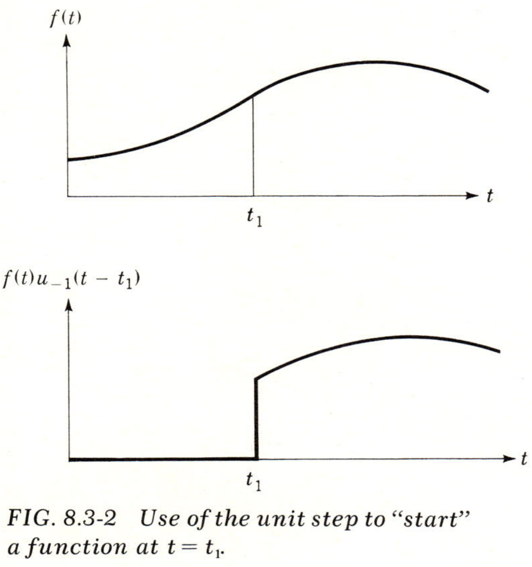

FIG. 8.3-2 Use of the unit step to "start" a function at $t = t_1$.

The singularity functions are a family of functions denoted by

$$u_k(t - t_1) \tag{4}$$

(where k is an integer—positive, negative, or zero) which are related as follows:

$$\int_{-\infty}^{t} u_k(\tau - t_1)\, d\tau = u_{k-1}(t - t_1) \quad \text{for all } k \text{ and } t \tag{5}$$

For example, if $k=-1$, then Eq. (5) becomes

$$\int_{-\infty}^{t} u_{-1}(\tau - t_1)\, d\tau = u_{-2}(t - t_1) \tag{6}$$

Thus the singularity function $u_{-2}(t-t_1)$ is the integral of the unit step. It is easy to see that

$$u_{-2}(t - t_1) = \begin{cases} 0 & \text{for } t < t_1 \\ t - t_1 & \text{for } t \geq t_1 \end{cases} \tag{7}$$

The singularity function $u_{-2}(t-t_1)$ is shown in Fig. 8.3-3 and is called the *unit*

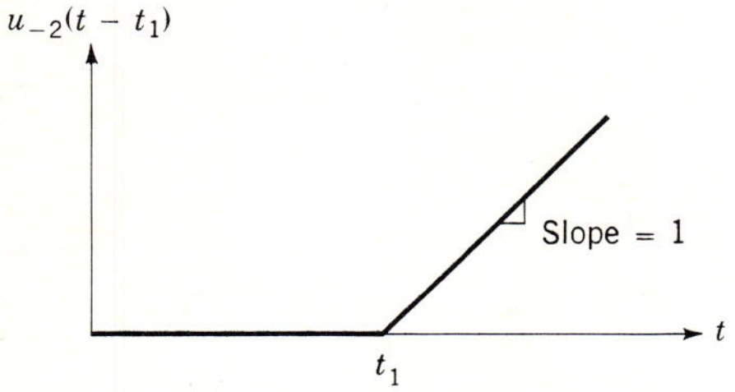

FIG. 8.3-3 The unit ramp.

ramp. In general,

For $k \geq 1$

$$u_{-k}(t - t_1) = \begin{cases} 0 & \text{for } t < t_1 \\ \dfrac{(t - t_1)^{k-1}}{(k-1)!} & \text{for } t \geq t_1 \end{cases} \tag{8}$$

If we consider Eq. (5) for $k=0$ we have

$$\int_{-\infty}^{t} u_0(\tau - t_1)\, d\tau = u_{-1}(t - t_1) \tag{9}$$

Equation (9) states that the unit step $u_{-1}(t-t_1)$ is the time integral of the singularity function $u_0(t-t_1)$, which is called the *unit impulse*.† If the function of Fig. 8.3-1*b* is considered an approximation to the unit step, then $u_0(t-t_1)$ should be the idealization of the derivative of Fig. 8.3-1*b*. The derivative of the waveform of Fig. 8.3-1*b* is a narrow but smooth pulse of very small duration, large height, and unit area, as shown in Fig. 8.3-4*a*.‡ The unit impulse $u_0(t-t_1)$, shown as it is usually drawn, in Fig. 8.3-4*b*, is a "spike" of zero width, infinite height, and unit area. From a mathematical point of view it is not a function,§ and from an engineering standpoint, one cannot generate an impulse.

†$u_0(t-t_1)$ is often written as $\delta(t-t_1)$ and is sometimes called the Dirac delta function.

‡Another waveform, the limit of a narrow pulse, was used informally in Chap. 7 as an approximation to the impulse.

§It is often called a generalized function or a distribution.

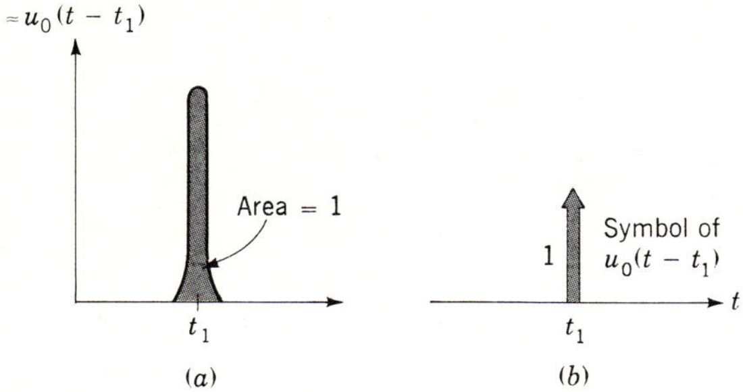

FIG. 8.3-4 (a) *An approximation to the unit impulse, and* (b) *the corresponding symbol.*

Nonetheless, as long as we think of these waveforms as idealizations of smooth functions, our intuition will not lead us astray. We will not have reason to refer to $u_k(t)$ for $k \geq 1$, although we mention that $u_1(t)$ is called a *unit doublet*.

Although it makes little formal sense to talk about the derivative of a unit step and of a unit impulse, it is nonetheless often useful to think of the unit impulse as the time derivative of the unit step, and of the unit doublet as the time derivative of the unit impulse. Thus, in conjunction with the normal interpretation of an equation like Eq. (5) we formally write

$$\frac{d}{dt}u_{k-1}(t - t_1) = u_k(t - t_1) \qquad \text{for all } k \text{ and } t \tag{10}$$

If we continue our analogy with conventional mathematics, we have

$$\frac{d}{dt}[u_{k-1}(t - t_1)f(t)] = u_{k-1}(t - t_1)\frac{df}{dt} + f(t)u_k(t - t_1) \tag{11}$$

In particular, let us consider

$$\frac{d}{dt}[u_{-1}(t - t_1)f(t)] = u_{-1}(t - t_1)\frac{df}{dt} + f(t)u_0(t - t_1) \tag{12}$$

Observe that the term $f(t)u_0(t - t_1) = f(t_1)u_0(t - t_1)$, since $u_0(t - t_1)$ is zero for $t \neq t_1$. Hence, Eq. (12) can be written as

$$\frac{d}{dt}[u_{-1}(t - t_1)f(t)] = u_{-1}(t - t_1)\frac{df}{dt} + f(t_1)u_0(t - t_1) \tag{13}$$

As an example, if $f(t) = 2 \cos t$ and $t_1 = 0$ in Eq. (13), then

$$\frac{d}{dt}[(2 \cos t)u_{-1}(t)] = (-2 \sin t)u_{-1}(t) + 2u_0(t) \tag{14}$$

as illustrated in Fig. 8.3-5*a* and *b*. A constant times $u_0(t)$, such as $f(0)u_0(t)$, is an impulse with area $f(0)$. The area is usually written alongside the impulse as shown in Fig. 8.3-5*b*.

We shall make extensive use of the singularity functions $u_0(t)$ and $u_{-1}(t)$ in the following sections.

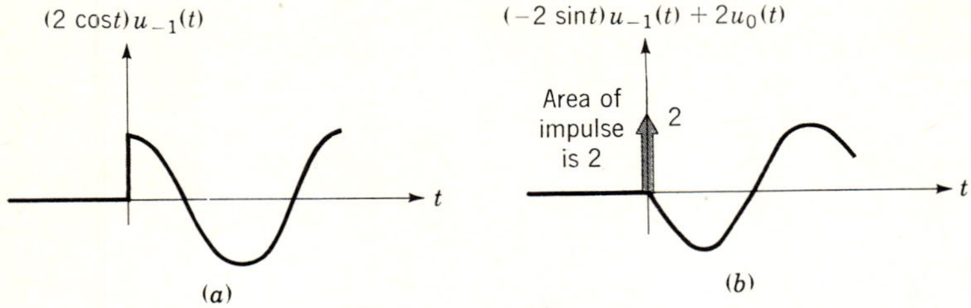

FIG. 8.3-5 (a) *A function of time and* (b) *its time derivative.*

8.4 SOLUTION OF LINEAR DIFFERENTIAL EQUATIONS

We shall consider first the solution (for $t \geq 0$) of the differential equation

$$\frac{d^2y}{dt^2} + a_1\frac{dy}{dt} + a_0y = u(t) \tag{1}$$

We shall assume that at $t = 0$ the initial (or boundary) conditions are

$$y(0) = \alpha \qquad \frac{dy}{dt}(0) = \beta \tag{2}$$

and that the input $u(t)$ is a complex exponential time function,†

$$u(t) = Ue^{s_0t} \qquad t \geq 0 \tag{3}$$

In accordance with our definition, the relation of Eq. (1) is called a *second-order linear differential equation with constant coefficients*. In Sec. 7.6 where we dealt with a first-order linear differential equation, it was necessary to specify only the value of the solution at a given time as an initial condition. Because this is a second-order equation, we will see that we must specify both the value and the first derivative of the solution at a given time.

We have chosen the forcing function Ue^{s_0t} because the complex exponential can represent the types of signals we will be interested in throughout this chapter.

We will assert that all differential equations in this section have a unique solution. This assertion is an extension of Theorem 7.5-1 for first-order equations.

The development used in Chap. 7 can be paralleled to a large extent to solve Eq. (1). Let us first solve the equation when $U = 0$, i.e., let us obtain the *unforced* or *zero-input solution*. Under zero input, Eq. (1) yields:

$$\begin{aligned} &\frac{d^2y}{dt^2} + a_1\frac{dy}{dt} + a_0y = 0 \\ &y(0) = \alpha \qquad \text{and} \qquad \frac{dy}{dt}(0) = \beta \end{aligned} \tag{4}$$

†The appendix discusses complex numbers and exponentials.

While Eq. (4) cannot be easily solved by direct integration, as was done with first-order equations in Sec. 7.6, we can use the zero-input solution to the first-order equation as the basis for an educated guess at the solution of Eqs. (4). Recalling that the first-order solution was an exponential, we hypothesize that the solution to Eqs. (4) has the form

$$y(t) = C_1 e^{s_1 t} + C_2 e^{s_2 t} \qquad t \geq 0 \tag{5}$$

To determine whether the constants in Eq. (5) can be chosen to satisfy the constraints of Eqs. (4), we substitute Eq. (5) into Eqs. (4). This yields

$$s_1{}^2 C_1 e^{s_1 t} + s_2{}^2 C_2 e^{s_2 t} + a_1 s_1 C_1 e^{s_1 t} + a_1 s_2 C_2 e^{s_2 t} + a_0 C_1 e^{s_1 t} + a_0 C_2 e^{s_2 t} = 0 \tag{6}$$

and

$$\begin{aligned} C_1 + C_2 &= \alpha \\ s_1 C_1 + s_2 C_2 &= \beta \end{aligned} \tag{7}$$

Equation (6) can be rewritten to emphasize its form as

$$(s_1{}^2 + a_1 s_1 + a_0) C_1 e^{s_1 t} + (s_2{}^2 + a_1 s_2 + a_0) C_2 e^{s_2 t} = 0 \tag{8}$$

If we assume that C_1 and C_2 are not always zero, which is reasonable in view of Eqs. (7), then Eq. (8) must be satisfied by having

$$s_1{}^2 + a_1 s_1 + a_0 = 0$$

and

$$s_2{}^2 + a_1 s_2 + a_0 = 0 \tag{9}$$

This implies that s_1 and s_2 are the two roots of the polynomial $s^2 + a_1 s + a_0$. If we look back at Eq. (1) we see that this polynomial is independent of the forcing function. For this reason it is called the *characteristic polynomial.* As long as $s_1 \neq s_2$, Eq. (8) can only be satisfied by having s_1 and s_2 roots of the characteristic polynomial

$$\boxed{p(s) = s^2 + a_1 s + a_0} \tag{10}$$

If s_1 and s_2 satisfy $p(s_1) = p(s_2) = 0$, Eq. (8) is satisfied, regardless of the values of C_1 and C_2. The initial conditions can then be met by choosing C_1 and C_2 to satisfy Eqs. (7).

This is consistent with the first-order result of Chap. 7, where we found the exponential term a function of the equation, and the coefficient a function of the boundary condition.

Example 8.4-1 If we take the network of Fig. 8.2-2 with the current source $i_s(t) = 0$, we have the network of Fig. 8.4-1.

The second-order equation for $i_c(t)$ is given by Eq. (11) [See Eq. (12) of Sec. 8.2]. We will assume that $G = 3$, $L = 1/2$, and $C = 1$.

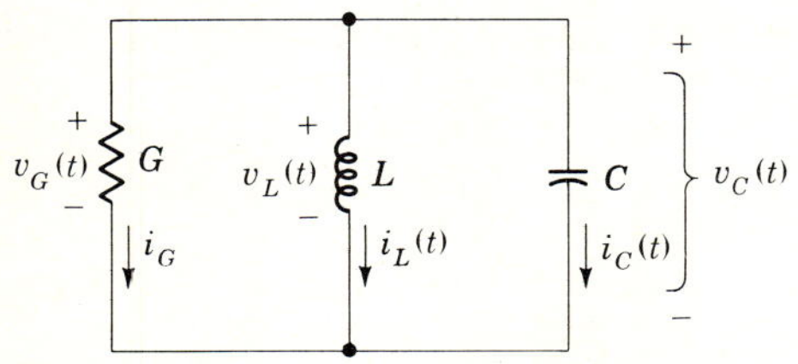

FIG. 8.4-1 The network of Fig. 8.2-2 with $i_s(t) = 0$.

$$\frac{d^2 i_c(t)}{dt} + \frac{G}{C}\frac{di_c(t)}{dt} + \frac{1}{LC} i_c(t) = 0 \tag{11}$$

This gives the characteristic polynomial

$$p(s) = s^2 + 3s + 2$$

which has the roots $s_1 = -1$ and $s_2 = -2$. Now, without knowing anything about initial conditions for the network, we know that $i_c(t)$ will have the form

$$i_c(t) = C_1 e^{-t} + C_2 e^{-2t} \tag{12}$$

Again, we stress that the network, and the network alone, determines Eq. (12) and that the initial conditions can only affect the coefficients C_1 and C_2.

Now, assume that

$$i_c(0) = 3 \qquad \frac{di_c}{dt}(0) = 6$$

Equation (7) requires that

$$C_1 + C_2 = 3 \qquad -C_1 - 2C_2 = 6$$

which gives

$$C_1 = 12 \qquad C_2 = -9$$

The zero-input solution is then

$$i_c(t) = 12e^{-t} - 9e^{-2t} \tag{13}$$

Since we are investigating the response of the circuit of Fig. 8.4-1 caused by nonzero initial conditions, we will typically know $i_L(0)$ and $v_C(0)$, as opposed to $i_c(0)$ and $(di_c/dt)(0)$. These quantities, however, can be related using the element constraints and the Kirchhoff voltage and current laws. First, KVL requires

$$Gv_C(0) = i_G(0)$$

and KCL requires

$$i_G(0) + i_L(0) + i_c(0) = 0$$

giving

$$i_c(0) = -i_L(0) - Gv_C(0)$$

Thus we have $i_c(0)$ as a function of $i_L(0)$ and $v_C(0)$. To get $(di_c/dt)(0)$, we proceed as follows: First,

$$\frac{di_c}{dt} = -\frac{di_G}{dt} - \frac{di_L}{dt}$$

which, because $v_C = v_L = L\,di_L/dt$, can be written as

$$\frac{di_c}{dt} = -\frac{di_G}{dt} - \frac{v_C}{L}$$

Now

$$\frac{di_G}{dt} = G\frac{dv_G}{dt} = G\frac{dv_C}{dt} = \frac{G}{C}i_C$$

Hence

$$\frac{di_c}{dt}(0) = -\frac{G}{C}i_c(0) - \frac{v_C(0)}{L}$$

But, since $i_c(0) = -i_L(0) - Gv_C(0)$, we have

$$\frac{di_c}{dt}(0) = \frac{G}{C}(i_L(0) + Gv_C(0)) - \frac{v_C(0)}{L}$$

$$= v_C(0)\left[\frac{G^2}{C} - \frac{1}{L}\right] + \frac{G}{C}i_L(0)$$

Thus, the initial conditions $i_c(0)$ and $(di_c/dt)(0)$ can be expressed in terms of $i_L(0)$ and $v_C(0)$ as

$$i_c(0) = -i_L(0) - Gv_C(0)$$

and

$$\frac{di_c}{dt}(0) = \frac{G}{C}i_L(0) + v_C(0)\left[\frac{G^2}{C} - \frac{1}{L}\right]$$

As this example demonstrates, it is sometimes tedious to translate from initial conditions specified in one fashion to the specification necessary for solving the given differential equation.

Now we can return to the problem of solving the forced case,

$$\frac{d^2y}{dt^2} + a_1\frac{dy}{dt} + a_0y = Ue^{s_0t}$$

with

$$y(0) = \alpha \qquad \frac{dy}{dt}(0) = \beta$$

We will assume that s_1 and s_2 are roots of $p(s) = s^2 + a_1s + a_0$ and that $s_1 \neq s_2$. To solve this equation, let us search for a function that can be substituted into the left-hand side to yield the right-hand side, temporarily ignoring the initial conditions. Again, we take direction from Chap. 7 and assume an exponential of the form

$$y(t) = C_0 e^{s_0 t} \tag{14}$$

Substitution of $y(t)$ gives

$$s_0^2 C_0 e^{s_0 t} + a_1 s_0 C_0 e^{s_0 t} + a_0 C_0 e^{s_0 t} = U e^{s_0 t}$$

which can be rewritten as

$$[s_0^2 + a_1 s_0 + a_0] C_0 e^{s_0 t} = U e^{s_0 t} \tag{15}$$

We recognize the coefficient of $C_0 e^{s_0 t}$ as $p(s_0)$, and we assume that $p(s_0) \neq 0$. Solving Eq. (15) for C_0 gives

$$C_0 = \frac{U}{p(s_0)} \tag{16}$$

Now, although $y(t) = [U/p(s_0)]\, e^{s_0 t}$ satisfies Eq. (1), it does not, in general, satisfy the initial conditions. If, however, we consider the function

$$\boxed{y(t) = \frac{U}{p(s_0)} e^{s_0 t} + C_1 e^{s_1 t} + C_2 e^{s_2 t}} \tag{17}$$

we see that it has certain desirable properties. The fact that s_1 and s_2 are roots of the characteristic polynomial ensures that when Eq. (17) is substituted into Eq. (1), the result is the same as substituting $[U/p(s_0)]\, e^{s_0 t}$. If we examine $y(0)$ and $\frac{dy}{dt}(0)$ for Eq. (17), we see that

$$y(0) = \frac{U}{p(s_0)} + C_1 + C_2$$

$$\frac{dy}{dt}(0) = \frac{U s_0}{p(s_0)} + s_1 C_1 + s_2 C_2$$

Thus, C_1 and C_2 can be chosen to meet the constraints given below, which are imposed by $y(0) = \alpha$ and $\frac{dy}{dt}(0) = \beta$.

$$\begin{aligned} \alpha - \frac{U}{p(s_0)} &= C_1 + C_2 \\ \beta - \frac{U s_0}{p(s_0)} &= s_1 C_1 + s_2 C_2 \end{aligned} \tag{18}$$

Hence, Eq. (17) gives the complete solution to Eqs. (1), (2), and (3).

Example 8.4-2 In this example, we take the second-order equation of Example 8.4-1 and add a forcing function as shown below:

$$\frac{d^2y}{dt^2} + 3\frac{dy}{dt} + 2y = 4e^{-3t}$$

$$y(0) = 11 \qquad \frac{dy}{dt}(0) = -20$$

Since $p(s) = s^2 + 3s + 2$, with roots $s_1 = -1$, $s_2 = -2$, and $p(s_0) = p(-3) = 2$, we have from Eq. (17) a solution of the form

$$y(t) = \frac{4}{2}e^{-3t} + C_1 e^{-t} + C_2 e^{-2t} \tag{19}$$

To find C_1 and C_2, we employ, according to Eq. (18), the constraints

$$11 - 2 = C_1 + C_2 \qquad -20 + 6 = -C_1 - 2C_2$$

which yield $C_1 = 4$, $C_2 = 5$.

Hence, the desired solution is

$$y(t) = 2e^{-3t} + 4e^{-t} + 5e^{-2t} \qquad t \geq 0 \tag{20}$$

A portion of this solution is similar in form to the zero-input solution, and the remainder is due particularly to the forcing function used. The zero-input solution characterizes the natural or undriven behavior of the system modeled by the differential equation. For this reason the roots of the characteristic polynomial are called *natural frequencies* of the equation. Because the zero-input solution satisfies the differential equation when the right-hand side is zero, it is also called the *homogeneous solution*. That portion of Eq. (20) which has the same exponential form as the forcing function is called the *particular solution*.

We see that *a forcing function will usually excite, or cause to be present in the solution, a response at the natural frequencies of the system.* The reason for this can be seen most clearly from examination of Eq. (18). If

$$\alpha = \frac{U}{p(s_0)} \qquad \text{and} \qquad \beta = \frac{Us_0}{p(s_0)}$$

then $C_1 = C_2 = 0$, and no homogeneous solution is present. But the above equation requires that the imposed initial conditions be identical to those of the particular solution. The role of the homogeneous solution is to "make up the difference" between the demands of the forcing function and the unrelated initial conditions.

So far we have required that *the two roots of* $p(s)$ *be distinct and that the complex driving frequency not correspond to a root of* $p(s)$, i.e., that $p(s) \neq 0$.

Double Roots

To remove the restriction of distinct roots, we can focus on the zero-input solution and look at a limiting situation. Assume for some small δ that

$$p(s) = (s - a)[s - (a + \delta)]$$

and

$$y(0) = \alpha \qquad \frac{dy}{dt}(0) = \beta$$

The homogeneous solution is

$$y(t) = C_1e^{at} + C_2e^{(a+\delta)t}$$

and

$$\alpha = C_1 + C_2 \qquad \beta = aC_1 + (a+\delta)C_2$$

or

$$C_1 = \alpha + \frac{a\alpha - \beta}{\delta} \qquad C_2 = -\frac{a\alpha - \beta}{\delta}$$

The complete solution is

$$\begin{aligned} y(t) &= \left[\alpha + \frac{a\alpha-\beta}{\delta}\right]e^{at} - \left[\frac{a\alpha-\beta}{\delta}\right]e^{(a+\delta)t} \\ &= e^{at}\left[\alpha + \frac{a\alpha-\beta}{\delta} - \left(\frac{a\alpha-\beta}{\delta}\right)\left\{1 + \delta t + \frac{(\delta t)^2}{2\,!} + \cdots\right\}\right] \\ &= e^{at}\left[\alpha - (a\alpha-\beta)t + \left(\frac{a\alpha-\beta}{\delta}\right)\left\{\frac{(\delta t)^2}{2\,!} + \cdots\right\}\right] \end{aligned}$$

where we have used the series expansion of $e^{\delta t}$. Now, if we take the limit as $\delta \to 0$, we cause the two roots of $p(s)$ to coincide, and the solution becomes

$$\lim_{\delta\to 0} y(t) = e^{at}[\alpha - (a\alpha - \beta)t]$$

We can rewrite this as

$$y(t) = e^{at}[K_1 + K_2t] \tag{21}$$

where

$$K_1 = \alpha \qquad K_2 = -(a\alpha - \beta)$$

and the solution is complete.

Example 8.4-3 Find the solution of

$$\frac{d^2y}{dt^2} + 4\frac{dy}{dt} + 4y = 0 \tag{22}$$

with

$$y(0) = \frac{dy}{dt}(0) = 1$$

The characteristic polynomial is $s^2 + 4s + 4$ with a double root at $s = -2$. Hence the solution is, by Eq. (21),

$$y(t) = e^{-2t}[1 + 3t]$$

Note that an appropriate choice of initial conditions can make the solution, Eq.

(21), either

$$y(t) = Ce^{-2t}$$

or

$$y(t) = Cte^{-2t}$$

Note further that since we know how to find the homogeneous solution with repeated roots, we should have no problem in finding the complete solution of an equation such as Eq. (22) for an exponential forcing function.

Excitation at a Natural Frequency

To see what happens when the forcing-function frequency s_0 is a natural frequency, we can guess that having s_0 coincide with s_1 is no different than having s_1 and s_2 coincide, as in the previous case. This assumption would point to a solution of the form

$$y(t) = e^{s_0 t}[C_1 + C_2 t] + C_3 e^{s_1 t} \tag{23}$$

Assume that the differential equation is

$$\frac{d^2y}{dt^2} + a_1\frac{dy}{dt} + a_0 y = Ue^{s_0 t} \tag{24}$$

with

$$y(0) = \alpha \qquad \text{and} \qquad \frac{dy}{dt}(0) = \beta$$

Our hypothesis can be verified by substitution. Since the characteristic polynomial has roots s_0 and s_1 (assuming $s_0 \neq s_1$), we know that $C_1e^{s_0 t} + C_3e^{s_1 t}$ is the homogeneous solution to Eq. (23). It remains only to verify that $C_2te^{s_0 t}$ is a particular solution. Substitution of $C_2te^{s_0 t}$ in Eq. (24) gives

$$C_2te^{s_0 t}[s_0^2 + a_1s_0 + a_0] + C_2e^{s_0 t}[2s_0 + a_1] = Ue^{s_0 t}$$

But $s_0^2 + a_1s_0 + a_0 = 0$; so we have

$$C_2e^{s_0 t}[2s_0 + a_1] = Ue^{s_0 t}$$

or finally

$$C_2 = \frac{U}{2s_0 + a_1}$$

Thus the solution is

$$\boxed{y(t) = e^{s_0 t}\left(C_1 + \frac{U}{2s_0 + a_1}t\right) + C_3e^{s_1 t}}$$

and exciting a system at a natural frequency is essentially the same as having a system in which two natural frequencies coincide.

To summarize, we can solve the equation

$$\frac{d^2y}{dt^2} + a_1\frac{dy}{dt} + a_0 = Ue^{s_0t} \qquad y(0) = \alpha \qquad \frac{dy}{dt}(0) = \beta \tag{25}$$

under a variety of circumstances. We have defined the characteristic polynomial as

$$p(s) = s^2 + a_1s + a_0$$

We know that the roots of $p(s)$ are the natural frequencies of the system, which determine the zero-input or homogeneous solution within multiplicative constants determined by the initial conditions. The particular solution is determined completely by $p(s_0)$ and U, and the sum of the particular and homogeneous solutions forms the total solution to Eq. (25).

Higher-order Equations

These ideas carry over unchanged to higher-order equations. If we consider the equation

$$\frac{d^3y}{dt^3} + 7\frac{d^2y}{dt} + 17\frac{dy}{dt^2} + 15y = e^{-t}$$

with

$$y(0) = \frac{dy}{dt}(0) = \frac{d^2y}{dt^2}(0) = 0$$

we can exhibit the extension. By analogy, we define the characteristic polynomial

$$p(s) = s^3 + 7s^2 + 17s + 15$$

which has roots

$$s_1 = -3$$

$$s_2 = -2 + j$$

$$s_3 = -2 - j = s_2^*$$

(Here * means complex conjugate as discussed in the appendix.)

Nothing prohibits the natural frequencies from being complex, although since the coefficients of the characteristic polynomial are real, the complex roots will occur in complex conjugate pairs. As we will see later in this chapter, many interesting systems have complex roots.

The particular solution will be

$$y_p(t) = \frac{U}{p(s_0)}e^{s_0t}$$

which in this case is

$$y_p(t) = \frac{1}{p(-1)}\, e^{-t} = \tfrac{1}{4} e^{-t}$$

Thus, the total solution is of the form

$$y(t) = \tfrac{1}{4} e^{-t} + C_1 e^{-3t} + C_2 e^{(-2+j)t} + C_3 e^{(-2-j)t}$$

Instead of giving an involved set of three linear equations to evaluate C_1, C_2, and C_3, we observe that this set can be derived by taking the appropriate derivative of $y(t)$, evaluating it for $t=0$, and using the stated initial conditions. Doing this yields

$$y(t) = \tfrac{1}{4} e^{-t} + \tfrac{1}{12} e^{-3t} + (-\tfrac{1}{6} + j\tfrac{1}{12})\, e^{(-2+j)t} + (-\tfrac{1}{6} - j\tfrac{1}{12})\, e^{(-2-j)t}$$

which can be reduced to the more convenient form

$$y(t) = \tfrac{1}{12}[3e^{-t} + e^{-3t} - 2e^{-2t}(2 \cos t - \sin t)] \qquad t \geq 0$$

Although this solution procedure becomes more tedious as the order of the equation increases, it is completely straightforward. From this example, we see that even though the initial conditions were all zero, the homogeneous solution still appears in the total solution. A solution derived with all zero initial conditions corresponds to the *zero-state response* of Sec. 7.6. We also see that the appearance of *complex natural frequencies* has introduced sinusoidal variations into the response, a situation which is not possible with first-order D.E.'s.

Sudden Excitations

We are now in a position to use the step and impulse to create "suddenly applied"excitations. In particular, we wish to solve equations of the form

$$\frac{d^n y}{dt^n} + a_{n-1}\frac{d^{n-1} y}{dt^{n-1}} + \cdots + a_1 \frac{dy}{dt} + a_0 y = U e^{s_0 t} u_{-1}(t)$$

with (26)

$$y(0) = \frac{dy}{dt}(0) = \cdots \frac{d^{n-1}y}{dt^{n-1}}(0) = 0$$

The forcing function is "suddenly applied" because prior to $t=0$ it is zero. Moreover, the system is at rest† prior to $t=0$ because of the initial conditions. Since all our results will extend easily to nth-order equations, we can work with the easily managed second-order case.

Consider the equation

$$\frac{d^2 y}{dt^2} + 3\frac{dy}{dt} + 2y = 4e^{-3t} u_{-1}(t) \qquad (27)$$

†We say that the system is at rest if $y(t)$ and all of its time derivatives are zero for $t \leq 0$.

which is identical to that of Example 8.4-2, except for the unit step multiplying the exponential part of the forcing function. With zero initial conditions, the solution to Example 8.4-2 is easily found to be

$$w(t) = 2e^{-3t} + 2e^{-t} - 4e^{-2t}$$

For $t > 0$, this is certainly a solution of Eq. (27). Thus we might conjecture that

$$y(t) = [2e^{-3t} + 2e^{-t} - 4e^{-2t}]u_{-1}(t) = w(t)u_{-1}(t) \tag{28}$$

is the solution to Eq. (27). If we substitute Eq. (28) into the left side of Eq. (27), we have

$$2wu_{-1}(t) + 3\left[\frac{dw}{dt}u_{-1}(t) + wu_0(t)\right]$$

$$+ \frac{d^2w}{dt^2}u_{-1}(t) + \frac{dw}{dt}u_0(t) + \frac{dw}{dt}u_0(t) + wu_1(t)$$

$$= \left[\frac{d^2w}{dt^2} + 3\frac{dw}{dt} + 2w\right]u_{-1}(t) + \left[2\frac{dw}{dt} + 3w\right]u_0(t)$$

$$+ wu_1(t)$$

This last expression must equal $4e^{-3t}u_{-1}(t)$ in order for $y(t)$ to be the solution. Because $w(t)$ is a solution to the equation of Example 8.4-2 with zero initial conditions, we know that

$$\left[\frac{d^2w}{dt^2} + 3\frac{dw}{dt} + 2w\right] = 4e^{-3t} \qquad \text{for } t \geq 0$$

The term

$$\left[2\frac{dw}{dt} + 3w\right]u_0(t)$$

can be written as

$$\left[2\frac{dw}{dt}(0) + 3w(0)\right]u_0(t)$$

This term is identically zero because $w(t)$ satisfies the constraint that

$$\frac{dw}{dt}(0) = w(0) = 0$$

Using similar reasoning,

$$wu_1(t) = 0$$

Thus

$$y(t) = [2e^{-3t} + 2e^{-t} - 4e^{-2t}]u_{-1}(t)$$

is the solution to Eq. (27). This expression and $w(t)$ are identical for $t > 0$, and the process of going to the solution for a suddenly applied exponential excitation requires no significant extensions.

Example 8.4-4 *Singularity Matching* Find the initial-rest solution of

$$\frac{d^2y}{dt^2} + 5\frac{dy}{dt} + 6y = e^{-t}u_{-1}(t) \tag{29}$$

The characteristic polynomial is $s^2 + 5s + 6$, with roots $s_1 = -2$, $s_2 = -3$. Since $p(s_0) = p(-1) = 2$, the form of the solution is

$$y(t) = [\tfrac{1}{2}e^{-t} + C_1e^{-2t} + C_2e^{-3t}]u_{-1}(t) \tag{30}$$

The constants C_1 and C_2 can be evaluated directly by means described earlier, or by substituting into Eq. (29) and "matching" singularity functions, as described below:

$$\begin{aligned}
&\frac{d^2y}{dt^2} + 5\frac{dy}{dt} + 6y \\
&= 6[\tfrac{1}{2}e^{-t} + C_1e^{-2t} + C_2e^{-3t}]u_{-1}(t) && \leftarrow \text{(the term } 6y\text{)} \\
&\left.\begin{aligned}&+ 5[-\tfrac{1}{2}e^{-t} - 2C_1e^{-2t} - 3C_2e^{-3t}]u_{-1}(t) \\ &+ 5[\tfrac{1}{2} + C_1 + C_2]u_0(t)\end{aligned}\right\} && \leftarrow \left(\text{the term } 5\frac{dy}{dt} \text{ simplified}\right)^\dagger \\
&\left.\begin{aligned}&+ [+\tfrac{1}{2}e^{-t} + 4C_1e^{-2t} + 9C_2e^{-3t}]u_{-1}(t) \\ &+ [-\tfrac{1}{2} - 2C_1 - 3C_2]u_0(t) \\ &+ [\tfrac{1}{2} + C_1 + C_2]u_1(t)\end{aligned}\right\} && \leftarrow \left(\text{the term } \frac{d^2y}{dt^2} \text{ simplified}\right)^\dagger \\
&= [e^{-t}]u_{-1}(t) + (2 + 3C_1 + 2C_2)u_0(t) \\
&+ [\tfrac{1}{2} + C_1 + C_2]u_1(t)
\end{aligned} \tag{31}$$

In order for Eq. (31) to equal the excitation, the coefficients of the impulse and doublet must both be zero, since there are no corresponding terms in the excitation. This in turn requires that

$$2 + 3C_1 + 2C_2 = 0 \qquad \tfrac{1}{2} + C_1 + C_2 = 0 \tag{32}$$

giving

$$C_1 = -1 \qquad C_2 = \tfrac{1}{2}$$

Although this approach is needlessly complex, it was used to illustrate the concept of singularity matching. This concept can be stated simply as the only thing that matches an impulse, or in general $[u_i(t), i \geq 0]$, is an impulse, or $[u_i(t), i \geq 0]$.

Observe finally that the conditions of Eq. (32) can be obtained by taking Eq. (30) and requiring

$$y(0^+) = 0 \qquad \text{or} \qquad \tfrac{1}{2} + C_1 + C_2 = 0$$

$$\frac{dy}{dt}(0^+) = 0 \qquad \text{or} \qquad -\tfrac{1}{2} - 2C_1 - 3C_2 = 0$$

†Here, simplification involves the step $f(t)u_0(t) = f(0)u_0(t)$.

Although this set of equations differs in appearance from Eqs. (32), the solution is the same.

Forcing-function and Solution Derivatives

As we did in Chap. 7 for first-order equations, we can find the *step response* of the system being modeled by using a suddenly applied exponential forcing function with zero frequency, i.e., $s_0 = 0$. The step response of the system modelled by Eq. (29) is the solution to the equation

$$\frac{d^2y}{dt^2} + 5\frac{dy}{dt} + 6y = u_{-1}(t)$$

All initial conditions are zero, since the step response is, by convention, a zero-initial-condition or zero-state response. For this example, $p(0) = 6$, and the solution is of the form

$$y(t) = [1/6 + C_1e^{-2t} + C_2e^{-3t}]u_{-1}(t)$$

with

$$1/6 + C_1 + C_2 = 0 \qquad -2C_1 - 3C_2 = 0$$

giving

$$y(t) = [1/6 - 1/2e^{-2t} + 1/3e^{-3t}]u_{-1}(t)$$

as the desired step response.

From the step response, we can obtain the impulse response through the following approach.

LEMMA 8.4-1 Differentiation Consider the differential equation

$$\frac{d^ny}{dt^n} + \cdots + a_1\frac{dy}{dt} + a_0y = v(t)u_{-1}(t) \tag{33}$$

with initial rest, i.e.,

For all k, and for $t \le 0$,

$$\frac{d^ky}{dt^k}(t) = 0$$

Let $\phi(t)$ denote the initial-rest solution of this equation. Now consider the same equation with the forcing function differentiated, i.e.,

$$\frac{d^ny}{dt^n} + \cdots + a_1\frac{dy}{dt} + a_0y = \frac{d}{dt}[v(t)u_{-1}(t)] \tag{34}$$

and let $\psi(t)$ denote the initial-rest solution to Eq. (34). Then

$$\psi(t) = \frac{d}{dt}[\phi(t)]$$

PROOF: The lemma states that the response to the derivative of a forcing function is the derivative of the response to the forcing function. Since,

$$\frac{d^n\phi}{dt^n} + \cdots + a_1\frac{d\phi}{dt} + a_0\phi = v(t)u_{-1}(t)$$

it follows, by differentiating both sides, that

$$\frac{d^{n+1}\phi}{dt^{n+1}} + \cdots + a_1 \frac{d^2\phi}{dt^2} + a_0 \frac{d\phi}{dt} = \frac{d}{dt}[v(t)u_{-1}(t)]$$

which can be rewritten as

$$\frac{d^n}{dt^n}\left[\frac{d\phi}{dt}\right] + \cdots + a_1 \frac{d}{dt}\left[\frac{d\phi}{dt}\right] + a_0\left[\frac{d\phi}{dt}\right] = \frac{d}{dt}[v(t)u_{-1}(t)]$$

Clearly $d\phi/dt$ meets the initial-rest requirement, and is therefore the solution of Eq. (34). Q.E.D.

Let us apply the result we just derived to the step response of Example 8.4-4. The initial-rest solution of

$$\frac{d^2y}{dt^2} + 5\frac{dy}{dt} + 6y = u_{-1}(t)$$

was found to be

$$y(t) = [1/6 - 1/2e^{-2t} + 1/3e^{-3t}]u_{-1}(t)$$

Differentiation of $y(t)$ yields

$$\psi(t) = \frac{dy}{dt} = [e^{-2t} - e^{-3t}]u_{-1}(t) + [y(0)]u_0(t)$$

$$= [e^{-2t} - e^{-3t}]u_{-1}(t) \tag{35}$$

By Lemma 8.4-1, Eq. (35) gives the initial-rest *impulse response,* i.e., the solution to the equation

$$\frac{d^2y}{dt^2} + 5\frac{dy}{dt} + 6y = u_0(t)$$

Alternatively, the impulse response can be found by first identifying it as the homogeneous solution, and by then selecting the constants to satisfy the singularity matching constraints. This approach is illustrated below, using the previous example.

Example 8.4-5 Find the initial-rest solution of

$$\frac{d^2y}{dt^2} + 5\frac{dy}{dt} + 6y = u_0(t) \tag{36}$$

The homogeneous solution, hence the impulse response, is of the form

$$[Ae^{-2t} + Be^{-3t}]u_{-1}(t)$$

Substitution of this equation into Eq. (36) yields

$$6[Ae^{-2t} + Be^{-3t}]u_{-1}(t)$$

$$+ 5[-2Ae^{-2t} - 3Be^{-3t}]u_{-1}(t) + 5[A + B]u_0(t)$$

$$+ [4Ae^{-2t} + 9Be^{-3t}]u_{-1}(t) + [-2A - 3B]u_0 t + [A + B]u_1(t) = u_0(t)$$

We need not be concerned with the step terms because the assumed solution is the homogeneous solution. The fact that no doublet appears on the right-hand side of Eq. (36) requires that the coefficient of the left-hand doublet be zero

$$A + B = 0$$

Likewise, the fact that the forcing function is a unit impulse requires that the sum of the coefficients of the left-hand impulses be unity

$$3A + 2B = 1$$

Solution of these two equations gives the impulse response as

$$[e^{-2t} - e^{-3t}]u_{-1}(t)$$

which agrees with Eq. (35). Notice that the constraint $A + B = 0$ requires that the impulse response be continuous at $t = 0$. If this were not so, its second derivative at $t = 0$ would have contained a doublet, which cannot be matched by the impulse forcing function.

Superposition

In Sec. 8.2 we derived a differential equation (7) relating the input (or forcing function) $u(t)$ and output $y(t)$ that had the form

$$\frac{d^3y}{dt^3} + 2\frac{d^2y}{dt^2} + 3\frac{dy}{dt} + y = \frac{d^2u}{dt^2} + 2\frac{du}{dt} + u$$

We have also indicated in Eq. (13) of Sec. 8.2 the most general equation that we are interested in solving. Lemma 8.4-2, stated and proved below, in conjunction with previous results, will permit us to solve these equations.

LEMMA 8.4-2 Superposition Consider the differential equation

$$\frac{d^ny}{dt^n} + a_n\frac{d^{n-1}y}{dt^{n-1}} + \cdots + a_1\frac{dy}{dt} + a_0y = v(t)$$

with initial rest, i.e., for $t \leq 0$,

$$\frac{d^{n-1}y}{dt^{n-1}}(t) = \frac{d^{n-2}y}{dt^{n-2}}(t) = \cdots = \frac{dy}{dt}(t) = y(t) = 0 \tag{37}$$

If $\phi_1(t)$ is the initial-rest solution when $v(t) = v_1(t)u_{-1}(t)$ and if $\phi_2(t)$ is the initial-rest solution when $v(t) = v_2(t)u_{-1}(t)$, then the initial-rest solution when $v(t) = [\beta_1v_1(t) + \beta_2v_2(t)]u_{-1}(t)$ is given by

$$\phi(t) = \beta_1\phi_1(t) + \beta_2\phi_2(t)$$

where β, and β_2 are constants.

PROOF: Since $\phi_1(t)$ and $\phi_2(t)$ satisfy the following equations:

$$\frac{d^n\phi_1}{dt^n} + \cdots + a_1\frac{d\phi_1}{dt} + a_0\phi_1 = v_1(t)u_{-1}(t)$$

$$\frac{d^n\phi_2}{dt^n} + \cdots + a_1 \frac{d\phi_2}{dt} + a_0\phi_2 = v_2(t)u_{-1}(t)$$

it follows that

$$\frac{d^n}{dt^n}(\beta_1\phi_1 + \beta_2\phi_2) + \cdots + a_1 \frac{d}{dt}(\beta_1\phi_1 + \beta_2\phi_2)$$

$$+ a_0(\beta_1\phi_1 + \beta_2\phi_2) = [\beta_1 v_1(t) + \beta_2 v_2(t)]u_{-1}(t)$$

Finally, since ϕ_1 and ϕ_2 are both initial-rest solutions, of their corresponding excitations, it is clear that $\beta_1\phi_1(t) + \beta_2\phi_2(t)$ is the initial-rest solution of the superposed excitations.
Q.E.D.

More generally, if $\phi_1(t)$ and $\phi_2(t)$ are *not* initial-rest solutions, then $\beta_1\phi_1(t) + \beta_2\phi_2(t)$ is the solution with initial conditions that are the sums of the initial conditions associated with $\phi_1(t)$ and $\phi_2(t)$, weighted by β_1 and β_2, respectively, as we have indicated in Chap. 7.

The use of Lemmas 8.4-1 and 8.4-2 is illustrated in the following example.

Example 8.4-6 Find the initial-rest solution to

$$\frac{d^2y}{dt^2} + 5\frac{dy}{dt} + 6y = 2\frac{du}{dt} + u \tag{38}$$

with

$$u(t) = e^{-t}u_{-1}(t)$$

From Example 8.4-4, we know that the initial-rest solution to

$$\frac{d^2y}{dt^2} + 5\frac{dy}{dt} + 6y = e^{-t}u_{-1}(t)$$

is

$$\phi(t) = [\tfrac{1}{2}e^{-t} - e^{-2t} + \tfrac{1}{2}e^{-3t}]u_{-1}(t)$$

Lemma 8.4-1 states that $\phi_1(t)$, the initial-rest solution corresponding to the derivative of the excitation, i.e., the solution of

$$\frac{d^2y}{dt^2} + 5\frac{dy}{dt} + 6y = \frac{d}{dt}[e^{-t}u_{-1}(t)]$$

is $\phi_1(t) = \dfrac{d}{dt}[\phi(t)]$ or

$$\phi_1(t) = [-\tfrac{1}{2}e^{-t} + 2e^{-2t} - \tfrac{3}{2}e^{-3t}]u_{-1}(t) + [0]u_0(t)$$

Finally, Lemma 8.4-2 states that the desired solution to Eq. (38) is, by superposition,

$$2\phi_1(t) + \phi(t) = [-\tfrac{1}{2}e^{-t} + 3e^{-2t} - \tfrac{5}{2}e^{-3t}]u_{-1}(t)$$

Generalization

In the remainder of this section, we will generalize the solutions to higher-order equations, and we will use Lemmas 8.4-1 and 8.4-2 where applicable. Consider again the general form of the input-output nth-order D.E.,

$$\frac{d^n y}{dt^n} + a_{n-1}\frac{d^{n-1}y}{dt^{n-1}} + \cdots + a_1\frac{dy}{dt} + a_0 y = b_m\frac{d^m u}{dt^m} + \cdots + b_1\frac{du}{dt} + b_0 u \tag{39}$$

with $n \geq m$, and a forcing function $u(t) = Ue^{s_0 t}u_{-1}(t)$. Using the experience gained in Example 8.4-6 we can predict what the initial-rest solution will look like.

First observe that the initial-rest solution to

$$\frac{d^n y}{dt^n} + a_{n-1}\frac{d^{n-1}}{dt^{n-1}} + \cdots + a_1\frac{dy}{dt} + a_0 y = Ue^{s_0 t}u_{-1}(t)$$

has the form

$$\phi(t) = \left[\frac{U}{p(s_0)}e^{s_0 t} + C_1 e^{s_1 t} + \cdots + C_n e^{s_n t}\right]u_{-1}(t)$$

where

$$p(s) = s^n + a_{n-1}s^{n-1} + \cdots + a_1 s + a_0$$

and $s_1, \ldots, s_n$ are the roots of $p(s)$. We have assumed no repeated roots for ease of presentation. Observe next that the initial-rest solution to

$$\frac{d^n y}{dt^n} + a_{n-1}\frac{d^{n-1}y}{dt^{n-1}} + \cdots + a_1\frac{dy}{dt} + a_0 y = b_i\frac{d^i}{dt^i}[Ue^{s_0 t}u_{-1}(t)]$$

is, from Lemmas 8.4-1 and 8.4-2 and from the above,

$$b_i\frac{d^i}{dt^i}\phi(t) = b_i\left[\frac{Us_0{}^i}{p(s_0)}e^{s_0 t} + C_1 s_1{}^i e^{s_1 t} + \cdots + C_n s_n{}^i e^{s_n t}\right]u_{-1}(t) \tag{40}$$

We can be sure that no impulses or higher-order singularity functions appear in the ith derivative of $\phi(t)$ because it is an initial-rest solution and hence $\phi(0)$ and its first $n-1$ derivatives evaluated at $t=0$ are zero. This, in conjunction with the fact that $i \leq m \leq n$ ensures the validity of Eq. (40).

The total solution of Eq. (39) has, by superposition, the form

$$\sum_{i=0}^{m} b_i\frac{d^i}{dt^i}\phi(t)$$

which can be written as

$$\left[\frac{U}{p(s_0)}e^{s_0 t}\sum_{i=0}^{m} b_i s_0{}^i + C_1 e^{s_1 t}\sum_{i=0}^{m} b_i s_1{}^i + \cdots + C_n e^{s_n t}\sum_{i=0}^{m} b_i s_n{}^i\right]u_{-1}(t)$$

If we define a polynomial associated with the right-hand side of the equation

$$q(s) \triangleq b_m s^m + b_{m-1}s^{m-1} + \cdots + b_1 s + b_0$$

we can rewrite the solution more compactly as

$$\left[\frac{q(s_0)}{p(s_0)} Ue^{s_0 t} + q(s_1)C_1 e^{s_1 t} + \cdots + q(s_n)C_n e^{s_n t}\right]u_{-1}(t) \tag{41}$$

This solution has the familiar form of being composed of a particular solution and a homogeneous solution. Because $\phi(t)$ is an initial-rest solution, we can state that the C_is in this solution satisfy the following constraints [obtained by successive differentiation of $\phi(t)$]:

$$\left.\begin{array}{l} \dfrac{U}{p(s_0)} + C_1 + C_2 + \cdots + C_n = 0 \\[2ex] \dfrac{Us_0}{p(s_0)} + s_1C_1 + s_2C_2 + \cdots + s_nC_n = 0 \\[2ex] \cdots\cdots\cdots\cdots\cdots\cdots\cdots\cdots\cdots \\[2ex] \dfrac{U}{p(s_0)}\, s_0^{\,n-1} + s_1^{\,n-1}C_1 + s_2^{\,n-1}C_2 + \cdots + s_n^{\,n-1}C_n = 0 \end{array}\right\} \; n \text{ equations}$$

Hence, the C_i can be evaluated through solution of the above system of equations.

We proceed now with an illustration of the above approach.

Example 8.4-7 Find the initial-rest solution of

$$\frac{d^2y}{dt^2} + 5\frac{dy}{dt} + 6y = 3\frac{du}{dt} + 2u \tag{42}$$

with

$$u(t) = 4e^{-t}u_{-1}(t)$$

We have

$$\begin{array}{ll} U = 4 & q(s) = 3s + 2 \\ p(s) = s^2 + 5s + 6 & p(s_0) = 2 \\ s_0 = -1 & q(s_0) = -1 \\ s_1 = -2 & q(s_1) = -4 \\ s_2 = -3 & q(s_2) = -7 \end{array}$$

Thus C_1 and C_2 must satisfy

$$4/2 + C_1 + C_2 = 0$$

$$\frac{4(-1)}{2} - 2C_1 - 3C_2 = 0$$

or

$$C_1 = -4 \qquad C_2 = 2$$

The final solution can be written from Eq. (41) as

$$\phi(t) = \left[\frac{-1}{2}(4)e^{-t} + (-4)(-4)e^{-2t} + (-7)(2)e^{-3t}\right]u_{-1}(t)$$

or, more compactly, as

$$\phi(t) = [-2e^{-t} + 16e^{-2t} - 14e^{-3t}]u_{-1}(t)$$

Note that though this is an initial-rest solution, its first derivative contains a step at $t=0$. This is as it should be, since if we look at Eq. (42), the right-hand side contains an impulse (the derivative of the forcing function), and hence the second derivative of the solution must also contain an impulse, implying that the first derivative must contain a step.

We can take advantage of Lemmas 8.4-1 and 8.4-2 to find the response when $u(t) = u_0(t)$. Since $\phi(t)$ is the response when

$$u(t) = 4e^{-t}u_{-1}(t)$$

then, by Lemma 8.4-1, $d\phi/dt$ is the response when

$$u(t) = \frac{d}{dt}[4e^{-t}u_{-1}(t)] = -4e^{-t}u_{-1}(t) + 4u_0 t$$

By Lemma 8.4-2, the function

$$\phi + \frac{d\phi}{dt}$$

is the solution when

$$u(t) = 4e^{-t}u_{-1}(t) - 4e^{-t}u_{-1}(t) + 4u_0(t) = 4u_0 t$$

Hence,

$$\frac{1}{4}\left(\phi + \frac{d\phi}{dt}\right)$$

is the impulse response.

In the next section we shall focus on those situations in which the particular solution is of primary interest. This investigation will lead to a compact and useful representation called the system function.

8.5 STEADY-STATE SOLUTION AND THE SYSTEM FUNCTION

In this section, we shall continue to investigate equations of the form of Eq. (39) in Sec. 8.4:

$$\frac{d^n y}{dt^n} + a_{n-1}\frac{d^{n-1}y}{dt^{n-1}} + \cdots + a_1\frac{dy}{dt} + a_0 y = b_m\frac{d^m u}{dt^m} + \cdots + b_1\frac{du}{dt} + b_0 u \qquad (1)$$

where $n \geq m$, $u(t)$ is the forcing function, and $y(t)$ is the response. In what follows, we shall be concerned with the behavior of solutions at very large values of time. Hence, the effect of singularities at $t = 0$ will be unimportant.

Recall that in Sec. 8.4 we defined the characteristic polynomial for Eq. (1) as

$$p(s) = s^n + a_{n-1}s^{n-1} + \cdots + a_1 s + a_0$$

and later in the same section we defined a similar polynomial $q(s)$, associated with the right-hand side of Eq. (1), as

$$q(s) = b_m s^m + b_{m-1}s^{m-1} + \cdots + b_1 s + b_0$$

Moreover, if the forcing function is of the form

$$u(t) = Ue^{s_0 t}u_{-1}(t) \qquad U \neq 0$$

we saw from Eq. (41) of Sec. 8.4 that the response is of the form

$$y(t) = \left[\frac{q(s_0)}{p(s_0)}\,Ue^{s_0 t} + q(s_1)C_1 e^{s_1 t} + \cdots + q(s_n)C_n e^{s_n t}\right]u_{-1}(t) \tag{2}$$

where $s_1, \ldots, s_n$ are roots of $p(s)$ and we have assumed $s_0 \neq s_1 \neq \cdots \neq s_n$.

Now assume that

$$\text{Re}\,[s_0] > \text{Re}\,[s_k] \qquad \text{for } k = 1, 2, \ldots, n \tag{3}$$

Then, with reference to Eq. (2) as t becomes large, i.e., $t \to \infty$, the particular solution will *dominate* all other terms in Eq. (2). By "dominate" we mean that

$$\left|\frac{q(s_0)}{p(s_0)}\,Ue^{s_0 t}\right| >> \left|q(s_k)C_k e^{s_k t}\right| \qquad \text{for } k = 1, 2, \ldots, n \text{ and } t \to \infty$$

Thus, for large t, the total response is, for all practical purposes, the particular solution, i.e.,

$$y(t) \cong \frac{q(s_0)}{p(s_0)}\,Ue^{s_0 t} \qquad \text{as } t \to \infty \tag{4}$$

If Eq. (3) holds, we define the *steady-state solution* or *response* as

$$\boxed{y_{ss}(t) \triangleq \frac{q(s_0)}{p(s_0)}\,Ue^{s_0 t}} \tag{5}$$

We will see that the steady-state response plays an important role in the analysis of systems.

First, observe that *in the steady-state solution, the output is essentially the input multiplied by* $q(s_0)/p(s_0)$. This quantity, which is evaluated at the driving frequency to determine $y_{ss}(t)$, is in form a function of the system. It is for this reason that we define the *system function* for Eq. (1) as

$$\boxed{H(s) \triangleq \frac{q(s)}{p(s)}} \tag{6}$$

$H(s)$, when evaluated at a driving frequency s_0, assuming $p(s_0) \neq 0$, determines the particular response. In the steady state, for large t, $H(s)$ also determines the total response.

The roots of $p(s)$ are the *natural frequencies* of the system and determine the exponents in the homogeneous solution. As s approaches a root of $p(s)$, the magnitude of $H(s)$ becomes unboundedly large. For this reason the roots of $p(s)$ are called *poles* of the system function. The roots of $q(s)$ are called *zeros* of the system function, since if s is a root of $q(s)$, then $H(s)=0$. Note that the system function and the concept of the particular solution dominating the homogeneous solution are independent. It is the importance of the steady state, however, that causes us to focus on the system function.

Example 8.5-1 Consider the differential equation

$$\frac{d^2y}{dt^2}+5\frac{dy}{dt}+6y=2\frac{du}{dt}+u \tag{7}$$

The system function is, by inspection of Eq. (7),

$$H(s)=\frac{2s+1}{s^2+5s+6}$$

which, to emphasize the poles and zeros, can be written as

$$H(s)=\frac{2(s+\frac{1}{2})}{(s+3)(s+2)}$$

Thus, $H(s)$ has a zero at $s=-\frac{1}{2}$ and two poles, at $s=-3$ and $s=-2$. For a forcing function of the form

$$u(t)=Ue^{s_0t}u_{-1}(t)$$

a steady-state condition exists† for RE $[s_0] > -2$. Thus, if we had

$$u(t)=4e^{-t}u_{-1}(t)$$

the steady-state response would be

$$y(t)=H(-1)4e^{-t}=-2e^{-t} \qquad \text{for } t>>0$$

If instead, the forcing function is

$$u(t)=[e^{jt}+e^{-jt}]u_{-1}(t)=[2\cos t]u_{-1}(t) \tag{8}$$

a steady-state condition exists and the steady-state response is, by superposition,

$$\begin{aligned} y(t) &= H(j)e^{jt}+H(-j)e^{-jt} \\ &= \frac{2j+1}{-1+5j+6}e^{jt}+\frac{-2j+1}{-1-5j+6}e^{-jt} \\ &= \frac{3+j}{10}e^{jt}+\frac{3-j}{10}e^{-jt} \\ &= 2\,\text{Re}\left[\frac{3+j}{10}e^{jt}\right] \\ &= \frac{\sqrt{10}}{5}\cos\left[t+\tan^{-1}(\tfrac{1}{3})\right] \end{aligned}$$

†That is, the particular solution dominates the homogeneous solution as t becomes large.

Thus, for large t, the output is the input cosine, *modified in magnitude and phase,* a characteristic of all linear systems that we will discuss in the next section.

The system function can be used instead of the differential equation as a concise description of the system. That is, we can completely describe a system, for purposes of input-output analysis, by giving the system function. For example, if we have

$$H(s) = \frac{3(s+2)}{(s+3)(s+4)}$$

the associated differential equation can be reconstructed, by inspection, as

$$\frac{d^2y}{dt^2} + 7\frac{dy}{dt} + 12y = 3\frac{du}{dt} + 6u$$

8.6 IMPEDANCE: SYSTEM FUNCTIONS FOR NETWORKS

In this section we shall utilize the concept of the system function to represent linear time-invariant network elements (resistors, capacitors, and inductors) by their *impedance*. As we shall see, the impedance of a network element or of a network is a system function $Z(s)$ which relates the current through the network element to the voltage across it. The replacement of network elements by their impedance greatly facilitates derivation of describing equations because the network becomes effectively a "resistive" network that can be analyzed by algebraic rather than by differential equations.

An *RLC* Network Example

Consider the *RLC* network shown in Fig. 8.6-1. Assume that the network elements are linear and time-invariant. Let $v(t)$ represent the applied voltage and $i(t)$ the current.

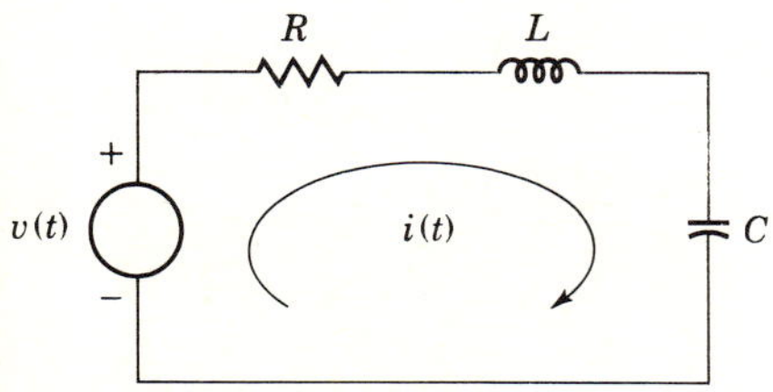

FIG. 8.6-1 Series RLC network.

First we will derive the second-order differential equation relating $v(t)$ and $i(t)$. Using KVL and the element-constraint equations, we see that the current $i(t)$ and the voltage $v(t)$ are related by the following equation:

$$\frac{d^2i}{dt^2} + \frac{R}{L}\frac{di}{dt} + \frac{1}{LC}i = \frac{1}{L}\frac{dv}{dt} \tag{1}$$

Suppose that the voltage $v(t)$ is the complex exponential†

$$v(t) = Ve^{s_0 t} \qquad \text{for } t > 0 \tag{2}$$

Then we know that the solution $i(t)$ to Eq. (1) will contain a term [the particular solution $i_p(t)$ of the differential equation] of the form

$$i_p(t) = H(s_0)Ve^{s_0 t} \tag{3}$$

where $H(s_0)$ is the complex number obtained by evaluating the system function $H(s)$ given by

$$H(s) = \frac{Cs}{LCs^2 + RCs + 1} \tag{4}$$

at the excitation frequency s_0. We assume that $H(s_0)$ is defined. If we write $i_p(t)$ as shown in Eq. (5),

$$i_p(t) = Ie^{s_0 t} \tag{5}$$

we see by comparing Eqs. (3) and (5) that the coefficients I and V are related by

$$V = \frac{LCs_0^2 + RCs_0 + 1}{Cs} I \tag{6}$$

Since Eq. (6) looks like Ohm's law, *we may think of the complex function* $Z(s)$, *called an impedance,*

$$Z(s) = \frac{LCs^2 + RCs + 1}{Cs} \tag{7}$$

as a *resistance whose value changes with the excitation frequency* s_0 *and which relates the coefficients of the complex voltage and current as shown in Eq.* (6). Observe that $Z(s)$ is a system function, specified by the network topology and the element values.

Next, we shall outline a technique, which can be used to derive Eq. (6) *by inspection of the network.* Consider again the network of Fig. 8.6-1.

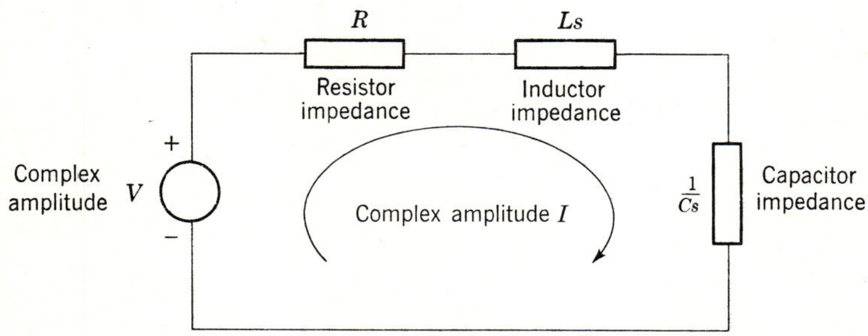

FIG. 8.6-2 Replacing the network elements of Fig. 8.6-1 by impedances.

†Complex exponentials are discussed in the appendix. No real-life voltage or current is a complex exponential. Complex exponentials are, however, a useful *tool* for finding network and system responses to real excitations, as we shall see in Sec. 8.8.

From this network let us construct a related network (with identical topology) as shown in Fig. 8.6-2.

In Fig. 8.6-2, we have replaced the elements of Fig. 8.6-1 by three generalized elements whose resistances are R, Ls, and $1/Cs$, as shown. By KVL, we can write for the network of Fig. 8.6-2

$$V = RI + (Ls)I + \left(\frac{1}{sC}\right)I \tag{8}$$

which can be rearranged into the following equation:

$$V = \frac{LCs^2 + RCs + 1}{sC} I \qquad \text{or } V = Z(s)I \tag{9}$$

Comparing Eq. (9) with Eq. (7), we see that *this purely mechanical method has* enabled us to find in a simple manner the impedance $Z(s)$. Since the differential equation can be constructed from the system function, we have also derived Eq. (1) by the above approach. The remainder of this section will make these ideas more precise.

General Considerations

In a network containing linear elements we can relate any current or voltage $y(t)$ to any other voltage or current $x(t)$ by eliminating from the differential equations obtained from KCL, KVL, and the branch constraint relations all variables except $y(t)$ and $x(t)$. When this is done, $y(t)$ and $x(t)$ will be related by a differential equation of the familiar form.

$$a_n \frac{d^n y}{dt^n} + a_{n-1} \frac{d^{n-1}y}{dt^{n-1}} + \cdots + a_0 y = b_m \frac{d^m x}{dt^m} + b_{m-1} \frac{d^{m-1}x}{dt^{m-1}} + \cdots + b_0 x \tag{10}$$

Let us recall that for $x(t) = Xe^{s_0 t}$ the solution of this equation takes the form

$$y(t) = C_0 e^{s_0 t} + C_1 e^{s_1 t} + \cdots + C_n e^{s_q t}$$

where $s_1, s_2, \ldots, s_q$ are the poles of the system function.

Hence, in a linear network containing a single source with excitation $Xe^{s_0 t}$, *every* current and every voltage can be expressed in the general form

$$y(t) = C_0 e^{s_0 t} + C_1 e^{s_1 t} + \cdots + C_q e^{s_q t} + \cdots + C_n e^{s_n t} \tag{11}$$

where $s_1, s_2, \ldots, s_n$ are *all* the natural frequencies of the network, and in general some of the C_i will be zero for every particular current or voltage $y(t)$.

Let us suppose that $i_1(t), \ldots, i_m(t)$ represent m currents of the network at a node, as shown in Fig. 8.6-3.
By KCL, we require that

$$\sum_{k=1}^{m} i_k(t) = 0 \qquad \text{for all } t \tag{12}$$

If the network is excited by a source of the form $Xe^{s_0 t}$, then from Eq. (11) we know

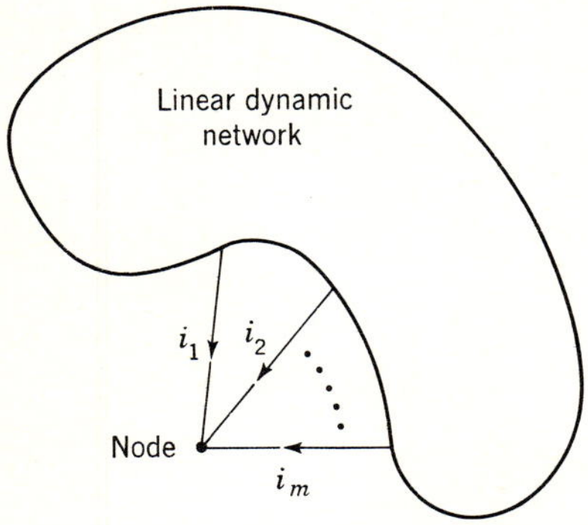

FIG. 8.6-3 A linear network with an isolated node.

that *every* current $i_k(t)$ in Fig. 8.6-3 will be of the form

For $k = 1, 2, \ldots, m$,

$$i_k(t) = \sum_{j=0}^{n} I_{kj} e^{s_j t} \tag{13}$$

where, for some k and j, I_{kj} may be zero; that is, some frequencies may not be present in some currents. Combining Eqs. (12) and (13) yields

$$\sum_{k=1}^{m} \sum_{j=0}^{n} I_{kj} e^{s_j t} = 0 \tag{14}$$

Interchanging the order of summation over k and j, we obtain

$$\sum_{j=0}^{n} (I_{1j} + I_{2j} + \cdots + I_{mj}) e^{s_j t} = 0 \tag{15}$$

For nonzero t, Eq. (15) can hold if and only if for each j, the quantity in parentheses is zero, i.e.,

For $j = 0, 1, \ldots, m$,

$$I_{1j} + I_{2j} + \cdots + I_{mj} = 0 \tag{16}$$

In other words, if we focus our attention at any frequency s_j, then *the complex amplitudes of the current components at that frequency will satisfy KCL at the node of Fig. 8.6-3.*

By the same argument, we can show that KCL will be satisfied *everywhere* in the network for frequency s_j. By a parallel argument, we can show that KVL will be satisfied by network voltages of frequency s_j around every loop in the network. Moreover, both KCL and KVL will be satisfied for *every* j, i.e., for every frequency present in the solution, regardless of whether or not s_j arises from a pole of the system function or from an excitation frequency. This result is fundamental to the algebraic analysis of networks using impedances.

Consider next the linear resistor, capacitor, and inductor. If E and

I are the complex amplitudes of voltage and current at a frequency s, then reasoning identical to the above requires that E and I be related as follows:

For a resistor of resistance R,

$$e(t) = Ri(t) \qquad \text{hence } Ee^{st} = RIe^{st}$$

Therefore

$$E = RI \tag{17}$$

For an inductor of inductance L,

$$e(t) = L\,\frac{di(t)}{dt} \qquad \text{hence } Ee^{st} = LIse^{st}$$

Therefore

$$E = (Ls)I \tag{18}$$

For a capacitor of capacitance C,

$$i(t) = C\,\frac{de(t)}{dt} \qquad \text{hence } Ie^{st} = CsEe^{st} \tag{19}$$

Therefore

$$E = \left(\frac{1}{sC}\right)I$$

The multiplicative quantity that relates the complex amplitude I to the complex amplitude E is called the *impedance* of the element. These element impedances are summarized below:

Resistor R	$Z(s) = R$	
Inductor L	$Z(s) = Ls$	(20)
Capacitor C	$Z(s) = \dfrac{1}{Cs}$	

Use of Element Impedances

The significance of impedance is that it can transform any linear RLC network into an equivalent "resistive" network whose network topology is identical to the original one, and in which R, L, C are replaced by their impedances R, Ls, and $1/Cs$. The desirable feature of the transformed network is that it permits the derivation of any system function associated with it, by algebraic manipulation. This, in turn, is possible because of two reasons:

First, a network system function relates the complex amplitudes of two signals of the general form Xe^{st}. Second, we have argued that the complex amplitudes of voltages and currents of this form obey KCL, KVL, and element constraints of the form shown in Eqs. (17) to (19). Hence, deriving arbitrary system functions from the transformed network is identical to solving a linear resistor network for the relation between two arbitrary voltages and/or currents. The derivation of such a relation involves purely algebraic and linear techniques; hence it is amenable to superposition and to the helpful equivalences developed in Chap. 4 for resistive networks.

The following example illustrates the use of impedances.

Example 8.6-1 Consider the network of Fig. 8.6-4. By inspection, it can be

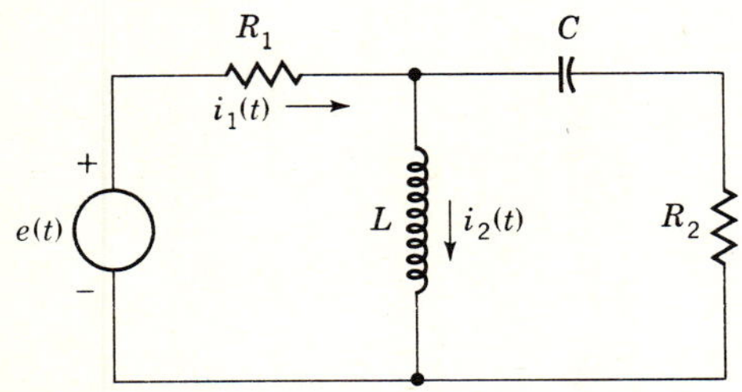

FIG. 8.6-4 An RLC network.

transformed into the network of impedances shown in Fig. 8.6-5.

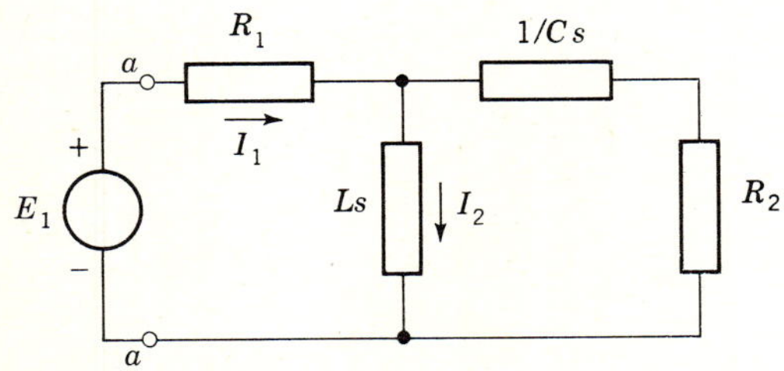

FIG. 8.6-5 The transformed network of Fig. 8.6-4.

If, for example, $e(t)$ had the form $e(t) = Ee^{st}$, then the system function relating E to the complex amplitude I of the component $i(t)$ at the frequency s, is

$$\frac{I_2}{E_1} = \frac{1/Cs + R_2}{R_1(1/Cs + R_2 + Ls) + Ls(1/Cs + R_2)} \tag{21}$$

Equation (21) is easily obtained by treating the network of Fig. 8.6-5 as a resistive network and by applying the voltage- and current-divider rules developed in Chap. 4.

If instead of the above, we wish to obtain the system function from I_1 to E_1,† i.e., the ratio E_1/I_1, then we use the same approach to write

$$\frac{E_1}{I_1} = R_1 + \frac{(R_2 + 1/Cs)Ls}{R_2 + 1/Cs + Ls} \tag{22}$$

†Here, we assume that the voltage source is removed and E is the complex amplitude of the voltage across terminals aa (Fig. 8.6-5).

Note again that derivation is by inspection; i.e., Eq. (22) is obtained by recognizing that R_1 is in series with an impedance which, in turn, is the parallel combination of two impedances—Ls and the series combination of $1/Cs$ and R_2.

From our discussion of equivalence in Sec 4.6, and from Eq. (22), it follows that the network of Fig. 8.6-5, excluding the voltage source, and with respect to terminals *aa*, is equivalent to a single "resistor," more precisely *an impedance*, of value

$$R_1 + \frac{(R_2 + 1/Cs)Ls}{R_2 + 1/Cs + Ls} \tag{23}$$

Besides having obtained the two system functions I_2/E_1 and E_1/I_1, by inspection, we can also write, by inspection of the system function, the differential equations relating $i_2(t)$ to $e_1(t)$ or $e_1(t)$ to $i_1(t)$, respectively. To do this, we must manipulate the appropriate system function into the ratio of two polynomials. For example, Eq. (22) can be manipulated as follows:

$$\begin{aligned}\frac{E_1}{I_1} &= \frac{R_1R_2 + R_1/Cs + R_1Ls + R_2Ls + L/C}{R_2 + 1/Cs + Ls} \\ &= \frac{(R_1 + R_2)Ls^2 + (R_1R_2 + L/C)s + R_1/C}{Ls^2 + R_2s + 1/C}\end{aligned} \tag{24}$$

From Eq. (24) we can now write, by inspection, the desired D.E. which relates $i_1(t)$ to $e(t)$:

$$L\frac{d^2e_1}{dt^2} + R_2\frac{de_1}{dt} + \frac{1}{C}e_1 = (R_1 + R_2)L\frac{d^2i_1}{dt^2} + \left(R_1R_2 + \frac{L}{C}\right)\frac{di_1}{dt} + \frac{R_1}{C}i_1 \tag{25}$$

A system function such as that of Eq. (24) which equals the ratio E/I of a complex voltage amplitude to a complex current amplitude, is called an *impedance*. When the system function represents the ratio I/E of a complex current amplitude to a complex voltage amplitude [Eq. (21)], it is called *an admittance*. When E and I are at the same terminal pair, as in Fig. 8.6-6*a*, the system function is referred to as a *driving-point impedance* or a *driving-point admittance*, according to whether it equals E/I or I/E, respectively. Hence,

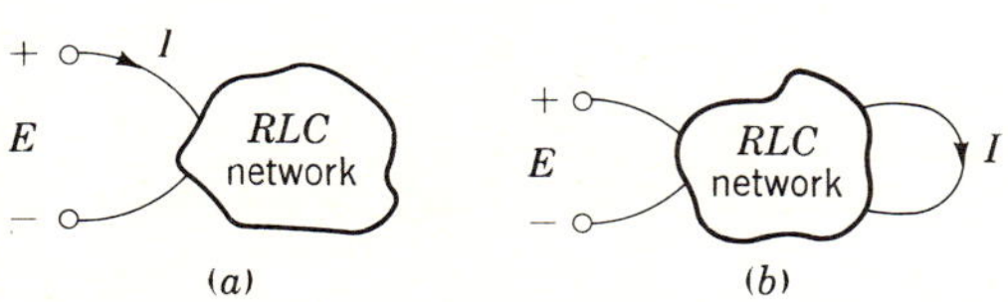

FIG. 8.6-6 (a) Driving-point and (b) transfer impedance.

Eq. (22) gives the driving-point impedance E_1/I_1. When E and I are not at the same terminal pair as in Fig. 8.6-6*b*, the corresponding system function is called a *transfer impedance* or a *transfer admittance*. Such is the case with the transfer admittance of Eq. (21). Finally, if the system function relates two voltages or two currents in the network, it is called a *voltage-transfer function* or a *current-transfer function,* respectively.

Now that we have established a convenient method for writing the system function that relates any two variables in a network, it is appropriate that we use our approach to study the behavior of *RLC* networks under certain specific forms of excitation—this is the subject of the following section.

8.7 THE s PLANE: GRAPHICAL EVALUATION OF $H(s)$

The locations of the poles and zeros of a system function $H(s)$ can be indicated on a complex plane, usually called the *s plane.* The coordinates of the s plane are the real part, σ, and the complex part, $j\omega$, of s. Hence, a complex number $s = \sigma_1 + j\omega_1$ corresponds to the point $[\sigma_1, \omega_1]$ on the s plane. Conventionally, pole locations are indicated with a cross, x, and zero locations with a zero, 0. As an example, consider the system function

$$H(s) = \frac{(s+1)(s+3)}{(s+2)(s+4)}$$

The pole-zero plot for $H(s)$ is shown in Fig. 8.7-1.

Observe that a pole-zero plot does not represent the system as completely as the system function, in that any multiplicative constant is ignored. Thus Fig. 8.7-1 is also the pole-zero plot for the following system function:

$$H(s) = \frac{-3(s+1)(s+3)}{(s+2)(s+4)}$$

Let us assume that we have a system function $H(s)$, with both the numerator and denominator polynomials factored, as shown in Eq. (1).

$$H(s) = \frac{b_m(s-z_1)(s-z_2)\cdots(s-z_m)}{(s-s_1)(s-s_2)\cdots(s-s_n)} \tag{1}$$

Our goal is to find a meaningful graphical interpretation for the evaluation of $H(s_0)$, where s_0 is the complex frequency associated with a drive of the form $e^{s_0 t}$. Substituting s_0 into Eq. (1) gives $H(s_0)$, as shown in Eq. (2).

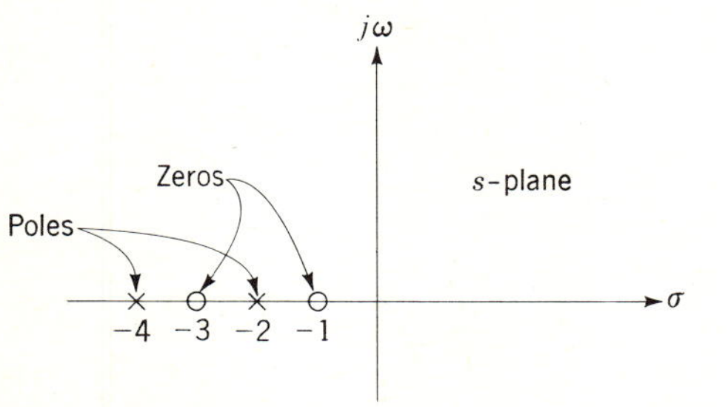

FIG. 8.7-1 Poles and zeros of H(s) on the s plane.

$$H(s_0) = \frac{b_m(s_0 - z_1)(s_0 - z_2) \cdots (s_0 - z_m)}{(s_0 - s_1)(s_0 - s_2) \cdots (s_0 - s_n)} \tag{2}$$

A typical term $(s_0 - z_i)$ in the numerator will be examined. We shall represent s_0 and z_i on the complex plane as vectors from the origin to the coordinates of the number, as shown in Fig. 8.7-2. Further, let $s_0 = k + z_i$ so that $k = s_0 - z_i$; thus k is

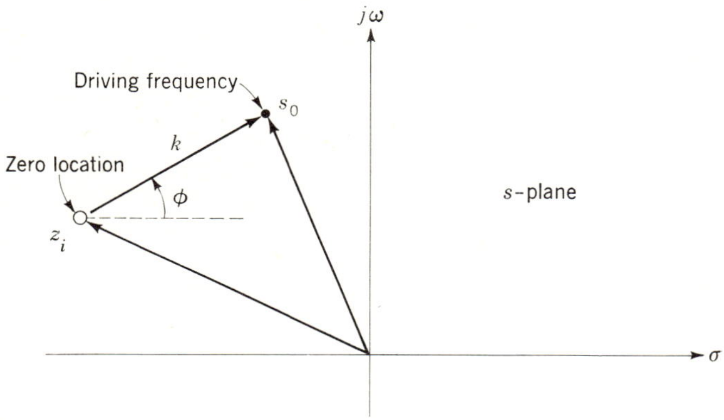

FIG. 8.7-2 A term of H(s) represented as a vector.

the complex number represented by a vector from the zero location to the coordinates of the driving frequency. The magnitude of k is the length of the vector, and the angle ϕ of k is measured *counterclockwise* from the positive real direction, as shown. This same interpretation can be used for all the terms in both the numerator and denominator of $H(s_0)$. This representation in polar form† gives $H(s_0)$ the form shown in Eq. (3), where k_i and ϕ_i have been used for the numerator and l_j and γ_j for the corresponding quantities of the denominator.

$$H(s_0) = \frac{(k_1 e^{j\phi_1})(k_2 e^{j\phi_2}) \cdots (k_m e^{j\phi_m})}{(l_1 e^{j\gamma_1})(l_2 e^{j\gamma_2}) \cdots (l_n e^{j\gamma_n})} \tag{3}$$

Equation (3) can now be rewritten as

$$H(s_0) = \frac{k_1 k_2 \cdots k_m}{l_1 l_2 \cdots l_n} \exp\left[j\left(\sum_{i=1}^{m} \phi_i - \sum_{j=1}^{n} \gamma_j\right)\right] \tag{4}$$

Since $H(s_0)$ is, in general, a complex number, it can be described by its magnitude $|H(s_0)|$ and its angle $\angle H(s_0)$. From Eq. (4), we can write the corresponding expressions given in Eqs. (5) and (6).

$$|H(s_0)| = \frac{k_1 k_2 \cdots k_m}{l_1 l_2 \cdots l_n} \tag{5}$$

$$\angle H(s_0) = \sum_{i=1}^{m} \phi_i - \sum_{j=1}^{n} \gamma_j \tag{6}$$

†Polar form is also discussed in the appendix.

Equations (5) and (6) will prove useful in the *qualitative* evaluation of $H(s_0)$. The *quantitative* evaluation of $H(s_0)$ is best handled by computer, but we can deduce a lot about how a system behaves without having numerical values for $H(s_0)$. We proceed with a qualitative example.

Example 8.7-1 Let us examine the pole-zero plot for the system function of Example 8.5-1, which was

$$H(s) = \frac{2(s + 1/2)}{(s+3)(s+2)} \tag{7}$$

The pole-zero plot is shown in Fig. 8.7-3.

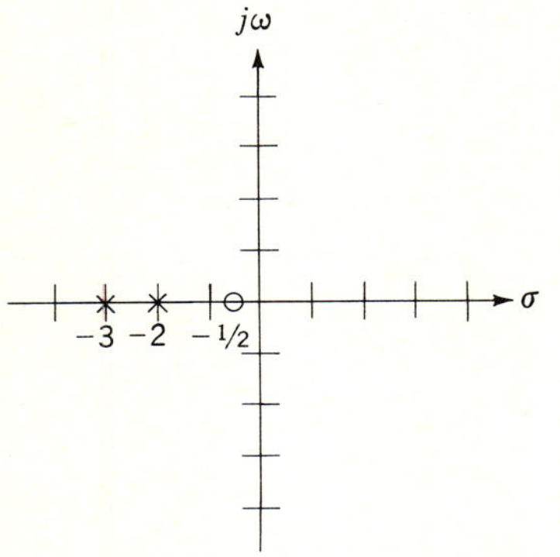

FIG. 8.7-3 Pole-zero plot of H(s).

In Example 8.5-1 we drove the system [Eq. (8) of Sec. 8.5] with a function of the form

$$u(t) = [e^{j\omega t} + e^{-j\omega t}]u_{-1}(t) = [2\cos\omega t]u_{-1}(t) \tag{8}$$

where we chose $\omega = 1$. To derive a qualitative feeling for the response of the system for various values of ω, let us examine $H(j\omega)$. For $\omega = 0$, we have the situation depicted in Fig. 8.7-4.

We can easily see from Eq. (5) that the magnitude of $H(j0)$ is

$$|H(j0)| = 2\,\frac{1/2}{2\cdot 3} = \frac{1}{6},$$

where the initial multiplier is the scale factor in Eq. (7). The angle of $H(j\omega)$ is,

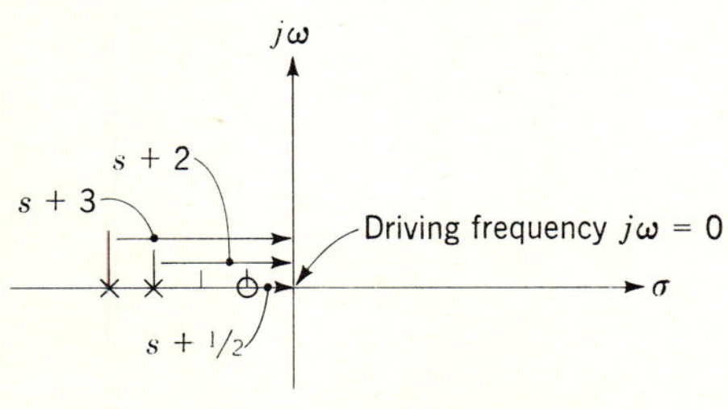

FIG. 8.7-4 Use of vectors to evaluate H(j0).

from Eq. (6), zero, since the angle contributed by each vector is zero. As ω increases in the positive direction, the magnitude of $H(j\omega)$ decreases, and for large ω, it goes as ω^{-1}. The same is true for ω increasing in the negative direction, for as long as the pole-zero pattern is symmetric about the real axis, $|H(\sigma + j\omega)| = |H(\sigma - j\omega)|$. This symmetry of the pole-zero pattern will exist whenever the coefficients of the numerator and denominator polynomials are real, a circumstance that will hold for all systems considered in this text. Thus we can qualitatively state that the magnitude of $H(j\omega)$ behaves approximately as shown in Fig. 8.7-5.

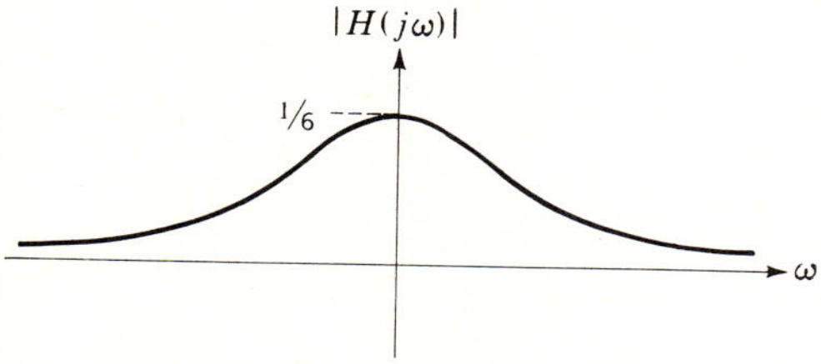

FIG. 8.7-5 Magnitude of H(jω) versus frequency.

The angle of $H(j\omega)$ is zero for $\omega = 0$ and becomes positive for small ω, as can be seen from Fig. 8.7-6, where $\omega = 1$, $\phi_1 \simeq 63°$, $\gamma_1 \simeq 26°$, and $\gamma_2 \simeq 18°$. As ω

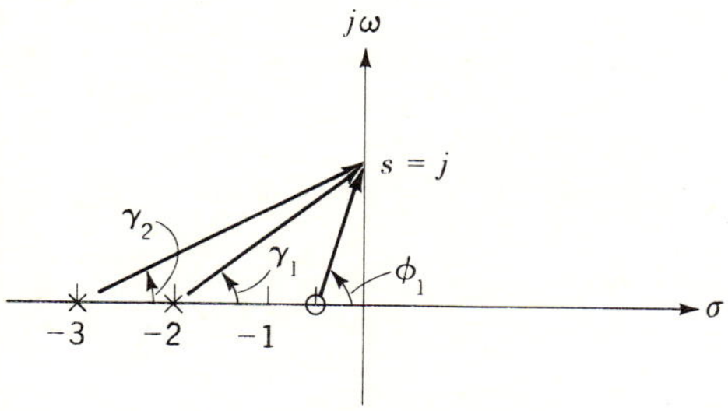

FIG. 8.7-6 Use of vectors to evaluate the angle of H(j).

continues to increase, the angle of $H(j\omega)$ goes through 0°; and for large ω, $\angle H(j\omega) \simeq -90°$. For ω less than zero, if the pole-zero plot is symmetric about the real axis, $\angle H(\sigma + j\omega) = -\angle H(\sigma - j\omega)$. To crudely estimate when $H(j\omega) \simeq 0$, we might assume† that for $\omega > 1$, $\phi_1 \simeq 90°$ and that $\gamma_1 = \gamma_2 = \tan^{-1}(2\omega/5)$. Using these assumptions, we want $2\tan^{-1}(2\omega/5) = 90°$, giving $\omega = 5/2$. The actual value of $\angle H(j5/2)$ is $-13°$. Thus we can conclude that $\angle H(j\omega)$ behaves approximately as shown in Fig. 8.7-7.

Before concluding this example, the reader should try to figure out how to use the information regarding $H(j\omega)$ to determine the steady-state response to the excitation given in Eq. (7).

The purpose of this example was to illustrate how graphical inter-

†From the geometry of Fig. 8.7-6.

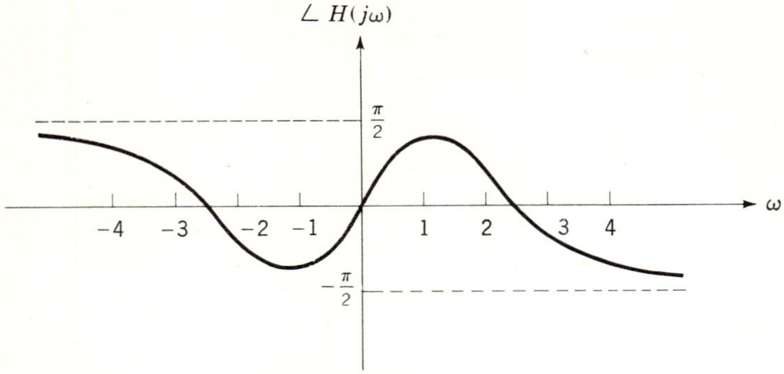

FIG. 8.7-7 *Angle of H(jω) versus frequency.*

pretation, via the pole-zero pattern, can help develop a qualitative feeling for the behavior of system functions. In the next section, we will concentrate on the steady-state behavior of linear systems and networks under *sinusoidal* inputs.

8.8 SINUSOIDAL STEADY-STATE ANALYSIS

General Considerations

The objective of this section is to analyze the steady-state output behavior of linear systems and networks when the input is a sinusoid.

Practical applications of this approach include *electronic filters*, used for selecting or supressing certain frequencies, e.g., in the audio range for hi-fi work or at higher frequencies for radio detection or remote control. Filters are quite fundamental to electronic systems, since they generally separate desirable information-carrying signals from undesirable noise.

Recall from our previous discussion that if a linear network or dynamic system has a system function $H(s)$, then any response to a drive of the form

$$x(t) = Xe^{s_0 t}u_{-1}(t) \tag{1}$$

will be as follows:

$$y(t) = (A_1 e^{s_1 t} + \cdots + A_n e^{s_n t} + XH(s_0)e^{s_0 t})u_{-1}(t) \tag{2}$$

where $s_1, \ldots, s_n$ are the poles of $H(s)$, hence the natural frequencies arising from the zero-input solution; and $H(s_0)$ is the system function evaluated at the excitation frequency.

In Sec. 8.5, we indicated that if the excitation dominates the natural frequencies, i.e., if

For $j = 1, 2, \ldots, n$,

$$\text{Re}\,[s_0] > \text{Re}\,[s_j] \tag{3}$$

then for sufficiently large values of t, i.e., at steady state, the response is given by the term

$$y(t) = XH(s_0)e^{s_0 t} \tag{4}$$

From an intuitive point of view, the condition for dominance [Eq. (3)] requires that the magnitude of the excitation grow faster (or decay slower) than the magnitude of the zero-input solution, so that after enough time has passed, the component of the solution due to the natural frequencies will be insignificant.

From a graphical point of view, dominance is assured, if s_0 (a complex number) *lies to the right of the rightmost pole of* $H(s)$, as shown in Fig. 8.8-1.

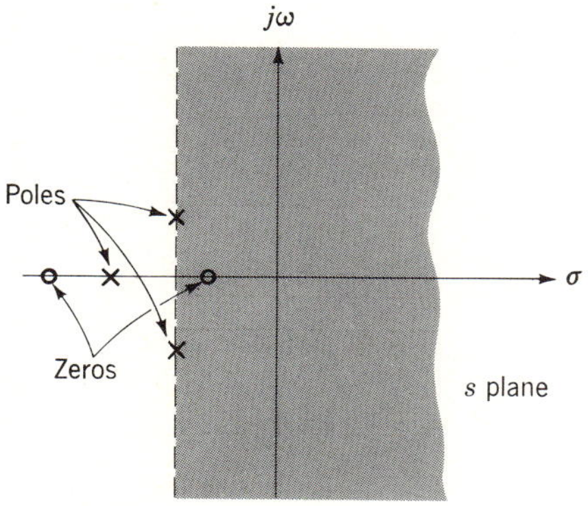

FIG. 8.8-1 Region of dominance of particular solution: s_0 *may be anywhere in shaded region for a dominant particular solution.*

Networks made up of positive resistors, inductors, and capacitors have the property of ultimately dissipating any energy which may have been initially stored in the capacitors and inductors. Hence, they are characterized by *decaying* zero-input solutions, i.e., by *negative real parts* for all their natural frequencies:

For $j = 1, 2, \ldots, n,$

$$\text{Re}\,[s_j] < 0$$

Most practical networks and systems also have this property, since in its absence the zero-input response would grow without bound, and would be of little use (as in the common whistling "feedback" encountered in public address systems). For such systems and networks, it follows that dominance is satisfied if $\text{Re}\,[s_0] = 0$, i.e., if the excitation *is a sinusoid of constant amplitude*—hence our preoccupation with sinusoidal steady-state analysis.

Complex and Real Excitations

Before we proceed with our analysis, we recall from our discussion of the appendix that excitations of the form of Eq. (1) are generally *complex numbers* which vary in time. We have seen that differential equations, when excited by complex-number excitations, exhibit complex-number solutions, e.g., Eq. (4) with X, H, and s_0 complex. *We have no physical counterpart, how-*

ever, to complex excitations, since the excitations to all our systems and networks are real numbers that vary with time. The reason that we are dealing with *complex* excitations and responses is strictly one of manipulative convenience—it is simpler to obtain the responses to real excitations by first obtaining the responses to complex excitations, through the use of the following theorem:

THEOREM 8.8-1 If the initial-rest response to (complex) $x(t)$ is (complex) $y(t)$, then the response to Re $[x(t)]$ is Re $[y(t)]$.

PROOF: By the premise of the theorem,

$$a_n \frac{d^n y}{dt^n} + \cdots + a_0 y = b_m \frac{d^m x}{dt^m} + \cdots + b_0 x \tag{5}$$

Since x and y are assumed to be complex, we can substitute the rectangular forms

$$x = x_r + jx_i \qquad \text{and} \qquad y = y_r + jy_i$$

in the above equation. The result, after collecting real and imaginary terms, is

$$\begin{aligned} &\left(a_n \frac{d^n y_r}{dt^n} + \cdots + a_0 y_r\right) + j\left(a_n \frac{d^n y_i}{dt^n} + \cdots + a_0 y_i\right) \\ &= \left(b_m \frac{d^m x_r}{dt^m} + \cdots + b_0 x_r\right) + j\left(b_m \frac{d^m x_i}{dt^m} + \cdots + b_0 x_i\right) \end{aligned} \tag{6}$$

The only way that this equation can be satisfied, for all t, is that the real part of the left side equal the real part of the right side, and the imaginary part of the left side equal the imaginary part of the right side. For the real parts, this means

$$a_n \frac{d^n y_r}{dt^n} + \cdots + a_0 y_r = b_m \frac{d^m x_r}{dt^m} + \cdots + b_0 x_r \tag{7}$$

which completes the proof, since y_r is by Eq. (7) the response to x_r. Q.E.D.

In particular, if we want to know the response of a linear system, with system function $H(s)$ to the excitation cos ωt, we recognize that

$$\cos \omega t = \text{Re}\,[e^{j\omega t}]$$

and can therefore assert that the response will be

$$\text{Re}\,[H(j\omega)e^{j\omega t}]$$

We proceed with an appropriate illustration.

Example 8.8-1 *The RL Network* We wish to find the sinusoidal steady-state behavior of $e(t)$ in the network of Fig. 8.8-2.

First observe that the excitation is $i_s(t) = \text{Re}\,[e^{j2\pi t}]u_{-1}(t)$. The driving-point impedance "seen" by the current source is, by inspection,

$$Z(s) = \frac{E}{I} = R + Ls = 1 + j\frac{2\pi}{2\pi} = 1 + j1 \qquad \text{for } s = j2\pi \tag{8}$$

For large t, it follows from Theorem 8.8-1 that

$$e(t) = \text{Re}\,[(1 + j1)e^{j2\pi t}] \tag{9}$$

which can be rearranged to yield

$$e(t) = \text{Re}\,[\sqrt{2}\,e^{j\pi/4}e^{j2\pi t}] = \sqrt{2}\,\text{Re}\,[e^{j(2\pi t+\pi/4)}]$$
$$= \sqrt{2}\cos\left(2\pi t + \frac{\pi}{4}\right) \tag{10}$$

This waveform is shown in Fig. 8.8-3; note that the steady-state voltage $e(t)$ is the input cosine, *time-shifted* by $\pi/4$ and *scaled* by $\sqrt{2}$. The response to this complex excitation could have also been obtained more directly, by first expressing the impedance E/I in polar form as follows:

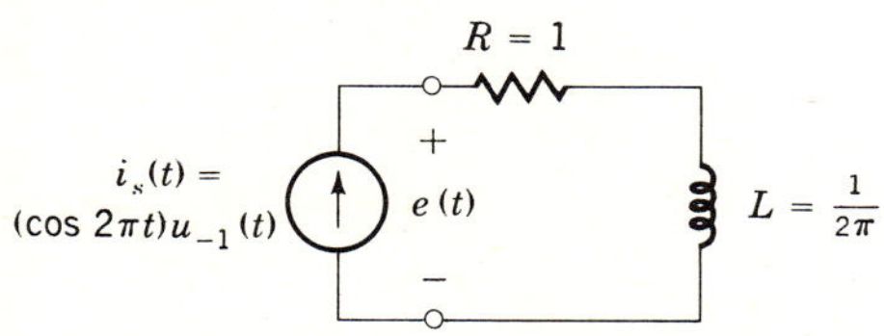

FIG. 8.8-2 An RL network.

$$E/I = 1 + j1 = \sqrt{2}\,e^{j\pi/4} \tag{11}$$

This is illustrated in Fig. 8.8-4a as a vector in the plane of complex numbers.

The complex excitation amplitude I, i.e., 1 in this example, is also represented as a vector in the plane of complex numbers (Fig. 8.8-4b). The complex amplitude of the response E is given by the product of the two complex numbers of Fig. 8.8-4. Recall that this product is another complex number with *magnitude equal to the product of the magnitudes and angle equal to the sum of angles* of the two vectors being multiplied. This result is shown in Fig. 8.8-4c, from which it can be seen that the complex amplitude of the response is the complex amplitude of the excitation scaled by $\sqrt{2}$ and shifted in angle by $\pi/4$ rad.

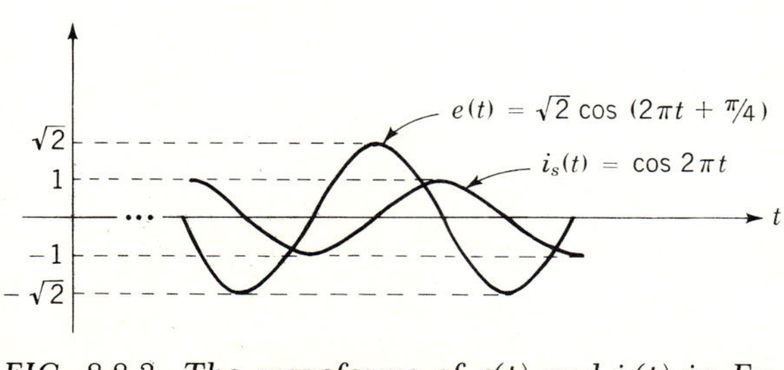

FIG. 8.8-3 The waveforms of $e(t)$ and $i_s(t)$ in Example 8.8-1.

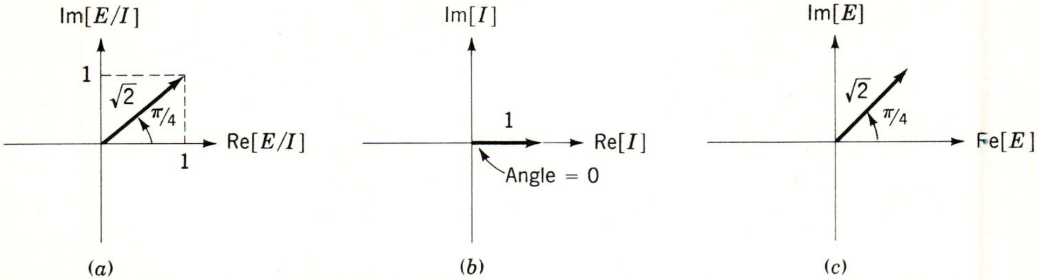

FIG. 8.8-4 Vector representation of complex amplitudes of (a) *system-function,* (b) *excitation, and* (c) *response.*

Rotating Vectors

In order to understand the shifting in angle (or *phase shift,* as it is called) between excitation and response, consider the following. The term $e^{j\omega t}$ is a complex number; hence it may be regarded as a vector in the plane of complex numbers. Its magnitude is unity, since

$$\text{Re}^2\,[e^{j\omega t}] + \text{Im}^2\,[e^{j\omega t}] = \cos^2 \omega t + \sin^2 \omega t = 1 \qquad \text{for all } \omega \text{ and } t \qquad (12)$$

Its angle is ωt; that is, the angle varies with t. As t increases, the angle ωt also increases. We may, therefore, visualize the vector $e^{j\omega t}$ as a unit vector rotating *counterclockwise* in the plane of complex numbers, with angular velocity ω, as shown in Fig. 8.8-5.

Returning now to Example 8.8-1 and Fig. 8.8-4*c*, we see that the complete response will be the complex amplitude E shown in that figure, times $e^{j\omega t}$. The result is a vector of magnitude $\sqrt{2}$ and of angle $\omega t + \pi/4$. That is, as time increases, that vector rotates with the same angular velocity as $e^{j\omega t}$, but is displaced from the latter by $\pi/4$ rad, as shown in Fig. 8.8-6. The real parts of these two rotating vectors, which are their projections on the real axis (and are also shown in Fig. 8.8-6), represent the actual excitation and the response.

In summary, we can interpret any excitation of the form $Xe^{j\omega t}$ as a rotating vector. The effect of the system function on this excitation vector is to scale *the excitation vector magnitude by the magnitude of the system function, and advance the angle of the excitation vector by the angle of the system function;* the angle of the system function may, of course, be negative. The

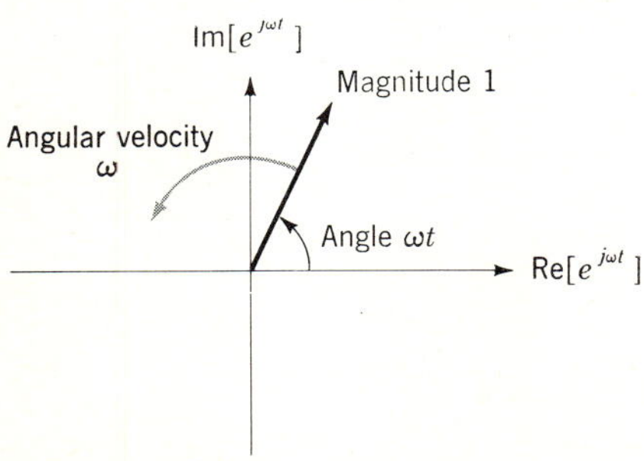

FIG. 8.8-5 Representation of $e^{j\omega t}$ *as a rotating vector.*

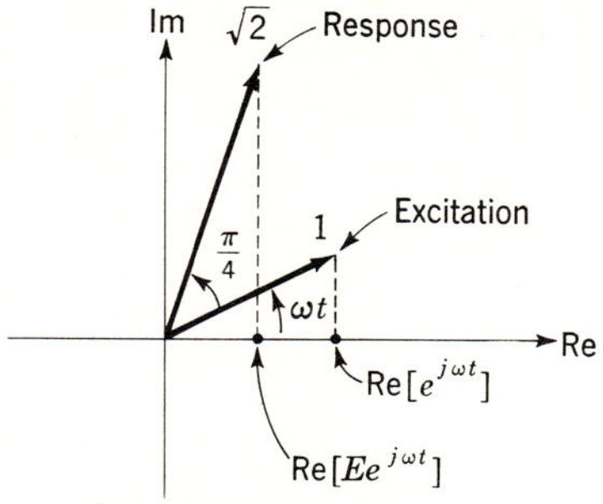

FIG. 8.8-6 Excitation and response of Example 8.8-1 viewed as rotating vectors.

resultant vector represents the complex response $Ye^{j\omega t}$ and rotates counterclockwise with the *same angular frequency* as the excitation vector. Real or imaginary parts of either excitation or response may be obtained by projecting these vectors on the appropriate axis. As the excitation and response vectors rotate, their projections on either axis undergo sinusoidal variations which are phase-shifted from each other by the angle of the system function.

Evaluation of Frequency Response

Besides the need for evaluation of the system function at some specific frequency, it is often desirable to know the behavior of the system function as the excitation frequency is varied. This is illustrated in the following two examples.

Example 8.8-2 *Frequency Response of RC Network: Low-pass Filter* In the network of Fig. 8.8-7, we wish to determine the steady-state output E_2 as the frequency ω of the input sinusoid varies.

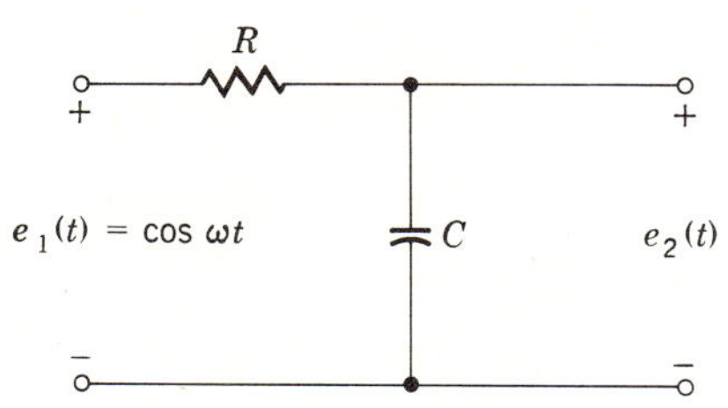

FIG. 8.8-7 An RC network.

By inspection, the transfer function is

$$\frac{E_2}{E_1}(s) = \frac{1/Cs}{R + 1/Cs} = \frac{1}{\tau s + 1} \qquad \tau = RC \tag{13}$$

Since $e_1(t) = \text{Re}\,[e^{j\omega t}]$, it follows that for very large t,

$$e_2(t) = \text{Re}\left[\frac{E_2}{E_1}(j\omega)e^{j\omega t}\right]$$

As an aside, we remark that dominance is guaranteed, since the solution has only one natural frequency, the pole of the transfer function, Eq. (13), at $s_1 = -1/\tau$; clearly, Re $[j\omega t] = 0 >$ Re $[s_1] = -1/\tau$, for positive R and C. Substitution of Eq. (13) in the above equation and some rearrangement yield the following:

$$e_2(t) = \text{Re}\left[\frac{1}{\sqrt{(\omega\tau)^2+1}} e^{j\arctan(-\omega\tau)} e^{j\omega t}\right]$$

$$= \frac{1}{\sqrt{(\omega\tau)^2+1}} \cos(\omega t - \arctan \omega\tau) \tag{14}$$

Equation (14)† expresses the desired result. It is a cosine waveform scaled by $1/\sqrt{(\omega\tau)^2+1}$, and phase-shifted by $-\arctan \omega\tau$. Qualitatively, as $\omega\tau$ increases from zero, the output magnitude decreases from 1 to 0, while the phase shift decreases from 0 to $-90°$. Moreover, at $\omega = (1/\tau)$ the magnitude is ½ and the phase is $-45°$. These results are shown in Fig. 8.8-8.

Figure 8.8-8 discloses that the *RC* network, viewed as an input-output system, *passes* low (relative to $1/\tau$) frequency signals and *rejects* high (relative to $1/\tau$) frequencies. Thus, if a signal is made up of two terms, e.g.,

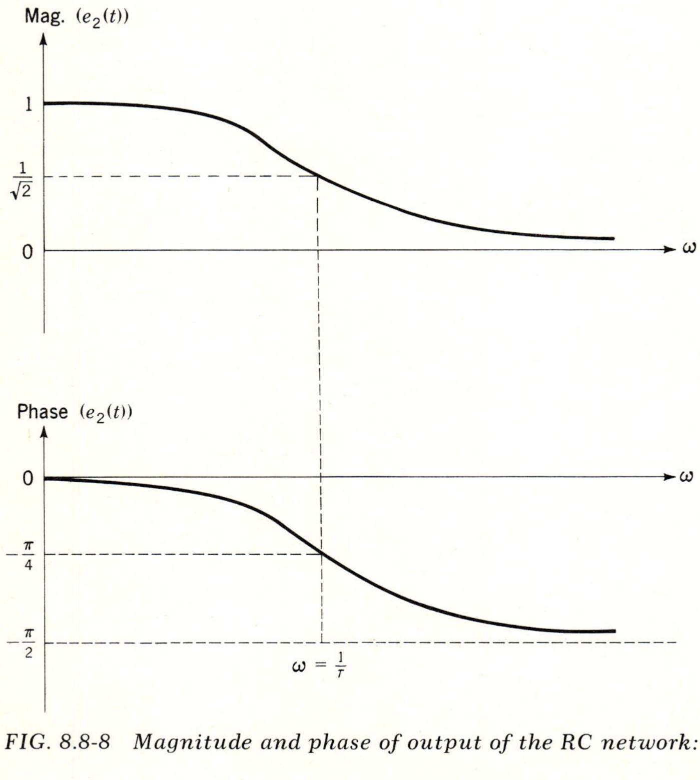

FIG. 8.8-8 Magnitude and phase of output of the RC network:

†arctan ϕ is another way of writing $\tan^{-1}\phi$, i.e., the angle whose tangent is ϕ.

$$e_1(t) = A \cos 10^{-3}t + B \cos 10^3 t$$

then, assuming $\tau = 1$, the steady-state output will be, from Eq. (14) and by superposition,

$$e_2(t) \cong 0.9999995\, A \cos (10^{-3}t - 0.001) + 0.001\, B \cos \left[10^3 t - \left(\frac{\pi}{4} - 0.001\right)\right]$$

In other words, the low-frequency signal $A \cos 10^{-3}t$ has "passed" with very nearly the same magnitude and phase (within 10^{-6} and 10^{-3}, respectively), while the high-frequency signal $B \cos 10^3 t$ has been attenuated by three orders of magnitude and has been phase-shifted by nearly 45°. Thus, if the former signal were desirable and the latter undesirable, the RC network acts as the appropriate filter.

Example 8.8-3 *The Series RLC Network: Resonance* Consider the series connection of elements shown in Fig. 8.8-9.

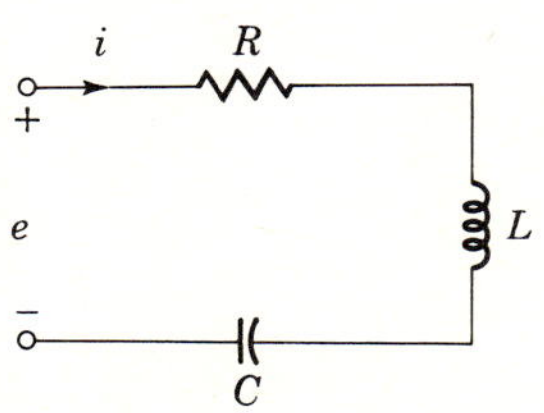

FIG. 8.8-9 The series RLC network.

By inspection, the driving-point impedance is

$$\frac{E}{I}(s) = Z(s) = sL + R + \frac{1}{Cs} = \frac{(LC)s^2 + (RC)s + 1}{Cs}$$

The driving-point admittance $Y(s)$ is the inverse of the driving-point impedance, or

$$\frac{I}{E}(s) = Y(s) = \frac{Cs}{(LC)s^2 + (RC)s + 1}$$

Recalling the definition of poles and zeros, we see that this admittance has one zero at $s = 0 + j0$ [where the numerator of $Y(s)$ vanishes], and two poles. The poles are evaluated by setting the denominator of $Y(s)$ equal to 0. Hence, the poles are at

$$s_1, s_2 = -\frac{R}{2L} \pm \sqrt{(R/2L)^2 - (1/LC)} \tag{15}$$

Observe that these poles are located in the *left half plane* of s (i.e., have negative real part), since R, L, and C are non-negative. Observe further that, if $R/2L < 1/\sqrt{LC}$, then the roots are complex conjugates.

As a specific example, let $R = 1$, $L = 1/2$, and $C = 2/101$. From Eq. (15) the poles are at

$$s_1 = -1 + j10 \qquad s_2 = -1 - j10 \tag{16}$$

and Y(s) can be rewritten as follows:

$$Y(s) = \frac{(1/L)s}{s^2 + (R/L) + (1/LC)} = \frac{2s}{(s+1-j10)(s+1+j10)}$$
$$= \frac{(1/L)(s - s_0)}{(s - s_1)(s - s_2)} \tag{17}$$

Here we have denoted the zero by s_0 ($s_0 = 0$), and the two poles by s_1 and s_2. For $s = j\omega$, i.e., a sinusoidal excitation, the admittance can be expressed as a complex number of magnitude

$$|Y| = \frac{(1/L)A_0}{B_1 B_2} \tag{18}$$

and angle

$$\angle Y = \phi_0 - \phi_1 - \phi_2 \tag{19}$$

where A_0, B_1, B_2 and ϕ_0, ϕ_1, ϕ_2 are respectively the magnitudes and angles of the vectors $s - s_0$, $s - s_1$, and $s - s_2$ shown in Fig. 8.8-10. From Eq. (18) the magnitude of $Y(j\omega)$ is given by

$$|Y(j\omega)| = \frac{(2)|\omega|}{(\sqrt{(\omega-10)^2+1})(\sqrt{(\omega+10)^2+1})} \tag{20}$$

and from Eq. (19) the angle of $Y(j\omega)$ is given by

$$\angle Y(j\omega) = \frac{\pi}{2} - \tan^{-1}(\omega - 10) - \tan^{-1}(\omega + 10) \qquad \omega > 0 \tag{21}$$

We shall generalize the system function of Eq. (17) as follows:

$$H(s) = \frac{Ks}{(s - s_1)(s - s_2)} \tag{22}$$

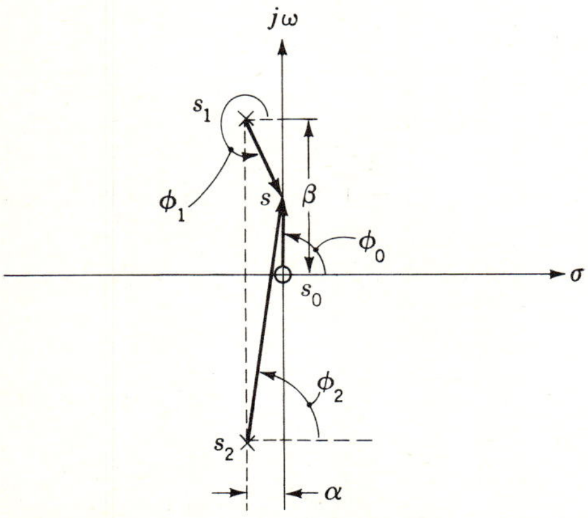

FIG. 8.8-10 The three vectors that make up Y(s).

where $s_1 = s_2^*$, $s_1 = -\alpha + j\beta$, $\alpha,\beta > 0$. We will also assume that $\alpha \ll \beta$, as is the case with our specific example. $H(s)$ is shown in Fig. 8.8-11. Our goal is to estimate qualitatively and with some accuracy the magnitude and phase of

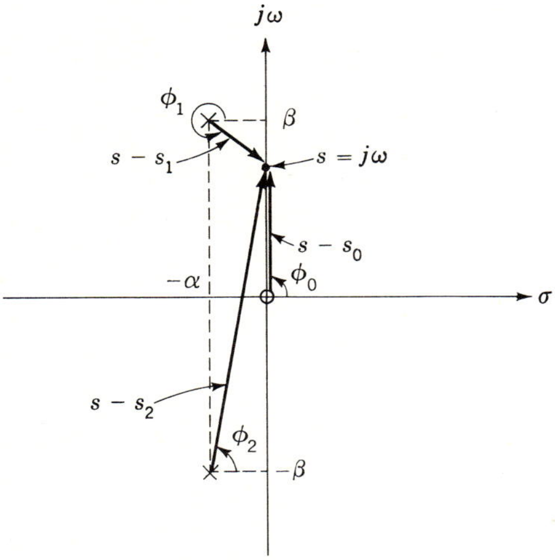

FIG. 8.8-11 Pole-zero plot of H(s), and the associated vectors for $s = j\omega$.

$H(s)$ for $s = j\omega$ as ω varies. To this purpose, we shall use the graphical approach of Sec. 8.7.

For $\omega \simeq 0$ and $\omega \to \pm\infty$, we see that $|H(j\omega)| = 0$. Particularly, for large ω, $|H(j\omega)| \simeq 1/\omega$. As $|\omega|$ goes from $0 \to \infty$, we expect $|H(j\omega)|$ to become maximum around $\omega = \pm\beta$, since these points are closest to the poles of the system function. Furthermore, if $\alpha \ll \beta$, as assumed, we expect $H(j\omega)$ to undergo relatively rapid changes on either side of $\omega = +\beta$ (also $\omega = -\beta$), since large percentage changes are occurring in the length of, say $s - s_1$, while s and $s - s_2$ are relatively constant. Observe next that the angle of $H(j\omega)$ will go from $\pi/2$ to $-\pi/2$ as ω goes from 0 to ∞. This can be seen (Fig. 8.8-11) by observing that for small positive ω, $\phi_0 = \pi/2$, $\phi_1 \cong -\pi/2$, and $\phi_2 \cong \pi/2$. From Eq. (19), it follows that the angle of $H(j\omega)$ is $\pi/2$. To see approximately where $\angle\, H(j\omega)$ goes through zero, observe that for $\omega = \beta$, $\phi_1 = 0$, $\phi_2 \cong \pi/2$, and $\phi_0 = \pi/2$. This gives $\angle\, H(j\omega)$ as approximately zero. Finally, observe that since the zero of $H(s)$ is real, and the poles occur in a complex conjugate pair, the magnitude of $H(j\omega)$ will be symmetric about $\omega = 0$, and the phase or angle of $H(j\omega)$ will be antisymmetric† about $\omega = 0$. Figures 8.8-12*a* and *b* are plots of $|H(j\omega)|$ and $\angle\, H(j\omega)$ which summarize all the foregoing observations.

We can estimate the peak value of $|H(j\omega)|$ by observing that for $\omega = \beta$, $|H(j\omega)| \simeq K\beta/2\beta\alpha = K/2\alpha$. Hence, as α becomes smaller, with β fixed, the peak in $|H(j\omega)|$ becomes larger. In addition, it becomes narrower. To verify the

†Antisymmetric here means that $\angle\, H(j\omega) = -\angle\, H(-j\omega)$.

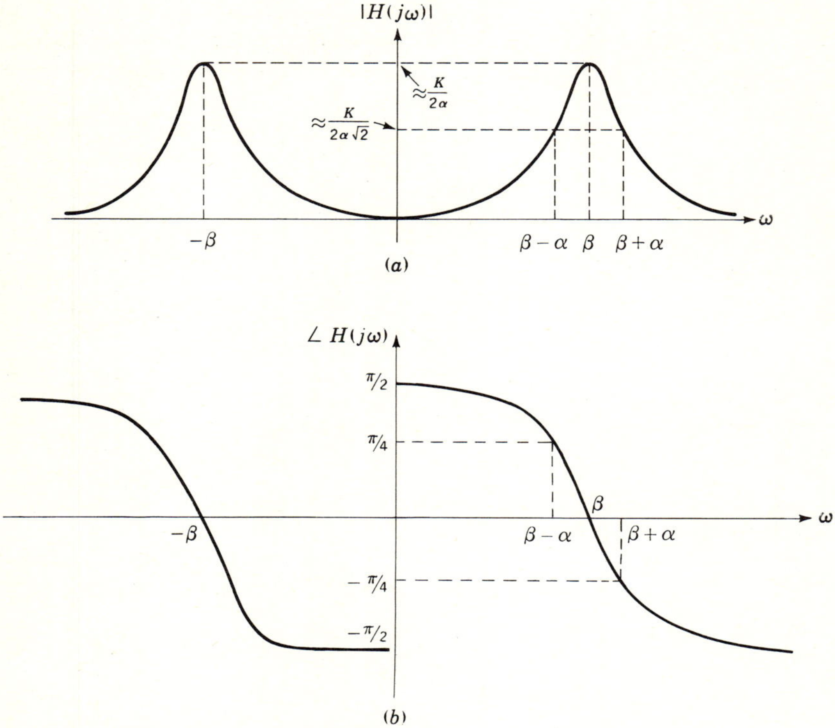

FIG. 8.8-12 Magnitude and phase of H(jω).

latter, let us calculate the width of the peak at some convenient point. Assume that as ω varies in the vicinity of $\omega = \beta$, the magnitude and phase contributed by s and $s - s_2$ are constant. Thus, when $\phi_1 = +\pi/4$, the value of ω is $\beta + \alpha$, the magnitude of $s - s_1$ is $\alpha\sqrt{2}$, and $|H(j\omega)| \simeq K\beta/2\beta\alpha\sqrt{2} = K/2\alpha\sqrt{2}$. Moreover, under these assumptions, $\angle\ H(j\omega) \cong -\pi/4$. For $\phi_1 = -\pi/4$, the value of ω is $\beta - \alpha$, the magnitude of $s - s_1$ is $\alpha\sqrt{2}$ and $|H(j\omega)| \simeq K/2\alpha\sqrt{2}$, while $\angle\ H(j\omega) \cong \pi/4$. These additional observations are also indicated in Fig. 8.8-12.

This phenomenon of a rapidly peaking magnitude is known as *resonance.* Every reader has experienced resonance in one form or another. for example, in rocking a car out of snow, there is a rhythm which seems to work best. The excitation frequency which the experienced rocker is using is one which corresponds to $\omega \approx \beta$ in the above example, and hence the response of the system (in this case the oscillation of the car) is maximum.

The network of Fig. 8.8-9 need not exhibit resonance. If $R/2L > 1/\sqrt{LC}$, both poles are real, and $|Y(j\omega)|$ and $\angle\ Y(j\omega)$ are quite different than shown in Fig. 8.8-12. (See Prob. 8.4-6.)

Viewed as a filter, the system function of Fig. 8.8-12 would be called a *band-pass filter,* because in the sinusoidal steady state it passes, relatively unchanged, sinusoids in a band of width 2α, about frequency β, while rejecting sinusoids outside this band.

Fourier theory permits the description of a large class of waveforms as summations of sinusoids of various frequencies, amplitudes, and relative phases. If it is desired to modify or filter the frequency content of a signal, especially to reduce the content at some frequencies relative to others, a filter is designed with the appropriate transfer function. Two important properties of a filter can be stated as questions: (1) Are the undesired frequencies sufficiently attenuated? and (2) Are the magnitude and phase suitable for the range of signals to be passed? A significant body of knowledge exists concerning filter design. The use of computer-aided network synthesis techniques has made it relatively easy to produce custom filters with almost any achievable characteristics.

8.9 IMPEDANCE AND NATURAL FREQUENCIES

In this section we shall discuss some relations between impedance and the homogeneous solution of the differential equation corresponding to such impedance.

Consider the linear network shown in Fig. 8.9-1*a*. Let the driving-

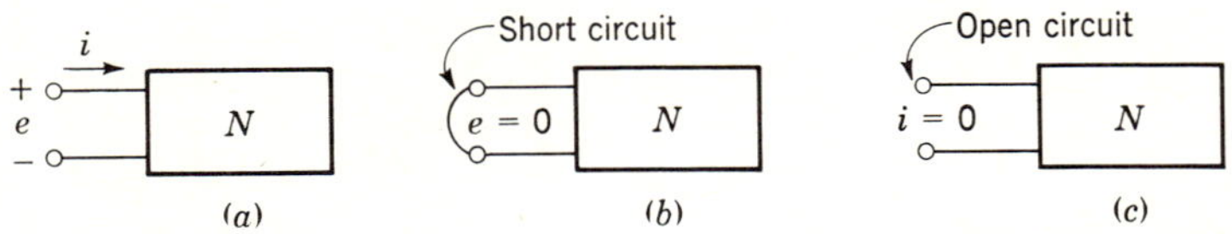

FIG. 8.9-1 (a) Network N with (b) terminals short- and (c) open-circuited.

point impedance at the terminal-pair be

$$Z(s) = \frac{A(s)}{B(s)} \tag{1}$$

where $A(s)$ and $B(s)$ are of the form

$$\begin{aligned} A(s) &= a_n s^n + a_{n-1} s^{n-1} + \cdots + a_1 s + a_0 \\ B(s) &= b_m s^m + b_{m-1} s^{m-1} + \cdots + b_1 s + b_0 \end{aligned} \tag{2}$$

It follows that the differential equation relating the voltage $e(t)$ to the current $i(t)$ is

$$a_n \frac{d^n i}{dt^n} + \cdots + a_1 \frac{di}{dt} + a_0 i = b_m \frac{d^m e}{dt^m} + \cdots + b_1 \frac{de}{dt} + b_0 e \tag{3}$$

If we imagine the experiment of short-circuiting or connecting the two terminals, as shown in Fig. 8.9-1*b*, then by KVL

$$e(t) = 0 \qquad \text{for all } t \tag{4}$$

Equation (3) then becomes

$$a_n \frac{d^n i}{dt^n} + \cdots + a_1 \frac{di}{dt} + a_0 i = 0 \tag{5}$$

and any current $i(t)$, due to initial conditions in the network, is of the form

$$i(t) = A_1 e^{s_1 t} + A_2 e^{s_2 t} + \cdots + A_n e^{s_n t} \tag{6}$$

where $s_1, \ldots, s_n$ ($s_i \neq s_j$) are the roots of the characteristic polynomial $A(s)$. But from Eq. (1), $A(s)$ is the numerator of the impedance, and $s_1, \ldots, s_n$ are the *zeros* of $Z(s)$.

Similarly, if we open-circuit the terminals of N (Fig. 8.9-1*c*), then by KCL

$$i(t) = 0 \qquad \text{for all } t \tag{7}$$

Hence, the differential equation becomes

$$0 = b_m \frac{d^m e}{dt^m} + \cdots + b_1 \frac{de}{dt} + b_0 e \tag{8}$$

and the only voltage $e(t)$ that can exist is of the form

$$e(t) = B_1 e^{z_1 t} + B_2 e^{z_2 t} + \cdots + B_m e^{z_m t} \tag{9}$$

where $z_1, z_2, \ldots, z_m$ ($z_i \neq z_j$) are the roots of the characteristic polynomial $B(s)$, which is the denominator of the impedance. Hence $z_1, z_2, \ldots, z_m$ are the *poles of the impedance.*

We refer to $s_1, \ldots, s_n$ as the *short-circuit natural frequencies* and to $z_1, z_2, \ldots, z_m$ as the *open-circuit natural frequencies.* These frequencies are called *natural* because they are a property of the network and not of the excitation. Of course, in order for these natural frequencies to be excited with nonzero amplitude in some short-circuit current or open-circuit voltage, we assume that energy has been stored in the capacitors and inductors of the network. If, while the network is at rest, an excitation is applied at the terminals, then certain natural frequencies (depending on the nature of the excitation) of the network will be "excited" by energy supplied to the network by the excitation. These frequencies appear as the homogeneous portion of the complete solution.

To summarize, we have shown that *the poles of a driving-point impedance* are the open-circuit natural frequencies of the network, while the zeros are the short-circuit natural frequencies. These observations lead to an easy way of determining impedance for simple networks, as the following examples indicate.

Example 8.9-1 Consider the "network" of Fig. 8.9-2. If we short-circuit the terminals, i.e., $e(t) = 0$, and ask at what frequencies could a current $i(t)$ exist, we see that the answer is "none," since the short-circuit forces $de/dt = 0$. Hence, there are no short-circuit natural frequencies, and therefore no zeros of the impedance. If we open the terminals, then the only voltage that can exist is a constant k, since only if $e(t) = k$, is the open-circuit constraint $i(t) = 0$ satisfied. Hence, there is one open-circuit natural frequency, namely, zero, since a con-

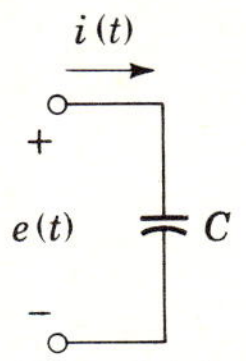

FIG. 8.9-2 A capacitor viewed as a two-terminal network.

stant is an exponential of zero frequency. Thus, we expect one pole at the origin and no zeros. Indeed, the impedance of a capacitor, derived by other means, is $1/Cs$, which has only one pole at the origin.

Example 8.9-2 If in the network of Fig. 8.9-3 we short-circuit the terminals,

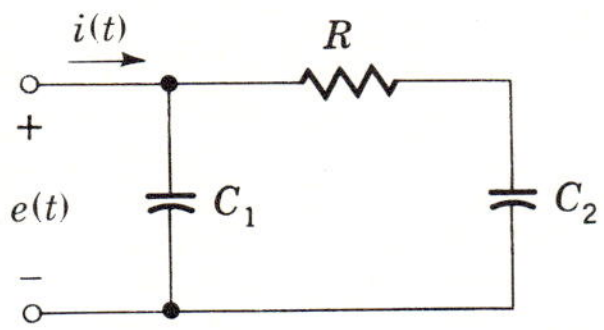

FIG. 8.9-3 An RC network.

we obtain the network of Fig. 8.9-4*a* where the only natural frequency that can exist is $1/RC_2$, i.e., $i(t) = Ie^{-t/RC_2}$. Hence $-1/RC_2$ is a zero of the driving-point

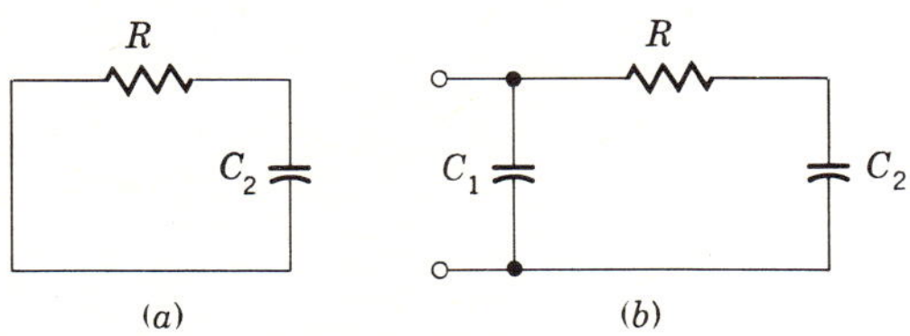

FIG. 8.9-4 Equivalent short-circuit and open-circuit networks for Example 8.9-2.

impedance at the terminals of Fig. 8.9-3. If we open-circuit the network, then we have (Fig. 8.9-4*b*) two capacitors, C_1 and C_2, and a resistor, R, in series. The only natural frequencies, i.e., the only poles of the impedance that can now exist are 0 and $-1/RC$, where C is net capacitance of the two series capacitors, i.e., $C = C_1C_2/(C_1 + C_2)$.† Zero frequency is possible here since the open-circuit voltage may be a constant. We expect then an impedance of the form

$$Z = \frac{k(s + 1/RC_2)}{s[s + (C_1 + C_2)/RC_1C_2]} \tag{10}$$

where k is a constant.

†This transformation is allowable even though $e(t)$ is the voltage across only C_1. The reason is that if a frequency is observed across C_1, it must also be observed everywhere in the network because KVL must be satisfied at that frequency.

By inspection of Fig. 8.9-3, the impedance Z is $1/C_1 s$ in parallel with $R + 1/C_2 s$ or

$$Z = \frac{(1/C_1 s)(R + 1/C_2 s)}{1/C_1 s + R + 1/C_2 s} = \frac{(1/C_1)(s + 1/RC_2)}{s[s + (C_1 + C_2)/RC_1C_2]} \tag{11}$$

Equation (11) is identical to Eq. (10) with $k = 1/C_1$.

PROBLEMS

8.2-1 *Formulation of Input-Output Equations.* Determine the differential equation relating the indicated response $y(t)$ to the excitation $x(t)$ in each of the networks and systems shown in the figure.

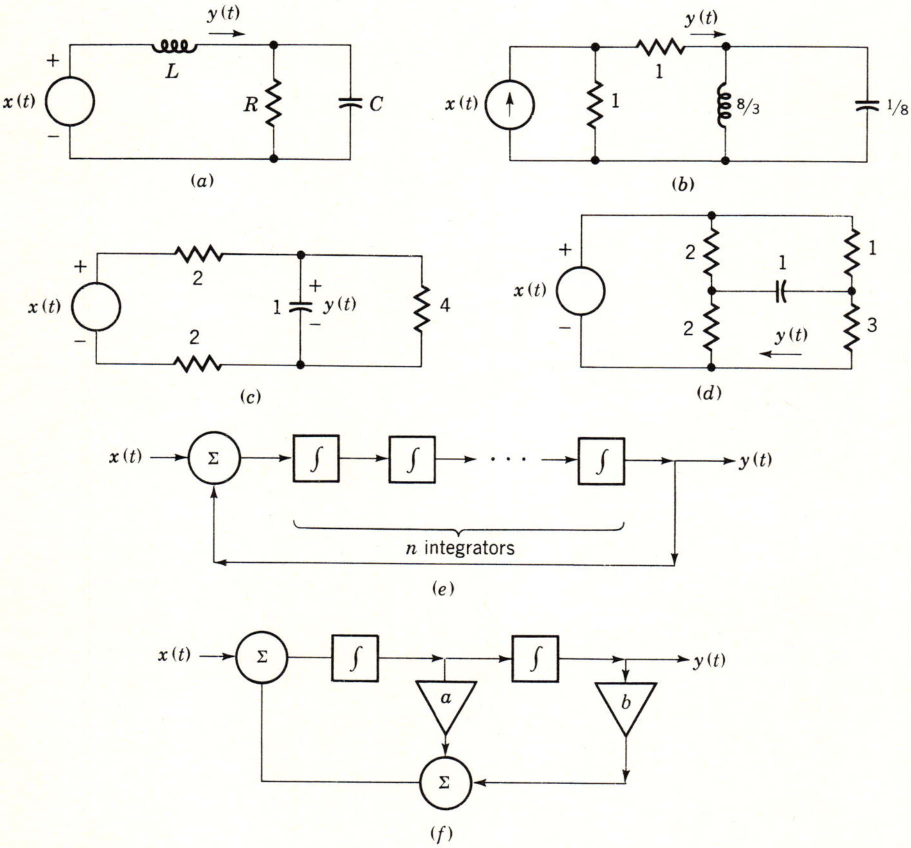

PROB. FIG. 8.2-1

8.2-2 *From State to Input-Output Equations.* Suppose that a second-order system can be described by the state equations

$$\dot{x}_1 = -4x_1 + 3x_2 \qquad x_1(0) = \xi_1$$

$$\dot{x}_2 = x_1 - 2x_2 + u \qquad x_2(0) = \xi_2$$

$$y = x_1 + x_2$$

(*a*) Determine the relations between $y(0)$ and $\dot{y}(0)$, the suitable initial conditions for an input-output description, and the given values of the initial state.

(*b*) Determine the input-output differential equation relating $y(t)$ to $u(t)$.

8.2-3 *From State to Input-Output Equations: Generalization.* Show that if a linear system with a single input u can be described by a set of n first-order state equations, then for any output $y(t)$, an input-output equation can always be derived in the form

$$a_n \frac{d^n y}{dt^n} + a_{n-1} \frac{d^{n-1} y}{dt^{n-1}} + \cdots + a_0 y = b_m \frac{d^m u}{dt^m} + b_{m-1} \frac{d^{m-1} u}{dt^{m-1}} + \cdots + b_0 u$$

where $m \leq n$, and all the a_i and b_i are constants.

8.3-1 Assuming that $u_0(t)$, as shown in the figure is a continuous and many-times differentiable function, sketch qualitatively the waveforms: (*a*) $u_{-1}(t)$, $u_{-2}(t)$, and $u_{-3}(t)$ through successive integration of $u_0(t)$; and (*b*) u_1, u_2 through successive differentiation.

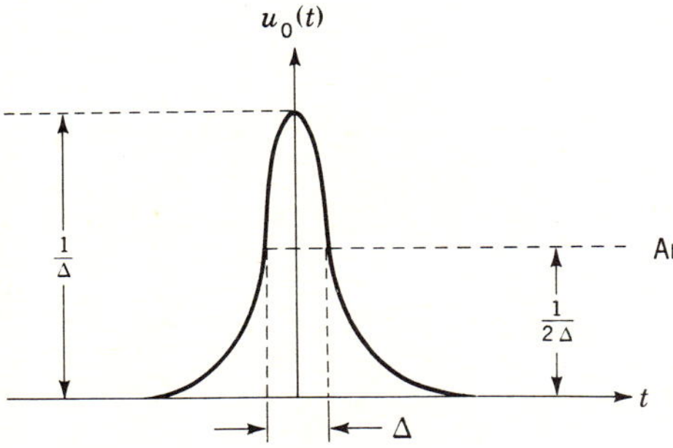

PROB. FIG. 8.3-1

8.3-2 Consider the pulse function $u_\Delta(t)$ shown in the figure, which will be used to represent the unit impulse in the limit as $\Delta \to 0$.

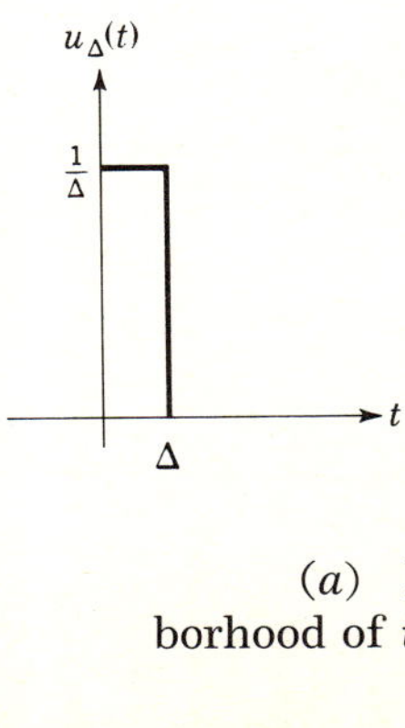

PROB. FIG. 8.3-2

(*a*) Let $f(t)$ be a time function which is continuous in the neighborhood of $t = 0$, and finite for all t. Show that

$$\lim_{\Delta\to 0} \int_{-\infty}^{t} f(\tau)u(\tau)\, d\tau = f(0) \int_{-\infty}^{t} u_0(\tau)\, d\tau$$

(*b*) Justify the statement that if $f(t)$ is continuous at $t = 0$, then

$$f(t)u_0(t) = f(0)u_0(t)$$

8.4-1 Consider the following differential equation:

$$\frac{d}{dt}\, y(t) + y(t) = u(t)$$

where $y(t) = 0$ for $t < 0$.

(*a*) Determine and express $y(t)$ when $u(t) = Ue^{s_p t}u_{-1}(t)$, where U and s_p are complex constants.

(*b*) Use the results of (*a*) to determine $y(t)$ for each of the following special cases of inputs:

$$u(t) = \frac{d}{dt}[Ue^{s_p t}u_{-1}(t)] \qquad (1)$$

$$u(t) = 110\sqrt{2}\,\sin\left(377t + \frac{\pi}{3}\right)u_{-1}(t) \qquad (2)$$

$$u(t) = e^{-t}\cos(t)u_{-1}(t) \qquad (3)$$

$$u(t) = \cosh(2t)u_{-1}(t) \quad \text{(hyperbolic cosine)} \qquad (4)$$

$$u(t) = u_{-1}(t) \quad \text{(unit step)} \qquad (5)$$

(*c*) Determine $y(t)$ when $u(t) = u_0(t)$, a unit impulse.

8.4-2 The solution to a differential equation of the form

$$\frac{dx}{dt}(t) = ax(t) + u(t) \qquad x(0) = q$$

is for $t \geq 0$

$$x(t) = e^{-t/2} + 4t$$

Find a, q, and $u(t)$.

8.4-3 Consider the differential equation

$$\frac{d^2y}{dt^2} + 3\frac{dy}{dt} + 2y = e^t u_{-1}(t)$$

where for $t < 0$,

$$\frac{d^2y}{dt^2}(t) = \frac{dy}{dt}(t) = y(t) = 0$$

(*a*) Give the characteristic polynomial, and express the homogeneous solution.

(*b*) Express the complete solution.

(*c*) Sketch under one another each of the terms d^2y/dt^2, $3dy/dt$, $2y$, and $e^t u_{-1}(t)$, showing all singularities. Does the D.E. hold for all t? Briefly explain, using your sketches.

8.4-4 *Double Roots and Excitation at the Roots.* Consider the differential equation

For $t \geq 0$,

$$\frac{d^2y}{dt} + 2\frac{dy}{dt} + y = e^{-t} \qquad y(0) = 1 \qquad \frac{dy}{dt}(0) = 1$$

(*a*) Identify the roots and express the zero-input solution for $t \geq 0$.
(*b*) Express the overall solution for $t \geq 0$.

8.4-5 Consider the network given in the figure.

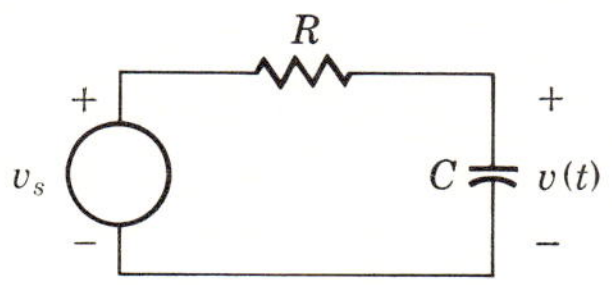

PROB. FIG. 8.4-5

where

$$v_s(t) = \sin\left(\frac{t}{RC}\right)u_{-1}(t) \qquad v(0) = 7 \qquad (1)$$

(*a*) Write the linear differential equation relating $v(t)$ to $v_s(t)$.
(*b*) Find the zero-input solution.
(*c*) Find the particular solution, using a complex exponential as drive, and then taking the real or imaginary part, as required (see also Theorem 8.8-1 for this part.)
(*d*) Knowing $v(0)$, find $v(t)$ for $t \geq 0$.
(*e*) Sketch $v(t)$ versus t for $t \geq 0$, as found in (*d*) above.
(*f*) If $v(0) = 0$, what is $v(t)$?
(*g*) Note that the difference between (*d*) and (*f*) is due to the initial condition on $v(t)$. What is the difference? That is, find $w(t) = v_d(t) - v_f(t)$ where subscripts refer to (*d*) and (*f*).
(*h*) To the network of (*f*), zero initial condition, add an independent source so that $v(t)$ is now the same $v(t)$ as found in (*d*). *Hint:* Use the results of (*g*) and superposition.

8.4-6 *Step Response of a Resonant Circuit.* In the network shown in the figure;

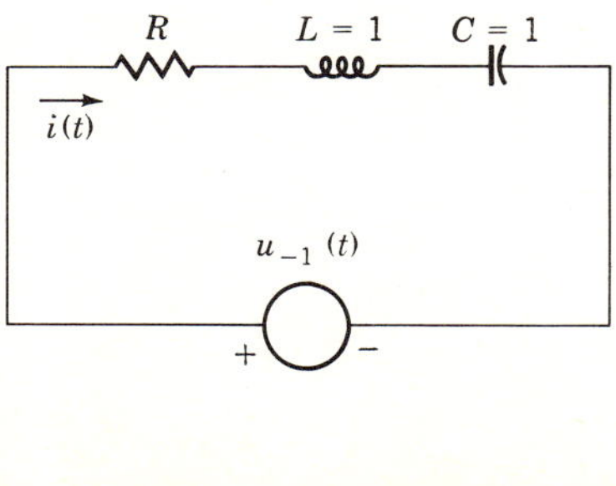

PROB. FIG. 8.4-6

$$i(t) = \frac{di}{dt}(t) = 0 \qquad \text{for } t \leq 0$$

Express and sketch $i(t)$ for each of the cases (a), (b), and (c), below:

(a) $R = 0.16$ $\left(\text{Note here that } \frac{R}{2L} < \frac{1}{\sqrt{LC}}\right)$

(b) $R = 2$ $\left(\frac{R}{2L} = \frac{1}{\sqrt{LC}}\right)$

(c) $R = 50.01$ $\left(\frac{R}{2L} > \frac{1}{\sqrt{LC}}\right)$

(d) The factor $1/\sqrt{LC}$ is called the *resonant frequency* ω_R, and $R/2L\omega_R$ the *damping factor*. Briefly explain, in view of your results, why these names are appropriate, and why the three responses to (a), (b), and (c) above are known as *underdamped, critically damped,* and *overdamped,* respectively. [*Suggestion:* Evaluate your expressions for $i(t)$ by computer so as to obtain accurate sketches.]

8.4-7 *An Automatic Antenna Positioning Servo.* To a first approximation, the relationship between the voltage $e(t)$ applied to an electric motor and the angular velocity $\frac{d\theta}{dt}(t)$ of the motor shaft is

$$\frac{d^2\theta}{dt^2} = -\frac{1}{\tau}\frac{d\theta}{dt} + \frac{k}{\tau}e$$

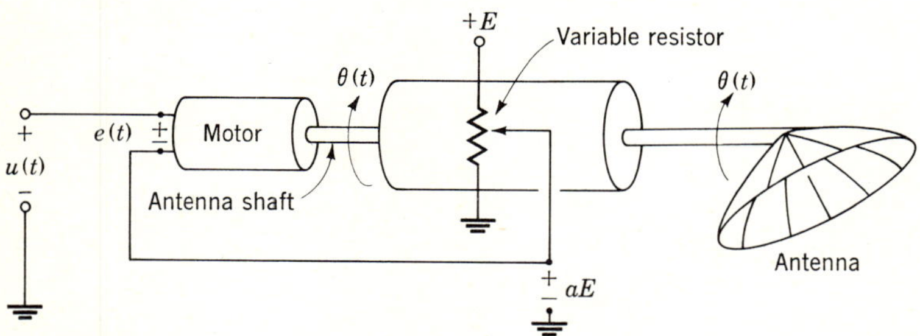

PROB. FIG. 8.4-7

where k is the constant linking voltage to top speed at that voltage, i.e., k has the units "radians per second per volt," and τ is the motor time constant (the time it takes to reach 63 percent of top speed). The motor is connected into a servo system as shown in the figure. The variable resistor driven by the motor has the property that $a = \theta(t)/2\pi$; i.e., it converts angular position $\theta(t)$, with 0.1 percent accuracy, to a proportional voltage $[\theta(t)/2\pi]E$.

(a) Draw a block diagram for the system, using *only integrators, scalers, and summers,* and treating $u(t)$ as the input and $\theta(t)$ as the output.

(*b*) Write the input-output differential equation for this system, and solve for the roots of the characteristic polynomial.

(*c*) In the light of the preceding problem, discuss what $\theta(t)$ will be, a long time after $u(t)$ has assumed a constant value $[0 \leq u(t) \leq E]$.

(*d*) We want the antenna to assume its new orientation, as commanded by a step of $u(t)$, as soon as possible, without oscillating about its final position. This is the case of critical damping we discussed in the preceding problem. Find E so that the system is critically damped.

8.4-8 The system in the figure can have any arbitrary initial (real) values at the four integrator outputs. That is,

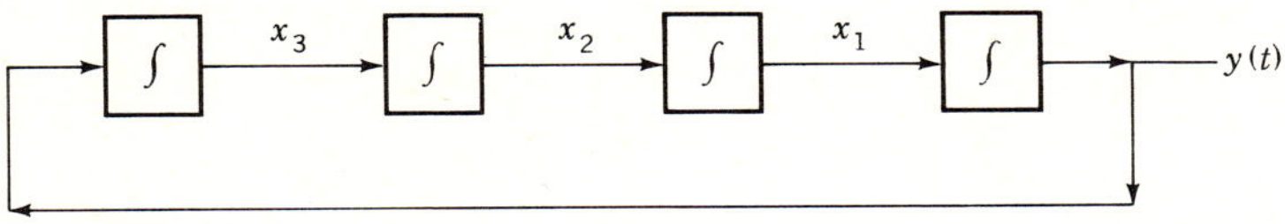

PROB. FIG. 8.4-8

$$y(0) = A \qquad x_1(0) = B \qquad x_2(0) = C \qquad x_3(0) = D$$

(*a*) Write a single D.E. on variable $y(t)$ and its derivatives that characterizes the above system.

(*b*) Find the roots of the characteristic polynomial and give an expression for $y(t)$, using unknown coefficients a, b, c, d as necessary. Your expression *must not* include j.

(*c*) Write the system of equations that relates a, b, c, d to A, B, C, D.

8.4-9 *Derivatives and Superposition.* Consider the typical RC network with $y(t) = 0$ for $t < 0$.

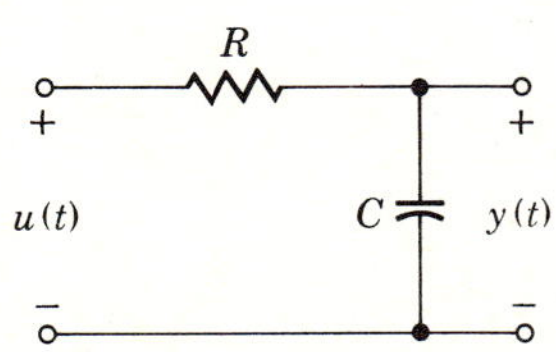

PROB. FIG. 8.4-9

(*a*) Write the D.E. relating $u(t)$ to $y(t)$, and express the solution $y(t)$ when $u(t) = u_{-1}(t)$, i.e., a unit step.

(*b*) Using only the result of (*a*) above, and without solving the D.E. again, write (by inspection) the exact solution $y(t)$ when the input is given by

$$u(t) = u_{-2}(t) + 2u_0(t) + 3u_1(t)$$

(*c*) Sketch $u(t)$, $y(t)$, and $\frac{dy}{dt}(t)$ under vertical alignment, and convince yourself that the D.E. is satisfied for all t $(-\infty \leq t \leq \infty.)$

8.5-1 For each of the following cases, find $z(t)$, where $z(t) = Xe^{st}$, and X and s are complex.

(i) $\text{Re}\,[z] = \cos \omega t$
(ii) $\text{Im}\,[z] = \sin \omega t$
(iii) $\text{Re}\,[z] = \sin \omega t$
(iv) $z + z^* = \cos \omega t$
(v) $\text{Re}\,[z] = Ae^{st} \cos(\omega t + \phi)$

8.5-2 (*a*) Each of the following differential equations relates the input $u(t)$ and output $y(t)$ of a linear system. Specify the corresponding system functions.

$$\frac{dy}{dt} + 2y = u \tag{1}$$

$$\frac{d^2y}{dt^2} + \frac{dy}{dt} + y = 3\frac{du}{dt} + u \tag{2}$$

(*b*) Determine the differential equation relating the input $u(t)$ and output $y(t)$ for each of the following system functions:

$$\frac{s+1}{s^2+s+2} \tag{3}$$

$$\frac{(s+1+j)(s+1-j)}{(s+1)(s+2)(s+3)} \tag{4}$$

$$\frac{s}{(s+j\sqrt{2})\,(s-j\sqrt{2})} \tag{5}$$

(*c*) In each of the above cases, (1) through (5), specify the homogeneous solution to the differential equation in terms of arbitrary constants.

8.5-3 Find the system function representing the quotient of output complex amplitude to input complex amplitude when the input is of the form Xe^{st}.

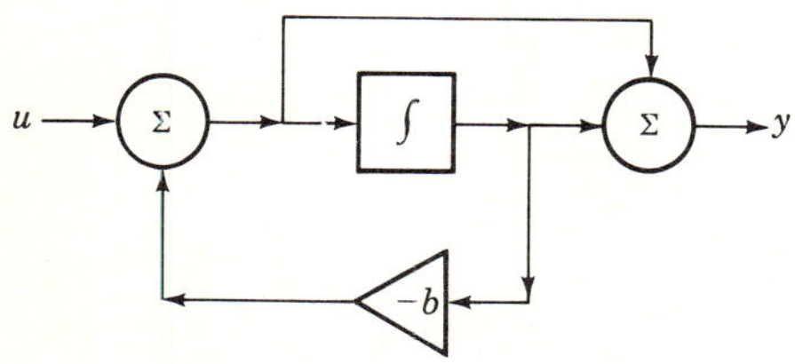

PROB. FIG. 8.5-3

8.6-1 Determine by inspection the system function relating the indicated response to the excitation.

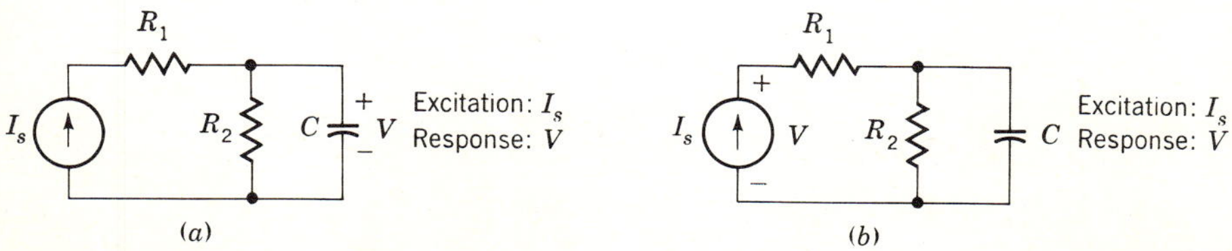

PROB. FIG. 8.6-1

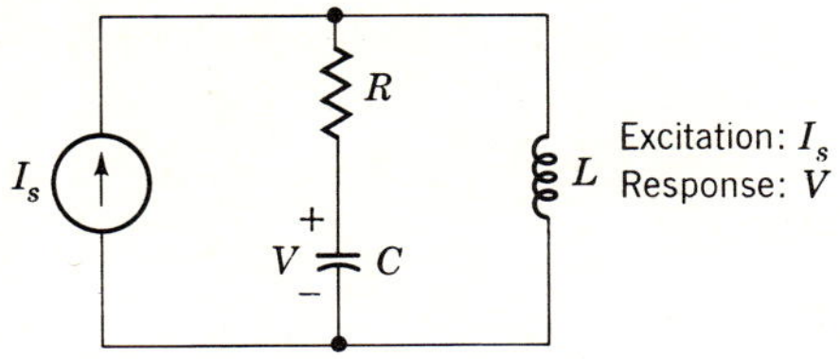

(c)

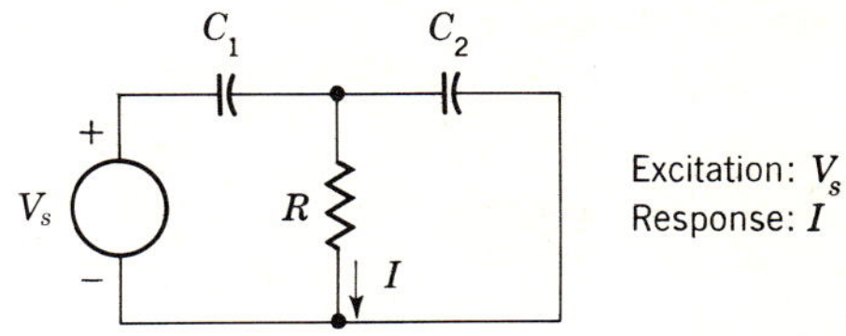

(d)

PROB. FIG. 8.6-1 (cont.)

8.6-2 (*a*) The roots of the characteristic polynomial of a certain linear system are plotted in the complex *s* plane as shown in part I of the figure. Indicate those regions on the *s* plane where the input frequencies s_0 should be located such that for large time,

(i) The particular solution dominates the homogeneous solution.
(ii) The homogeneous solution dominates the particular solution.

(*b*) For each network shown in parts II to IV clearly indicate the position in the *s* plane of natural frequencies s_0 of the input. In each case, indicate the *form* of the solution for large time, i.e., the dominant solution. (You do not need to determine the solution.)

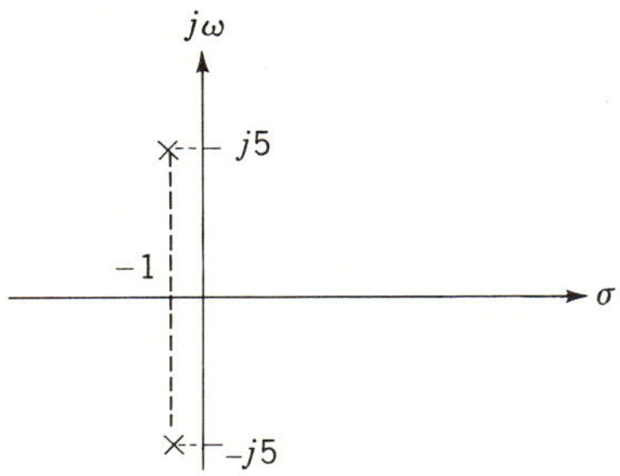

I

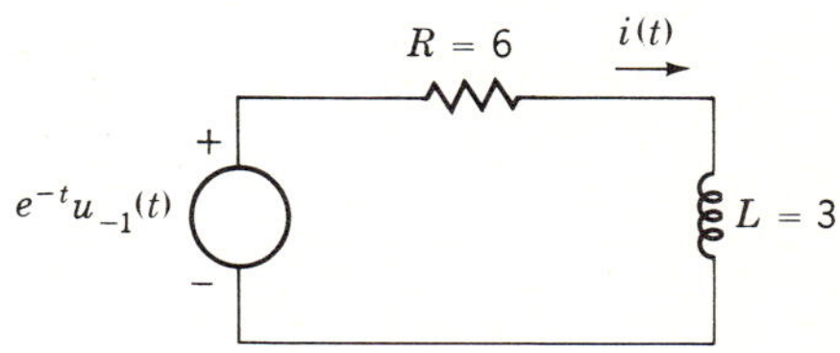

II

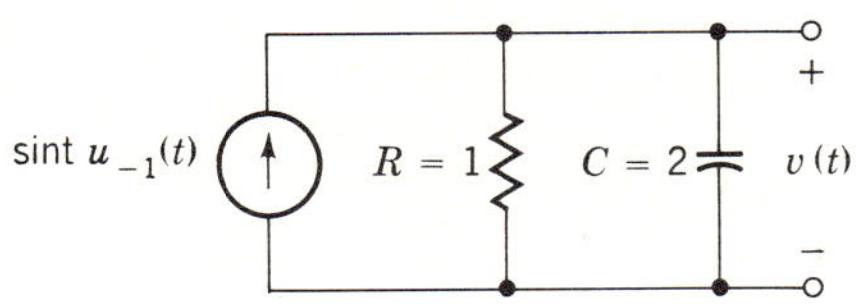

III

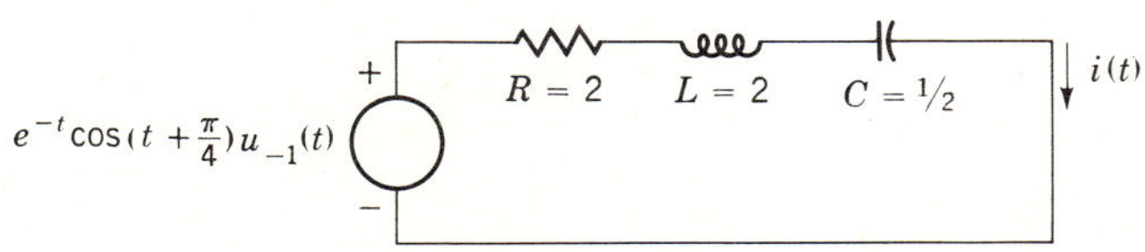

IV

PROB. FIG. 8.6-2

8.6-3 The network shown in the figure is composed of linear time-invariant R, L, C with values indicated in ohms, henrys, and farads, respectively.

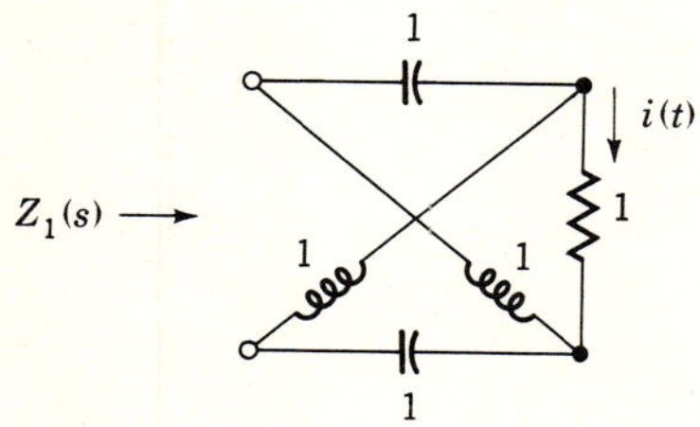

PROB. FIG. 8.6-3

Determine:

(*a*) The input impedance $Z_1(s)$.

(*b*) The transfer impedance $Z_t(s)$ between the input voltage and the current $i(t)$ indicated.

(*c*) The differential equation relating the input voltage $u(t)$ to the input current and the current $i(t)$ indicated.

8.6-4 Three experiments are performed on the linear network N which has an input impedance $Z(s)$. Determine $Z(s)$ so that the experiments are consistent.

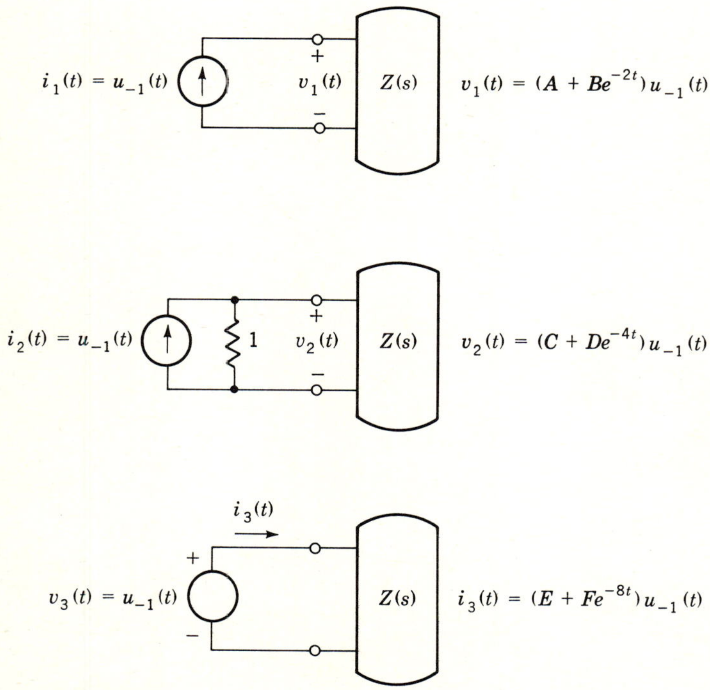

PROB. FIG. 8.6-4

8.7-1 The linear time-invariant circuit shown in the figure is called the Maxwell bridge. It can be used to measure the inductance L and series resistance

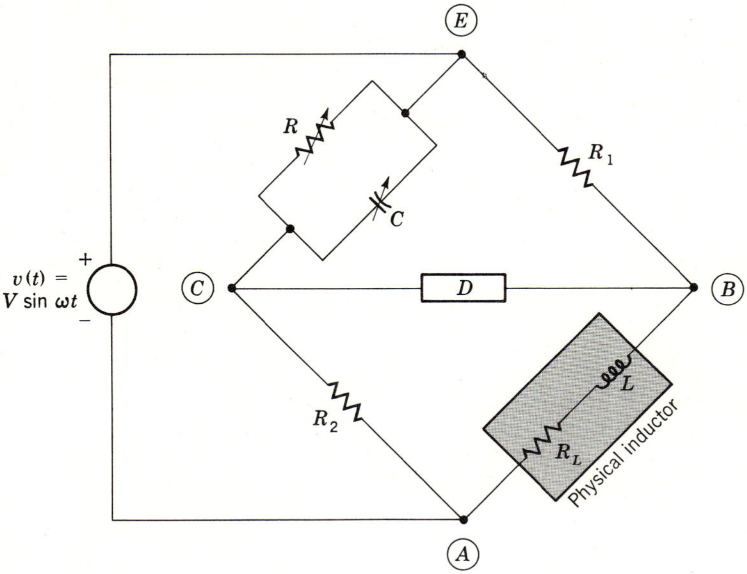

PROB. FIG. 8.7-1

R_L of an inductor connected at the terminals A and B. The measurement is made by adjusting the resistance R and the capacitance C so that, at sinusoidal steady state, the current through the resistive detector D is zero; that is, the bridge is balanced. Show that when the bridge is balanced,

$$L = R_1 R_2 C \qquad R_L = \frac{R_1 R_2}{R}$$

8.7-2 *Neatly* sketch the magnitude versus frequency and phase versus frequency for the pole-zero plots shown in the figure.

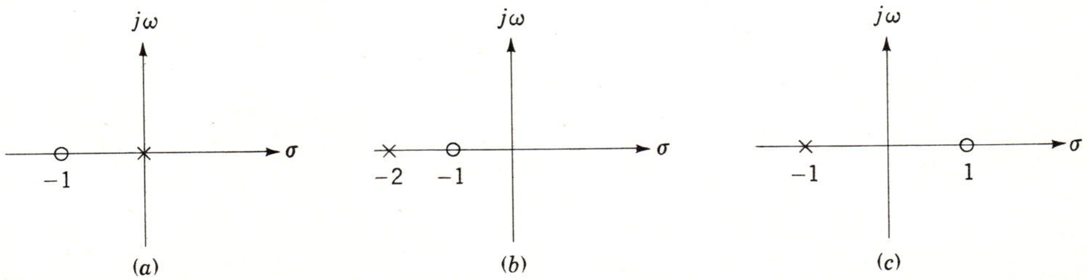

PROB. FIG. 8.7-2

8.7-3 Find ω such that the steady-state resistor current is zero.

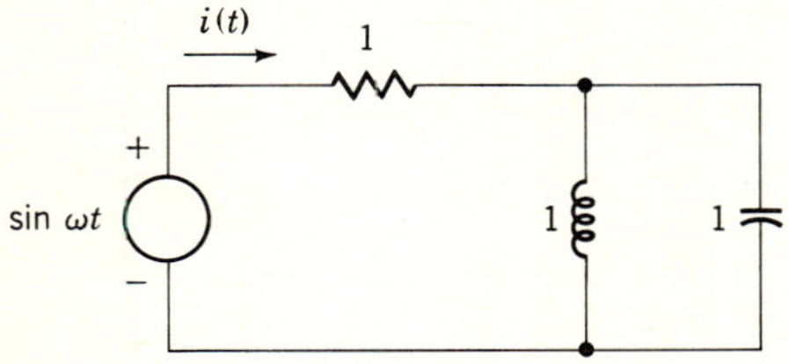

PROB. FIG. 8.7-3

8.7-4 In the following equations $y(t)$ represents the output and $u(t)$ the input for a linear time-invariant dynamic system. It is assumed that the system is initially at rest.

(*a*) Determine the complete solution of the differential equations.

(*b*) State whether or not the input excitation frequency is dominant.

(*c*) Determine in each case the system function $H(s)$ from input to output.

(*d*) Plot the poles and zeros of the system function in the *s* plane.

(*e*) Sketch $|H(j\omega)|$ and $\angle H(j\omega)$ for $\omega > 0$ (rough but accurate). [*Hint:* At least one of the natural frequencies is a simple number.]

$$\frac{d^3y}{dt^3} + 6\frac{d^2y}{dt^2} + 11\frac{dy}{dt} + 6y = u \qquad u(t) = u_{-1}(t) \tag{1}$$

$$\frac{d^3y}{dt^3} + 6\frac{d^2y}{dt^2} + 11\frac{dy}{dt} + 6y = \frac{d^2u}{dt^2} + u \qquad u(t) = u_{-1}(t) \tag{2}$$

$$\frac{d^3y}{dt^3} + 2\frac{d^2y}{dt^2} + 5\frac{dy}{dt} = u \qquad u(t) = \cos(t)u_{-1}(t) \tag{3}$$

$$\frac{d^3y}{dt^3} + 2\frac{d^2y}{dt^2} + 5\frac{dy}{dt} = \frac{d^2u}{dt^2} - u \qquad u(t) = \cos(t)u_{-1}(t) \tag{4}$$

8.7-5 *Twin-T Filter.* The network in the figure is in the form of a so-called

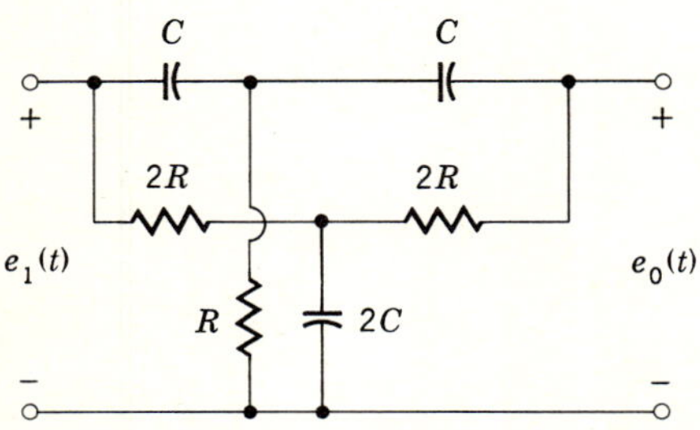

PROB. FIG. 8.7-5

twin-T filter. Calculate and sketch the magnitude and phase of the system function $\frac{E_0}{E_1}(s)$ when $s = j\omega$, and ω varies from 0 to ∞.

chapter 9
Computational Solution of Differential Equations

This chapter is the memory counterpart of iteration (Chap. 5). In the same way that iteration is fundamental to the computational solution of equations arising from memoryless systems and networks, this chapter is fundamental to the computational solution of ordinary differential equations arising from dynamic systems and networks. The chapter concentrates first on numerical integration, then on the solution of first-order ordinary differential equations by open and closed methods, and finally on the computational errors associated with these approaches.

9.1 INTRODUCTION

In this chapter we are concerned with the solution of problems that arise in the analysis of continuous systems with memory: specifically systems that contain, or can be modeled with integrators. Our strategy will be to introduce and study the fundamental computational issues in terms of one-state-variable systems, postponing the treatment of multivariable memory systems to the second volume.

The first fundamental numerical problem to be considered is the *numerical integration* of functions. Numerical integration is applicable to the analysis of explicit systems with memory (Fig. 9.1-1*a*), i.e., systems that have no memory loops. Numerical integration is also useful in evaluating the response of *linear* systems to *arbitrary* drives—through evaluation of the convolution integral [Eq. (11) of Sec. 7.6].

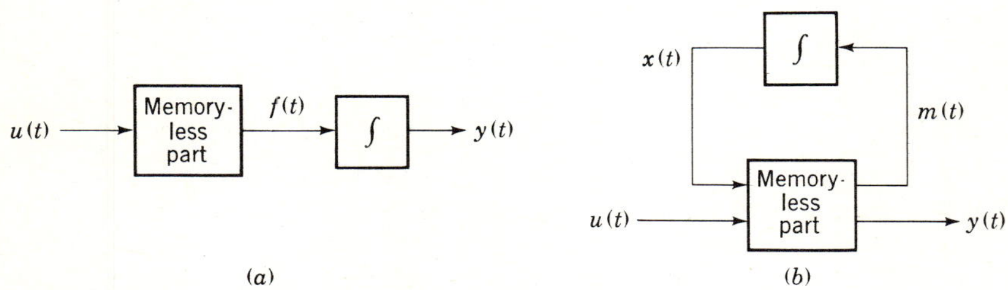

FIG. 9.1-1 Explicit and implicit memory systems.

The second fundamental problem of interest to us is the *solution of ordinary differential equations.* Our motivation here is the analysis of systems which are in state model form, and whose integrator output is fed back to the memoryless part of the system, thus forming a memory loop (Fig. 9.1-1*b*).

We start the following section with the first of these problems—the computation of definite integrals.

9.2 COMPUTATION OF DEFINITE INTEGRALS

Given a function $f(x)$, defined on an interval $I = [a,b]$ of the real line, we wish to evaluate the definite integral

$$\int_{x_0}^{x_n} f(x)\,dx \tag{1}$$

where x_0, x_n are in I.

The fundamental question which must be raised both now and later in the chapter is the information available about f. In calculus, f is usually in the form of an expression, and the basic problem is to find an expression for an integral of f. For example, if $f(x) = k_1x + k_2x^2$, where k_1 and k_2 are real, then we can integrate f explicitly since we know that

$$\int x^n \, dx = \frac{1}{n+1} x^{n+1} \qquad \text{for } n \neq -1 \tag{1a}$$

$$\int kg(x) \, dx = k \int g(x) \, dx \qquad \text{where } k \text{ is real} \tag{1b}$$

$$\int [g(x) + f(x)] \, dx = \int g(x) \, dx + \int f(x) \, dx \tag{1c}$$

Although finding an expression for the integral of f is important, we will not treat it here in any detail.

A common situation is that f is not known everywhere, but only for certain values of its arguments. This occurs frequently in the analysis of systems, because the values of $f(x)$ may arise from either sampled measurements or other calculations. To illustrate this point, consider the following example.

Example 9.2-1 *A Network That Requires Numerical Integration* The non-

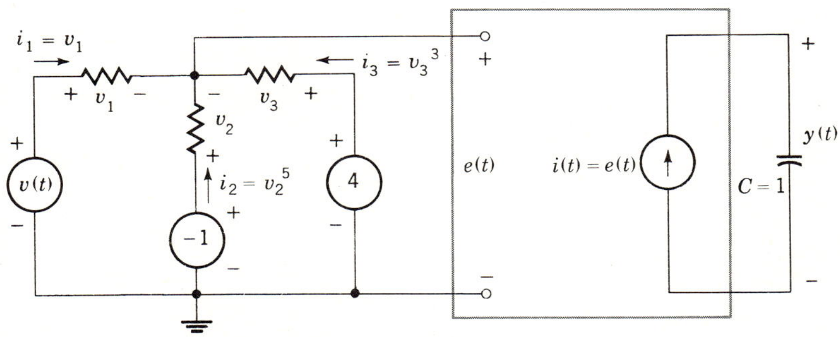

FIG. 9.2-1 The network of Example 9.2-1.

linear three-resistor network of Fig. 9.2-1 is identical to the network of Ex. 5.2-1, with the exception that the leftmost source of that network has been replaced here by a time-varying voltage source $v(t)$. The capacitor voltage $y(t)$ is given by

$$y(t) = \int e(t) \, dt$$

where to simplify matters we have dropped the limits of integration. But the value of $e(t)$, at any time t, depends only on the value of $v(t)$ at that time, since the three-resistor network is memoryless. Accordingly, we can write

$$e = f(v) \qquad \text{hence } y(t) = \int f(v(t)) \, dt$$

or

$$y(t) = \int h(t) \, dt \tag{2}$$

where h is a new function representing the composition of f with v. The function f, however, and hence h, is not known over all values of v; yet it can be evaluated, by iteration, for any *given* value of v. If, for example, at some time $t = t_1$, $v(t_1) = 6$, then we have the problem of Ex. 5.2-1 where by iteration we obtained $e = f(6) = 1$ [or $e = h(t_1) = 1$]. Thus, because of the nature of the problem, we can evaluate h only at discrete points and must therefore resort to some approximation of the desired integral.

In order to solve problems such as the one shown in Fig. 9.2-1, we must be willing to approximate Eq. (1) by evaluating the integrand function at a *finite* number of points over the interval I, giving an expression which approximates the integral to any required accuracy in terms of these values. For ease of discussion, we will restrict these evaluations to equidistant† points x_i, where $x_i = ih$ $(i = 0, 1, \ldots)$, and where h will be called the *step size*. With reference to Fig. 9.2-2, we wish to approximate the area under the continuous and unknown curve $f(x)$, using only the known values $f(x_i)$ (shown as small circles on the graph of Fig. 9.2-2).

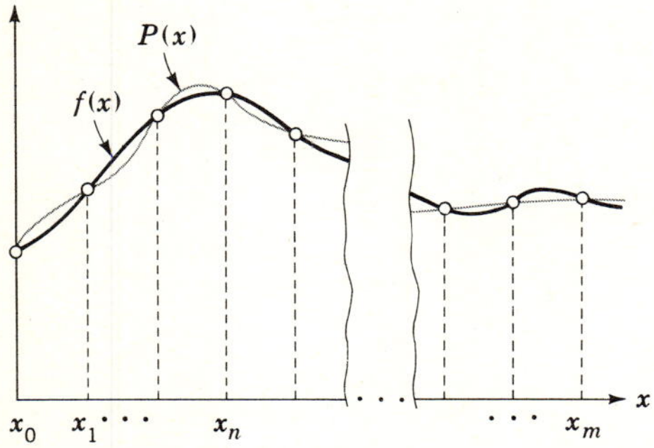

FIG. 9.2-2 Approximation of $f(x)$ by $P(x)$.

Our basic strategy will be to choose a known function $P(x)$ which (1) is specified over all x, (2) passes through the known points $(x_i, f(x_i))$, and (3) is easy to integrate explicitly. At this point, the reader might question the accuracy of any such scheme. Certainly, if we know only the sample values $f(x_i)$, we have no reason for choosing any one approximation $P(x)$ over any other. Furthermore, the resulting error between $P(x)$ and $f(x)$ could be arbitrarily large. In practice, however, either the physical source of $f(x)$ or some constraints on it will be known. For example, f may be known to be continuous, or second order, and steplike. As we shall see, information of this type will be used to influence the choice of the approximating function, so that the ultimate accuracy is acceptable.

The Interpolating Polynomial

Let the approximating function $P(x)$ be an nth-order polynomial in x of the form

$$P(x) = a_n x^n + a_{n-1} x^{n-1} + \cdots + a_1 x + a_0 \tag{3a}$$

To specify the $n+1$ quantities $a_0, \ldots, a_n$, we shall require that $P(x)$ pass through

†This restriction is not essential. Often unequal distances are used to increase the efficiency of the integration process. However, the above restrictions cause no significant loss of generality in understanding the fundamentals of numerical integration.

the first $n+1$ points $f(x_i)$. That is,

for $i=0, 1, \ldots, n$,

$$P(x_i) = f(x_i) \tag{3b}$$

Normally, n will be small, i.e., 2, 3, or 4, so that we will "fit" $P(x)$ through the first three, four, or five points $f(x_i)$. The m samples $f(x_i)$, over which we wish to integrate, will be more than n—typically several hundred (see Fig. 9.2-2). To cope with this situation, we have several alternatives. Some typical approaches are as follows:

1. If all we want is the definite integral from x_0 to x_m, i.e., the area under the curve from x_0 to x_m, we shall integrate first the function $P(x)$ as given by Eq. (3a) from x_0 to x_n. We shall then fit the next $n+1$ points, $i=n, \ldots, 2n$, by respecifying the a_i, and we shall integrate the function $P(x)$ from x_n to x_{2n}. The process of shifting to the right by $n+1$ points, fitting, and integrating will continue until we get close to m, i.e., until an integer k is reached such that $x_{m-n} < x_{kn} \leq x_m$. The integral of the remaining samples $f(x_{kn}) \cdots f(x_m)$ can then be evaluated by a lower-order polynomial. Finally, the sum of the above integrals constitutes the desired result, namely the "area under the curve" from x_0 to x_m.
2. If we want the integral function at every point, i.e., if we wish to obtain the function $F(x_i)$ such that

   for $i=0, 1, \ldots m$,

   $$F(x_i) = \int_{x_0}^{x_i} f(u)\, du$$

   then we can rewrite the integral as follows:

   $$F(x_i) = F(x_{i-n}) + \int_{x_{i-n}}^{x_i} f(u)\, du$$

   We can now approximate the right-hand integral by the nth-order polynomial $P(x)$ over the $n+1$ points, $x_{i-n}, \ldots, x_i$; the approximate value for $F(x_i)$ will then be the sum of this integral and the value of $F(x_{i-n})$. An approximation to the latter, however, is obtained from the same equation after the indices are shifted back by $n+1$ points. Repetitive use of this approach permits approximate evaluation of $F(x_k)$ for every k, as long as $k \geq n$. The first $n+1$ values, $F(x_0), F(x_1), \ldots, F(x_n)$, are assumed to be known or can be obtained by approximate integration of lower order.

Without loss of generality and regardless of which of the above approaches we shall use, we can concentrate now on our main problem—the integration of $f(x)$ over the first $n+1$ points, i.e., the approximate integration of

$$\int_{x_0}^{x_n} f(x)\, dx$$

In order to establish a polynomial $P(x)$, as given by Eq. (3*a*), which passes through the first $n+1$ points, $f(x_i)$, as required by Eq. (3*b*), we shall define $P(x)$ as a sum of product terms. In particular,

$$P(x) \triangleq \sum_{i=0}^{n} A_i(x) f(x_i) \tag{4a}$$

where

$$A_i(x) \triangleq \frac{(x-x_0)(x-x_1)\cdots(x-x_{i-1})(x-x_{i+1})\cdots(x-x_n)}{(x_i-x_0)(x_i-x_1)\cdots(x_i-x_{i-1})(x_i-x_{i+1})\cdots(x_i-x_n)} \tag{4b}$$

Observe that $P(x)$ is an nth-order polynomial in x, and has the desired properties (3*b*), since for each i, $A_i(x_i)=1$: while for $i \neq j$, $A_i(x_j)=0$. $P(x)$ is called an *interpolating polynomial,* since for values of x other than x_i it "interpolates" the value $f(x)$, *assuming that f behaves like an nth-order polynomial.* The polynomial $P(x)$ is shown in Fig. 9.2-3*a* and *b* for $n=1$ and 2, respectively.

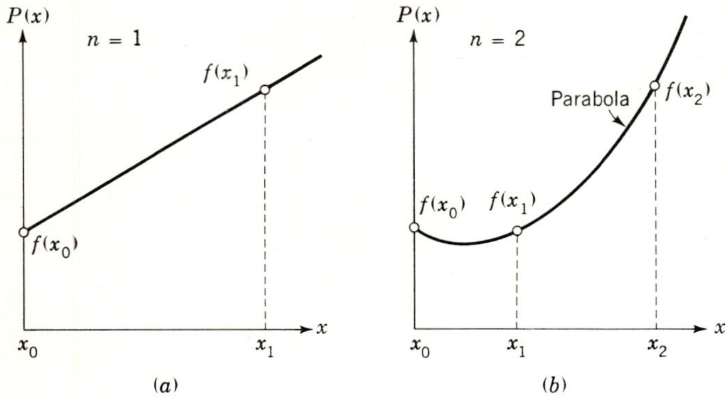

FIG. 9.2-3 $P(x)$ for (a) $n=1$ *and* (b) $n=2$.

The definite integral $\int_{x_0}^{x_n} f(x)\,dx$ can now be *approximated* by the integral of the interpolating polynomial. That is,

$$\int_{x_0}^{x_n} f(x)\,dx \cong \int_{x_0}^{x_n} P(x)\,dx \tag{5}$$

To integrate $P(x)$ explicitly, observe that

$$x_i = x_0 + ih \qquad \text{where } i = 0, 1, 2, \ldots, n \tag{6a}$$

and h is the step size. Observe further that for every x $(x_0 \le x \le x_n)$,

$$x = x_0 + ph \tag{6b}$$

where p is a real number such that $0 \le p \le n$. Substitution of Eq. (6*b*) into Eq. (4*b*) yields a function of p as follows:

$$A_i(x) = B_i(p) = \frac{p(p-1)\cdots(p-i+1)(p-i-1)\cdots(p-n)}{i(i-1)\cdots(1)(-1)\cdots(i-n)} \tag{7}$$

Use of Eqs. (4a) and (7) yields the following:

$$\int_{x_0}^{x_n} P(x)\,dx = \int_{x_0}^{x_n} \sum_{i=0}^{n} A_i(x) f(x_i)\,dx = \sum_{i=0}^{n} f(x_i) \int_{x_0}^{x_n} A_i(x)\,dx$$

$$= h \sum_{i=0}^{n} f(x_i) \int_0^n B_i(p)\,dp \qquad (8)$$

where we have made use of Eq. (6b) to establish that $dx = h\,dp$, and to change the limits of integration from x to p. Observe now that the term

$$\int_0^n B_i(p)\,dp$$

depends only on n and i and is independent of f; *it may therefore be computed separately to yield a set of coefficients.* To preserve convention, we define these $n+1$ coefficients, $C_{i,n}$, called *Cotes numbers*, as follows:

$$C_{i,n} = \frac{1}{n} \int_0^n B_i(p)\,dp \qquad \text{for } i = 0, 1, \ldots, n \qquad (9)$$

Use of Eq. (9) in Eq. (8) yields the desired result:

$$\int_{x_0}^{x_n} P(x)\,dx = nh \sum_{i=0}^{n} C_{i,n} f(x_i) \qquad (10)$$

Observe that in Eq. (10), nh is the width of the interval $[x_0,x_n]$; hence the weighted sum $\sum_{i=0}^{n} C_{i,n} f(x_i)$ represents the "average value" of $P(x)$ in the interval $x_0 \le x \le x_n$.

It should be emphasized again that the $C_{i,n}$ are, for any given n, $n+1$ coefficients which are used, along with the known values $f(x_i)$, to compute the integral of $P(x)$ through Eq. (10). We will now investigate some well-known integration schemes that result when n is chosen small.

Specialization to Integration Methods

For $n = 1$, we have what is known as *trapezoidal integration*, where

$$C_{0,1} = C_{1,1} = 1/2$$

and Eq. (10) becomes

$$\int_{x_0}^{x_1} P(x)\,dx = (x_1 - x_0)\,[1/2\, f(x_0) + 1/2\, f(x_1)] \qquad (11)$$

Since $n = 1$, $P(x)$ fits $f(x)$ at two points. Figure 9.2-4a shows graphically that the area under $f(x)$ is approximated by the area of the shaded trapezoid. To develop some feeling as to how f influences the accuracy of the approximation, we observe that if $f(x)$ has small second and higher-order derivatives in the interval $[x_0,x_1]$, then the approximation of Eq. (11) should be quite accurate. In fact, if these derivatives are zero, i.e., if $f(x)$ is a straight line, then it is easily seen that Eq. (11) gives the *exact* value of the integral of $f(x)$.

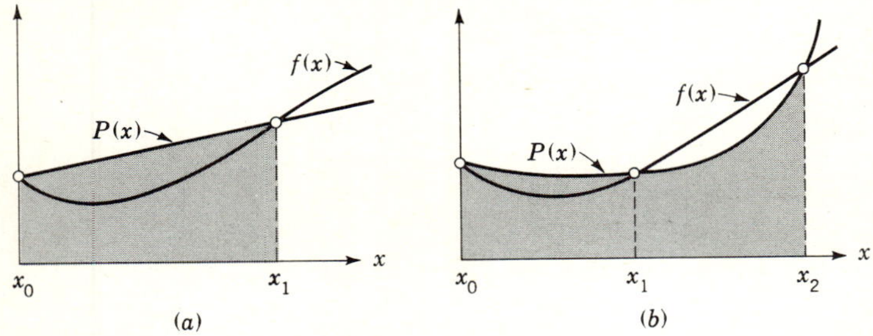

FIG. 9.2-4 (a) *Trapezoidal* ($n = 1$) *and* (b) *Simpson* ($n = 2$) *integration.*

For $n = 2$, the Cotes numbers are

$$C_{0,2} = C_{2,2} = 1/6 \qquad C_{1,2} = 2/3$$

and we have the well-known *Simpson's formula* or *Simpson's rule;* that is, Eq. (10) becomes

$$\int_{x_0}^{x_2} P(x)\, dx = (x_2 - x_0)[1/6\, f(x_0) + 2/3\, f(x_1) + 1/6\, f(x_2)] \tag{12}$$

In this case the interpolating polynomial is a parabola passing through the three points $(x_0,f(x_0))$, $(x_1,f(x_1))$, and $(x_2,f(x_2))$, and the computed area is shown shaded in Fig. 9.2-4*b*. For $n = 2$, if the third and higher derivatives are small, we expect that the approximation error will be small; in fact, if these derivatives are zero, then Simpson's rule gives the exact value of the integral of $f(x)$. It follows then that Simpson's rule evaluates exactly the integrals of a *larger class* of functions as compared to the trapezoidal rule.

Extending this approach to higher n, we see that the resulting integration methods of successively higher order will integrate exactly a pro-

TABLE 9.2-1 Cotes Coefficients $C_{i,n}$ for $1 \le n \le 5$

n \ i	0	1	2	3	4	5
1	1/2	1/2				
2	1/6	2/3	1/6			
3	1/8	3/8	3/8	1/8		
4	7/90	16/45	6/45	16/45	7/90	
5	19/288	75/288	25/144	25/144	75/288	19/288

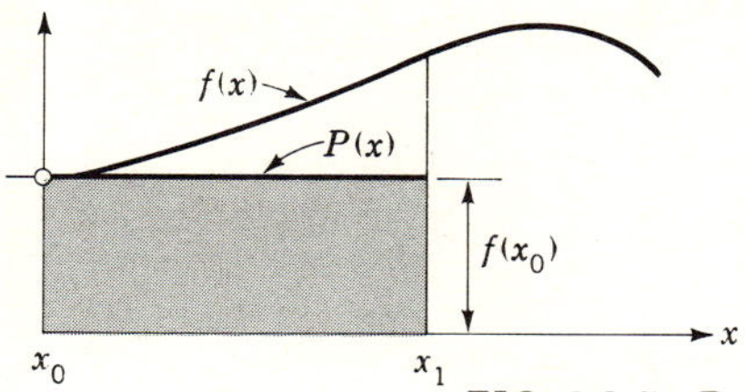

FIG. 9.2-5 Rectangular or Euler integration.

gressively larger class of functions, at the expense of more computation. The coefficients $C_{i,n}$ for $1 \le n \le 5$ are given in Table 9.2-1.

Finally, note that for $n = 0$ the integration interval is 0, and we cannot obtain a meaningful coefficient $C_{0,0}$. If we take a different point of view, however, and we follow the progression: $n = 2$, $P(x)$ is a polynomial of degree two; $n = 1$, $P(x)$ is a polynomial of degree one; with $n = 0$, $P(x)$ is a polynomial of degree zero, hence a constant; and if we retain the interval of integration as h, then we have what is known as *rectangular* or *Euler integration*. The resulting approximation to the integral is shown in Fig. 9.2-5 as the shaded area under the zero-degree polynomial $P(x)$ of constant value $f(x_0)$.

Example 9.2-2 *Evaluation of Convolution Integral* Consider the RC

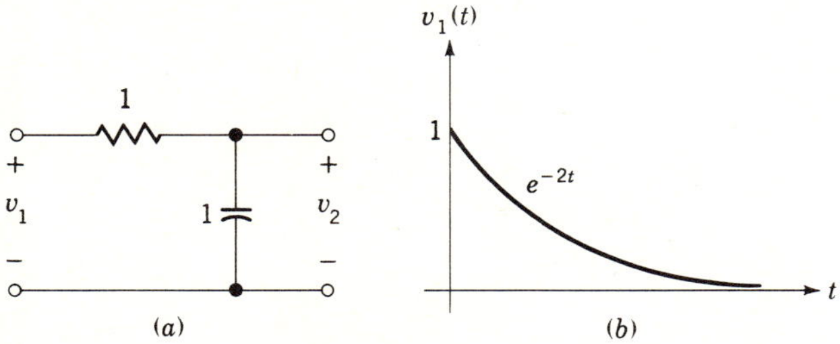

FIG. 9.2-6 (a) *RC network and* (b) *input waveform.*

network of Fig. 9.2-6*a*. We wish to determine $v_2(t)$, when $v_1(t)$ is as given in Fig. 9.2-6*b* and $v_2(0) = 2$. By inspection of the network, we can write the input-output D.E.:

$$\frac{dv_2}{dt} = v_1 - v_2 \qquad v_2(0) = 2 \tag{13}$$

From Eq. (11) of Sec. 7.6 we can write the solution as follows:

$$v_2(t) = e^{-t}[2 + \int_0^t e^{\tau} v_1(\tau)\, d\tau] \tag{14}$$

Substitution of $v_1(\tau) = e^{-2\tau}$ yields

$$v_2(t) = e^{-t}(2 + \int_0^t e^{-\tau}\, d\tau) \tag{15}$$

Equation (15) can be integrated directly to give an explicit expression, so that the exact solution is given by

$$v_2(t) = 3e^{-t} - e^{-2t} \tag{16}$$

Here, we have chosen an example whose exact solution is known, so that we may compare the numerical results. We proceed, hence, with the numerical computation of the integral of Eq. (15), using rectangular, trapezoidal, and Simpson integration. We begin by dividing time, quite arbitrarily, into segments of width 0.01 (one hundredth of the time constant). The integral function will be evaluated over the first 24 time points. From the preceding discussion of integration methods we list, by inspection, the four corresponding formulas as they apply to this example:

1. Rectangular integration ($n = 0$):

$$\text{rec}_0 = 0$$

For $i = 0, 1, 2, \ldots,$

$$\text{rec}_{i+1} = \text{rec}_i + (0.01)e^{-(0.01)i} \tag{17}$$

where rec_i is the value of the numerical integral of e^{-t} at time $t = 0.01i$.

2. Trapezoidal integration ($n = 1$):

$$\text{trap}_0 = 0$$

For $i = 0,1,2, \ldots,$

$$\text{trap}_{i+1} = \text{trap}_i + 0.005(e^{-0.01i} + e^{-0.01(i+1)}) \tag{18}$$

where trap_i is analogous to rec_i for trapezoidal integration.

3. Simpson integration ($n = 2$):

$$\text{simp}_0 = 0$$

For $i = 0,1,2, \ldots,$

$$\text{simp}_{i+2} = \text{simp}_i + \frac{0.02}{6}(e^{-0.01i} + 4e^{-0.01(i+1)} + e^{-0.01(i+2)}) \tag{19}$$

The errors between the above three types of numerical integrals and the exact integral are shown in Table 9.2-2 for each of the 24 time points. It can be seen from this table that the most accurate integral is that obtained through Simpson's method.

The above example motivates the natural question as to the choice of spacing h and integration order n. This is the topic of the following section.

Integration Error

In this section we shall concentrate on the approximation error

$$\epsilon(x) = f(x) - P(x) \tag{20}$$

and we shall assume that computation is otherwise error-free.

TABLE 9.2-2 Integration Errors: Numerical$_i$−Exact (ih)

Rectangular	*Trapezoidal*	*Simpson*
0.000000000E0	0.000000000E0	0.000000000E0
4.983374917E⁻5	8.291791389E⁻8	—
9.917164425E⁻5	1.650107808E⁻7	1.100066219E⁻12
1.480186191E⁻4	2.462868099E⁻7	—
1.963795584E⁻4	3.267541291E⁻7	2.178332167E⁻12
2.442592983E⁻4	4.064207851E⁻7	—
2.916626268E⁻4	4.852947446E⁻7	3.235258415E⁻12
3.385942844E⁻4	5.633838951E⁻7	—
3.850589641E⁻4	6.406960456E⁻7	4.27125002E⁻12
4.310613126E⁻4	7.172389273E⁻7	—
4.766059300E⁻4	7.930201946E⁻7	5.286729388E⁻12
5.216973709E⁻4	8.680474257E⁻7	—
5.663401445E⁻4	9.423281234E⁻7	6.282085963E⁻12
6.105387151E⁻4	1.015869716E⁻6	—
6.542975026E⁻4	1.088679557E⁻6	7.257749957E⁻12
6.976208828E⁻4	1.160764928E⁻6	—
7.405131882E⁻4	1.232133038E⁻6	8.214082192E⁻12
7.829787080E⁻4	1.302791024E⁻6	—
8.250216889E⁻4	1.372745950E⁻6	9.151485125E⁻12
8.666463351E⁻4	1.442004814E⁻6	—
9.078568092E⁻4	1.510574540E⁻6	1.007031958E⁻11
9.486572321E⁻4	1.578461986E⁻6	—
9.890516841E⁻4	1.645673941E⁻6	1.097097413E⁻11
1.029044205E⁻3	1.712217125E⁻6	—

At one extreme, if we only know the values $f(x_i)$ at equidistant points, $x_0, x_1, \ldots, x_n$, and we know nothing else about f, then there is little that can be said about $\epsilon(x)$, since for any given set of $f(x_i)$ we could choose f so that $\epsilon(x)$ attains any desired value when $x \neq x_i$. At the other extreme, if f has a vanishing derivative of order $n + 1$, then the function and its integral may be represented exactly by $P(x)$ and $\int P(x)\,dx$, respectively, thus resulting in zero approximation error. We are therefore tempted to classify functions in terms of the magnitude of their high-order derivatives, in order to study the associated error of integration.

Let us assume that f is continuously differentiable $n + 1$ times in the interval $[x_0, x_n]$. The following theorem provides an expression for $\epsilon(x)$.

THEOREM 4.2-1 For any fixed x $(x_0 \leq x \leq x_n)$,

$$\epsilon(x) = \frac{1}{(n+1)!} \prod_{i=0}^{n} (x - x_i)\, f^{(n+1)}(\xi) \tag{21}$$

where Π means the product of, $f^{(n+1)}$ is the $n + 1$st derivative of f, and ξ is a value which depends on x and is in the same interval, i.e., $x_0 \leq \xi \leq x_n$.

PROOF: For $x = x_j$, $j = 0, 1, \ldots, n$, it follows by inspection of Eq. (21) that $\epsilon(x_j) = 0$, as is required by the construction of $P(x)$. For any *fixed* value x_f $(x_0 < x_f < x_n)$ we define the following auxiliary function $g(p)$:

$$g(p) \triangleq \epsilon(p) - \frac{\epsilon(x_f)}{\prod_{i=0}^{n}(x_f - x_i)} \prod_{i=0}^{n}(p - x_i) \tag{22}$$

Observe that $g(p)$ equals zero for $p = x_k$, $k = 0, 1, \ldots, n$, and for $p = x_f$. Since g has these $n+2$ distinct zeros in the range $x_0 \leq p \leq x_n$, it follows by Rolle's theorem that g' has at least $n+1$ zeros, and, recursively, $g^{(n+1)}$ has at least one zero, say at $p = \xi$. Differentiation of Eq. (22) $n+1$ times and evaluation of both sides at $p = \xi$ yields

$$0 = \epsilon^{(n+1)}(\xi) - \frac{\epsilon(x_f)}{\prod_{i=0}^{n}(x_f - x_i)}(n+1)!$$

Observe here that the term $(n+1)!$ is the $(n+1)$st derivative of the highest power of p in $\prod_{i=0}^{n}(p - x_i)$, i.e., of p^{n+1}; all the other (smaller) powers of p vanish after $n+1$ differentiations. But $\epsilon(x) = f(x) - P(x)$ and since $P(x)$ is a polynomial of degree n, its $(n+1)$st derivative vanishes. Hence $\epsilon^{(n+1)}(\xi) = f^{(n+1)}(\xi)$. Thus, the above equation becomes

$$0 = f^{(n+1)}(\xi) - \frac{\epsilon(x_f)}{\prod_{i=0}^{n}(x_f - x_i)}(n+1)!$$

or after some rearranging;

$$\epsilon(x_f) = \frac{\prod_{i=0}^{n}(x_f - x_i)}{(n+1)!} f^{(n+1)}(\xi) \tag{23}$$

for any x_f such that $x_0 \leq x_f \leq x_n$. Q.E.D.

We see from Eq. (23) that the error function $\epsilon(x_f)$ equals the product of two terms, the first dependent on n and independent of f, and the second dependent on $f^{(n+1)}$. The first term attains a maximum which is approximately equal to h^{n+1} so that for a given n, $\epsilon(x)$ is bounded by $h^{n+1}f^{(n+1)}(\xi)$.

Equation (23) is an expression for the error resulting in the approximation of $f(x)$ by $P(x)$. What we are really interested in is the error resulting from the approximation of the desired integral of $f(x)$ by the integral of $P(x)$. This *integration error*, as it is called, is defined as follows:

$$\left|E_n\right| = \int_{x_0}^{x_n} f(x)\,dx - \int_{x_0}^{x_n} P(x)\,dx \triangleq \int_{x_0}^{x_n} \epsilon(x)\,dx$$

Assuming that the largest magnitude of $f^{(n+1)}(\xi)$ for $x_0 \leq \xi \leq x_n$ is $f_m^{(n+1)}$, we can obtain from Eq. (23) the following inequality:

$$\left|E_n\right| = \left|\int_{x_0}^{x_n} \epsilon(x)\,dx\right| \leq \frac{f_m^{(n+1)}}{(n+1)!}\left|\int_{x_0}^{x_n} \prod_{i=0}^{n}(x - x_i)\,dx\right| \tag{24}$$

This inequality, in turn, can be reduced to the general form

$$\left|E_n\right| \leq kh^{n+2} f_m^{(n+1)} \tag{25}$$

where k is constant. From Eq. (25) it can be seen that the nth-order integration error is bounded by the $(n + 2)$nd power of h and the $(n + 1)$st derivative of f. Since h is very small, it follows that the smaller h is, the higher the integration accuracy. The computation† needed, however, for nth-order integration grows at least as

$$\frac{2n + 1}{h} \tag{26}$$

since the number of points in an interval of interest grows as $1/h$; and at each point evaluation of the integral involves $n + 1$ multiplications and n additions.

For a given n, a more accurate estimate of the error E_n can be obtained by direct integration of Eq. (24). Thus, for trapezoidal integration ($n = 1$), a change of variables, $x = x_0 + ph$ (where $0 < p < 1$), converts Eq. (24) as follows:

$$\left|E_1\right| \leq \frac{h^3 f_m^{(2)}}{2} \left| \int_0^1 p(p - 1)\, dp \right| = \left(\frac{h^3}{12}\right) f_m^{(2)} \tag{27}$$

Example 9.2-3 Consider again the numerical integration of e^{-t}, discussed in the preceding example.

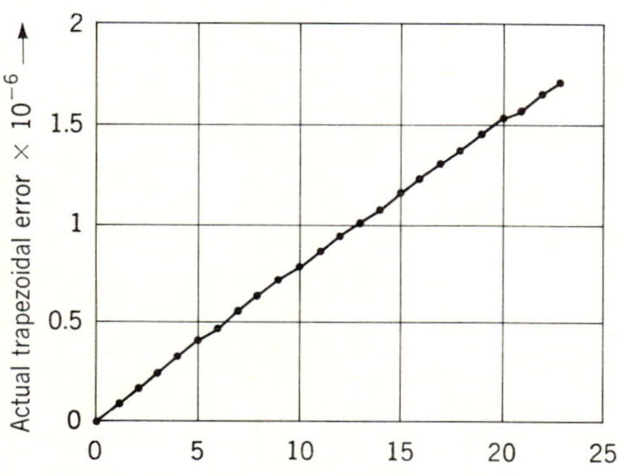

FIG. 9.2-7 Progress of trapezoidal error in the integration of e^{-t}.

Application of Eq. (27) to the parameters of that example ($h = 0.01$ and $f_m^{(2)} = 1$) yields

$$|E_1| \leq 10^{-7}$$

Observing that Eq. (27) is integrated over *one time interval h*, we see that the results of Table 9.2-2, column 2, are certainly within this error for the first time

†Number of addition and multiplication operations.

interval. As integration progresses over the designated time points, the error of Table 9.2-2 accumulates over successive intervals (see Fig. 9.2-7). As a rough check we might expect that over 24 time points the final error should be of the order $24|E_1|$ (or 2.4×10^{-6}). Indeed, the final error is $1.7 + 10^{-6}$, confirming our analysis.

9.3 COMPUTATIONAL SOLUTION OF ORDINARY DIFFERENTIAL EQUATIONS

In the preceding section we dealt with the numerical evaluation of explicit integrals, that is, integrals of the form

$$y(t) = \int_{t_0}^{t} x(\tau)\, d\tau$$

In this section, we shall treat the numerical solution of the ordinary differential equation

$$\frac{dx}{dt}(t) = f(x,t) \qquad \text{with } x(t_0) \text{ and } f \text{ given} \tag{1}$$

This equation can be rewritten as the integral equation,

$$x(t) = x(t_0) + \int_{t_0}^{t} f(x(\tau), \tau)\, d\tau \tag{2}$$

with $x(t_0)$ the initial value. Observe that this equation is in implicit form, since the unknown function $x(t)$ appears both on the left side of the equation and on the right as an argument of the integrand.

Our motivation for solving Eq. (2) numerically is based on the fact that continuous dynamic systems can be represented in state model form, and Eq. (2) is one of the simplest nontrivial forms that such a state model can take. Equation (2) characterizes the closed-loop portion of the dynamic system shown in Fig. 9.3-1. To demonstrate that this is so, observe that in Fig. 9.3-1 the state

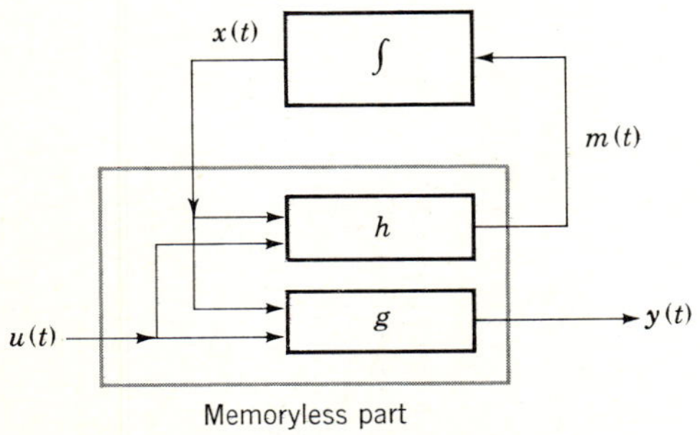

FIG. 9.3-1 State model of first-order dynamic system for numerical solution.

update $m(t)$ is related to state $x(t)$ and input $u(t)$ by a function h as follows:

For all t,

$$m(t) = h(x(t),u(t)) \tag{3}$$

Since u is a known function of time, h may be rewritten in terms of a new function f, which is defined as follows:

For all x and t,

$$f(x,t) = h(x,u(t)) \tag{4}$$

Next observe that we may write from the state model of Fig. 9.3-1 the following equations:

For all $t \geq t_0$,

$$x(t) = x(t_0) + \int_{t_0}^{t} m(\tau)\, d\tau$$

Substitution of Eqs. (3) and (4) into this equation yields

For all $t \geq t_0$,

$$x(t) = x(t_0) + \int_{t_0}^{t} f(x(\tau),\tau)\, d\tau$$

which is Eq. (2).

As we have pointed out in Chap. 7, after the function $x(t)$ of Fig. 9.3-1 has been numerically evaluated, it is a straightforward matter to evaluate the output as follows:

For all t,

$$y(t) = g(x(t),u(t)) \tag{5}$$

Classification of Methods

Our approach to the numerical solution of Eq. (2) will be to approximate the integral by the area under an interpolating polynomial, as we did in Sec. 9.2. Again we will be interested in values of $x(t)$ at a set of sample points. In contrast to plain integration, however, our approximation will give rise to two types of equations, according to whether or not the interpolating polynomial passes through the point $x(t_n)$ being evaluated. These two classes, called *open* and *closed* or *explicit* and *implicit* methods, respectively, will be specialized to encompass several well-known methods for the solution of ordinary differential equations. We start with an example that demonstrates the general idea.

Example 9.3-1 *A Nonlinear RC Network* In the network of Fig. 9.3-2*a*, the capacitor is linear and the resistor is nonlinear with the current equal to the

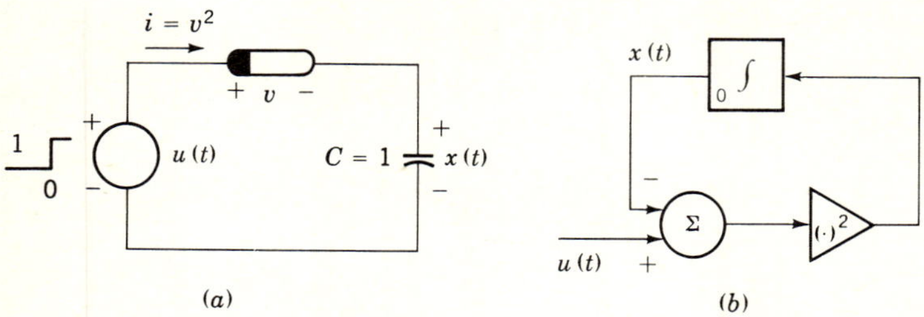

FIG. 9.3-2 Nonlinear network and corresponding block diagram system.

square of the voltage. The voltage source is a unit step occurring at $t=0$, and the capacitor voltage is initially zero. The state model for the network, shown in Fig. 9.3-2*b*, yields the relation

$$x(t) = \int_0^t [1 - x(\tau)]^2 \, d\tau \qquad \text{for } t \geq 0 \tag{6}$$

Here, for $t \geq 0$ the integrand is a function of x and not of t (as will be sometimes the case). Hence, the function f of Eq. (2) is given by

$$f(x) = (1 - x)^2 \tag{7}$$

where x is the value of $x(t)$ at any time $t \geq 0$.

Recall from the preceding section that a first-order interpolating polynomial (trapezoidal integration) yields the approximation

$$\int_{t_p}^{t_{p+1}} g(\tau) \, d\tau \cong \frac{h}{2} [g(t_{p+1}) + g(t_p)] \tag{8}$$

If we write Eq. (6) as

$$\begin{aligned} x(t_{n+1}) &= \int_0^{t_{n+1}} [1 - x(\tau)]^2 \, d\tau \\ &= \int_0^{t_n} [1 - x(\tau)]^2 \, d\tau + \int_{t_n}^{t_{n+1}} [1 - x(\tau)]^2 \, d\tau \\ &= x(t_n) + \int_{t_n}^{t_{n+1}} [1 - x(\tau)]^2 \, d\tau \end{aligned} \tag{9}$$

we can use the approximation of Eq. (8) to form an estimate of $x(t_{n+1})$. This gives

$$x(t_{n+1}) \cong x(t_n) + \frac{h}{2} \{[1 - x(t_{n+1})]^2 + [1 - x(t_n)]^2\} \tag{10}$$

Equation (10) is *closed* or *implicit* since $x(t_{n+1})$, the desired quantity, appears on both the left- and the right-hand sides. It can, however, be treated with the iterative methods discussed in Chap. 5.

An alternative to this procedure is to use an approximation for $x(t_{n+1})$. Specifically, we can use our knowledge of $x(t_n)$ and of the slope of $[1 - x(t_n)]^2$. Thus, using the approximation

$$x(t_{n+1}) \cong x(t_n) + h[1 - x(t_n)]^2 \tag{11}$$

on the right side of Eq. (10) we can rewrite the latter as follows:

$$x(t_{n+1}) \cong x(t_n) + \frac{h}{2}\{[1 - x(t_n) - h[1 - x(t_n)]^2]^2 + [1 - x(t_n)]^2\} \tag{12}$$

This approximation is *open* or *explicit* in that $x(t_{n+1})$ may be computed directly from knowledge of $x(t_n)$.

Closed methods are also called *multistep* methods since several iteration steps are generally used to compute each value of x_n, with the process repeated for each point. The explicit methods are also called *single-step* methods because only a single explicit calculation is required for each point. We proceed now with a more detailed discussion of open and closed methods.

Open Methods

Recall that we wish to solve the differential equation

$$\dot{x}(t) = f(x(t),t) \qquad \text{with initial value } x(t_0)$$

We assume a step size h, i.e., $t_i = t_0 + ih$, and integrate the above equation to relate $x(t_{j+1})$ to $x(t_j)$. Thus

$$x(t_{j+1}) = x(t_j) + \int_{t_j}^{t_{j+1}} f(x(\tau),\tau)\, d\tau \tag{13}$$

The next step is to approximate the integral in Eq. (13) by a weighted sum, as we did in Sec. 9.2. This yields the following equation, where we have focused on finding $x(t_1)$ with no loss of generality. Thus Eq. (13) becomes

$$x(t_1) - x(t_0) = \int_0^h f(x(\tau),\tau)\, d\tau \cong \sum_{k=0}^{n} A_k f_k \tag{14}$$

The A_k are suitable weights from the interpolating polynomial, and ideally the f_k should be exact values of the integrand at $n + 1$ points between $t = 0$ and $t = h$. Unfortunately, it is not possible to compute exact values for f_k since that would entail knowing the argument x_k of f_k, and this would mean that Eq. (13) had been solved exactly to yield values for x between $t = 0$ and $t = h$. We can clearly not assume that we have available what we are trying to find, namely, $x(t_1)$. We have entered then, an implicit "situation which can be resolved by resorting to a second "layer" of approximation; that is, we will approximate the arguments x_k at which we will evaluate f_k, and we will then use these approximate values of f_k to approximate new x_k, repeating this process until we finally obtain $x(t_1)$. The fact that these methods are open means that we must establish

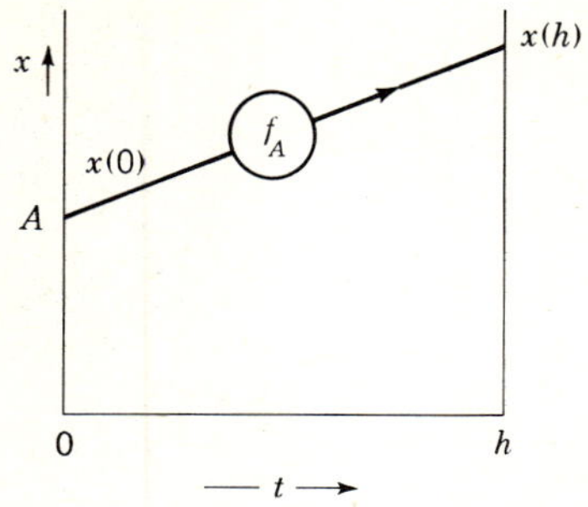

$$x(h) = x(0) + hf_A$$
$$f_A = f(x(0),0)$$

FIG. 9.3-3 *Euler's method.*

an explicit "chain" of approximating x_k and f_k, starting from $x(t_0)$ and continuing until $x(t_1)$ has been computed.

In order to specialize the above discussion to several well-known open methods, we will show the explicit chain of approximations employed by each such method. This is demonstrated graphically in the x-t plane in Figs. 9.3-3 to 9.3-5 for the Euler, Heun, and Runge-Kutta methods, along with the appropriate sequences of equations. To "read" these figures start from point A, which is the point $(x(0),0)$ in the x-t plane, and proceed with the construction of points B, C, . . . in alphabetical order, until the desired last point $(x(h),h)$ has been computed. These constructions involve the evaluation of f at several points of the x-t plane. Accordingly, we denote in these figures the value of f at a point k of the x-t plane by f_k [thus $f_A = f(x(0),0)$]. Observe further, that f denotes, by the defining differential equation, a slope in the x-t plane. In the figures, lines drawn with slope f_k show f_k encircled "within" the line. Thus, in Fig. 9.3-3, the desired value $x(h)$ is obtained as the intersection of a line of slope f_A, drawn from point A, with the line $t = h$; meaning that $x(h) = x(0) + hf(x(0),0)$. This is the *Euler method* that we used in Example 7.6-2.

In Fig. 9.3-4 point B is located as in Fig. 9.3-3, and the desired value $x(h)$ is obtained by drawing a line from A, with a slope which is the average of the slopes at A and B, until it intersects the line $t = h$. This is the method that we have used in Example 9.3-1, Eq. (12), called *Heun's method.*

Finally in Fig. 9.3-5, the point $(x(h),h)$ is obtained by drawing a line which is the weighted sum of four slopes which are established successively, starting from point A and constructing points B, C, and D, as shown in the figure.

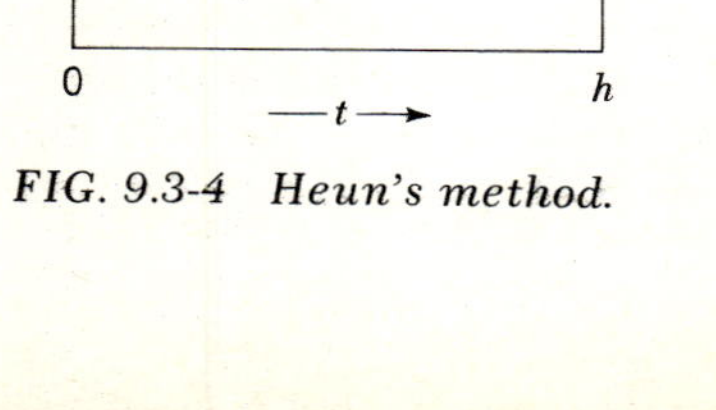

$$x(h) = x(0) + \frac{h}{2}(f_A + f_B)$$

$$f_A = f(x(0),0)$$
$$f_B = f(x(0) + hf_A,h)$$

FIG. 9.3-4 *Heun's method.*

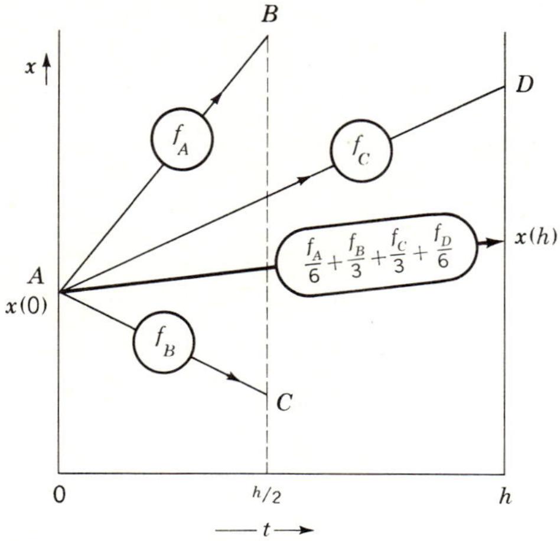

$$x(h) = x(0) + h(\tfrac{1}{6} f_A + \tfrac{1}{3} f_B + \tfrac{1}{3} f_C + \tfrac{1}{6} f_D)$$

$$f_A = f(x(0),0)$$

$$f_B = f(x(0) + \frac{h}{2} f_A, \frac{h}{2})$$

$$f_C = f(x(0) + \frac{h}{2} f_B, \frac{h}{2})$$

$$f_D = f(x(0) + hf_C, h)$$

FIG. 9.3-5 Fourth-order Runge-Kutta method.

This method, known as the *fourth-order Runge-Kutta method* of integration, is widely used in practice.

Closed Methods

We will now consider methods in which the desired value for $x(t_{j+1})$ is given in terms of itself as well as $x(t_j)$, and previous values of x. That is, we shall be dealing with equations of the form

$$x(t_{j+1}) = g(x(t_{j+1}), x(t_k)) \qquad \text{for } k \leq j \tag{15}$$

Following our earlier discussion, we need to resort to iteration, in order to solve the above equation for $x(t_{j+1})$.

Before we can begin this iteration process, however, we need an initial† estimate $x^\circ(t_{j+1})$ for $x(t_{j+1})$. This estimate is usually obtained from calculated past values by an explicit integration method, such as the ones studied earlier, i.e.,

$$x_{j+1}^\circ = x_{j-q} + h \sum_{k=j-q+1}^{j} A_k f_k \tag{16}$$

where we denote by x_p the calculated value for $x(t_p)$ and by f_p the value of $f(x_p,t_p)$. Such a formula, in this context, is called a *predictor*.

We will then establish an equation implicit in x_{j+1} which is an approximation to the integral. This equation, known as the *corrector*, will be of the form

$$x_{j+1} = x_{j-m} + h \sum_{k=j-m}^{j} b_k f_k + hb_{j+1} f(x_{j+1}, t_{j+1}) \tag{17}$$

where x_{j+1} is the unknown and appears on both sides of Eq. (17). The value of x_{j+1}° obtained from Eq. (16) will be used as a starting value in the following itera-

†This is initial in the dimension along which we shall iterate. It is not initial in time.

tion rule [obtained from Eq. (17)] where the iteration index appears as a superscript.

$$x_{j+1}^{i+1} = x_{j-m} + h \sum_{k=j-m}^{j} b_k f_k + h b_{j+1} f(x_{j+1}^{i}, t_{j+1}) \tag{18}$$

In actual practice, if the predictor is fairly accurate, it is necessary to iterate only once; i.e., x_{j+1}^{1} is used as the desired value for $x(t_{j+1})$, assuming that x_{j-1}° is "very close" to $x(t_{j+1})$.

Two well-known predictor-corrector methods are shown graphically in Figs. 9.3-6 and 9.3-7. The *modified Euler method* of Fig. 9.3-6 establishes, on the basis of the already computed value x_j, the predicted value x_{j+1}° as follows (Fig. 9.3-6*a*):

$$x_{j+1}^{\circ} = x_j + hf(x_j, t_j) \tag{19a}$$

The corrector is

$$x_{j+1} = x_j + h[\tfrac{1}{2}f(x_j, t_j) + \tfrac{1}{2}f(x_{j+1}, t_{j+1})] \tag{19b}$$

where the value x_{j+1}° of Eq. (19*a*) is used on the right of Eq. (19*b*) as a "starting value" for the iteration process.

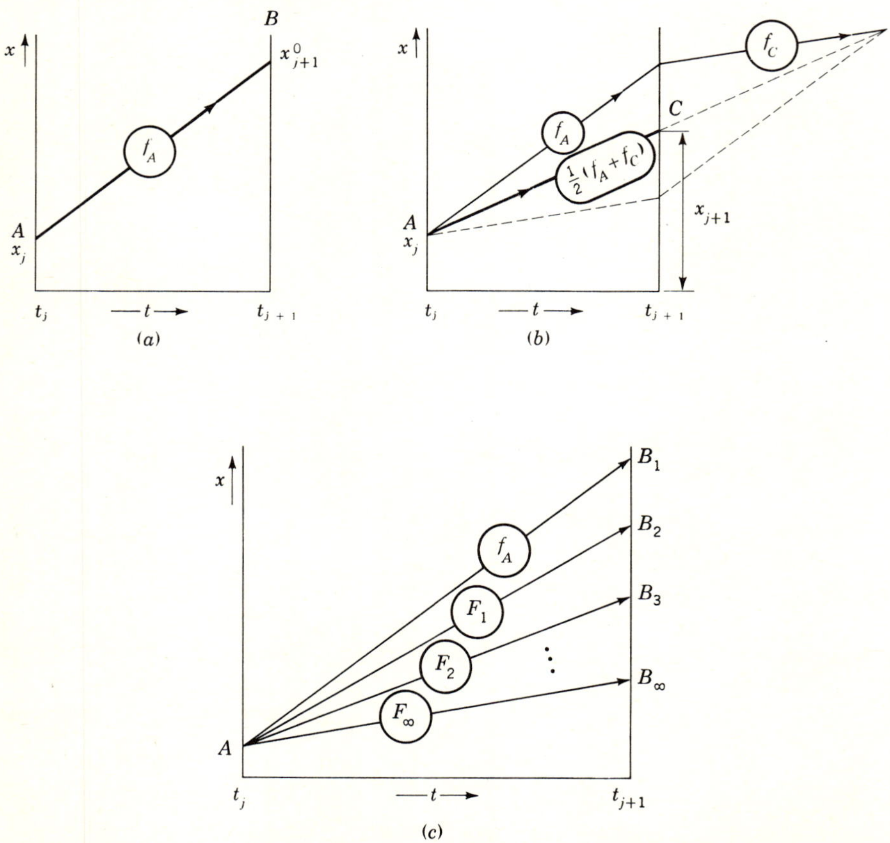

FIG. 9.3-6 Modified Euler predictor-corrector method. (a) *Predictor.* (b) *Corrector.* (c) *Iteration with* $F_i = \frac{1}{2}(f_A + f_{Bi})$.

The iteration rule, given by Eq. (19*b*), corresponds (Fig. 9.3-6*b*) to searching for a point C such that a line of slope $\frac{1}{2}f_A + \frac{1}{2}f_C$ passes from point A to C. The implicit nature of the corrector is now vividly demonstrated, for we cannot draw the line that establishes C unless we know f_C and hence C. The predictor-corrector process of Eqs. (19*a*) and (19*b*) corresponds, as shown in Fig. 9.3-6*c*, to drawing from A a line of slope f_A, thus establishing the "predicted" or initial point B_1. Thereafter points B_{i+1} ($i = 1, 2, 3$), are successively and explicitly established† by drawing from A a line of slope $\frac{1}{2}(f_A + f_{B_i})$ and finding where that line intersects the line $t = t_{j+1}$. As we stated earlier, the corrector is usually employed only once, since normally the predicted value is fairly close to the value that results after a great number of iterations.

Figure 9.3-7 describes another predictor-corrector, *the Milne method*. Notice that the predictor requires four previous values of x_j to compute the estimate x°_{j+1}. In such a situation, techniques requiring fewer previous points must be used to start the process, i.e., to accumulate four points so that the predictor of Fig. 9.3-7*a* can be used. The associated corrector (Fig. 9.3-7*b*) requires only two previous values. The number of prior points required is called the *order* of the technique.

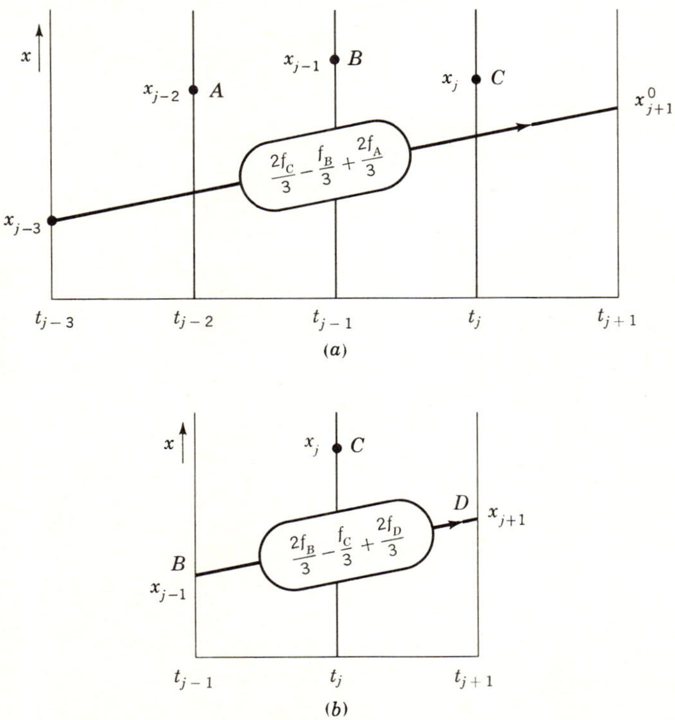

FIG. 9.3-7 Milne predictor-corrector method. (a) *Predictor.* (b) *Corrector.*

†Here we expect that the sequence $B_1, B_2, \ldots$ converges to point C. For further discussion of this convergence see the following section.

9.4 ERRORS OF OPEN AND CLOSED METHODS

In Sec. 9.3, we described explicit and implicit integration methods. In this section, we will briefly consider the relationship between these methods and a Taylor series expansion of the solution. From this discussion we will gain insight into the nature of the errors and a way of "designing" integration rules. To simplify this task we assume that in the differential equation $\dot{x} = f(x,t)$, f is a function of only x and not of t, i.e., $\dot{x} = f(x)$. The Taylor series for $x(t)$ about $t = t_j$ is

$$x(t) = x_j + \dot{x}_j(t - t_j) + \ddot{x}_j \frac{(t - t_j)^2}{2!} + \cdots \tag{1}$$

where $x_j = x(t_j)$, $\dot{x}_j = \dot{x}(t_j)$, etc. If we assume that x_j is an "exact" solution point, and we evaluate Eq. (1) for $t = t_j + h$, we have

$$x(t_j + h) = x_j + h\dot{x}_j + \frac{h^2}{2!}\ddot{x}_j + \frac{h^3}{3!}\dddot{x}_j + \cdots \tag{2}$$

A common property of explicit integration methods is that they match a number of the first terms of Eq. (2). For example, if we look at the Euler method of Fig. 9.3-3 which, in our current terminology, is

$$\begin{aligned} x_{j+1} &= x_j + hf(x_j) \\ &= x_j + h\dot{x}_j \end{aligned}$$

we see that it matches the first two terms of the Taylor series [Eq. (2)]. Heun's method (Fig. 9.3-4), which is

$$\begin{aligned} x_{j+1} &= x_j + \frac{h}{2}[f(x_j) + f(x_j + hf(x_j))] \\ &= x_j + \frac{h}{2}[\dot{x}_j + f(x_j + h\dot{x}_j)] \end{aligned} \tag{3}$$

can be rewritten, by further expanding $f(x_j + h\dot{x}_j)$ in a Taylor series about x_j. This yields for Eq. (3)

$$\begin{aligned} x_{j+1} &= x_j + \frac{h}{2}\dot{x}_j + \frac{h}{2}[f(x_j) + f'(x_j)h\dot{x}_j + \cdots] \\ &= x_j + \frac{h}{2}\dot{x}_j + \frac{h}{2}(\dot{x}_j + h\ddot{x}_j + \cdots) \\ &= x_j + h\dot{x}_j + \frac{h^2}{2}\ddot{x}_j + \cdots \end{aligned} \tag{4}$$

Evidently, this method matches the Taylor series to terms of order h^2, so that we expect an error of order h^3.

We will now use Heun's method to illustrate the derivation of an integration rule. Assume that we wish to use a rule of the form

$$x_{j+1} = x_j + \lambda_1 hf(x_j) + \lambda_2 hf(x_j + \mu hf(x_j)) \tag{5}$$

and we want to pick the real constants λ_1, λ_2, and μ to make the approximation match as many terms of the Taylor series as possible. Equation (5) can be written as

$$\begin{aligned} x_{j+1} &= x_j + \lambda_1 h\dot{x}_j + \lambda_2 h[f(x_j) + f'(x_j)\mu h\dot{x}_j + \cdots] \\ &= x_j + (\lambda_1 + \lambda_2)h\dot{x}_j + h^2\lambda_2\mu\ddot{x}_j + \cdots \end{aligned} \tag{6}$$

If we are going to match Eq. (2) we must have

$$\lambda_1 + \lambda_2 = 1 \qquad \lambda_2\mu = 1/2 \tag{7}$$

Equation (7) has many solutions. The one chosen for Heun's method, Eq. (3), is

$$\lambda_1 = 1/2 \qquad \lambda_2 = 1/2 \qquad \mu = 1 \tag{8}$$

which makes the computation easy to implement. An equally valid method, with accuracy equal to that of Heun's method, is achieved with

$$\lambda_1 = 0 \qquad \lambda_2 = 1 \qquad \text{and} \qquad \mu = 1/2$$

giving

$$x_{j+1} = x_j + hf(x_j + \frac{h}{2}f(x_j)) \tag{9}$$

To examine the error in predictor-corrector methods, we will use the modified Euler technique, given in Fig. 9.3-6 as an example. Our first question concerns the convergence of the corrector, which is given by

$$x_{j+1} = x_j + \frac{h}{2}\dot{x}_j + \frac{h}{2}f(x_{j+1}) \tag{10}$$

Since Eq. (10) is in the form $x_{j+1} = g(x_{j+1})$, we can examine g for conditions that ensure convergence. We assert that these conditions (Chap. 5) are satisfied if f is continuous and if h is sufficiently small. Our reasoning is as follows:

1. Continuity of f implies continuity of g.
2. Choose an interval $I = [x_j - \epsilon, x_j + \epsilon]$ where $\epsilon > 0$. Choose sufficiently small h, so that $\left|\frac{h}{2}f(x_j) + \frac{h}{2}f(x_{j+1})\right| < \epsilon$ for all x_{j+1} in I. This guarantees for all x_{j+1} in I, that $g(x_{j+1})$ is also in I.
3. Choose h sufficiently small (even smaller than the value chosen in 2 above, if necessary) so that for all u_1, u_2 in I

   $$\frac{h}{2}|f(u_1) - f(u_2)| < L|u_1 - u_2| \qquad \text{where } L < 1$$

   This satisfies the Lipschitz condition on g.

Therefore, the iteration rule given by Eq. (10) converges on a solution x_{j+1}. Even though convergence is assured, the solution is not generally the exact solution of the differential equation. The reader is invited to think why this is the case. Observe further that if only *one* iteration is used, i.e., if the corrector is used only once with the predicted value of x_{j+1} as initial value, then this predictor-corrector approach becomes Heun's method, which was shown to have an error of order h^3.

The question naturally arises as to which method is "best" for the solution of ordinary differential equations. As in all complex problems, there is no clear-cut answer. We can make several observations, however, which apply to explicit and implicit methods.

1. The explicit methods are "self-starting," since an initial value x_0 and $f(x,t)$ are all the information needed to compute $x_1, x_2, \ldots$.
2. The implicit methods are not self-starting, since they are generally dependent on *past values* of x_i.
3. The high-order explicit methods require several evaluations of f between x_j and x_{j+1}, before a new value x_{j+1} is established. These evaluations are used only once, since x_{j+2} requires an entirely new set of evaluations between x_{j+1} and x_{j+2}.
4. The implicit methods use past computed values of x_i in order to compute a new value, and do not therefore require intermediate evaluations of f, as in 3 above.

In practice, an explicit method is often used to get the solution started, i.e., to compute the first p values $x_1, x_2, \ldots, x_p$; then a predictor-corrector method which needs p past values of x is used to continue the solution process over the desired interval.

PROBLEMS

9.2-1 *Computer Problem.* The linear system shown in the figure is excited by

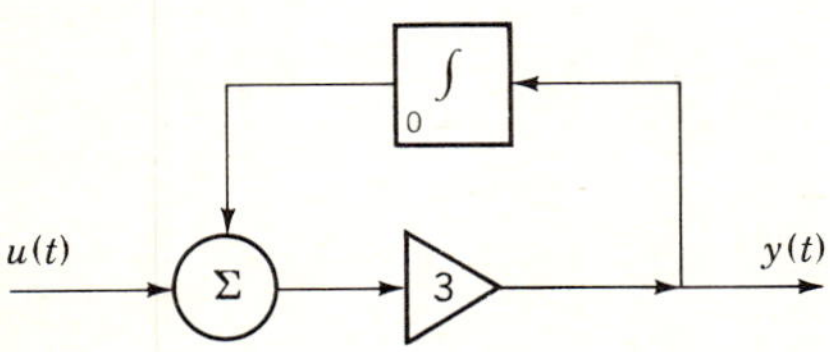

PROB. FIG. 9.2-1

an input waveform $u(t) = t^2$. Compute the response $y(t)$ for $0 \le t \le 2\pi$, using trapezoidal integration and $h = 2\pi/20$, $2\pi/60$, $2\pi/100$. Sketch the results obtained with $h = 2\pi/100$.

9.2-2 Choose the coefficients a, b, and c so that the numerical integration formula

$$2h[ay(-h) + by(0) + cy(h)]$$

gives the exact value of the integral

$$\int_{-h}^{h} y(t)\,dt$$

for each of the three functions

$$y(t) = 1$$

$$y(t) = t$$

$$y(t) = t^2$$

Comment on your result.

9.2-3 *Computer Problem.* A saturating amplifier has the input-output characteristic shown in part *a* of the figure and given by

For $u \geq 0$,

$$y = 1 - e^{-u}$$

For $u < 0$,

$$y = -1 + e^{u}$$

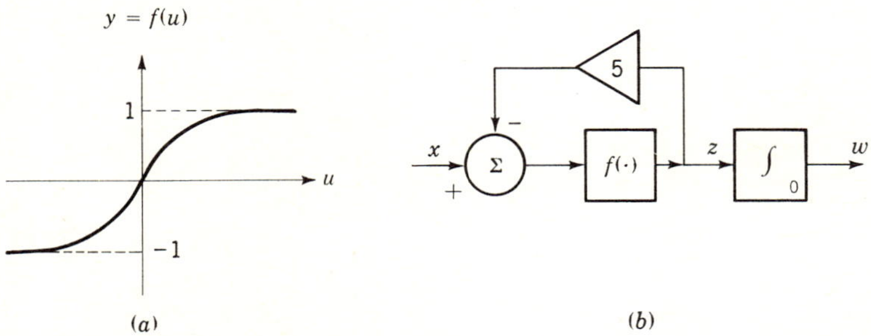

PROB. FIG. 9.2-3

The amplifier is connected as shown in part *b*. The input waveform is given by

$$x(t) = t$$

Compute and sketch accurately $w(t)$ for $0 \leq t \leq 10$.

9.2-4 *Computer Problem.* Compute the function $y(t) = \int_0^t \sin(\tau)\,d\tau$ over 20 points equally spaced in the interval $[0, 2\pi]$, using the following methods, and plot the errors for comparison:

(*a*) Euler.
(*b*) Trapezoidal.
(*c*) Simpson.
(*d*) $n = 4$.

9.2-5 *Computer Problem.* Evaluate the integral

$$\int_0^1 x^3\,dx$$

using for h the values 0.1, 0.01, and using each of the first three integration techniques, i.e., $n = 1$, $n = 2$, $n = 3$. Compare your six numerical results to the exact solution, and discuss the effect of the function x^3, the interval h, and the integration order n on the corresponding six errors.

9.2-6 *Design of an Integration Rule.* Extend the approach of Prob. 9.2-2 so that the formula

$$3h[ay(-\tfrac{3}{2}h) + by(-\tfrac{1}{2}h) + cy(\tfrac{1}{2}h) + dy(\tfrac{3}{2}h)]$$

gives the exact integral of the first four powers of t, i.e., 1, t, t^2, and t^3.

9.2-7 A manufacturer of varactors (a type of nonlinear capacitor) has the problem of characterizing his varactors in a simple yet meaningful way. It is relatively straightforward to measure the capacitance as a function of voltage, but for many purposes it is desired to know the elastance as a function of charge. The varactor is, of course, a nonlinear capacitor, so that "capacitance" means the incremental capacitance, $C = dq/dv$. Similarly, the "elastance" means the incremental elastance, $S = 1/C = dv/dq$. A certain varactor was measured at 0, 1, 2, 3, 4, and 5 V, with the following capacitances (in picofarads):

$$C(0) = 15.28 \qquad C(3) = 6.24$$

$$C(1) = 9.35 \qquad C(4) = 5.52$$

$$C(2) = 7.34 \qquad C(5) = 5.00$$

A plot of elastance versus charge can be made in the following way: For each voltage, find a corresponding charge by integrating the equation

$$dq = C(v)\,dv$$

and the corresponding elastance by $S(q) = 1/C(v)$. Set $q(0) = 0$. The only problem that remains is to decide what numerical integration scheme to use.

(*a*) Solve this problem using two schemes: rectangular (Euler) integration, and trapezoidal integration. Plot the results on the same graph for comparison.

(*b*) It was suggested that an analytical approximation to $C(v)$ be used instead. The following formula is based on semiconductor theory:

$$C(v) = C_{\min}\sqrt{\frac{v_B + \phi}{v + \phi}}$$

where $C_{\min}$ = capacitance at breakdown, v_B = breakdown voltage, and ϕ = "contact potential," which typically is less than 1 V. Note that v_B and ϕ are constant. Test this approximation by plotting $1/C^2(v)$ versus v and attempting to fit the points to a straight line (using a ruler). In the process determine the value of ϕ. You may take $v_B = 5$ V and $C_{\min} = 5$ pF. From this analytical approximation find an expression for $S(q)$, and plot this on your graph to compare with the numerical approximations.

9.3-1 Consider the linear *RC* network shown. The differential equation for charge is

$$\frac{dq}{dt} + q = 0$$

Suppose that $q = 1$ at $t = 0$. Our objective is to find $q(t)$ over 100 equally spaced integration points through several approaches.

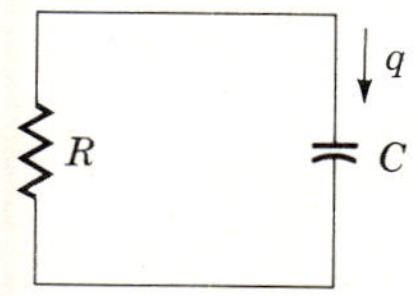

PROB. FIG. 9.3-1

(*a*) Solve the equation exactly to obtain an analytic solution.

(*b*) Solve the equation numerically by use of Euler's method, with the step size $h = 0.1$; repeat with $h = 0.01, 0.05, 0.1, 0.5$, and 3. In each case carry the numerical results on to at least five places, and compare with the exact solution.

(*c*) Repeat (*b*), using Heun's open method.

(*d*) Write a paragraph summarizing your conclusions from these calculations. You will want to compare the numerical results with the exact result, and perhaps express some opinion as to which method you prefer and how to select the step size.

9.3-2 *Exact Analysis of Rectifier: Computer Problem.* In the rectifier of alternating to direct current shown in the figure, $v(0) = 0$, and the diode characteristic is given by $i(v) = 10^{-5}\,(e^{40v} - 1)$

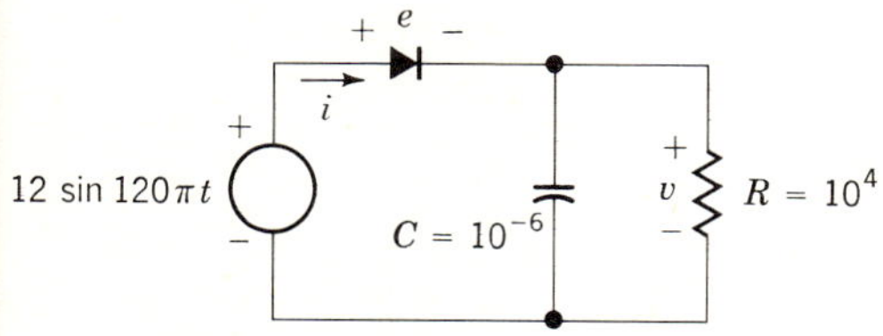

PROB. FIG. 9.3-2

(*a*) Compute and accurately sketch the output voltage $v(t)$ for the duration of the first four cycles, using 20 points per cycle and Heun's method.

(*b*) Write an expression for $v(t)$, assuming that the diode is ideal, and compare with the results of (*a*) above.

9.3-3 *Computer Problem.* Given the D.E.

$$\frac{dx}{dt} = -x^3 \qquad x(0) = 1$$

(*a*) Obtain an expression for the exact solution for $0 \leq t \leq 2$. Let $h = 0.1$, and compute over the first 20 points the solution using

(*b*) Euler's method.

(*c*) Heun's method.

(*d*) Plot the errors between (*a*) and (*b*) and between (*a*) and (*c*) above, and make any appropriate comments.

chapter 10
Finite-state Machines

The discrete counterparts of continuous dynamic systems and networks are known as sequential circuits, or more abstractly as finite-state machines. In structure, these systems have combinatorial networks (Chap. 6) for their memoryless part, and discrete storage elements, such as flip-flops, for memory. In behavior, they exhibit outputs which depend on the past history of the inputs. The chapter focuses on the input-output characterization, the logical capabilities, and the design of simple finite-state machines and sequential circuits.

10.1 INTRODUCTION

In this chapter, we will study systems that have a finite number of possible configurations or states. We will begin by examining some typical memory elements which form the basis for many digital systems. As we did in Chap. 6, we will restrict our attention to systems in which the signals are bivalued. We will then introduce and analyze a formal model of finite discrete systems—the *finite-state machine*. Finally, we will deal with the synthesis of circuits that have dynamic behavior and perform a desired input-output task.

The reasons for studying finite-state machines are many. All digital computers can be modeled by finite-state machines, although the techniques that we will discuss are, in practice, applied to less complex subsystems of large computers. Finite-state machines are also important in the study of pattern recognition problems, and information theory and coding, as well as in the study of some neurological processes. Everyday life provides many additional examples of discrete devices with dynamic behavior. Elevator controls, vending machine mechanisms, and traffic control devices are a few such examples.

10.2 DISCRETE MEMORY ELEMENTS

In our study of continuous dynamic systems, we have seen that the integrator was the fundamental memory element. The purpose of this section is to introduce the corresponding fundamental memory building block of discrete systems—the flip-flop.

The combinational gates of Chap. 6 can be used, along with some delay, which is naturally present in these gates, to construct a flip-flop. Consider, for example, the NOR gate, the input-output description of which is repeated in Fig. 10.2-1.

Consider next an interconnection of two NOR gates, as shown in Fig. 10.2-2. Here we have included two delay elements, each of delay δ, at the output of the NOR gates. Normally these delays are inherent to the physical NOR gates. In Fig. 10.2-2 we have isolated the delays from the ideal and memoryless NOR gates to facilitate our analysis.

The system of Fig. 10.2-2 will be analyzed by considering the four possible input combinations.

1. If $S = 1$ and $R = 0$, then after a time 2δ has elapsed, the only possible values for the remaining variables are $q' = Q' = 0$ and $q = Q = 1$, as the reader is invited to verify.

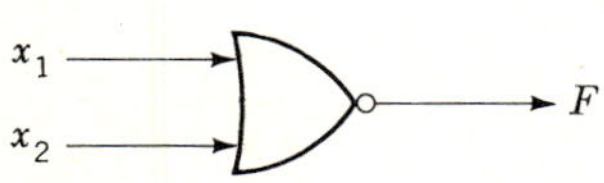

x_1	x_2	F
0	0	1
0	1	0
1	0	0
1	1	0

FIG. 10.2-1 NOR gate and input-output description.

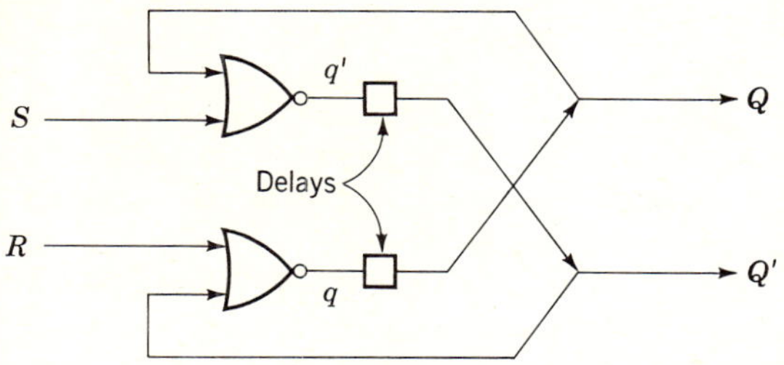

FIG. 10.2-2 Making a SET-RESET flip-flop out of NOR gates.

2. If $S = 0$ and $R = 1$, and since the system is symmetric, the only possible ultimate values are $q' = Q' = 1$ and $q = Q = 0$.
3. If $S = 1$ and $R = 1$, then $q' = Q' = q = Q = 0$.
4. If $S = 0$ and $R = 0$, then *either one* of two sets of values is ultimately possible:

$$q' = Q' = 1 \qquad \text{and} \qquad q = Q = 0 \tag{1}$$

$$q' = Q' = 0 \qquad \text{and} \qquad q = Q = 1 \tag{2}$$

It is this combination that makes the flip-flop useful, since the system will be at state (1) or (2) above, *depending on the past history* of the inputs S and R.

We illustrate this dependence of flip-flop memory on input past history as follows: Suppose that as we start a hypothetical experiment (at time

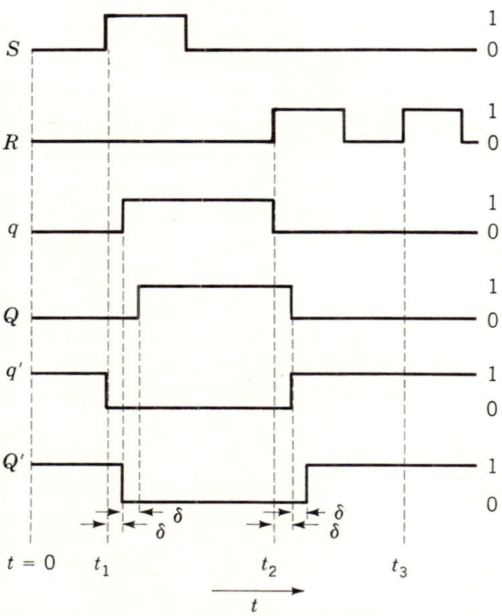

FIG. 10.2-3 Dynamic behavior of the flip-flop.

$t=0$), $S=0$ and $R=0$; suppose further that the flop-flop is in state (1) above, i.e., $q'=Q'=1$ and $q=Q=0$. All these assumptions can be seen in Fig. 10.2-3 at $t=0$.

At time t_1, input S undergoes a step from 0 to 1. This is immediately reflected at the output q' of the top NOR gate (Fig. 10.2-2), which is forced to 0. It follows that δ seconds later Q' is forced to 0; this, however, causes an immediate change of q to 1, since both inputs to the bottom NOR gate are now 0. This forces Q to 1 δ seconds later, and the flip-flop has entered state (2) above. At that time both inputs of the top gate are 1, so that q' remains at 0, and no further changes take place. When S returns to 0, there is still no change in the other variables, since Q, being equal to 1, holds the output of the top gate at 0, *regardless* of the value of S.

When at time t_2 the R input is raised to 1, a symmetric sequence of events takes place, forcing first q, and, δ seconds later, Q to 0. The latter causes q', and, δ seconds later, Q' to become 1, thereby "latching" the flip-flop in state (1). Any further change of value of R will not change the flip-flop state, as shown in Fig. 10.2-3 at time t_3. The only way the state can be changed is by raising and dropping input S.

Normally, S and R will be 0, and either one or the other will be rapidly (yet for a time greater than 2δ) raised and lowered. We will call such inputs *pulses*. Thus, the input-output behavior of the system of Fig. 10.2-2 can be summarized as follows:

The flip-flop will be in state $Q=1$, $Q'=0$ if the last pulse is at S. It will be in state $Q=0$, $Q'=1$ if the last pulse is at R. Such a flip-flop is called a SET-RESET flip-flop or *S-R flip-flop* and is given the symbol of Fig. 10.2-4. Observe that the variable Q' has been renamed $\overline{Q}$ since it is normally the logical complement of Q (this may not be the case for only a very short time when the flip-flop is changing state).

To summarize, only one of the inputs S and R is to be 1 at any time. The flip-flop remembers which line was most recently 1. If the S (or set) line was most recently 1, then $Q=1$ and $\overline{Q}=0$, and if the R (or reset) line was most recently 1, then $Q=0$ and $\overline{Q}=1$. The behavior of the flip-flop will not be specified if both S and R are simultaneously 1.

Combination Lock

In our analysis of *RLC* networks, we saw that, subject to certain constraints, the voltages on capacitors and the currents through inductors serve as state variables for those networks. That is, knowing these state variables at some time and knowing the external influences on the circuit, one can find the output or response of the circuit and the new values of the state variables. For *RLC* networks, the state variables are continuous; i.e., they can

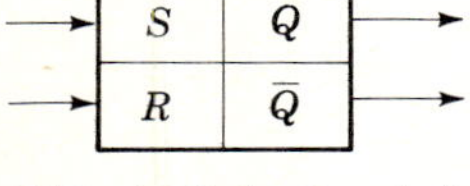

FIG. 10.2-4 Symbol of SET-RESET flip-flop.

assume any real value. Our purpose in this chapter is to analyze systems for which the state variables are constrained to take on a finite number of discrete values. Of primary interest will be the problem of determining how external inputs affect the state of the system and how the output can be determined from the state of the system.

As an example let us consider an idealized model of a combination lock, a common example of a discrete system with memory. Our lock has only three numbers 1, 2, and 3 on the dial, and the combination is R_3, L_1, R_2, i.e., the dial must be turned to the right to 3, then left to 1, and finally right to 2 in order to open the lock. At any time, there are six possible "inputs" that can be applied to the lock: R_1, R_2, R_3, L_1, L_2, L_3. The lock has two "outputs," locked and unlocked.

If a person were attempting to simulate the behavior of the lock, what would he need to remember? An initial answer might be the first, second, and third outputs. Additional thought, however, reveals that the actual value of the first input need not be remembered: the only information that is important with respect to producing the correct output is whether or not the input is R_3. We will first analyze the operation of the lock in terms of a "flow chart" of questions and answers. This chart is shown in Fig. 10.2-5. The initial question is "Is the input R_3?"

If the answer is *no,* the same question must be asked about the next input because the occurrence of R_3 must begin the input sequence that opens the lock. The response to the answer *no* is indicated by a returning arrow labeled *no.* If the answer is *yes,* the next question (about the next input) should be "Is the input L_1?" Note that if the answer to "Is the input L_1?" is *no,* the starting question must be asked again. Using similar reasoning, the remainder of the

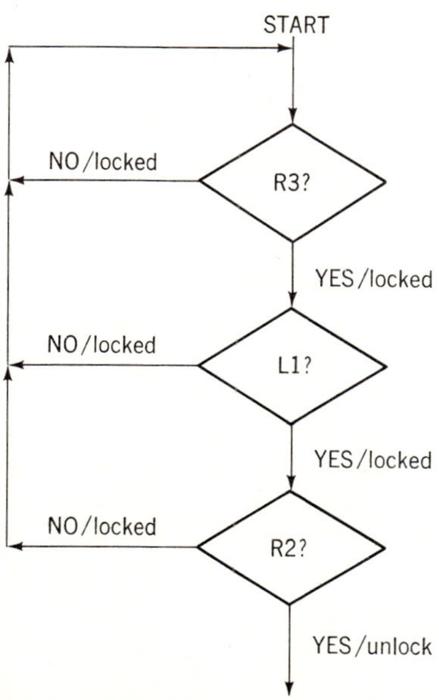

FIG. 10.2-5 Flow chart simulation of lock.

chart can be completed. In Fig. 10.2-5 we have indicated after each answer whether or not the lock is to be opened.

We are now in a position to informally interpret Fig. 10.2-5 in terms of states of the lock. On an intuitive basis, we see that the lock has three states, each state corresponding to one of the three questions of Fig. 10.2-5. For convenience we will call these state A, B, and C. We will also replace the labels "yes" and "no" by the corresponding input or set of inputs. For example, in the state "Is the input R_3?," or state A, all inputs except R_3 cause the *next state* to be A. Hence the self loop in Fig. 10.2-6 leading from A to A is labeled R_1, R_2, L_1, L_2, L_3. Input R_3 causes the next state to be B; hence the arrow from A to B is labeled R_3/ locked. Figure 10.2-6 gives the state chart that results when the substitutions are made. The reader is encouraged to verify that the input labels on the various arrows are correct.

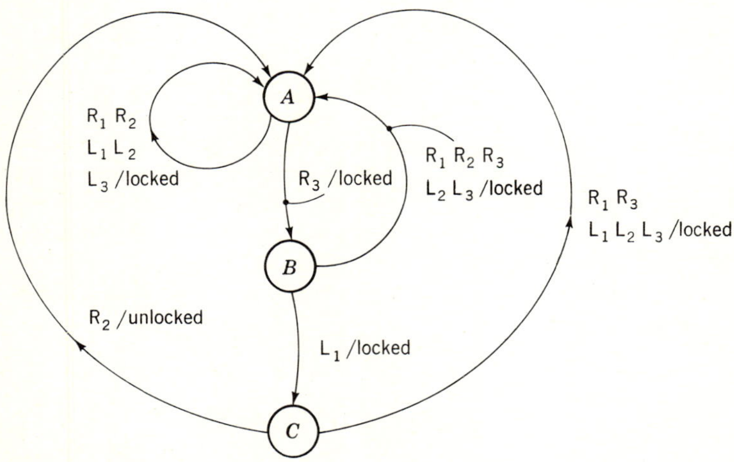

FIG. 10.2-6 Chart of states for the lock.

On an intuitive basis, it is clear that the lock being discussed can have no fewer than three distinct internal states, corresponding to A, B, and C. Using suitable bivalued memory elements such as the S-R flip-flops that we have introduced, we would need only two such elements for the total memory of the lock, since together these elements would make possible four states.

Viewed in the light of this example, the present chapter has three objectives: the first is to provide a formalism for models of the type shown in Fig. 10.2-6; the second is to present techniques for deriving such models from logic circuits; the third is to demonstrate how to derive logic circuits from the model.

The next section introduces a formal and abstract model called a finite-state machine, which will serve as a model of a class of useful digital systems.

10.3 DEFINITION OF A FINITE-STATE MACHINE

We have chosen as a basic definition of a finite-state machine one that is mathematical in nature, primarily in order to preserve generality.

Let $S = \{S_1, S_2, \ldots, S_n\}$ be a finite set called the *state set*. Let $I = \{I_1, I_2, \ldots, I_k\}$ and $O = \{O_1, O_2, \ldots, O_m\}$ be two finite sets. Let I be called the *input set* and O the *output set*. The elements of I are called the *symbols* (or characters) of the *input alphabet*, and the elements of O the symbols of the *output alphabet*.

DEFINITION 10.3-1 A finite-state machine M is an ordered set of five objects, or a *quintuple* $M = (S,I,O,f,g)$ where S, I, O are as above and

1. f is a function from $I \times S$ into S (next state function): i.e., given an input I_j and a state S_k, this function returns the next state.
2. g is a function from $I \times S$ into O (output function): i.e., given an input I_j and a state S_k, this function returns the current output.

The function f is called the *next state function* because it gives the next state of M as a function of the present state and the input. The function g is called an *output function*, because it gives the output in terms of the present state and the input. This form is exactly that of the state model of Chap. 2, and is analogous to what occurs in the state characterization of a continuous system, except that in the latter the next state is determined by a differential equation. The interpretation of f as a *next-state* function introduces implicitly another variable which does not appear in the above definition. It is perhaps natural to think of the variable to which "next" refers to as time. Finite-state machines, however, can serve as models of systems in which this variable is other than time, for example, space. For this reason, the specification of this implicit variable is assumed to take place when a particular system is modeled by a finite-state machine.

As an example of a finite-state machine, consider the quintuple $M = (S,I,O,f,g)$ specified in Fig. 10.3-1. In this figure, we have listed, for each ordered pair belonging to $I \times S$, the image under the functions f and g. This method of description is not always satisfactory, primarily because it is not concise. Two alternate forms of describing a finite-state machine are presented in Figs. 10.3-3 and 10.3-4.

$I \times S$	f
0, A	B
0, B	A
0, C	C
0, D	A
1, A	C
1, B	D
1, C	B
1, D	A

$I \times S$	g
0, A	0
0, B	0
0, C	1
0, D	1
1, A	0
1, B	1
1, C	1
1, D	0

FIG. 10.3-1 *Example of a finite-state machine description. State set* $S = \{A,B,C,D\}$. *Input set* $I = \{0,1\}$. *Output set* $O = \{0,1\}$.

The Combination Lock as a Finite-state Machine—State Tables and Transition Graphs

Let us formally interpret the combination lock of Fig. 10.2-6 as a finite-state machine. It is easy to see that the state set is $S = \{A,B,C\}$, the input set is $I = \{R_1,R_2,R_3,L_1,L_2,L_3\}$, and the output set is $O = \{L,U\}$ for locked and unlocked, respectively. The functions f and g are specified in Fig. 10.3-2. Observe that in these tables values of f and g appear as intersections of I and S members.

S \ I	R_1	R_2	R_3	L_1	L_2	L_3
A	A	A	B	A	A	A
B	A	A	A	C	A	A
C	A	A	A	A	A	A

f

S \ I	R_1	R_2	R_3	L_1	L_2	L_3
A	L	L	L	L	L	L
B	L	L	L	L	L	L
C	L	U	L	L	L	L

g

FIG. 10.3-2 Functions f and g for the lock.

Since, for a given machine, both functions f and g have as their domain the set $I \times S$, we can display the two functions in a single table called a *state table*. The table will have columns labeled with the inputs (elements of I) and rows labeled with the states (elements of S). Hence, the intersection of a "column" and a "row" represents an element of $I \times S$. Accordingly, the entry in the table at the intersection of column I_j and row S_k is the pair $f(I_j,S_k)$, $g(I_j,S_k)$. The state table for the machine of Fig. 10.3-1 is shown in Fig. 10.3-3. The state

State \ Input	0	1
A	B, 0	C, 0
B	A, 0	D, 1
C	C, 1	B, 1
D	A, 1	A, 0

Machine M_1

FIG. 10.3-3 State table for the machine of Fig. 10.3-1.

table allows one to easily determine the values of f and g. Thus the entry in input column 0, state row D, namely, A, 1 indicates that $f(0,D) = A$ and $g(0,D) = 1$. We will use the state table description extensively throughout this chapter.

Another graphic vehicle for describing a finite-state machine is the *transition graph*. A transition graph is a directed graph (i.e., a graph with directed branches) in which states are represented as nodes and the values of f and g as appropriately labeled directed branches. The transition graph for Machine M_1 of Fig. 10.3-1 is given in Fig. 10.3-4. Here, the directed branches are labeled "input/output." Thus, the branch from A to C, labeled 1/0, is interpreted as $f(1,A) = C$ and $g(1,A) = 0$. It is quite natural to think of the input label of a branch as causing the change of state and producing the associated output as the state changes.

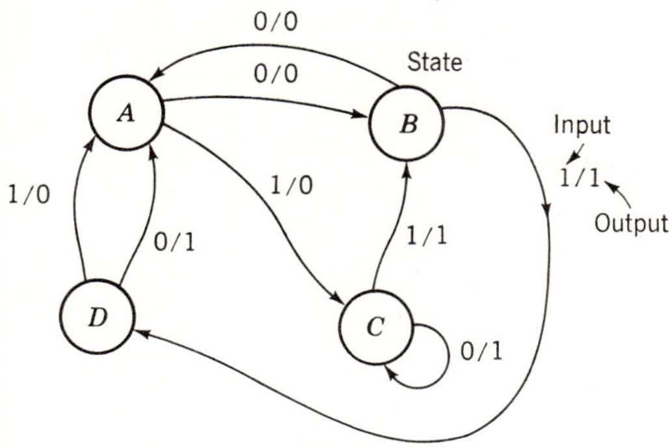

FIG. 10.3-4 Transition graph (or state diagram) for the lock.

Transition graphs, or *state diagrams*, as they are often called, are quite useful in describing the behavior of finite-state machines with few states. State tables can easily be coded into computer programs and are useful in the description of more complex machines.

Besides the state table and transition graph descriptions, finite-state machines can also be described by their input-output behavior. Input-output descriptions (Fig. 10.3-5), however, must deal with input and output *sequences*, since input values alone are not sufficient to establish output values. A more detailed discussion of input-output sequences is presented in the following section.

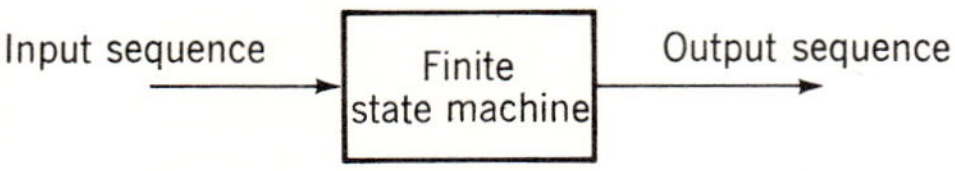

FIG. 10.3-5 Input-output description of finite-state machine.

10.4 INPUT-OUTPUT BEHAVIOR AND CHARACTERIZATION

Before we can describe the input-output behavior of a finite-state machine, we need some way of specifying the initial or starting state of the machine. Since it is clear that the behavior will depend on the starting state, we agree to always specify a starting state when we specify a machine.† *The starting state will always be the state listed first in the state table.* Thus the starting state of machine M_1 of Fig. 10.3-3 is assumed to be state A, whereas the starting state of the machine of Fig. 10.4-1*a* is B. When a machine is specified by its transition graph, the starting state will be identified by a *short unlabeled arrow,* as shown in Fig. 10.4-1*b* for the machine of Fig. 10.4-1*a*.

Viewed as a processor of input-output sequences, a finite machine is often called a *transducer*. By a sequence we mean a temporal series of sym-

†This is analogous to specifying an initial state (or condition) together with the differential equations for a continuous system.

	0	1
(Initial state) *B*	*C*,0	*B*,0
A	*B*,0	*A*,0
C	*A*,0	*B*,1

(*a*)

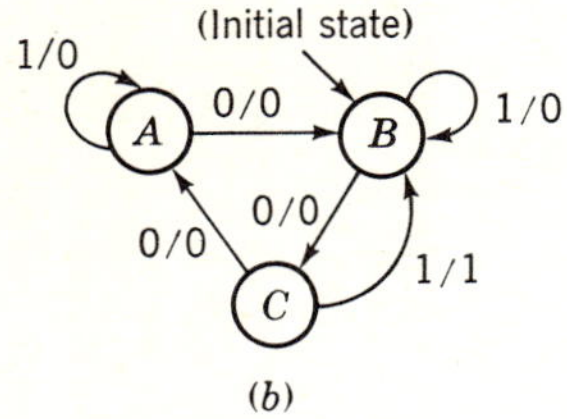

(*b*)

FIG. 10.4-1 Specification of a machine along with its initial state.

bols from the particular alphabet in question. Thus, if $I = \{0,1\}$, then 011010 is an input sequence in which we consider the first, or leftmost, symbol to have occurred earliest in time.

To illustrate the interpretation of a finite-state machine as a transducer, consider the machine of Fig. 10.4-2. If this machine is supplied with the input sequence 01111010111110, then the corresponding state sequence and output sequence will be as shown in Fig. 10.4-3.

	0	1
A	A, 0	B, 0
B	A, 0	C, 0
C	A, 0	D, 0
D	A, 0	A, 1

FIG. 10.4-2 State table of a machine.

Notice that we have written the states *between* inputs symbols, and the outputs coincident with the inputs. This is to emphasize the causal relationship among inputs, outputs, and state changes, or transition. The uniform spacing of the input symbols is not intended to reflect a uniform spacing in time. The only time relationshp required among input symbols is that they follow the ordering indicated by the input sequence, regardless of how much time elapses between adjacent symbols. Timing relations will become important when we consider physical realizations of finite-state machines in Sec. 10.6. Finally, note that the state sequence starts in state *A*, since that state appears first in the state table description of the machine, and hence is the initial state.

Suppose now that we wish to describe the output sequence produced by this machine for any arbitrary input sequence. Certainly, it is out of the question to list all the possibilities, since there is *an infinite numer* of finite-length input sequences. An alternative is to find a concise word statement that permits the derivation of the output sequence for a particular input sequence. Looking at the state table of Fig. 10.4-2, we see that an input of 0 always produces an output of 0. Thus, for any input sequence, the response to all the 0 inputs could be specified without knowledge of the state sequence. For 1 inputs the situation is somewhat more complicated. It is possible for an input 1 to

INPUT SEQUENCE: 0 1 1 1 1 0 1 0 1 1 1 1 1 0
STATE SEQUENCE: *A A B C D A A B A B C D A B A* → time
OUTPUT SEQUENCE: 0 0 0 0 1 0 0 0 0 0 0 1 0 0

FIG. 10.4-3 Input, state, and output sequences.

produce as output 1, but only if the state immediately prior to the occurrence of the input 1 is D. Continuing this line of reasoning, we see that if the input 1 is the fourth consecutive 1 after an input 0, then an output 1 will be produced. This would also be true if the input 1 were the eighth, twelfth, etc., consecutive 1 after an input 0. We conclude that *the output sequence associated with any input sequence has 1s coincident with every fourth consecutive input 1, immediately after an input 0, or immediately after the start of the sequence. The remaining outputs are all 0.*

An alternate method of input-output description, primarily useful for machines with only two outputs, will be described next. This method of description requires that the machine be interpreted as a *classifier* instead of a transducer.

Consider a finite-state machine with two symbols in its output alphabet, called for convenience 0 and 1, i.e., $0 = \{0,1\}$. Such a machine, regardless of the input alphabet, is called a *finite-state acceptor,* under the following interpretation:

DEFINITION 10.4-1 An input sequence is *accepted* by a finite-state acceptor if and only if the output produced by the *last* symbol of the input sequence is a 1. An input sequence that is not accepted is *rejected.*

DEFINITION 10.4-2 The set of input sequences accepted by a finite-state acceptor is the set of sequences *recognized* by the machine.

To see that specifying the recognized set is equivalent to specifying the input-output behavior of the machine, we argue as follows:

To specify the first output symbol, determine whether the sequence consisting of the first input symbol is accepted. To find the second output symbol, determine whether the sequence consisting of the first two input symbols is accepted. Continue this procedure in order to generate the entire output sequence.

As an example, we return to the machine of Fig. 10.4-3, which is a finite-state acceptor. The set of sequences recognized by this machine is the set of sequences *ending* with† 01^{4i} or sequences having the form 1^{4i} for $i = 1, 2, 3, \ldots$. For example, the sequences 1111, 0101111, 1011111111 are accepted, whereas 010, 0111, 01101 are not.

Now that we have seen how we can describe the input-output behavior of a given finite-state machine, we are ready to examine the "inverse" problem of designing a finite-state machine that has a desired input-output behavior. This is the topic of the following section.

10.5 DESIGN OF FINITE-STATE MACHINES

The design of discrete dynamic systems with finite memory can be divided into two parts: (1) the conversion of a desired input-output behavior into a finite-state machine, i.e., into a state table or transition graph; and (2) the realization of this state table by a network of logic gates and flip-flops—which we

†The notation 1^x is shorthand for the sequence $1\ 1 \ldots 1$, containing x 1s.

shall call *a sequential circuit.* This section discusses part 1 above. Sequential circuits are discussed in the following section.

The general problem of designing a finite-state machine that has a special input-output description is too complex to be treated here. For this reason, we shall take a more informal approach, and we shall illustrate through appropriate examples a few problems associated with the design of finite-state machines. A problem of particular interest is taking a given machine and determining whether there exists a machine with fewer states that has the same input-output behavior. This is an important problem, because it affects the complexity of the resultant sequential circuit, as we will see in Sec. 10.6.

Example 10.5-1 *Recognizing Three or More Ones* Our first synthesis example begins with the specification of the recognized set. This set consists of *exactly those sequences that contain at least three consecutive 1s.* The input set is {0,1}, and since the problem is stated in terms of recognition, we may assume that the output set is also {0,1}. Our general approach in attempting to construct the required state table will be to add states only when we must do so to obtain correct behavior, until no further states are necessary. The machine must start in some state, to which we give the name *A* (Fig. 10.5-1*a*).

	0	1
A	*A*, 0	*B*, 0

(*a*)

	0	1
A	*A*, 0	*B*, 0
B	*A*, 0	*C*, 0

(*b*)

	0	1
A	*A*, 0	*B*, 0
B	*A*, 0	*C*, 0
C	*A*, 0	*D*, 1

(*c*)

	0	1
A	*A*, 0	*B*, 0
B	*A*, 0	*C*, 0
C	*A*, 0	*D*, 1
D	*D*, 1	*D*, 1

(*d*)

FIG. 10.5-1 Progressive construction of a state table.

If the first input is a 0, the machine can remain in state *A* while producing an output of 0. This is possible because as long as only 0 inputs occur, the machine need not "remember" anything. If, however, the first input is a 1, the machine cannot remain in state *A* since such an action would ignore the information that an input 1 had occurred. Hence, under a 1 input we "send the machine" to a new state, and the corresponding entry is listed as *B*, 0 in the state table. Using similar reasoning, the entries for state *B* are filled as shown in Fig. 10.5-1*b*. At this point, we have had to add a third state *C* to "remember" the occurrence of a second consecutive 1. The entries for state *C* are shown in Fig. 10.5-1*c*. Note that state *D* remembers the occurrence of at least three consecutive 1s, as seen from Fig. 10.5-1*d*. The reader should verify that the machine of Fig. 10.5-1*d* recognizes exactly the set of sequences containing at least three consecutive 1s.

Example 10.5-2 *Recognizing Exactly Three Ones* Here, we require an output of 1 to be produced coincident with the first 0 following a block† of *exactly* three 1s. Using the same type of reasoning as before, we can easily develop the partial state table of Fig. 10.5-2*a*. This machine will enter state *D*, if exactly three 1s have occurred since the last 0, or since the beginning of the input sequence. If a 0 occurs when the machine is in state *D*, an output of 1 should be

(*a*)

	0	1
A	A, 0	B, 0
B	A, 0	C, 0
C	A, 0	D, 0

(*b*)

	0	1
A	A, 0	B, 0
B	A, 0	C, 0
C	A, 0	D, 0
D	A, 1	

(*c*)

	0	1
A	A, 0	B, 0
B	A, 0	C, 0
C	A, 0	D, 0
D	A, 1	E, 0
E	A, 0	E, 0

FIG. 10.5-2 State table for Example 10.5-2.

produced and the machine can return to state *A*, as shown in Fig. 10.5-2*b*. If while in state *D* an input of 1 occurs, the machine cannot go to any of the listed states, and so another state, *E*, must be added. The complete table is shown in Fig. 10.5-2*c*.

Example 10.5-3 *A Serial Adder* As a final example, consider a machine that has the input alphabet {00,01,10,11} and the output alphabet {0,1}. The machine is to implement addition, in the following sense. Consider the binary addition of

1	0	1	1	1	0	0	A
1	1	1	0	0	1	0	B
1	0	0	1	1	1	0	Z

FIG. 10.5-3 Sequences for Example 10.5-3.

two sequences, *A* and *B*, shown in Fig. 10.5-3. The desired machine should produce the *i*th bit of the sum, i.e., Z_i, when presented with the input symbol representing the *i*th bits of the two numbers being added, i.e., A_i, B_i. Thus, for the example of Fig. 10.5-3, the input-output sequences will be as shown in Fig. 10.5-4. Observe that the state table for this machine will have four columns,

INPUT:	00	01	10	10	11	01	11
OUTPUT:	0	1	1	1	0	0	1

→ time

FIG. 10.5-4 Input-output sequences from Fig. 10.5-3.

†A block of 1s is a continuous sequence of 1s occurring in an input sequence. If a block is of length *k*, then it consists of *k* consecutive 1s bounded on the left by either the beginning of the input sequence or a 0, and on the right by a 0 or the end of the input sequence.

one for each input symbol. A little thought reveals that all we have to remember, by being in the appropriate state, is whether or not a "carry" was generated by the previous addition. The starting state can correspond to a no-carry situation. Realizing this, the state table can be easily generated, and is shown in Fig. 10.5-5. This machine is called a *serial adder* and is a widely used component of digital

	00	01	10	11	
A	$A, 0$	$A, 1$	$A, 1$	$B, 0$	(no carry)
B	$A, 1$	$B, 0$	$B, 0$	$B, 1$	(carry)

FIG. 10.5-5 State table of serial adder.

systems. Note that if two n-bit numbers are added, an $(n + 1)$-bit number can result. Since the machine of Fig. 10.5-5 would produce only n output bits in this situation, we need to be able to determine what the $(n + 1)$st bit should be. This can be accomplished by either applying input 00 at the end of the sequence and by observing the output, or by observing the state of the machine after the generation of the first n output bits. If the state is A, the $(n + 1)$st output bit is 0, and if the state is B, the $(n + 1)$st output bit is 1.

We have seen in the preceding examples that our informal approach to the synthesis of a state table involves the following steps:

1. Let A be the starting state (first row).
2. From that state, consider every possible input, and determine whether the machine can (*a*) go to itself or to one of the states above it in the table, or (*b*) go to a new state. If the latter is necessary, designate new states, as required, and append them at the bottom of the state table.
3. Proceed to the next state (row) of the table and go to Step 2 above. If there are no further states to be considered, the procedure terminates.

State Equivalence

Although the preceding design examples were simple, the reader can imagine that a major problem encountered in the design of finite-state machines is deciding when the machine should go to a state already present in the partial table and when it should go to a new state. To cope with this problem, we shall allow the specification of redundant machines, i.e., machines with too many states, and we shall develop a test to determine whether some states can be eliminated from the table without changing the input-output behavior. As we shall see in Sec. 10.6, a reduction in the number of states of a finite-state machine generally leads to a reduction in the complexity of the corresponding sequential circuits.

Example 10.5-4 *Two Equivalent Machines* Consider the two machines shown in Fig. 10.5-6. The reader can verify that both machines recognize the same set, or equivalently, perform the same transduction; yet M_1 has four states

	0	1
A	B, 0	C, 0
B	D, 1	A, 0
C	D, 1	B, 1
D	D, 1	A, 0

M_1

(a)

	0	1
A	B, 0	C, 0
B	B, 1	A, 0
C	B, 1	B, 1

M_2

(b)

FIG. 10.5-6 Two equivalent machines.

while M_2 has only three. In the state table for M_1 we see that the entries for states B and D are identical, *in both their next states and their outputs*. This means that whenever machine M_1 is in state B, its future behavior for all time will be the same as though it were in state D. Carrying this reasoning a step further, we could replace every transition to state D with a transition to state B, without changing the observable input-output behavior of the machine. If this replacement is made, we obtain the state table shown in Fig. 10.5-7*a*. The transition diagram, Fig. 10.5-7*b*, for this machine reveals that there are no transitions *into*

	0	1
A	B, 0	C, 0
B	B, 1	A, 0
C	B, 1	B, 1
D	B, 1	A, 0

(a)

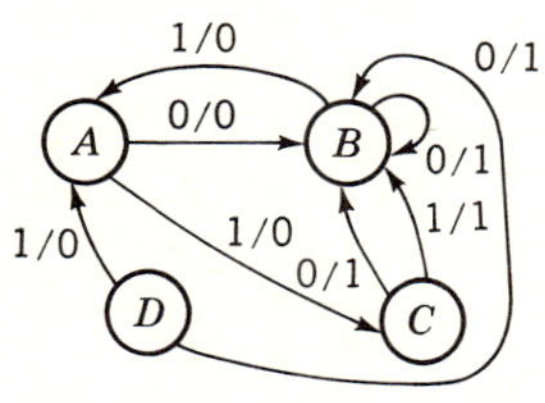

(b)

FIG. 10.5-7 (a) *State table, and* (b) *State diagram of* M_1 *with transitions to D replaced by transitions to B.*

state D. Since the starting state is A, the machine can *never* be in state D, and hence state D can be eliminated, resulting in machine M_2 of Fig. 10.5-6*b*.

The observation that led to the elimination of state D in the above example was that states B and D were the same as far as the input-output behavior of the machine was concerned. At this point we can formalize our reasoning and define state equivalence for finite-state machines.

DEFINITION 10.5-1 Two states S_i and S_j of a finite-state machine M are *equivalent* if and only if the input-output behavior of M, starting in state S_i, is identical to the behavior of M, starting in state S_j, under all input sequences. If S_i and S_j are equivalent, we write $S_i = S_j$.

If two states are equivalent, according to Definition 10.5-1, it should be clear that the substitution used in the previous example can be employed to eliminate either state from the machine, while retaining the same input-output behavior. From Definition 10.5-1, it follows that a necessary condition for two states to be equivalent is that their respective outputs be the same under each input. Thus, states A and B of M_2 in Fig. 10.5-6*b* are not equivalent because M_2 starting in A gives a 0 output in response to a 0 input, whereas M_2 starting in B gives a 1 output for a 0 input.

Example 10.5-5 For machine M_1 in Fig. 10.5-6*a*, it was easy to determine that states B and D were equivalent because their row entries were identical. The process of identifying equivalent states is not always as simple, as we shall see for the two machines shown in Fig. 10.5-8*a* and *b*. These two machines have identical input-output behavior. Yet it is not obvious that any states of M_1 can

	0	1
A	$B, 0$	$C, 1$
B	$B, 1$	$D, 0$
C	$C, 1$	$E, 0$
D	$A, 1$	$B, 1$
E	$A, 1$	$C, 1$

M_1
(*a*)

	0	1
A	$B, 0$	$B, 1$
B	$B, 1$	$C, 0$
C	$A, 1$	$B, 1$

M_2
(*b*)

FIG. 10.5-8 Equivalent machines of Example 10.5-5.

be eliminated. Certainly, no two rows have identical entries, so that such a simple procedure will not work. Focusing on the outputs† we see that B and C *could* be equivalent and D and E *could* be equivalent, but no state can be equivalent to A. Observe now that D and E would be equivalent if B and C were equivalent. Certainly there is no problem with an input of 0 since both D and E have identical entries. If the input is 1, D goes to B while E goes to C. Thus, if B and C were equivalent, we would be assured that D and E are equivalent. Focusing next our attention on states B and C, we see that they would be equivalent if B and C, and D and E were equivalent. At this point, we have achieved consistency, and we conclude that $B = C$ and $D = E$. The reader should verify that replacing all transitions to C with transitions to B, replacing all transitions to E with transitions to D, and eliminating states C and E gives machine M_2, shown in Fig. 10.5-8*b*.

The procedure used above can be summarized as follows. First, we reasoned that two states are equivalent if and only if (1) their respective outputs for each input are identical and (2) under all inputs they make transitions to the same *or* equivalent states. Then we assumed that $D = E$ and saw that this implied that $B = C$, which in turn implied that $B = C$ and $D = E$. Hence, the assumption that $D = E$ produces no contradictions. Thus, we are able to eliminate states C and E from the state table of M_1 without changing the input-output behavior. Before we can present a more efficient procedure for detecting state equivalence, it is necessary to introduce some additional terminology.

DEFINITION 10.5-2 Two states S_i and S_j of a finite-state machine are *K-equivalent* if and only if there exists no input sequence of length K or less that when applied to M starting in S_i, produces an output sequence different from that produced when the same input sequence is applied to M starting in S_j.

Using Definitions 10.5-1 and 10.5-2, we have the following self-evident theorem:

†That is, checking whether the outputs of two states are the same under a 0 input and under a 1 input.

THEOREM 10.5-1 Two states S_i and S_j of a finite-state machine are equivalent if and only if they are K-equivalent for all $K \geq 1$.

DEFINITION 10.5-3 Two states that are not equivalent will be called *distinguishable.*

It is clear from Definition 10.5-2 that if two states are K-equivalent, then they are l-equivalent for $l \leq K$.

We shall now present an equivalence testing procedure, using the machine of Fig. 10.5-9 as an example. The first step is to group together all states that are *one-equivalent.* It is clear that each state in every such group

	0	1
A	B, 0	C, 0
B	C, 0	E, 0
C	A, 1	F, 0
D	G, 0	F, 0
E	F, 0	G, 0
F	D, 1	C, 0
G	C, 0	B, 0

FIG. 10.5-9 Machine used for equivalence testing procedure.

must have the same output under a given input. This grouping, called G_1, is shown in Fig. 10.5-10. Observe that our machine has only two groups of states that are 1-equivalent.

G_1: [*ABDEG*] [*CF*]
G_2: [*AD*] [*BEG*] [*CF*]
G_3: [*AD*] [*BEG*] [*CF*]
Equivalence Groups

FIG. 10.5-10 The equivalence-testing procedure.

The next step is to form a similar grouping of all states that are 2-equivalent. This grouping is called G_2. A moment's reflection will reveal that *states are in the same group of G_2 if and only if they are in the same group of G_1 and under each input, they make transitions to members of the same group of G_1.* This is because if two states are 2-equivalent, then they themselves must be 1-equivalent and the states they make transitions to must also be 1-equivalent. Thus, states A and D are together in G_2 because they are together in G_1 and because B and G, and C and F are together in G_1. The reader can easily verify that the remaining groups satisfy the required condition.

Continuing this reasoning, we form G_3, in which states are in the same group if and only if they are 3-equivalent. This implies that they are together in G_2 and that for each input they make transitions to states in the same group of G_2. At this point we see that G_3 is identical to G_2. By our construction, G_4 will then be identical to G_3, G_5 to G_4, etc. Since the states grouped together in G_K represent precisely those states that are K-equivalent and since $G_K = G_2$

for $K \geq 2$, we deduce that $A = D$, $B = E = G$, and $C = F$. The reader should now be able to construct a machine with three states which has identical input-output behavior to that of the machine of Fig. 10.5-9.

In Prob. 10.5-9, the reader is asked to show that for an n-state machine, G_n must be identical to G_{n-1}. This fact implies that those states that appear together in G_{n-1} are precisely those states that are equivalent. The problem of finding equivalent states has then a simple and relatively compact solution. A finite-state machine for which no two states are equivalent is called a *reduced* machine.

Partially Specified Machines

When a state table or transition diagram is derived from a practical application, the symbols of the input alphabet represent combinations of physical variables. It may well be the case that the source of these input variables is not capable of producing certain sequences of input symbols. As an example, assume that input signals are being received indicating when a door has just opened or just closed. One would not expect an input sequence indicating that the door had opened twice in succession without ever closing in between. In this case the particular sequence cannot occur because of an obvious physical constraint. In more complex examples the input constraints can be of such a nature that they escape immediate detection.

How might such constraints reveal themselves in a state table? There will be certain states of the machine for which certain inputs cannot occur. This situation is indicated in Fig. 10.5-11. The missing entry for state B, input 0, means that if the machine is in state B, an input of 0 cannot occur.

	0	1
A	A, 0	B, 1
B	—	C, 0
C	A, 1	B, 0

FIG. 10.5-11 A partially specified machine.

Such an entry is called a *don't care*† because the person who specified the state table is not concerned about this particular transition. In other words, it is guaranteed that such a (state, input) combination will never be encountered when a realization of the machine is operated in the environment for which it was designed. A finite-state machine with one or more "don't cares" is called a *partially specified machine.*

We will see shortly that before any physical realization of the finite-state machine can be designed, it is necessary for the don't care entries to be specified. A reasonable way is to choose entries so as to minimize the "complexity" of the resulting machine. The only tractable measure of complexity is the number of states in a machine. The manner in which a don't care is specified

†We have encountered don't cares in Chap. 6.

can affect the number of states in the resulting reduced machine. If, for example, in the table of Fig. 10.5-11, the missing entry is specified as B, 0 the resulting machine has no equivalent states. If, on the other hand, the entry is chosen as A, 1, states B and C are equivalent and the machine can be reduced by one state. Finding a reduced partially specified machine is a very complex problem which will not be discussed here.

Properties of Reduced Machines

In this section we will state briefly some additional results that are important in the design of finite-state machine. Fully specified† machines that have identical input-output behavior are called equivalent machines. Two such equivalent machines cannot be essentially different. That is, if both machines are *reduced*, they can differ only in *the naming of their states*—they are then called *isomorphic*. The implication of this result is that *there is really only one way (state table) to accomplish every task that a finite-state machine is capable of doing*. The reader who continues to study automata theory will find that such algorithmic uniqueness is not a general property of computational processes or of machines.

The converse of the above result is, of course, also true. Specifically, if two machines are reduced and found to be isomorphic, then they must have the same input-output behavior. The only way in which we have to qualify this statement is with regard to starting states. That is, for any starting state of either machine, there exists a starting state of the other machine so that the two machines have the same input-output behavior.

At this point, we are ready to design physical realizations of finite-state machines. Though we will confine ourselves to circuit realizations, the reader should be aware that finite-state machines can also be realized by other means, e.g., mechanically, hydraulically, or by a computer program.

10.6 SEQUENTIAL CIRCUITS

The general form of the sequential circuits with which we will realize finite-state machines is shown in Fig. 10.6-1. It is the familiar form of the state model.

The memory is a subsystem capable of assuming a *finite* number of distinct configurations or states; the flip-flops, discussed in Sec. 10.2, are typical discrete memory elements. Information uniquely specifying the present state of memory is fed to the control portion of the realization by the lines labeled Y. The *control* is a memoryless system consisting of combinational logic. When an input occurs, the control circuit uses the input and the information about the "present state" (on the Y line) to produce outputs and, state-update signals y which determine the next state of the memory. Thus the control portion of the system implements the functions f and g of Definition 10.3-1.

†A machine without don't care entries is called a *fully specified* machine.

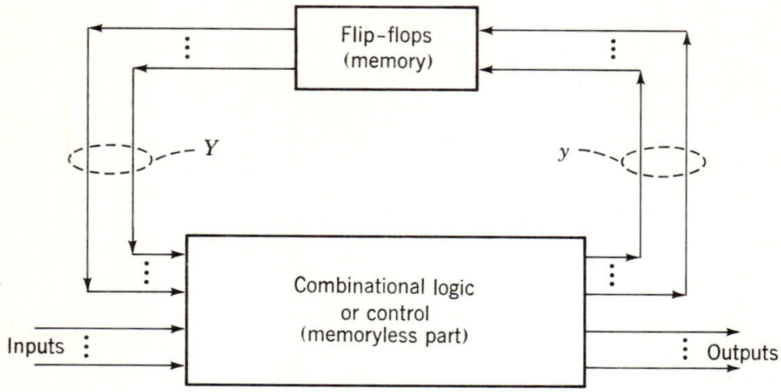

FIG. 10.6-1 General form of a sequential circuit.

In the remainder of this chapter, we will consider two specializations of Fig. 10.6-1. In the first of these, the input and output signals are pulses and the memory elements are SET-RESET flip-flops of the type discussed in Sec. 10.2. This specialization is easy to understand but not widely used in practice. In the second specialization, the input and output signals are conventional logic levels, discussed in Chap. 6, and the memory elements are the so-called *J-K* flip-flops, which will be introduced later in the section. These realizations represent a large segment of practical sequential circuits but are somewhat more involved than the previous type.

For the first specialization, the occurrence of an input will be represented by the appearance of a pulse on an associated input line. These pulses are not to be interpreted as impulses† and we have purposely not described the physical variable by which they are represented—this variable can be either voltage or current. Thus, if a finite-state machine description has two input columns X_1 and X_2 (or 0 and 1), a corresponding sequential circuit will have two input lines, X_1 and X_2. A pulse on the X_1 line will represent an input of X_1, and a pulse on the X_2 line an input of X_2. Thus, the input sequence $X_1X_1X_2X_1X_2$ would be represented as shown in Fig. 10.6-2. Note that *we do not allow two input pulses to occur simultaneously,* because slight delays between them could result in difficulties within the control logic. If a finite-state machine has two output values, say 0 and 1, then these values will be interpreted as follows: An output

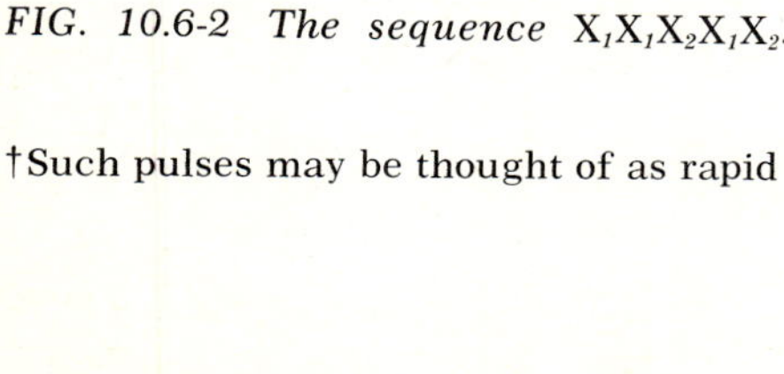

FIG. 10.6-2 The sequence $X_1X_1X_2X_1X_2$.

†Such pulses may be thought of as rapid transitions of a signal from 0 to 1 and back to 0.

value of 0 will indicate that *no pulse is to appear on the output line,* while an output value of 1 will indicate that *a pulse is to appear on the output line.* For machines with a larger output alphabet, we can think of coding the output symbols in a binary fashion, having as many output lines in the realization as there are binary digits, and applying the above interpretation to each line.

The control logic will then receive pulse inputs from the input lines and level inputs from the memory (the flip-flop outputs). This logic must, in turn, generate pulses on the appropriate output line (or lines) and state-update pulses which cause the memory to change to the required next state. Such a logic circuit which combines pulses and levels is usually called *pulse steering logic,* because, as we will see, the level inputs from the memory cause various input pulses to be "steered" to the appropriate output lines. At the logical design level, steering logic is described and implemented, using the gates of Chap. 6. The specific gates that we will use in this section will be ANDs, ORs, and INVERTERS. We introduce next some restrictions on the way in which these gates may be interconnected when their inputs are a mixture of pulse and level signals. While a pulse is simply a rapid succession of two level changes, care must be exercised in the logical processing of pulse signals. For example, if we attempt to AND two or more pulse signals, we would expect a pulse to occur at the gate output only if there were pulses on all the gate inputs. The problem arises in situations where logically there are pulses on all input lines, but because of delay differences prior to such AND gates, the pulses arrive at slightly different times. These time differences may be enough to prevent the occurrence of a pulse on the output line, even though logically there should have been one. *For this reason, two or more pulse lines are never ANDed together. Two or more pulses can be ORed and a single pulse line can be ANDed with an arbitrary number of level signals.* In the former case an output pulse occurs when a pulse occurs on either or both input lines; and in the latter case an output pulse occurs only when all level inputs are high and an input pulse occurs.

To summarize, in a sequential circuit containing *S-R* flip-flops, the state variables are *levels,* while the output and state-update variables are *pulses* which are coincident with the input pulses.

With the above restrictions specified, we can now analyze the operation of an actual circuit. Figure 10.6-3 shows a sequential circuit with two pulse inputs X_1 and X_2 and a single output Z. First, observe that the circuit has

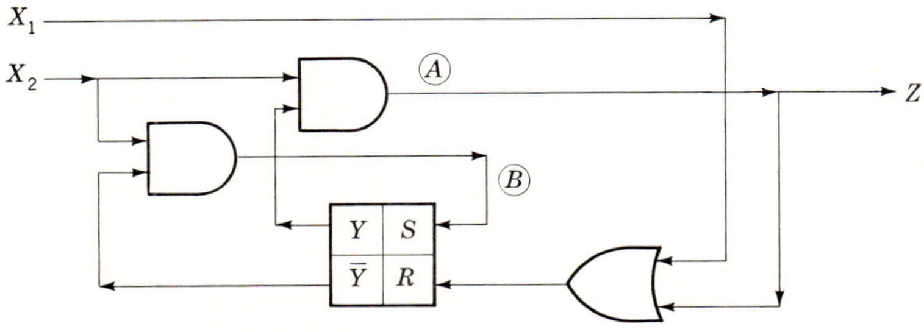

FIG. 10.6-3 A one-flip-flop sequential circuit.

the form shown in Fig. 10.6-1 and that the interconnections meet all the specified restrictions. The behavior of the circuit when an input occurs will generally depend on the "state" of the flip-flop. Recall that the state of the flip-flop will be SET if the S line was pulsed last or RESET if the R line was pulsed last. Thus SET corresponds to $Y = 1$ and RESET to $Y = 0$. The effect of an input pulse can be divided into two parts: (1) does it change the state of the flip-flop; and (2) does it produce an output pulse? We will analyze the circuit by answering these questions for both possible states of the flip-flop, under all possible inputs.

If the flip-flop is RESET ($Y = 0$) and an X_1 pulse occurs, then this X_1 pulse appears directly at the reset input of the flip-flop and does not produce an output pulse. Since the flip-flop is already RESET, no state change will occur. This is summarized in the entry for state RESET with an X_1 input in the table of Fig. 10.6-4. If an X_2 input pulse occurs, we must determine which, if any, of the

	INPUT	
	X_1	X_2
RESET	RESET/NO	SET/NO
SET	RESET/NO	RESET/YES

FIG. 10.6-4 State table for the sequential circuit of Fig. 10.6-3.

AND gate outputs exhibits a pulse. Since the flip-flop is RESET, the Y output is 0, and no pulse appears at point A. The $\overline{Y}$ output is 1, however, and a pulse does appear at point B, causing the flip-flop to change to the SET state. Observe now that if the flip-flop changes state in zero time, difficulties would be encountered. This is the case because the $\overline{Y}$ output would change to 0 instantaneously, although it was assumed to be 1, in order to allow the pulse to appear at point B. Fortunately, actual flip-flops do require a small nonzero time to change state, so that the difficulty is avoided. With this assumption, we can now enter SET/NO for state RESET under input X_2 in Fig. 10.6-4.

If we now assume that the flip-flop is SET and that an X_1 input pulse occurs, we see that the behavior is the same as though the flip-flop were RESET, namely, RESET/NO. Finally, for an X_2 input pulse, we see that the flip-flop is RESET and a pulse occurs on the output line. This completes the table of Fig. 10.6-4, which completely describes the behavior of the circuit of Fig. 10.6-3.

Since the names SET and RESET simply describe conditions of the flip-flop, they may be renamed as A and B. In addition, 0 can be substituted for NO, and 1 for YES, for the output specification. These changes give rise to the state table of Fig. 10.6-5, which is now in the familiar finite-state machine form.

	X_1	X_2
A	$A/0$	$B/0$
B	$A/0$	$A/1$

FIG. 10.6-5 State table for the circuit of Fig. 10.6-3.

The sequential circuit of Fig. 10.6-3 is called a *realization* of the finite-state machine shown in Fig. 10.6-5. Other state names could have been chosen, or A could have been associated with RESET and B with SET. All state tables so derived would be, nevertheless, isomorphic to the state table of Fig. 10.6-5; i.e., they could be obtained from Fig. 10.6-5 by a renaming of symbols.

The above example illustrates that the general form of Fig. 10.6-1 can be refined as shown in Fig. 10.6-6.

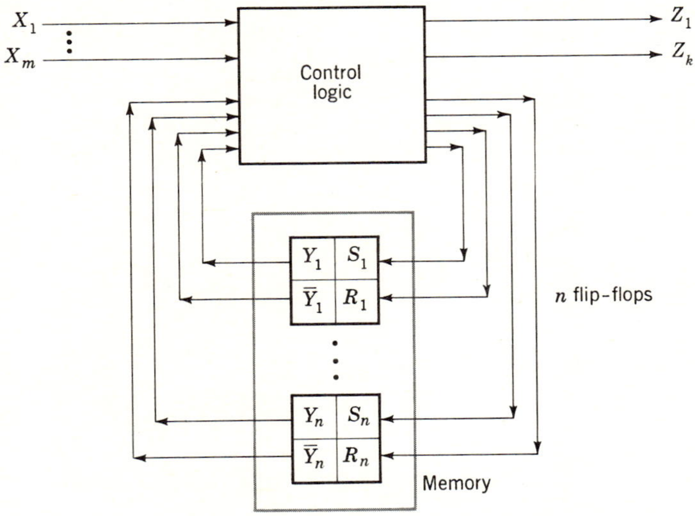

FIG. 10.6-6 General form of sequential circuit with SET-RESET flip-flop memory.

Since the sequential circuit of Fig. 10.6-6 has n flip-flops, it follows that it has 2^n states. In principle, any such circuit could be analyzed in exactly the same way as the simple circuit of Fig. 10.6-3. However, as the number of states increases, this method becomes very tedious. To indicate how we can avoid exhaustive state-by-state analysis, and at the same time to provide some tools for sequential circuit synthesis, we shall present a more formal analysis approach through the circuit of Fig. 10.6-7.

Our first objective in analyzing this circuit is to determine under what circumstances pulses will appear at inputs S_1, R_1, S_2, R_2, and Z. This information is enough to specify the state transitions and output of the circuit. To keep our representations concise, we shall use Boolean expressions as follows: Consider the input R_2; whether or not an X_1 pulse causes a pulse at R_2 is determined only by the present values of Y_1 and Y_2. In particular, we can write a Boolean function $f_1(Y_1,Y_2)$ that is 1 if and only if an X_1 pulse causes a pulse at R_2. Examination of the circuit reveals that $f_1(Y_1,Y_2)=Y_1$. A similar function, $f_2(Y_1,Y_2)$, can be specified in order to relate occurrences of X_2 pulses to R_2 pulses. For the circuit of Fig. 10.6-7, $f_2(Y_1,Y_2) = 1$; i.e., an X_2 pulse *always* causes a pulse at R_2. We shall interpret the Boolean expression

$$R_2 = X_1Y_1 + X_2$$

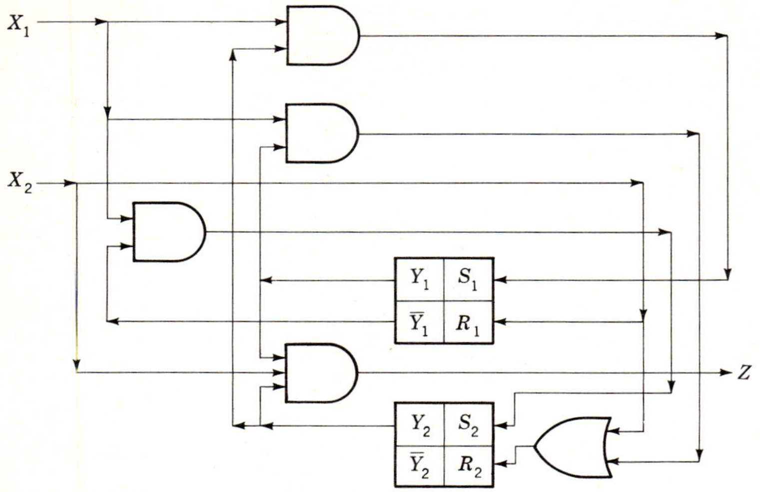

FIG. 10.6-7. Sequential circuit to be analyzed in terms of Boolean expressions.

as representing the circumstances under which input pulses appear at R_2. If the presence of a pulse on X_1 is interpreted logically as $X_1 = 1$ and the expression for R_2 as a Boolean expression, then when $X_1 = 1$ and $X_2 = 0$, R_2 will be logical one if and only if a pulse appears at Y_1. The constraint in the above expression is that, at most, one of the X_1 and X_2 can be 1. In a similar fashion, we can write for the remaining variables of interest the following expressions:

$$
\begin{aligned}
S_1 &= X_1Y_2 \\
R_1 &= X_2 \\
S_2 &= X_1\overline{Y}_1 \qquad (1) \\
R_2 &= X_1Y_1 + X_2 \\
Z &= X_2Y_1Y_2
\end{aligned}
$$

In terms of the general form of a sequential circuit (Fig. 10.6-6), each of the above expressions assumes the general form

$$S_j = X_1 f_1{}^j(Y_1, \ldots, Y_n) + X_2 f_2{}^j(Y_1, \ldots, Y_n) + \cdots + X_m f_m{}^j(Y_1, \ldots, Y_n) \qquad (2)$$

where the $f_1{}^j(Y_1, \ldots, Y_n)$ are Boolean expressions in the above sense. The important characteristic of such an equation is its form—a sum of terms, each of which is an X_i ANDed with a function involving only Y variables.

The expressions in Eqs. (1) can be compactly represented, using a Karnaugh map, as shown in Fig. 10.6-8.

In the X_1 column the functions mapped are S_1, R_1, S_2, R_2 and Z, each with the constraint that $X_1 = 1$ and $X_2 = 0$. In the X_2 column the same functions are mapped with the constraint that $X_1 = 0$, $X_2 = 1$. Looking at the general form of Eq. (2), we see that the portion of S_j mapped in Column X_k will be S_j constrained by $X_i = 0$, for all $i \neq k$, and $X_k = 1$, i.e., $f_k{}^j(Y_1, \ldots, Y_n)$, the *steering function* between input X_k and S_j. The table of Fig. 10.6-8 is called an *excitation ma-*

	Y_1Y_2	X_1	X_2
(A)	0 0	00, 10, 0	01, 01, 0
(B)	0 1	10, 10, 0	01, 01, 0
(C)	1 1	10, 01, 0	01, 01, 1
(D)	1 0	00, 01, 0	01, 01, 0

S_1R_1, S_2R_2, Z

FIG. 10.6-8 Excitation matrix of the sequential circuit of Fig. 10.6-7.

trix because it indicates how the flip-flop inputs and system outputs are excited.

The entry in the map of Fig. 10.6-8 for $Y_1Y_2 = 11$ and X_1, is interpreted as follows: If both flip-flops are SET ($Y_1Y_2 = 11$) and an X_1 pulse occurs, a pulse will appear at S_1 and at R_2, and no pulse will appear at the output Z. This will leave flip-flop number 1 in the SET state and flip-flop number 2 in the RESET state, giving as a next state $Y_1Y_2 = 10$. Figure 10.6-8 also allows us to test the consistency of the realization. If an entry in the table were $R_j = 1$ and $S_j = 1$, this would mean that both the set and reset inputs of flip-flop j were pulsed simultaneously, a situation that is not permitted, by the definition of the set-reset flip-flop.

The next step in our analysis procedure is to convert this map into a state table and thus to an abstract finite-state machine description of the behavior of the circuit of Fig. 10.6-7. To do this, we determine the next state caused by each input pulse for each present state. To simplify the procedure further, it is convenient to rename the states. This is done to the left of the Y_1Y_2 column in Fig. 10.6-8. Thus, for the entry discussed in the previous paragraph, namely, $Y_1Y_2 = 11$, or state C and input X_1, we have a new state of $Y_1Y_2 = 10$, or state D, and no output pulse. This is shown in Fig. 10.6-9. As another example,

	X_1	X_2
A	B, 0	A, 0
B	C, 0	A, 0
C	D, 0	A, 1
D	D, 0	A, 0

FIG. 10.6-9 Finite-state machine description of Fig. 10.6-7.

we take $Y_1Y_2 = 00$ or state A, and X_2. The table of Fig. 10.6-8 indicates that both R_1 and R_2 will be pulsed, giving a new state of $Y_1Y_2 = 00$, or state A; the output is not pulsed. The reader is encouraged to verify the remaining entries in Fig. 10.6-9. Notice that in going from Fig. 10.6-8 to Fig. 10.6-9, we have dropped all references to Y_1 and Y_2, and hence to the exact nature of the realization. We can now use the state table of Fig. 10.6-9 to predict the input-output behavior of the circuit of Fig. 10.6-7, provided that we specify the initial values of Y_1 and Y_2, i.e., the initial state.

This analysis procedure is quite general and gives us the ability to analyze any circuit in the prescribed form. To summarize, the procedure steps are as follows:

1. Determine expressions for every set, reset, and output line.
2. Construct the excitation matrix.
3. Derive the state table, using flip-flop output values.
4. Form the abstract state table by renaming states and outputs.

Synthesis of Sequential Circuits

We are now ready to synthesize sequential circuit realizations of finite-state machines. We have already seen that a certain form of sequential circuit, that of Fig. 10.6-6, always has a representation as a finite-state machine. In this section we shall see that every finite-state machine can be realized by a sequential circuit. Observe, however, that any machine derived from a sequential circuit with *binary* storage elements must have 2^k states. Since there are finite-state machines that have other than 2^k states, it follows that a sequential circuit corresponding to such a finite-state machine will have more states than the machine. In particular, any sequential circuit realization of a given finite-state machine would certainly be adequate if for every state of the finite-state machine the state table of the sequential circuit has an equivalent state. There would be nothing wrong if, in addition, the circuit state table has states that are not equivalent to any in the finite-state machine. In such a case we will say that the state table of the sequential circuit *contains* the finite-state machine. *The synthesis problem is to find a sequential circuit that contains the specified finite-state machine.*

As an example, consider the state table of Fig. 10.6-10. Any real-

	X_1	X_2
A	B, 0	C, 0
B	C, 1	A, 0
C	C, 1	B, 1

FIG. 10.6-10 A finite-state machine.

ization of this state table must have at least two flip-flops. In the interests of economy, we shall attempt to find a realization with just this minimum number of flip-flops. If we imagine following the analysis procedure of the preceding section backward, the next step would be to *assign states* of the circuit flip-flops to states of the state table. As we will see, the manner in which this assignment is made affects only the complexity, and not the existence, of a realization. Let us arbitrarily choose the *state assignment* shown in Fig. 10.6-11*a*. That is,

	Y_1Y_2
A	0 0
B	0 1
C	1 1
—	1 0

(*a*)

	Y_1Y_2	X_1	X_2
(*A*)	0 0	0∅, 10, 0	10, 10, 0
(*B*)	0 1	10, ∅0, 1	0∅, 01, 0
(*C*)	1 1	∅0, ∅0, 1	01, ∅0, 1

S_1R_1, S_2R_2, Z

(*b*)

FIG. 10.6-11 State assignment and excitation matrix for the machine of Fig. 10.6-10.

$Y_1Y_2 = 00$ is assigned to state A, and so forth. Observe that the state $Y_1Y_2 = 10$ of the circuit flip-flops is "left over." Temporarily ignoring this fact, we shall construct an excitation matrix based on this state assignment.

If the sequential circuit is in state A and the input X_1 occurs, then according to Fig. 10.6-10, a transition to state B is required. According to our state assignment, this requires flip-flop number 1 to remain reset, while flip-flop number 2 is to be set. Accordingly, the entry in the excitation matrix for S_2R_2 must be 10. The entry for S_1R_1, however, could be either 01 or 00, since either entry leaves the flip-flop reset. The choice of entries in such an instance is usually made to simplify the steering logic, and is therefore postponed until all other entries have been specified. To remind ourselves of the choice, we will enter $0\emptyset$, where $\emptyset$, the "don't care" familiar from Chap. 6, may be chosen as 0 or 1 without influencing the proper operation of the circuit. In order to facilitate our construction of the excitation matrix, we pause to examine what values should be specified for S_i and R_i in order to make each of the four possible changes of state Y_i of the ith flip-flop. The results are shown in Fig. 10.6-12 where an arrow denotes the desired change.

Y_i change	S_iR_i
$0 \rightarrow 0$	$0\,\emptyset$
$0 \rightarrow 1$	1 0
$1 \rightarrow 0$	0 1
$1 \rightarrow 1$	$\emptyset\,0$

FIG. 10.6-12 Excitations corresponding to desired state changes.

Using this table, the excitation matrix can be completed as shown in Fig. 10.6-11*b*. As an example of another entry, if the circuit is in state B and an input of X_2 occurs, the circuit state must change to A. This requires $Y_1: 0 \rightarrow 0$, $Y_2: 1 \rightarrow 0$, and $Z = 0$. This gives the entries $S_1R_1 = 00$, $S_2R_2 = 01$, and $Z = 0$. The reader is invited to check the remainder of the excitation matrix entries.

At this point, were it not for the state $Y_1Y_2 = 10$, we could develop steering functions for the various flip-flop inputs and the output. However, any realization that we obtain must have a well-defined behavior when it is in state $Y_1Y_2 = 10$; hence, we must consider this behavior. The reader should be aware

Y_1Y_2	X_1	X_2
0 0	$0\emptyset$, 10, 0	10, 10, 0
0 1	10, $\emptyset 0$, 1	$0\emptyset$, 01, 0
1 1	$\emptyset 0$, $\emptyset 0$, 1	01, $\emptyset 0$, 1
1 0	$\emptyset\emptyset$, $\emptyset\emptyset$, $\emptyset$	$\emptyset\emptyset$, $\emptyset\emptyset$, $\emptyset$

S_1R_1, S_2R_2, Z

FIG. 10.6-13 Excitation matrix containing all the circuit states.

that if the circuit starts in state A, B, or C, it will never enter the state $Y_1Y_2 = 10$. In this sense, we don't care what the entries are; we will agree, however, to retain the restriction that S_iR_i is never 11. The resulting excitation matrix for *all* the circuit states is shown in Fig. 10.6-13, where ØØ cannot be 11.

The synthesis problem is now reduced to one of Boolean minimization in the presence of don't cares and input constraints. The function S_1, whose general form is shown in Fig. 10.6-14*a*, will be used as an example. The two functions f_1^1 and f_2^1 can be obtained from the above excitation matrix and are shown in Fig. 10.6-14*b*.

(*a*) $S_1 = X_1 f_1^1(Y_1,Y_2) = X_2 f_2^1(Y_1,Y_2)$

$Y_2 \backslash Y_1$	0	1
0	0	0
1	1	0

f_1^1

$Y_2 \backslash Y_1$	0	1
0	1	0
1	0	0

f_2^1

(*b*)

FIG. 10.6-14 Minimizing the excitation S_1.

Since inputs X_1 and X_2 cannot coincide, the minimization of f_1^1 and f_2^1 must be accomplished independently. In practice, the minimization can be accomplished directly on the excitation matrix, remembering that the constraint imposed on the 00 entries must be observed. Figure 10.6-15 shows a reasonable choice for the various don't cares. The reader is encouraged to experiment with other choices.

Y_1Y_2	X_1	X_2
0 0	01, 10, 0	10, 10, 0
0 1	10, 10, 1	01, 01, 0
1 1	10, 10, 1	01, 00, 1
1 0	01, 10, 0	10, 10, 1

S_1R_1, S_1R_2, Z

FIG. 10.6-15 Excitation matrix with don't cares assigned.

The equations for S_1, R_1, S_2, R_2, and Z can now be written by inspection from the excitation matrix of Fig. 10.6-15, as shown in Eqs. (3).

$$\begin{aligned} S_1 &= X_1Y_2 + X_2\overline{Y}_2 \\ R_1 &= X_1\overline{Y}_2 + X_2Y_2 \\ S_2 &= X_1 + X_2\overline{Y}_2 \\ R_2 &= X_2\overline{Y}_1Y_2 \\ Z &= X_1Y_2 + X_2Y_1 \end{aligned} \tag{3}$$

From these equations, and from Chap. 6 background, a sequential circuit realization of the finite-state machine can be constructed by inspection. Such a circuit is shown in Fig. 10.6-16.

As a concluding step in the synthesis, we may wish to verify that

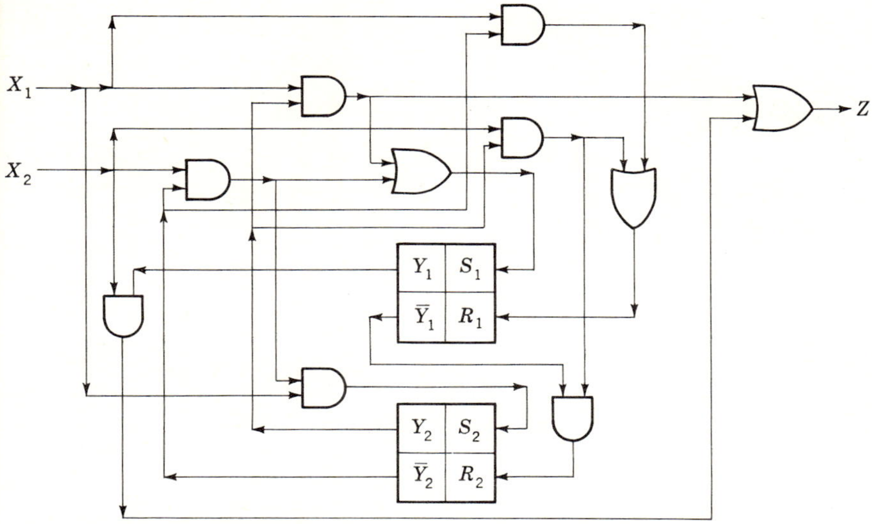

FIG. 10.6-16 A realization of the machine of Fig. 10.6-10.

the state table associated with this circuit contains the table of Fig. 10.6-10. Using $A = 00$, $B = 01$, $C = 11$, $D = 10$, we derive from Fig. 10.6-16 the state table shown in Fig. 10.6-17.

Observe that from states A, B, or C the machine cannot get into state D. It is easily seen that this table contains the given table of Fig. 10.6-10. Observe, further that if the assignment of Y_1Y_2 values to states A, B, and C in the table of Fig. 10.6-11*a* had been different, the excitation matrix would have been different, the steering functions would have been different, the circuit would have been different, but the state table of the resulting sequential circuit *would still contain the original table*. Observe finally that there is no general procedure for deciding ahead of time which assignment will result in the simplest final circuit. Trial and error plus experience, however, will usually yield good realizations.

The synthesis procedure just presented is perfectly general and will yield a sequential-circuit realization of *any finite-state machine*. The assignment of flip-flop states to rows of the table is arbitrary, and once this is accomplished, the table of Fig. 10.6-12 can be used to develop the excitation matrix.

	X_1	X_2
A	B, 0	C, 0
B	C, 1	A, 0
C	C, 1	B, 1
D	B, 0	C, 1

FIG. 10.6-17 State table of the sequential circuit of Fig. 10.6-16.

In summary, the synthesis procedure that we have discussed entails the following steps:

1. Choose a state assignment.
2. Develop the excitation matrix, with don't cares.
3. Assign the don't cares so as to minimize individual excitations; and obtain expressions for all excitations, S_iR_i, and for all outputs.
4. Design the sequential circuit, from the results of step 3 above.

Realizations Using *J-K* Flip-Flops

In many digital systems and almost all present-day computers, all logical events are synchronized with a periodic pulse signal called the clock. It is the occurrence of a clock pulse, CP, that initiates state changes, and the nature of these changes is determined by signal values just before the clock pulse. The fundamental memory element used in such circuits differs from the SET-RESET flip-flop and is called a *J-K* flip-flop.

Consider a binary storage element with two level inputs, *J* and *K*, and a single pulse input, CP, as shown in Fig. 10.6-18. The outputs Q and $\overline{Q}$ are complements and represent the state of the device. We will specify the behavior

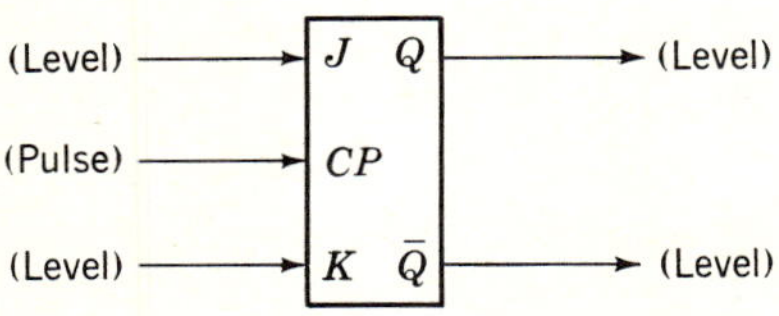

FIG. 10.6-18 The J-K flip-flop.

by giving the new state as a function of the present state and the values of *J* and *K*, when a clock pulse occurs. The state changes are initiated by the clock pulse, in fact, by the trailing edge of that pulse. If we consider Fig. 10.6-6 and assume that the Y_i inputs to the control logic are external inputs, that the clock pulse is the only X_i, and if we permit some Y_i to be external level outputs, we have the general characteristics of the device being discussed. This reasoning leads to the table of Fig. 10.6-19, where the values of *J* and *K* (columns) and the state Q (rows) determine the next state (intersection of row and column), *when the clock pulse is applied.*

Present State Q \ JK	00	01	11	10
0	0	0	1	1
1	1	0	0	1

Next State

FIG. 10.6-19 J-K flip-flop behavior. Indicated state transitions occur after CP has been applied.

Thus, if $JK = 00$, the state after CP will be the same as the state before CP. If $JK = 11$, the state after CP will be the complement of the state before CP. For $JK = 01$ and 10, the state after CP is independent of the state before CP. *Thus, by controlling the values of J and K, we can leave the state unchanged, complement it, or drive it to either value.* This flexibility is what makes the *J-K* flip-flop so popular.

An interesting exercise is to synthesize a *J-K* flip-flop, using SET-RESET flip-flops. Consider the excitation matrix shown in Fig. 10.6-20. Obvious

Q \ JK (CP)	00	01	11	10
0	0Ø	0Ø	10	10
1	Ø0	01	01	Ø0

SR

FIG. 10.6-20 Excitation matrix of J-K flip-flop.

choices for the 0's give

$$S = J\overline{Q}(CP)$$

$$R = KQ(CP)$$

which in turn yield the realization shown in Fig. 10.6-21. For a variety of reasons, actual *J-K* flip-flops are usually not realized, as shown in Fig. 10.6-21.

The general form of a sequential circuit employing *J-K* flip-flops as memory elements is shown in Fig. 10.6-22. All signals are levels, with the exception of CP, which is pulselike. As a specific example of such a realization consider the circuit of Fig. 10.6-23. As a first step in our analysis of this circuit, we will derive equations for J_1, K_1, J_2, K_2, and Z. These equations are given below.

$$J_1 = X$$

$$K_1 = \overline{X}Q_2 + X\overline{Q}_2$$

$$J_2 = X\overline{Q}_1$$

$$K_2 = 1$$

$$Z = Q_1 + \overline{X}Q_2$$

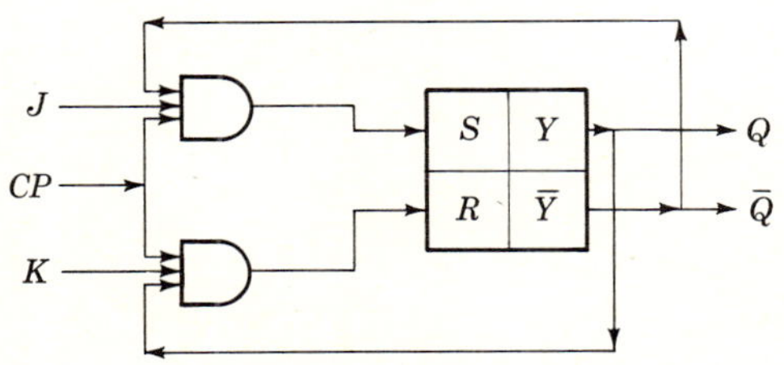

FIG. 10.6-21 S-R flip-flop realization of J-K flip-flop.

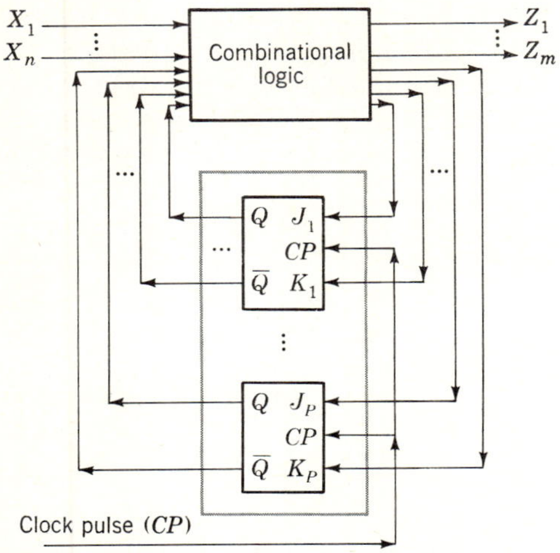

FIG. 10.6-22 General form of J-K flip-flop sequential circuit.

Note that none of these equations involves the clock pulse. The clock pulse and its effect on the circuit will be understood throughout the analysis. The functions for J_1, K_1, J_2, and K_2 are next mapped in the form of an excitation matrix (Fig. 10.6-24). Now assume that the circuit is in a particular state as specified by the values of Q_1 and Q_2, that X has a particular value, and that a clock pulse

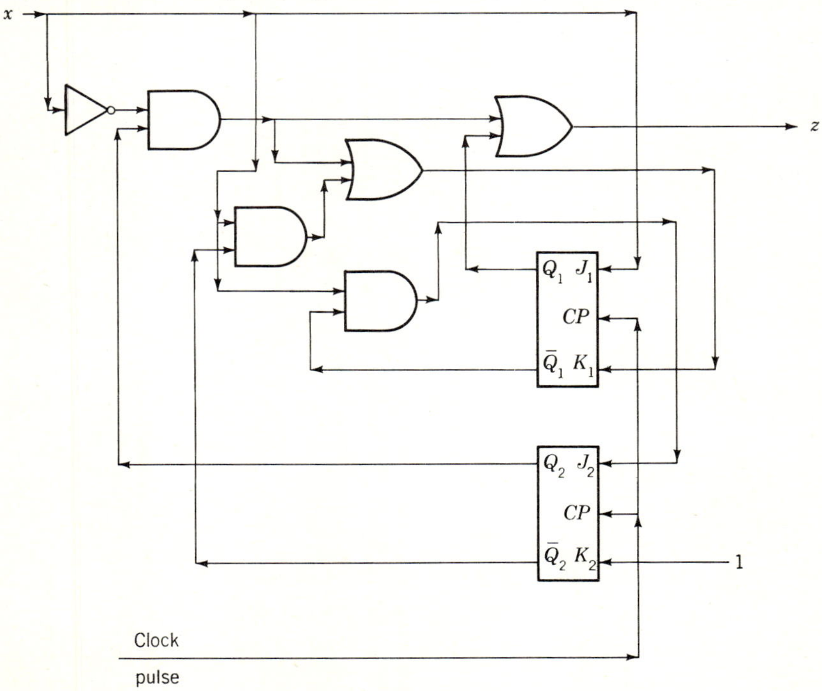

FIG. 10.6-23 A J-K flip-flop sequential circuit.

Q_1Q_2 \ X	0	1
00	0001	1111
01	0101	1011
11	0101	1001
10	0001	1101

$J_1K_1J_2K_2$

FIG. 10.6-24 Excitation matrix for the circuit of Fig. 10.6-23.

occurs. The values of Q_1, Q_2, and X determine J_1, K_1, J_2, and K_2 just before the clock pulse; hence the values of Q_1 and Q_2 immediately after the clock pulse are uniquely determined, and are shown in Fig. 10.6-25. To illustrate the derivation of this table, consider the $Q_1Q_2 = 01$, $X = 1$ entry.

	Q_1Q_2 \ X	0	1
(A)	00	00	11
(B)	01	00	10
(C)	11	00	10
(D)	10	10	00

Q_1Q_2 after clock pulse

FIG. 10.6-25 State table of circuit.

Figure 10.6-24 gives $J_1K_1 = 10$ and $J_2K_2 = 11$. Figure 10.6-19 indicates that the new value of Q_1 will be 1, and the new value of Q_2 will be the complement of its previous value, which gives the new value of Q_2 as 0. The reader should be able to verify the remaining entries. To complete the analysis we need only consider the output Z. In Fig. 10.6-26 we have coded the states as indicated on the left side of Fig. 10.6-25, and we have also introduced the output. Observe that in *J-K* flip-flop sequential circuits the outputs are *levels* and are therefore associated with every state and input combination.

State \ Input X	0	1
A	A, 0	C, 0
B	A, 1	D, 0
C	A, 1	D, 1
D	D, 1	A, 1

Next State, Output

Fig. 10.6-26 Final state table.

The interpretation of this state table is as follows: If the present state is A and the input is $X = 0$, the state *after the next clock pulse* is A. The output entry for state A and $X = 0$ is $Z = 0$. That is, there is a 0 on the Z line until the next clock pulse arrives. *Note that this is different from our earlier con-*

vention with S-R flip-flop circuits in which an output of 0 or 1 indicates the presence or absence of a pulse or an output line.

The analysis procedure just presented is general and can be applied to any circuit employing *J-K* flip-flops or memory elements. Furthermore, it can be applied to realizations using any memory element which is controlled by levels and triggered by pulselike signals.

The synthesis procedure for *J-K* flip-flop realizations is very similar to the technique presented earlier for SET-RESET flip-flops. There are only three differences. First, the equivalent of Fig. 10.6-12 must be modified to reflect the behavior of *J-K* flip-flops; this is done in the next paragraph. Second, since the input signals are levels and not pulses, there is no need to retain the special form given in Eq. (2). Third, since the outputs are levels, they are associated with combinations of state and input values.

We shall first derive a table for the *J-K* flip-flop transitions, as we did for the *S-R* flip-flop in Fig. 10.6-12. With reference to Fig. 10.6-19, assume that we want Q to undergo a 0 to 0 change. The values of J and K that can accomplish this change can be either 00, leaving the state unchanged, or 10 forcing the state to 0. Combining these alternatives gives $JK = 0\emptyset$, which is the first entry in the table of Fig. 10.6-27. The remainder of the table has been completed, using similar reasoning.

Q change	JK
$0 \rightarrow 0$	$0\emptyset$
$0 \rightarrow 1$	$1\emptyset$
$1 \rightarrow 0$	$\emptyset 1$
$1 \rightarrow 1$	$\emptyset 0$

FIG. 10.6-27 JK values corresponding to desired state changes.

As an initial synthesis example, we shall realize the state table of the serial adder, which is given in Fig. 10.5-5, and repeated in Fig. 10.6-28.† The table requires a single *J-K* flip-flop and we choose the state assignment $A = 0$,

	XY = 00	01	11	10
(0) A	A, 0	A, 1	B, 0	A, 1
(1) B	A, 1	B, 0	B, 1	B, 0

Next State, Output

FIG. 10.6-28 State table of serial adder.

Q \ XY	00	01	11	10
0	$0\emptyset$, 0	$0\emptyset$, 1	$1\emptyset$, 0	$0\emptyset$, 1
1	$\emptyset 1$, 1	$\emptyset 0$, 0	$\emptyset 0$, 1	$\emptyset 0$, 0

JK, Output

FIG. 10.6-29 Excitation for serial adder.

†With the last two columns permuted to match the Karnaugh map adjacency requirement.

(a) J

Q \ XY	00	01	11	10
0	0	0	1	0
1	0	0	1	0

(b) K

Q \ XY	00	01	11	10
0	1	0	0	0
1	1	0	0	0

FIG. 10.6-30 Maps for J and K excitations.

$B = 1$. Using Fig. 10.6-27, we can immediately derive the required entries for J and K. These values are shown in the table of Fig. 10.6-29. Separate maps for J and K with appropriate entries for the Ø's are given in Fig. 10.6-30*a* and *b*.

These maps give rise to the following expressions:

$$J = XY$$

$$K = \overline{X}\overline{Y}$$

and from Fig. 10.6-29 we have

$$Z = \overline{X}\overline{Y}Q + \overline{X}Y\overline{Q} + XYQ + X\overline{Y}\overline{Q}$$

These three equation specify the desired realization.

The synthesis procedure just presented is applicable to any state table and will yield a realization of that table. As before, the assignment of circuit state to the rows of the state table is arbitrary, but it should be kept in mind that certain assignments will result in more economical realizations than others.

Present State	(CP)
A	B, 00
B	C, 01
C	D, 10
D	A, 11
	Next State, Z_1Z_0

FIG. 10.6-31 State table of two-bit counter.

As another synthesis example, let us consider a two-bit binary counter, whose state table is shown in Fig. 10.6-31. If we make the assignment of $A = 00$, $B = 01$, $C = 10$, and $D = 11$, we get the excitation matrix of Fig. 10.6-32.

	Q_1Q_0	(CP)
(A)	00	0Ø, 1Ø, 00
(B)	01	1Ø, Ø1, 01
(D)	11	Ø1, Ø1, 11
(C)	10	Ø0, 1Ø, 10
		J_1K_1, J_0K_0, Z_1Z_0

FIG. 10.6-32 Excitation matrix for two-bit counter.

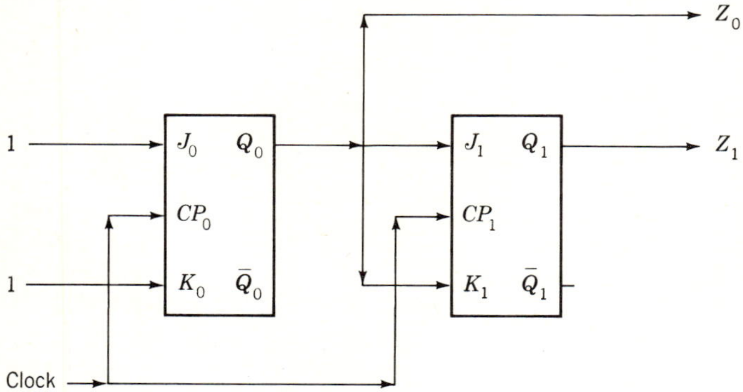

FIG. 10.6-33 J-K realization of two-bit counter.

Among the possible equations for J_1, K_1, J_0, K_0 we have

$$J_1 = Q_0 \qquad K_0 = 1$$

$$K_1 = Q_0 \qquad Z_1 = Q_1$$

$$J_0 = 1 \qquad Z_0 = Q_0$$

This set of equations yields the realization of Fig. 10.6-33, which requires no logic gates, indicating the flexibility of the *J-K* flip-flop.

In this section, we have seen that the finite-state machine serves as an abstract model of sequential circuits employing a variety of memory elements. There are other practical memory elements in use, and the synthesis procedure used here can be applied with little modification.

10.7 LIMITATIONS OF FINITE-STATE MACHINES

Intuitively, a finite-state machine is constrained by its ability to remember only a finite number of different things. The machine remembers, of course, by being in various states at various times.

As an example of something that a finite-state machine cannot do, we consider the set of symmetric sequences† in the alphabet {0,1}. This set includes exactly those sequences that are their own reverses. Thus, 0, 010, and 0110 are members of the set, while 011, 01, and 100 are not. *There is no finite-state machine that recognizes exactly this set of symmetric sequences.* The formal reasoning required to prove this result will not be presented. We will, however, indicate the line of reasoning used in the proof. After any K input symbols have occurred, the machine must be prepared to indicate acceptance if the appropriate additional K symbols are presented. The only way the machine can be prepared, however, is to be in a particular state. Furthermore, it must be

†Also called palindromes.

in a different state for each input sequence of input length K. If it were not, it could not exhibit the assumed behavior. This, however, requires that the number of states, N, be greater than or equal to 2^K, for all K—an impossibility.

There are many other, less obvious, examples for which it is correspondingly harder to prove that the associated set cannot be recognized. The existence of sets that cannot be recognized implies the existence of transformations that cannot be accomplished. This, in turn, implies that if certain transformations are desired, either the length of the input sequences must be restricted or the machines must be provided with unbounded memories. The first of these alternatives usually occurs in practice. Machines with unbounded memory, such as Turing machines, are used primarily in theoretical studies to determine the logical capabilities of computing machines.

There exists a precise algebraic method of describing *exactly* the sets that finite-state machines are capable of recognizing. The expressions that accomplish this are called *regular expressions*. For an excellent discussion of this and many other topics concerning finite and infinite machines, we refer the reader to "Computation: Finite and Infinite Machines" by M. Minsky.

PROBLEMS

10.3-1 Let M_1 and M_2 be the machines shown in part *a* of the figure. Let the machines be connected as shown in part *b*. Give the state table and transition graph of the resulting machine M.

	0	1
A	A, 1	B, 0
B	B, 1	A, 1

M_1

	0	1
C	D, 0	C, 1
D	D, 1	C, 0

M_2

(*a*)

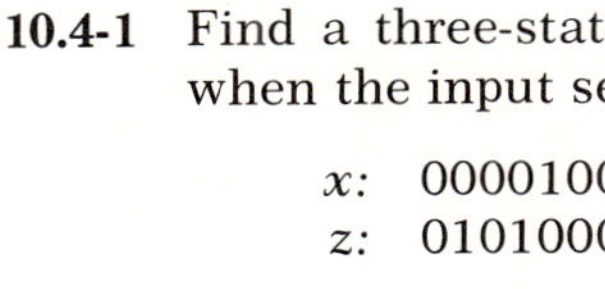

(*b*)

PROB. FIG. 10.3-1

10.4-1 Find a three-state machine that will produce the output sequence z, when the input sequence is x.

x: 00001000100010
z: 01010000101001

Are there any other (nonisomorphic) three-state machines that have the same behavior?

10.4-2 Give a word description of the set of sequences recognized by the machine in the figure.

	X_1	X_2
A	A, 0	B, 0
B	A, 0	C, 0
C	C, 0	D, 0
D	C, 0	E, 1
E	E, 1	F, 1
F	E, 1	F, 0

PROB. FIG. 10.4-2

10.4-3 As Prob. 10.3-1 indicates, finite-state machines can be interconnected to form other machines. A possible connection is shown in the figure.

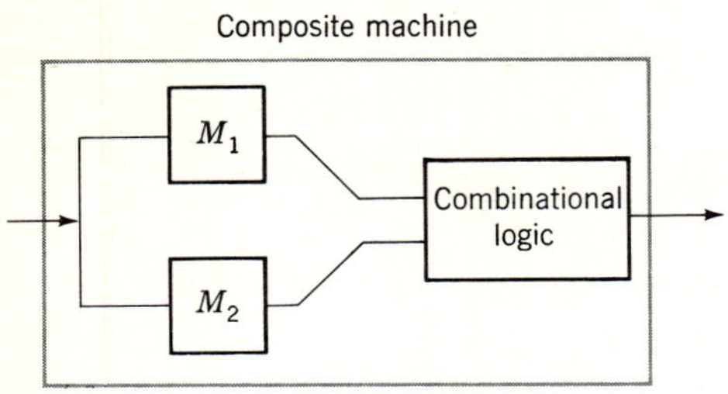

PROB. FIG. 10.4-3

Using this connection as a basis, show that if M_1 recognizes the set S_1, and M_2 recognizes the set S_2, then the following sets can also be recognized by a finite-state machine.

(i) $\overline{S_1}$
(ii) $S_1 \cup S_2$
(iii) $S_1 \cap S_2$

You must give a convincing argument that the result of the connection *is* a finite-state machine.

10.4-4 Prove that if a machine with n states recognizes a nonempty set, this set must contain at least one sequence of length n or less. (This is a nontrivial problem.)

10.5-1 (*a*) Find the state table for the transition diagram given in the figure.

(*b*) Indicate trivially why this is a reduced machine.

(*c*) Which of the following sequences are accepted by the machine?

(i)	01	(iv)	00101
(ii)	0110	(v)	111101
(iii)	110	(vi)	00001

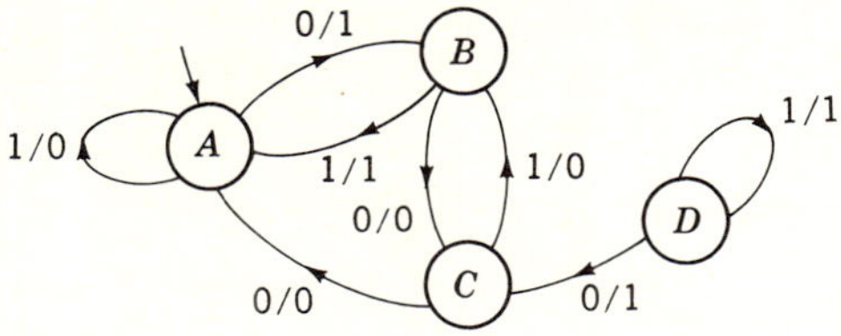

PROB. FIG. 10.5-1

10.5-2 (*a*) Find a reduced state table for a machine that accepts only the input sequence 011.

(*b*) Modify your state table so that the resulting machine recognizes the set

$$\{011, 011011, 011011011, \ldots\}$$

10.5-3 Design a finite-state machine to recognize the set of input sequences in which the number of 0s is even, or the number of 1s is odd, or both. (Consider zero to be an even number.) Use as few states as possible.

10.5-4 For each of (*a*) and (*b*) below, find a finite-state machine having the specified behavior. Use as few states as possible, and use *A* as the starting state. The input and output alphabets are {0,1}.

(*a*) An output of 1 is to be produced coincident with the first input 0 following a group of two or more consecutive input 1s. Outputs of 0 are to be produced in all other cases.

(*b*) The first two symbols in any input sequence are to result in outputs of 0. Thereafter the output should be equal to the second previous input.

10.5-5 Determine all equivalent states, and give the reduced-machine state table.

	X_1	X_2	X_3
A	B, 1	B, 0	E, 0
B	A, 0	D, 1	D, 1
C	B, 1	B, 0	E, 0
D	C, 0	B, 1	B, 1
E	F, 1	D, 0	C, 0
F	H, 0	J, 1	F, 1
G	F, 1	B, 0	H, 0
H	D, 1	D, 0	G, 0
J	G, 0	J, 1	G, 1

PROB. FIG. 10.5-5

10.5-6 Using the equivalence-testing method described in this chapter, find all the equivalent states in the machine M, and give the reduced machine.

	0	1
A	$B, 0$	$C, 0$
B	$D, 1$	$E, 0$
C	$A, 0$	$B, 1$
D	$E, 0$	$F, 1$
E	$F, 0$	$F, 0$
F	$C, 1$	$E, 0$

M *PROB. FIG. 10.5-6*

10.5-7 Consider the state table in the figure. Each blank entry indicates that for that particular state of the machine the indicated output can never occur.

	0	1
A	$C, 0$	$K, 0$
B	$I, 0$	$E, 0$
C	$F, 1$	$L, 0$
D	$I, 0$	$G, 1$
E	$J, 1$	$B, 1$
F	$A, 1$	$H, 1$
G	$B, 0$	$K, 1$
H	$D, 1$	$B, 0$
I	$B, 0$	—
J	—	$I, 0$
K	—	$D, 1$
L	$A, 1$	$J, 1$

PROB. FIG. 10.5-7

Assign next state and output values to each of the blanks, so that the reduced machine has a minimum number of states. Give the state table of the new machine.

10.5-8 A *palindrome* is a finite sequence of 0s and 1s that reads the same from right to left as from left to right. For example, 01010 is a palindrome of length five. 01111 is not a palindrome.

(a) We want to construct a pulse-sequential machine that detects palindromes of length exactly three (3). The machine is to have two pulse inputs u_0 (logical ZERO) and u_1 (logical ONE), and a single pulse output z. The machine produces an output pulse coincident with the 3d, 6th, 9th, 12th, . . . , etc., input pulse if and only if the three-pulse sequence ending at each such pulse forms a palindrome. A typical input-output behavior is shown in the figure. Give a transition diagram for the reduced machine (five states are sufficient: can you do better?).

10.5-9 Prove that application of the equivalence testing procedure of this chapter to any n-state finite machine is guaranteed to terminate in $n - 1$ steps; i.e., G_n is identical to G_{n-1}.

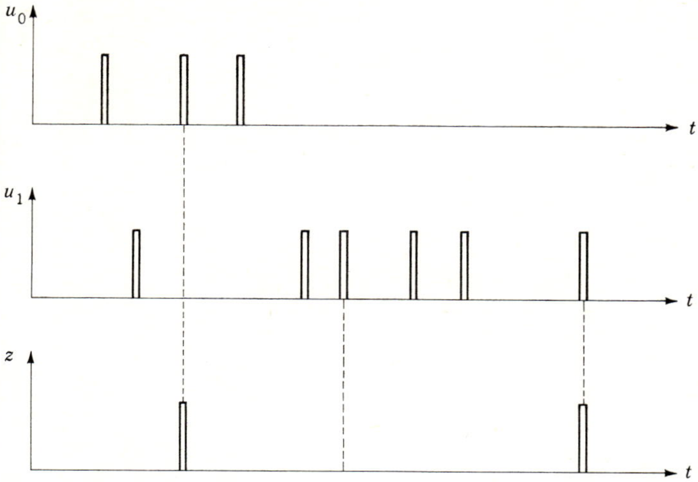

PROB. FIG. 10.5-8

10.6-1 Give a realization, using SET-RESET flip-flops, of the machine shown in the figure. Be sure to give all steps in your synthesis, and make all choices in a way that simplifies the resulting circuit.

	0	1
A	B, 0	A, 0
B	A, 0	C, 1
C	B, 0	C, 1

M PROB. FIG. 10.6-1

10.6-2 Binary numbers *A* and *B* are represented by time signals u_A and u_B, respectively. The numbers are presented least significant bit first. For example, if $A = 1110010$, then u_A is given by the plot in part *a* of the figure. A clock pulse generator *u* is available which produces pulses at half units of time, as shown in part *b*. Using SET-RESET flip-flops,

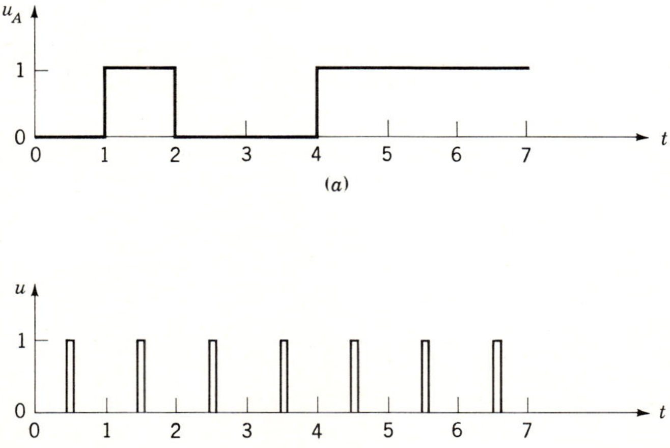

PROB. FIG. 10.6-2

design a circuit which has inputs u_A, u_B, and u, and output Z. It is required that Z have the same form as A and B (except for a delay of ½) and $Z = A + B$ (ordinary arithmetic sum).

10.6-3 Consider the *complementing flip-flop*, shown in the figure. By definition, this flip-flop changes state whenever the C line is pulsed.

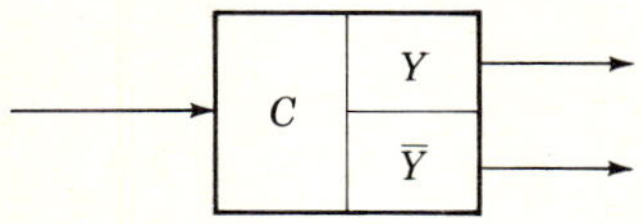

PROB. FIG. 10.6-3

(a) Describe a procedure which starting from a finite-state machine generates the steering logic of sequential circuits that use complementing flip-flops.

(b) Using complementing flip-flops and your procedure of (a) above, find a realization of the machine M of Prob. 10.6-1.

10.6-4 (See Prob. 10.6-3.)

(a) Construct a set-reset flip-flop using one complementing flip-flop, together with as few gates as possible.

(b) Construct a complementing flip-flop, using a SET-RESET flip-flop together with as few gates as possible.

10.6-5 Design a scale-of-five counter, i.e., a finite-state machine with input u and output y which issues a pulse on y upon the arrival of the fifth u pulse, since the last y output (or the start of the counting process). Use complementing flip-flops (see Prob. 10.6-3.), and carry out your design to a schematic of interconnected logic gates and flip-flops.

10.6-6 *Counter of Arbitrary Scale*

(a) Extend the circuit of Fig. 10.6-33 to a K-bit counter, which counts up with scale 2^K.

(b) A common procedure for the design of arbitrary-scale m, counters ($2^{K-1} < m < 2^K$)is as follows: Count every clock pulse as in (a) above *until* the flip-flops form the binary number $m-1$. Detect this (and only this) state of the system with a combinational network which on the next clock pulse (the mth since start) resets *all* flip-flops to zero. Use this approach to design a decade counter ($m = 10$), using J-K flip-flops.

10.6-7 *Up-Down Counter*

(a) Using J-K flip-flops, design a sequential circuit which has two inputs X_u and X_d and two outputs Z_u and Z_d. X_u and X_d are levels, constrained by $X_u \cdot X_d = 0$. Input-output behavior is as shown in the figure. (All transitions are under clock pulse, CP.) Note that as long as X_u is 1, the machine "counts up," making $Z_u = 1$ after every fourth clock pulse. Conversely, as long as X_d is 1, the machine "counts down" in exactly the opposite way. If both X_u and X_d are 0, then the machine stops counting.

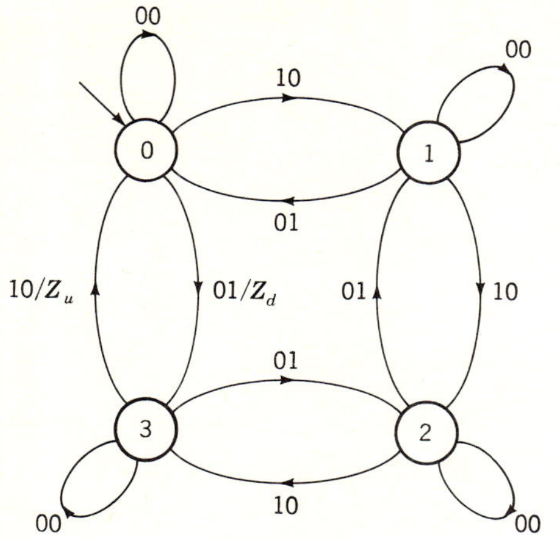

PROB. FIG. 10.6-7

(*b*) Extend the circuit you obtained in (*a*) above so as to give the general form of a scale, 2^K ($K = 1, 2, 3, \ldots$) up-down counter.

10.6-8 *Shift Register.* A register is an array of contiguous binary storage cells, e.g., *J-K* flip-flops. A shift register is capable of shifting (to the right or to the left) the contents of every flip-flop in the array upon receipt of the appropriate command.

(*a*) Without resorting to *any* formal procedures, connect several *J-K* flip-flops in a simple cascade (no logic gates needed) so as to form a shift register which shifts one position to the right at every clock pulse. The shift register input is a line x which is set to 1 or 0 between clock pulses.

(*b*) Let j be the (left-to-right) index of flip-flops in the register; Let the state of the jth flip-flop before a clock pulse be Q_j and after the clock pulse Q_j'. Using a more formal approach, write the equations for the jth flip-flop (j other than first or last) so that the register on every clock pulse (1) shifts to the right, (2) shifts to the left, (3) does not shift, (4) resets all flip-flops to zero according to whether two input lines $u_1 u_2$ are at (1) 01, (2) 10, (3) 00, (4) 11. Design the typical stage denoted by your equations.

10.7-1 We want a machine with two pulse inputs u_1, u_2 and one pulse output z. z is to be pulsed coincident with a u_1 or a u_2 pulse if and only if the two input lines have been pulsed *the same number of times* since the beginning of the experiment. If this task can be done by a finite-state machine, give the number of states; otherwise argue that it cannot be done with a finite number of states.

chapter 11

Ideal Delay and Convolution

In this chapter we develop general models, equivalences, and input-output relationships for those linear systems in which memory takes the form of ideal delay. A limiting case of a particular system in this class leads to the identification of a basic linear operation called convolution *with the aid of which the output waveform of any given linear system can be calculated directly for any input-signal waveform. Finally, linear systems having delay memory and integrator memory are compared, and close relationships are established.*

11.1 INTRODUCTION: A LINEAR TRANSMISSION SYSTEM UTILIZING NONLINEARITY AND IDEAL DELAY

Before we turn to the consideration of general models, the following purified design problem is introduced to illustrate the purposeful use of ideal delay, and thereby to whet the appetite for subsequent results.

Consider the schematic communications system shown in Fig. 11.1-1*a*. Suppose that the channel† is perfect for sufficiently small signal levels,

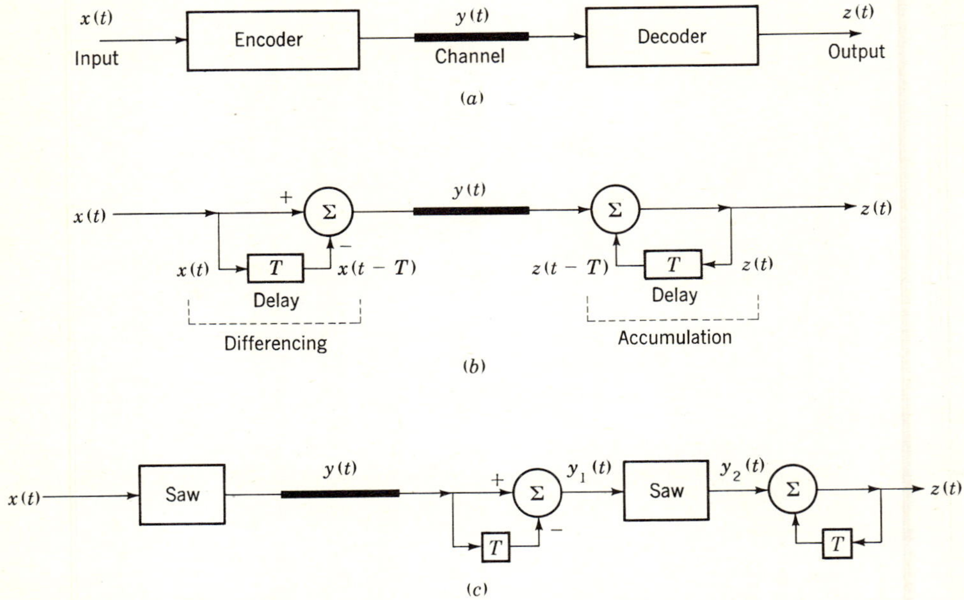

FIG. 11.1-1 Transmission of a limited-change signal over a saturable channel.

but that it produces distortion whenever the channel signal $y(t)$ has a magnitude in excess of some threshhold, say, 1.5. Suppose also that the signal $x(t)$ to be transmitted is unrestricted in value but known to have a "delay difference," $x(t) - x(t - T)$, whose magnitude is less than 1.5 for a specified relative delay T and for all time t. (An example of a signal having such properties is the analog output from an up-down counter, which can change by no more than one count value in the time required for one count.) The requirement on $y(t)$ and the known property of $x(t)$ are

$$|y(t)| < 1.5 \tag{1}$$

$$|x(t) - x(t - T)| < 1.5 \tag{2}$$

We wish to establish two block-diagram systems called *encoder* and *decoder* in Fig. 11.1-1, such that the output signal $z(t)$ is identical to the input signal $x(t)$ for every signal $x(t)$ obeying Eq. (2). One such encoder is that shown in Fig. 11.1-1*b*, characterized by the difference equation

†A *channel*, or *communications channel*, is the medium between a transmitter and a receiver of information. It can be a pair of wires or the atmosphere. In this context, a *perfect channel* is one that transmitts information faithfully, without change or distortion.

$$y(t) = x(t) - x(t - T) \tag{3}$$

The corresponding decoder is an "accumulator" described by the difference equation,

$$z(t) = y(t) + z(t - T) \tag{4}$$

Elimination of $y(t)$ between Eqs. (3) and (4) yields the input-output difference equation

$$z(t) - z(t - T) = x(t) - x(t - T) \tag{5}$$

If $x(t)$ and $z(t)$ were both zero over some past time interval of duration T, which is the same as saying that the accumulator delay element was cleared (zero set), over its storage time interval T, before the first application of nonzero x, then we can work our way forward along the time scale with the aid of Eq. (5) to prove that

$$z(t) = x(t) \tag{6}$$

for all time. Moreover, note that by Eq. (2) and (3), the constraint of Eq. (1) is met. The encoder is a quasi-differentiator, the decoder is a quasi-integrator, their effects cancel, and $x(t)$ is reproduced at the output $z(t)$. With respect to its input-output relationship, the system is linear and memoryless. Table 11.1*a* shows sample values taken at discrete time intervals, but the analysis applies, of course, to functions x, y, z defined over the continuous time domain t.

TABLE 11.1-1 Transmission-system Performance: (*a*) The System of Fig. 11.1-1*b*; (*b, c, d*) The System of Fig. 11.1-1*c*

	t/T	1	2	3	4	5	6	7	8	9
	x	0	1	2	3	4	4	3	2	1
(*a*)	y	0	1	1	1	1	0	−1	−1	−1
	z	0	1	2	3	4	4	3	2	1
	x	0	1	2	3	4	4	3	2	1
	y	0	1	−1	0	1	1	0	−1	1
(*b*)	y_1	0	1	−2	1	1	0	−1	−1	2
	y_2	0	1	1	1	1	0	−1	−1	−1
	z	0	1	2	3	4	4	3	2	1
		1.4	2.8	4.2	5.6	5.6	4.2	2.8	1.4	2.8
	y	1.4	−0.2	1.2	−0.4	−0.4	1.2	−0.2	1.4	−0.2
(*c*)	y_1	1.4	−1.6	1.4	−1.6	0	1.6	−1.4	1.6	−1.6
	y_2	1.4	1.4	1.4	1.4	0	−1.4	−1.4	−1.4	1.4
	z	1.4	2.8	4.2	5.6	5.6	4.2	2.8	1.4	2.8
	x	1.6	3.2	4.8	6.4	6.4	4.8	3.2	1.6	3.2
	y	−1.4	0.2	−1.2	0.4	0.4	−1.2	0.2	−1.4	0.2
(*d*)	y_1	−1.4	1.6	−1.4	1.6	0	−1.6	1.4	−1.6	1.6
	y_2	−1.4	−1.4	−1.4	−1.4	0	1.4	1.4	1.4	−1.4
	z	−1.4	−2.8	−4.2	−5.6	−5.6	−4.2	−2.8	−1.4	−2.8

If for some practical reason the encoder must be memoryless but may be nonlinear, then the system in Fig. 11.1-1*c* represents an alternative solution to this purified design problem. This solution is not obvious. Here the encoder is simply the function block whose curve is plotted in Fig. 11.1-2. The reader is encouraged to prove that for any two values a and b,

$$\text{saw}(a+b) = \text{saw}(\text{saw}(a) + \text{saw}(b)) \tag{7}$$

We will have use for this result in a moment. Now by inspection of Fig. 11.1-1*c*,

$$y(t) = \text{saw}(x(t)) \tag{8}$$

$$y_1(t) = y(t) - y(t-T) \tag{9}$$

$$y_2(t) = \text{saw}(y_1(t)) \tag{10}$$

With the aid of relation (2), Eqs. (7) to (10), and Fig. 11.1-2, we find that Eq. (1) is satisfied and that

$$\begin{aligned} y_2(t) &= \text{saw}(\text{saw}(x(t)) - \text{saw}(x(t-T))) \\ &= \text{saw}(x(t) - x(t-T)) \\ &= x(t) - x(t-T) \end{aligned} \tag{11}$$

Finally, comparison of Eqs. (3) and (11) shows that $y_2(t)$ in Fig. 11.1-1*c* is the same as $y(t)$ in Fig. 11.1-1*b*. and it follows that the same $z(t)$ is produced at the output of the two systems. Hence the alternative system is linear and memoryless in overall behavior in spite of, or perhaps we should say because of, the very special nonlinear memoryless encoder and the compatible nonlinear decoder. Parts *b*, *c*, and *d* of Table 11.1-1 give numerical examples of performance.

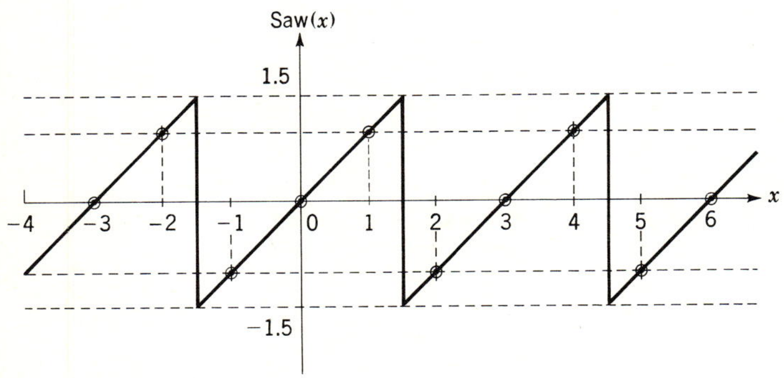

FIG. 11.1-2 The sawtooth function Saw(x) of period 3.

Questions for reader participation (some of these questions go beyond the present material and are not necessarily easy): How do you explain the inequality of $z(t)$ and $x(t)$ in part *d* of Table 11.1-1? With $x(t)$ restricted to integer values, as in part *b* of Table 11.1-1, can you offer a simple interpretation of the system? Is "exclusive OR" a linear operation? What are the practical effects of finite accuracy in the component operations of the systems in Fig. 11.1-1; in particular, subtraction? Why are differentiators said to be "noisy"? What have you learned from this whole example that is interesting to you?

11.2 IDEAL DELAY AND INTEGRATION

As additional background for subsequent material, we will now show that integration can be approximated by means of ideal delay, and vice versa, so that neither of these operations deserves to be treated as a more fundamental form of system memory than the other.

For a smooth (continuous derivative) curve $y(t)$, the extrapolation from t to some nearby time $t+A$ is given by the first two terms of a Taylor series,

$$y(t+A) \approx y(t) + A\frac{dy(t)}{dt} \tag{1}$$

The system in Fig. 11.2-1a has the associated difference equation,

$$y(t+A) = y(t) + Ax(t) \tag{2}$$

Comparison of Eqs. (1) and (2) shows that, for small A, the input $x(t)$ approximates the derivative of the output $y(t)$, so that the system evidently behaves as a good approximation to an ideal integrator. In particular, in the limit of small A,

$$x(t) = \frac{dy(t)}{dt} \tag{3}$$

so that

$$y(t) = \int_0^t x(\xi)\, d\xi + y(0) \tag{4}$$

where $y(0)$ is the initial condition of the integrator. If we make the common assumption that the integrator was clear (zero output) at some time in the past, prior to which the input was always zero, then Eq. (4) is equivalent to

$$y(t) = \int_{-\infty}^t x(\xi)\, d\xi \tag{5}$$

With a unit step input, that is, with $x=0$ for t negative and $x=1$ for t positive, we can deduce from Eq. (2) or by consideration of the accumulating circulation of signal around the delay loop in Fig. 11.2-1a that the step response is as shown in Fig. 11.2-1b. In the limit of small A, the step response

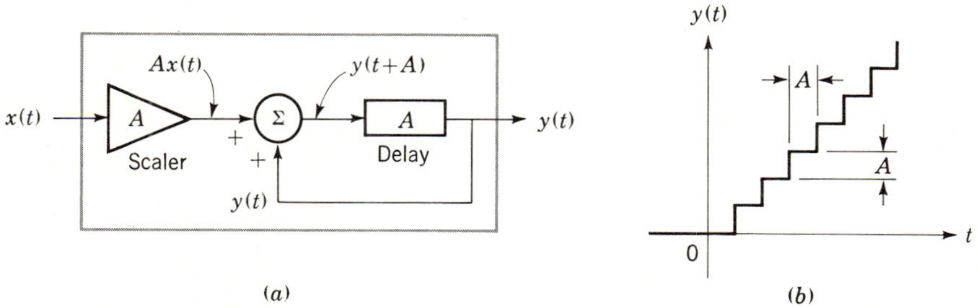

FIG. 11.2-1 Utilization of ideal delay to approximate an integrator. (a) *The system model.* (b) *The response to a unit step input approximates a unit ramp when A is made small.*

approximates the unit ramp that would be expected at the output of an ideal integrator.

The use of integration to approximate ideal delay is depicted in Fig. 11.2-2. The system model in Fig. 11.2-2*a* is exactly equivalent to a delay A, as can be verified by checking the signal levels indicated at internal points of the system. The innermost box approaches the behavior of an integrator for small A, and its replacement by an integrator block produces one section of the cascaded system shown in Fig. 11.2-2*b*. The scale factor $1/A$ in Fig. 11.2-2*a* is renamed n/T in Fig. 11.2-2*b* and lumped with the integrator, so that each section provides a small time delay T/n.

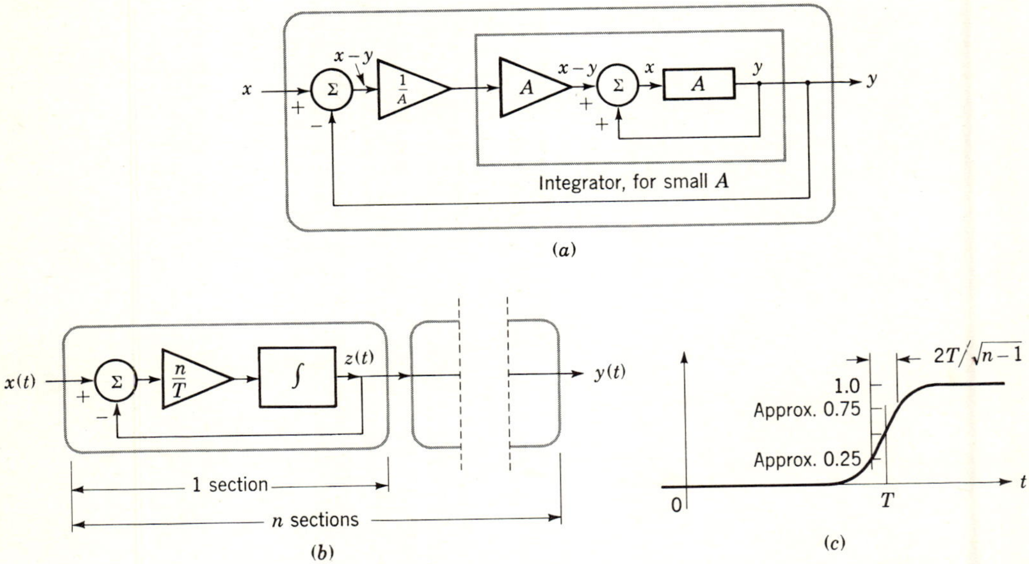

FIG. 11.2-2 Utilization of integrators to approximate ideal delay. (a) *Ideal delay* $y(t) = x(t - A)$. (b) *For large* n, $y(t) \to x(t - T)$. (c) *Step response of the approximated delay.*

Inspection of the first section in Fig. 11.2-2*b* yields

$$\frac{T}{n}\frac{dz(t)}{dt} = x(t) - z(t) \tag{6}$$

Solution of Eq. (6) for x gives

$$x(t) = z(t) + \frac{T}{n}\frac{dz(t)}{dt} \tag{7}$$

The right side of Eq. (7) is recognizable as the linear extrapolation of a smooth function $z(t)$ from t to $t + T/n$, so that in the limit of large n,

$$z\left(t + \frac{T}{n}\right) = z(t) + \frac{T}{n}\frac{dz(t)}{dt} \tag{8}$$

and comparison of Eqs. (7) and (8) yields

$$x(t) = z\left(t + \frac{T}{n}\right) \tag{9}$$

$$z(t) = x\left(t - \frac{T}{n}\right) \tag{10}$$

The cascade of n sections therefore provides a total delay T. The result is that we have approximated delay, using integrators as building blocks.

The plot in Fig. 11.2-2*c* indicates the detailed way in which the step response approaches a delayed step as n becomes large. This curve can be computed with methods developed later in this chapter and is inserted here without derivation.

11.3 THE STATE MODEL WITH MEMORY IN THE FORM OF IDEAL DELAY

Figure 11.3-1 shows the schematic model of a system in which ideal delay serves as the memory subsystem. We will introduce now, and continue to use throughout this chapter, the convenient notation,

$$\begin{aligned} &r^{(k)} \text{ means } r(t-kT) \\ &r^{(0)} \text{ or } r \text{ means } r(t) \end{aligned} \tag{1}$$

for any time-dependent system variable. In general, the system memory may consist of a number of ideal delay elements, and there may be multiple inputs and outputs. However, the approach to the general case is clearly revealed by a study of Fig. 11.3-1, where the memory consists of two ideal delay elements and there is a single input and a single output. The difference equations of the system are

$$y(t) = F(x(t), r_1(t-T), r_2(t-T)) \tag{2}$$

$$r_1(t) = G_1(x(t), r_1(t-T), r_2(t-T)) \tag{3}$$

$$r_2(t) = G_2(x(t), r_1(t-T), r_2(t-T)) \tag{4}$$

and the superscript notation lets us write them in the much more compact form,

$$y = F(x, r_1{}^{(1)}, r_2{}^{(1)}) \tag{5}$$

$$r_1 = G_1(x, r_1{}^{(1)}, r_2{}^{(1)}) \tag{6}$$

$$r_2 = G_2(x, r_1{}^{(1)}, r_2{}^{(1)}) \tag{7}$$

Observe that the above three equations are similar in form to the output [Eq. (5)], and state [Eqs. (6) and (7)] equations of Chap. 7. Nevertheless, the state of the system is no longer represented by the *values* of $r_1{}^{(1)}$ and $r_2{}^{(1)}$ but by two functions, as discussed next.

An ideal delay element of delay T effectively stores or "remembers"

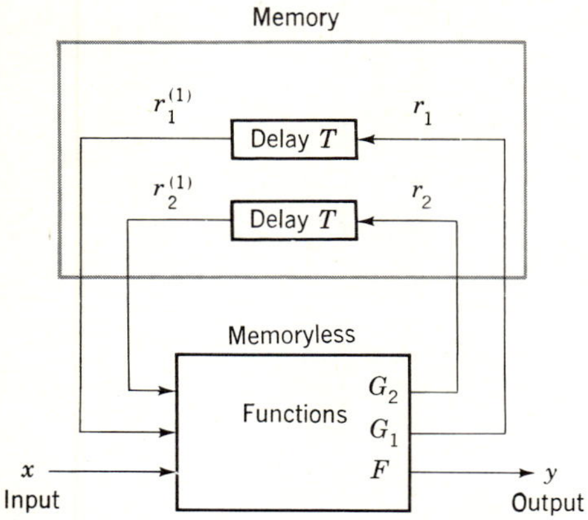

FIG. 11.3-1 The general schematic model of a system whose memory consists of two ideal delay elements.

the *waveform* representing the values assumed by its input variable r_k over the past T seconds of its history. *Specification of the "state" of the system therefore involves specifying this waveform for each of the delay elements comprising the system memory.* Knowledge of the present state of the system and the present and future values of the system input $x(t)$ allows us to compute the present and future values of the system output $y(t)$. In particular, suppose that $x(t)$ is specified for all positive t, and $r_1(t)$ and $r_2(t)$ are known only in the interval of t lying between 0 and $-T$. It is apparent that the values of $r_1(t)$ and $r_2(t)$ in the interval 0 to T may then be calculated directly from Eqs. (3) and (4), after which recursion of the process gives us $r_1(t)$ and $r_2(t)$ in the interval T to $2T$, and so on into the future. With $r_1(t)$ and $r_2(t)$ known for all positive t, Eq. (2) specifies $y(t)$.

A slightly different concept of "state," for systems containing ideal delay, is obtained by limiting attention to a discrete set of "sampling times": $-T, 0, 2T, 3T, \ldots$. Now, the sample values of the input $x(0), x(T), x(2T), \ldots$, together with the two initial state-variable values $r_1(-T)$ and $r_2(-T)$, permit us recursively to calculate $y(0), y(T), y(2T), \ldots$, from Eqs. (2) to (4). Had we started instead with the sampling times, $0.37 - T, 0.37, 0.37 + T, 0.37 + 2T, \ldots$, then the initial state of the system would have been represented by $r_1(0.37 - T)$, $r_2(0.37 - T)$; and we would have been able to calculate recursively $y(0.37)$, $y(0.37 + T), y(0.37 + 2T), \ldots$. The sampled-value viewpoint is different, but the result is the same; the future behavior is determined by the present state and the present and future excitation.

11.4 THE BASIC LINEAR MODEL: CASCADED DELAY WITH PARALLEL FEEDBACK

A system may be visualized as an operator acting upon an input waveform to produce an output waveform. Recall that *a linear* system is one for which the system operator is commutative under scaling and distributive over summing. In particular, the system (output waveform) = F (input waveform) is linear if and only if $F(a_1x_1 + a_2x_2) = a_1F(x_1) + a_2F(x_2)$ for all x_1 and x_2 in the allowable class of input waveforms and for all values of constants a_1 and a_2. Observe that this definition is identical to the definition of a linear memoryless system (Def. 3.6-1), with the important distinction that here x_1 and x_2 are two waveforms, i.e., two *functions on an interval of time*, rather than two values.

This definition of linearity obviously implies uniqueness of the output produced by a given input and also that a null input ($x = 0$ for all t) produces a null output. Therefore, if an otherwise linear system should contain an independent internal signal source, or storage in its memory, that produces "spontaneous" output in the absence of an externally applied input, then we must either formally classify that system as "not linear" or else subtract such spontaneous components of the total output before testing the system for linear character.

Specialization to linearity replaces the general functional equations (5) to (7) of Sec. 11.3 by the linear equations.

$$y = a_{00}x + a_{01}r_1^{(1)} + a_{02}r_2^{(1)} \tag{1}$$

$$r_1 = a_{10}x + a_{11}r_1^{(1)} + a_{12}r_2^{(1)} \tag{2}$$

$$r_2 = a_{20}x + a_{21}r_1^{(1)} + a_{22}r_2^{(1)} \tag{3}$$

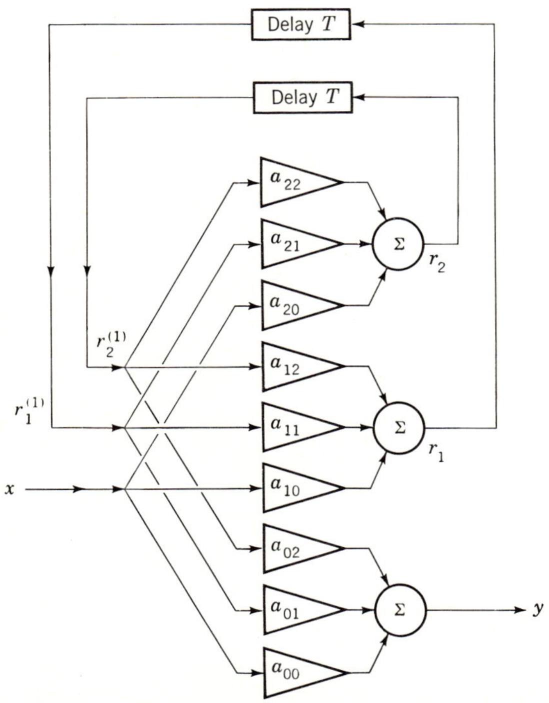

FIG. 11.4-1 *General model of a* linear *system whose memory consists of two ideal delays.*

with corresponding conversion of the system in Fig. 11.3-1 to the linear form detailed in Fig. 11.4-1. Equations (1) to (3) can be manipulated to eliminate the state variables r_1 and r_2, thereby obtaining a difference equation involving only the input and output variables. The procedure is straightforward but nontrivial: First solve Eqs. (2) and (3) for $r_1^{(1)} = L_1(x,r_1,r_2)$ and $r_2^{(1)} = L_2(x,r_1r_2)$, where $L(p,q)$ means "a linear combination of p and q." Second, substitute into Eq. (1) to obtain $y = L_3(x,r_1,r_2)$, which also implies $y^{(1)} = L_3(x^{(1)},r_1^{(1)},r_2^{(1)})$. Third, with the aid of the L_1 and L_2 equations, obtain $y^{(1)} = L_4(x,x^{(1)},r_1,r_2)$, which also implies $y^{(2)} = L_4(x^{(1)},x^{(2)},r_1^{(1)}r_2^{(1)})$. Fourth, again with the aid of the L_1 and L_2 equations, obtain $y^{(2)} = L_5(x,x^{(1)},x^{(2)},r_1,r_2)$. Fifth, solve the equations $y^{(1)} = L_4(x,x^{(1)},r_1,r_2)$ and $y^{(2)} = L_5(x,x^{(1)},x^{(2)},r_1,r_2)$ for r_1 and r_2 in terms of $y^{(1)}$, $y^{(2)}$, x, $x^{(1)}$, $x^{(2)}$, and substitute into $y = L_3(x,r_1,r_2)$ to obtain the *input-output difference equation,*

$$y^{(0)} - b_1y^{(1)} - b_2y^{(2)} = a_0x^{(1)} + a_1x^{(1)} + a_2x^{(2)} \tag{4}$$

wherein

$$a_0 = a_{00} \tag{5}$$

$$a_1 = a_{10}a_{01} + a_{20}a_{02} - a_{00}(a_{11} + a_{22}) \tag{6}$$

$$\begin{aligned} a_2 = a_{00}&(a_{11}a_{22} - a_{12}a_{21}) \\ &+a_{01}a_{12}a_{20} + a_{02}a_{21}a_{10} \\ &-(a_{01}a_{10}a_{22} + a_{02}a_{20}a_{11}) \end{aligned} \tag{7}$$

$$b_1 = a_{11} + a_{22} \tag{8}$$

$$b_2 = a_{12}a_{21} - a_{11}a_{22} \tag{9}$$

The reward for the labor of reducing the state equations (1) to (3) to the form of an input-output equation (4) is that we can now recognize Eq. (4) as the input-output equation of the much simpler system model shown in Fig. 11.5-1. The state equations of the basic cascade model in Fig. 11.5-1 are

$$y = a_0s + a_1s^{(1)} + a_2s^{(2)} \tag{10}$$

$$s = x + b_1s^{(1)} + b_2s^{(2)} \tag{11}$$

It follows directly that

$$y^{(0)} = a_0s^{(0)} + a_1s^{(1)} + a_2s^{(2)} \tag{12}$$

$$-b_1y^{(1)} = -b_1(a_0s^{(1)} + a_1s^{(2)} + a_2s^{(3)}) \tag{13}$$

$$-b_2y^{(2)} = -b_2(a_0s^{(2)} + a_1s^{(3)} + a_2s^{(4)}) \tag{14}$$

and

$$a_0x^{(0)} = a_0(s^{(0)} - b_1s^{(1)} - b_2s^{(2)}) \tag{15}$$

$$a_1x^{(1)} = a_1(s^{(1)} - b_1s^{(2)} - b_2s^{(3)}) \tag{16}$$

$$a_2x^{(2)} = a_2(s^{(2)} - b_1s^{(3)} - b_2s^{(4)}) \tag{17}$$

from which it is obvious that Eqs. (10) and (11) imply Eq. (4). Hence, the system models in Fig. 11.4-1 and Fig. 11.5-1 are entirely equivalent. Any input-output operation that can be accomplished with the general linear model can also be synthesized as a cascaded delay model with parallel feedback. The cascaded delay model is especially convenient because of the simple relationship between the parameters of the model and the coefficients of the input-output equation.

11.5 THE UNCOUPLED LINEAR MODEL: PARALLEL DELAY WITH LOCAL FEEDBACK

In the special case $a_{12} = a_{21} = 0$ (and $a_{10} = a_{20} = 1$), the general linear model in Fig. 11.4-1 reduces to the special form shown in Fig. 11.5-2. This model is said to be "uncoupled" or "diagonalized," because the cross coupling (a_{12} and a_{21}) between the memory elements is missing. Each memory element influences its own input through a local feedback path (c_1 or c_2) but does not affect the inputs of other memory elements.

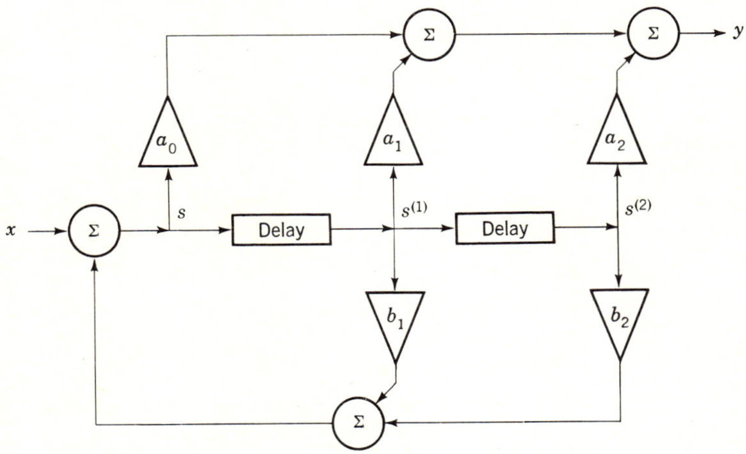

FIG. 11.5-1 The linear model with cascaded delay and parallel feedback.

The analysis of the uncoupled case is particularly simple, because the output of an nth-order (n-delay-element) uncoupled system is effectively the sum of the outputs of n different first-order (single-delay-element) systems operating independently in parallel. The uncoupled case is of special interest to us because for any particular basic linear model with cascaded delay and parallel feedback (Fig. 11.5-1), we can find a corresponding uncoupled model having exactly the same input-output equation, and therefore entirely equivalent operation.

By inspection of Fig. 11.5-2 the state equations of the uncoupled model are

$$y = k_0 x + k_1 v_1^{(1)} + k_2 v_2^{(1)} \tag{1}$$

$$v_1 = x + c_1 v_1^{(1)} \tag{2}$$

$$v_2 = x + c_2 v_2^{(1)} \tag{3}$$

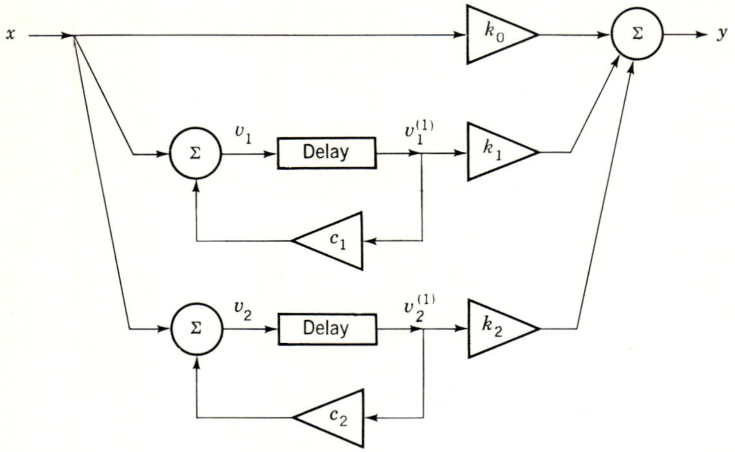

FIG. 11.5-2 The linear model with parallel delay and local feedback.

At this point, the bookkeeping can be simplified greatly by introduction of operational notation. Let multiplication of any system signal by the symbol "lambda" denote the operation of negative delay T upon that signal. It follows that multiplication by lambda obeys the laws of ordinary algebra. For example,

$$(\lambda - c_1)(\lambda - c_2)y^{(2)} = [\lambda^2 - (c_1 + c_2)\lambda + c_1 c_2]y^{(2)} \tag{4}$$

$$= \lambda^2 y^{(2)} - (c_1 + c_2)\lambda y^{(2)} + c_1 c_1 y^{(2)} \tag{5}$$

$$= y^{(0)} - (c_1 + c_2)y^{(1)} + c_1 c_2 y^{(2)} \tag{6}$$

Applying the linear operator in Eq. (4) to both sides of Eq. (1) after first delaying both sides by $2T$, we find

$$(\lambda - c_1)(\lambda - c_2)y^{(2)} = (\lambda - c_1)(\lambda - c_2)(k_0 x^{(2)} + k_1 v_1{}^{(3)} + k_2 v_2{}^{(3)}) \tag{7}$$

With operator notation, Eqs. (2) and (3) can be put in the form

$$(\lambda - c_1)v_1{}^{(3)} = x^{(2)} \tag{8}$$

$$(\lambda - c_2)v_2{}^{(3)} = x^{(2)} \tag{9}$$

By distributing the operator over the variables in Eq. (7) and then substituting Eqs. (8) and (9), we arrive at the input-output equation,

$$(\lambda - c_1)(\lambda - c_2)y^{(2)} = [(\lambda - c_1)(\lambda - c_2)k_0 + (\lambda - c_2)k_1 + (\lambda - c_1)k_2]x^{(2)} \tag{10}$$

Further convenience results if we now introduce the following auxiliary *rational function:*

$$f(\lambda) = \frac{a_0\lambda^2 + a_1\lambda + a_2}{\lambda^2 - b_1\lambda - b_2} \tag{11}$$

with the aid of which the parametric relationships between Figs. 11.5-1 and 11.5-2 can be clarified. A particular rational function is associated with a particular input-output difference equation. The denominator and numerator are respectively the linear operators applied to the input and output variables on the two sides of the input-output equation.† Function (11) evidently represents Eq. (4) of Sec. 11.4.

The rational function (a quotient of lambda polynomials) may be recast in an equivalent form called the *partial fraction expansion:*

$$f(\lambda) = k_0 + \frac{k_1}{\lambda - c_1} + \frac{k_2}{\lambda - c_2} \tag{12}$$

Equivalence of functional form is demonstrated by collection of the constant first term and the fractions over a common denominator,

$$f(\lambda) = \frac{k_0(\lambda - c_1)(\lambda - c_2) + k_1(\lambda - c_2) + k_2(\lambda - c_1)}{(\lambda - c_1)(\lambda - c_2)} \tag{13}$$

which produces numerator and denominator polynomials of the same degree as those in Eq. (11). The function of Eq. (13) may be made to represent the function of Eq. (11). Moreover, by inspection of Eqs. (11) and (12), the parameters of the expansion may be evaluated as follows:

$$c_1, c_2 = \text{roots of } (\lambda^2 - b_1\lambda - b_2 = 0) \tag{14}$$

$$k_0 = f(\infty) = a_0 \tag{15}$$

$$k_1 = \lim[(\lambda - c_1)f(\lambda)] \qquad \text{as } \lambda \to c_1 \tag{16}$$

$$k_2 = \lim[(\lambda - c_2)f(\lambda)] \qquad \text{as } \lambda \to c_2 \tag{17}$$

For example, with an inspection of the denominator of Eq. (13) and the numerator of Eq. (11), and a glance at Eq. (16), we see that k_1 has the value $(a_0c_1{}^2 + a_1c_1 + a_2)/(c_1 - c_2)$.

To summarize, the partial fraction expansion of the rational function of the system in Fig. 11.5-1 is just exactly the rearrangement that is needed to put Fig. 11.5-1 into the uncoupled form of Fig. 11.5-2.

Let us verify the form in two simple special cases of interest. First, set $k_0 = k_2 = 0$ in Eq. (13), and cancel a factor, $\lambda - c_2$, in numerator and denominator to obtain $k_1/(\lambda - c_1)$, the rational function associated with the second subsystem branch of the parallel system in Fig. 11.5-2. Second, set $a_0 = a_2 = b_2 = 0$ in Eq. (11), and cancel a lambda factor to obtain $a_1/(\lambda - b_1)$, the rational function for Fig. 11.5-1 when scalers a_0, a_2, b_2 are disconnected or erased. The effective system model structures and the function forms are the same in the two cases. The cancellation of an operator factor in numerator and denominator of the rational function implies cancellation of that operator from both sides of the associated input-output difference equation, which is permissible so long as a suitable assumption is made about initial conditions of the input and output

†This rational function is for input-output difference equations what the system function $H(s)$ of Chap. 8 is for input-output differential equations.

variables. A sufficient assumption is that the input and output were both zero prior to some finite past time.

11.6 PULSE RESPONSE OF THE PRIMITIVE FIRST-ORDER MODEL

Each delay subsystem in Fig. 11.5-2 is a primitive first-order system like that in Fig. 11.6-1*a*. If we can compute the pulse response (the output produced by a rectangular input pulse of duration T) of this system, then we can

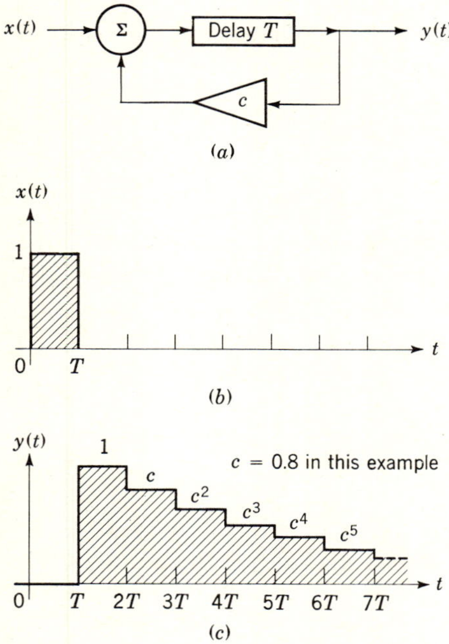

FIG. 11.6-1 A single delay with feedback. (a) *The model.* (b) *The test signal, a pulse of duration T.* (c) *The "pulse response."*

find the pulse response of the general nth-order system as the sum of n such first-order responses. By inspection of Fig. 11.6-1a the input-output equation is

$$y^{(k)} = x^{(k+1)} + cy^{(k+1)} \tag{1}$$

and, in particular,

$$y^{(0)} = x^{(1)} + cy^{(1)} \tag{2}$$

Substitution for y on the right of Eq. (2) gives

$$y^{(0)} = x^{(1)} + c(x^{(2)} + cy^{(2)}) \tag{3}$$

$$y^{(0)} = x^{(1)} + cx^{(2)} + c^2y^{(2)} \tag{4}$$

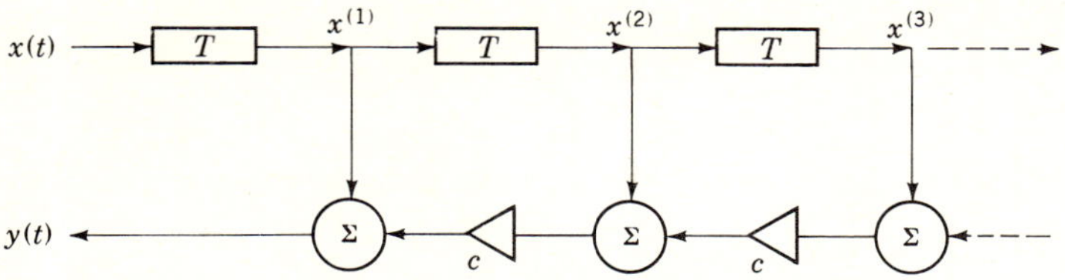

FIG. 11.6-2 Infinite non-feedback *model equivalent to the model of Fig. 11.6-1*a.

and continuation of this *recursive expansion* of the input-output equation leads to the infinite explicit form

$$y^{(0)} = x^{(1)} + cx^{(2)} + c^2x^{(3)} + c^3x^{(4)} + \cdots \tag{5}$$

In accordance with Eq. (5) the response to the input pulse in Fig. 11.6-1*b* is the *pulse response* plotted in Fig. 11.6-1*c*. Observe that the envelope of the stair-step response curve is exponential.

The transmission of a signal through the system in Fig. 11.6-1*a* can be visualized as the parallel combination of paths that go once, twice, three times, . . . , through the delay element, circulating through the feed back scaler before each repetition of passage through the delay. This viewpoint leads to the infinite nonfeedback system shown in Fig. 11.6-2 or its alternative form in Fig. 11.6-3.

The algebraic interpretation of this recursive expansion is offered by

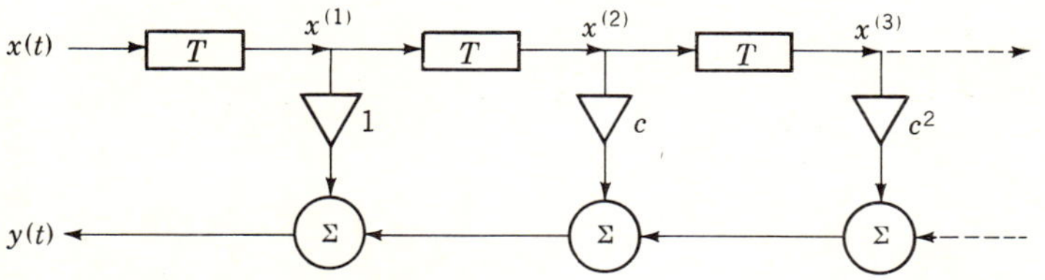

FIG. 11.6-3 An alternate equivalent form of the system in Fig. 11.6-2.

the infinite power-series expansion of the rational function associated with the primitive system:

$$\frac{1}{\lambda - c} = \frac{1/\lambda}{1 - c/\lambda} = \frac{1}{\lambda}\left[1 + \frac{c}{\lambda} + \left(\frac{c}{\lambda}\right)^2 + \left(\frac{c}{\lambda}\right)^3 + \cdots\right] \tag{6}$$

Remember that multiplication by lambda means negative delay, so that the reciprocal of lambda is associated with a positive delay T.

11.7 ILLUSTRATIVE EXAMPLES OF TRANSFORMATION FROM CASCADE TO PARALLEL DELAY

Example 11.7-1 The system in Fig. 11.7-1*a* is a particular member of the class whose analysis is offered by Figs. 11.5-1, 11.5-2, and 11.6-1; and Eqs. (11) through (17) of Sec. 11.5. By comparison with the standard form we see that

$$a_0 = a_1 = 0 \quad a_2 = 0.5 \tag{1}$$

$$b_1 = 1.5 \qquad b_2 = -0.5 \tag{2}$$

$$f(\lambda) = \frac{0.5}{\lambda^2 - 1.5\lambda + 0.5} \tag{3}$$

$$f(\lambda) = \frac{1}{\lambda - 1.0} - \frac{1}{\lambda - 0.5} \tag{4}$$

$$c_1 = 1.0 \qquad k_1 = 1.0 \tag{5}$$

$$c_2 = 0.5 \qquad k_2 = -1.0 \tag{6}$$

so that the pulse response is the superposition or summation of the two first-order responses plotted in Fig. 11.7-1*b*. The reader is invited to account for the values of the first few levels of the response by reasoning directly in terms of Fig. 11.7-1*a*, for comparison with the result shown in Fig. 11.7-1*b*.

Example 11.7-2 The system in Fig. 11.7-2*a* is a cascade of two identical first-order subsystems, whose input-output equations are

$$(\lambda - c)z^{(1)} = x^{(1)} \tag{7}$$

$$(\lambda - c)y^{(1)} = z^{(1)} \tag{8}$$

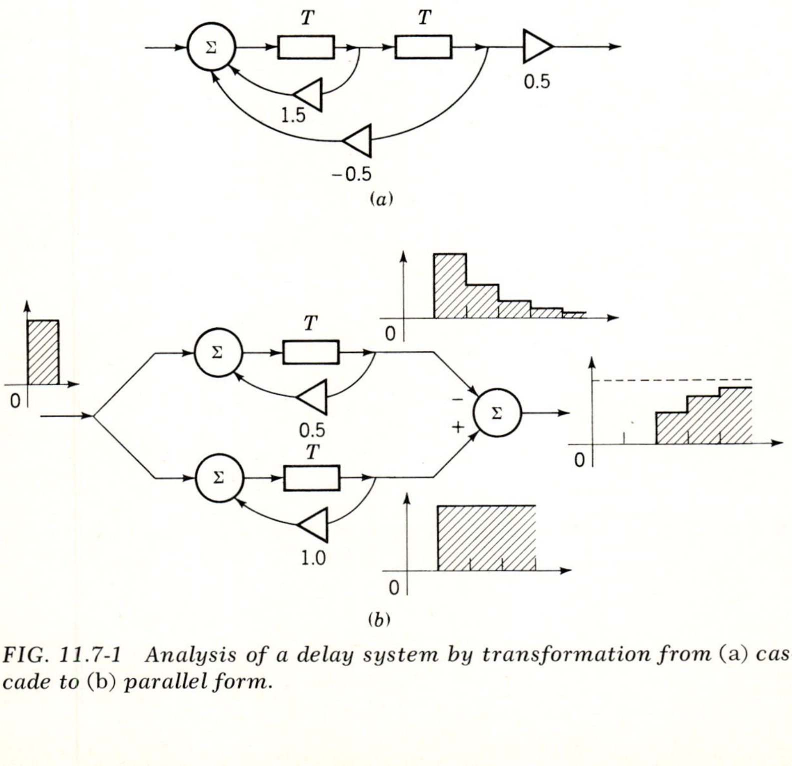

FIG. 11.7-1 Analysis of a delay system by transformation from (a) *cascade to* (b) *parallel form.*

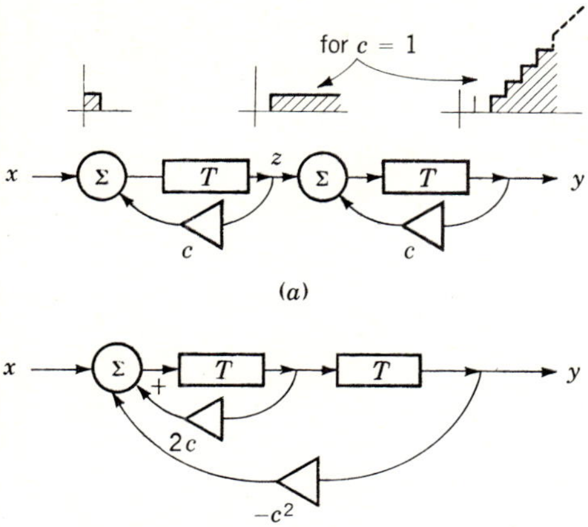

(a)

(b)

FIG. 11.7-2 A system having a multiple zero of the characteristic polynomial. (a) *The cascade of two identical subsystems.* (b) *The equivalent standard form.*

The pulse-response waveforms indicated in the figure follow directly from Figs. 11.2-1 and 11.6-1. Substitution of Eq. (8) into Eq. (7) yields the overall input-output equation,

$$(\lambda - c)^2 y^{(1)} = x^{(1)} \tag{9}$$

Expansion of the operator and injection of an additional delay T gives

$$(\lambda^2 - 2c\lambda + c^2)y^{(2)} = x^{(2)} \tag{10}$$

which means

$$y^{(0)} - 2cy^{(1)} + c^2 y^{(2)} = x^{(2)} \tag{11}$$

The input-output equation (11) is recognizable as that of the particular cascade delay system in Fig. 11.7-2*b*. The standard analysis of such a system is again offered by Figs. 11.5-1, 11.5-2, and 11.6-1; and Eqs. (11) through (17) of Sec. 11.5. However, in this case a difficulty arises in the evaluation of k_1 and k_2 because the characteristic lambda polynomial has two identical roots. We can surmount the difficulty by perturbing parameter b_2 slightly to the value $c^2 - \alpha^2$, in place of c^2, where α is small. The result is

$$a_0 = a_1 = 0 \qquad a_2 = 1 \tag{12}$$

$$b_1 = -2c \qquad b_2 = c^2 - \alpha^2 \tag{13}$$

$$f(\lambda) = \frac{1}{\lambda^2 - 2c\lambda + (c^2 - \alpha^2)} \tag{14}$$

$$f(\lambda) = \frac{1/2\alpha}{\lambda - (c + \alpha)} - \frac{1/2\alpha}{\lambda - (c - \alpha)} \tag{15}$$

$$y(1.5T + kT) = \frac{(c+\alpha)^k - (c-\alpha)^k}{2\alpha} \tag{18}$$

The numerator and denominator in Eq. (18) both vanish for zero alpha, so that the limit may be approached by first differentiating both numerator and denominator. In the limit of small alpha, we find

$$y(1.5T + kT) = kc^{k-1} \tag{19}$$

which is consistent with the output waveform plotted in Fig. 11.7-2*a*.

In the limit of small alpha, the rational function, Eq. (14) becomes a perfect square and may be expressed as a product of two rational functions,

$$f(\lambda) = \left(\frac{1}{\lambda - c}\right)\left(\frac{1}{\lambda - c}\right) \tag{20}$$

each of which is recognizable as the rational function of one of the cascaded subsystems in Fig. 11.7-2*a*. Just as paralleling of subsystems (Fig. 11.5-2) is associated with addition of rational functions [Eq. (57)], so cascading of subsystems is associated with multiplication of their rational functions to obtain the function for the entire system.

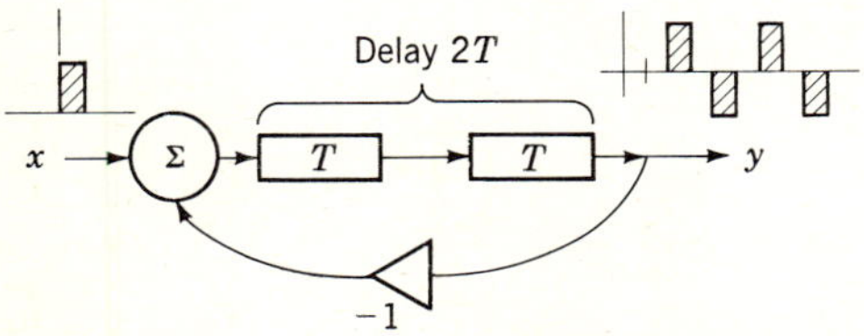

FIG. 11.7-3 A system exhibiting oscillatory pulse response.

Example 11.7-3 The system in Fig. 11.7-3 may be visualized as either a first-order system of delay $2T$ or a special case of the standard second-order system in Fig. 11.5-1. Taking the latter viewpoint, and returning once more to Figs. 11.5-1, 11.5-2, and 11.6-1; and Eqs. (11) through (17) of Sec. 11.5, we find

$$a_0 = a_1 = 0 \qquad a_2 = 1 \tag{21}$$

$$b_1 = 0 \qquad b_2 = -1 \tag{22}$$

$$f(\lambda) = \frac{1}{\lambda^2 + 1} \tag{23}$$

$$f(\lambda) = \frac{1}{(\lambda - j)(\lambda + j)} \tag{24}$$

$$f(\lambda) = \frac{1/2j}{\lambda - j} - \frac{1/2j}{\lambda + j} \tag{25}$$

$$c_1 = j \qquad k_1 = \frac{1}{2j} \tag{26}$$

$$c_2 = -j \qquad k_2 = -\frac{1}{2j} \tag{27}$$

Here all four parameters of the equivalent parallel system have imaginary values. The analysis accompanying Fig. 11.6-1 holds for complex values of c, so that for non-negative integers k,

$$y(1.5T + kT) = \frac{(j)^k - (-j)^k}{2j} \tag{28}$$

which is identifiable as

$$y(1.5T + kT) = \sin \frac{k\pi}{2} \tag{29}$$

The reader may wish to explain the output waveform in Fig. 11.7-3 directly in terms of the system model and compare the result with Eq. (29). The point is that the transformation from Fig. 11.5-1 to Fig. 11.5-2 leads, in general, to complex values of the scalers in Fig. 11.5-2, but the overall result is a real output signal, provided that the input signal and the scalers in the original system (Fig. 11.5-1) are real.

11.8 THE FINITE NONFEEDBACK MODEL AND ITS RESPONSE TO A UNIT PULSE

The infinite model in Fig. 11.6-2, and its pulse response (Fig. 11.6-1c) suggest the system in Fig. 11.8-1a as a useful general model that can realize any desired pulse-response behavior by choice of n and T and adjustment of the values of the parameters, $h_0, h_1, h_2, \ldots, h_n$, as indicated in Fig. 11.8-1c. The explicit input-output relationship is

$$y(t) = [h_0 x(t) + h_1 x(t - T) + h_2 x(t - 2T) + \cdots + h_n x(t - nT)]T \tag{1}$$

or in other notation

$$y(t) = \left[\sum_{k=0}^{n} h_k x(t - kT)\right] T \tag{2}$$

$$y^{(0)} = \left[\sum_{k=0}^{n} h_k x^{(k)}\right] T \tag{3}$$

The standard input pulse is here taken to be one having unit area, as in Fig. 11.8-1b, in preparation for subsequent passage to the limit of small T to obtain the so-called "superposition integral" form of Eq. (2). However, before we discuss the "superposition integral," there are certain important general properties of linear systems that can be examined as conveniently, and perhaps more conveniently, with T small but not infinitesimal.

The model in Fig. 11.8-1a is sometimes called a "tapped delay line." The cascade of delay elements may be visualized as a continuous wave-propagating structure, excited by the input signal and displaying dynamic samples of the past history of the input $x^{(k)}$ at the tapping or sampling points. The output is a weighted sum of these samples, with weighting coefficient $h_k T$.

To appreciate the generality of this linear model, we can return to the general state model described by Eqs. (5) to (7) of Sec. 11.3, and consider the

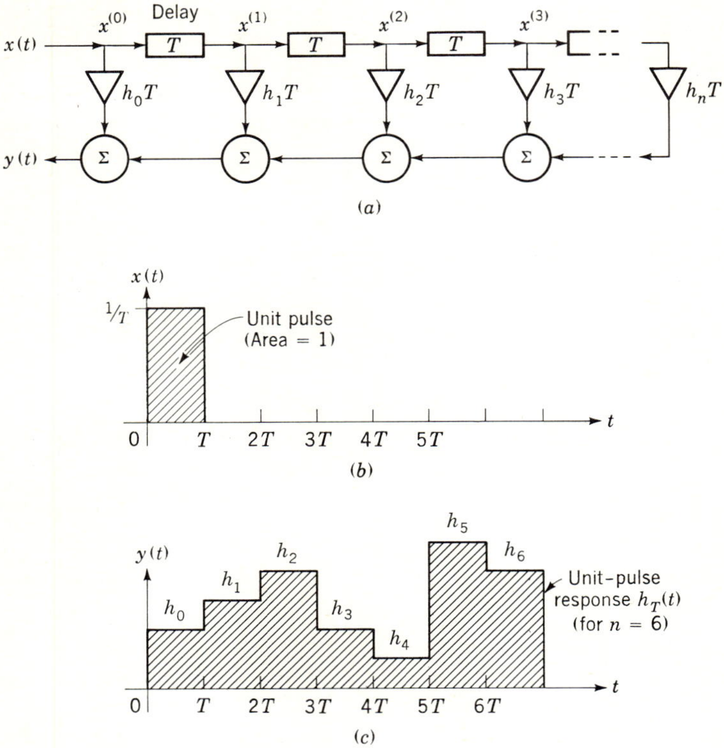

FIG. 11.8-1 The finite non-feedback system. (a) *The system model.* (b) *The "unit-pulse" input signal.* (c) *The "unit-pulse response."*

special case of n delay elements:

$$y^{(0)} = F(x^{(0)}, r_1^{(1)}, r_2^{(1)}, \ldots, r_n^{(1)}) \tag{4}$$

$$r_1^{(0)} = x^{(1)} \tag{5}$$

$$r_k^{(0)} = r_{k-1}^{(1)} \qquad \text{for } k = 2, 3, \ldots, n \tag{6}$$

Recursive substitution of Eqs. (5) and (6) into Eq. (4) gives

$$y^{(0)} = F(x^{(0)}, x^{(1)}, x^{(2)}, \ldots, x^{(n)}) \tag{7}$$

Equation (7) describes a system consisting of a tapped delay line feeding the $n + 1$ inputs of a memoryless system F. The system output is a function of the history of the system input, that history being made up of n samples separated by time intervals T. By choosing T small and nT large we can make the present output dependent upon an arbitrarily fine-grained picture of the input waveform extending over an arbitrarily long past time. It is apparent, therefore, that Eq. (7), containing the arbitrary function F, is a useful model for a very large class of systems and processes.

If we specialize to the linear case, then the nonlinear system represented by function F is replaced by the summers and scalers shown in

Fig. 11.8-1*a*, and Eq. (7) itself becomes Eq. (3). Thus Fig. 11.8-1*a* is a useful model for a very large class of *linear* systems and processes. Observe that once the unit-pulse response is known, the response of the system to any input waveform can be computed directly from Eq. (2).

11.9 THE MATCHED FILTER

Equation (2) of Sec. 11.8 may be interpreted as the dot product of two vectors, and such interpretation leads easily to a very important application called "matched filtering." The *dot product* of two vectors ($\mathbf{a} \cdot \mathbf{b}$) is the sum of the products of their corresponding components along the different space coordinates:

$$\mathbf{a} \cdot \mathbf{b} \text{ means } \sum_k a_k b_k \tag{1}$$

For an orthogonal coordinate system, the dot product is identifiable as the product of the lengths of the two vectors and the cosine of the angle between their two directions. If ϕ is a scalar and **a** and **b** are vectors, then it follows from the elementary properties of vector addition and scaling, with which we assume that the reader is conversant, and from the definition of the dot product, that

$$\mathbf{a} \cdot \mathbf{a} \geq 0 \tag{2}$$

$$(\phi \mathbf{a}) \cdot \mathbf{b} = \phi(\mathbf{a} \cdot \mathbf{b}) \tag{3}$$

$$(\mathbf{a} + \mathbf{b}) \cdot \mathbf{c} = (\mathbf{a} \cdot \mathbf{c}) + (\mathbf{b} \cdot \mathbf{c}) \tag{4}$$

$$\mathbf{a} \cdot \mathbf{b} = \mathbf{b} \cdot \mathbf{a} \tag{5}$$

With reference to Fig. 11.9-1 for geometrical interpretation only, we shall now prove a vector inequality that will be needed shortly. Because of inequality (2) we can write

$$(\mathbf{a} - \phi \mathbf{b}) \cdot (\mathbf{a} - \phi \mathbf{b}) \geq 0 \tag{6}$$

and with the help of Eqs. (3) through (5) this can be expanded to

$$(\mathbf{a} \cdot \mathbf{a}) - 2\phi(\mathbf{a} \cdot \mathbf{b}) + \phi^2(\mathbf{b} \cdot \mathbf{b}) \geq 0 \tag{7}$$

Differentiating the left side of inequality (7) with respect to ϕ and setting the derivative equal to zero, we find

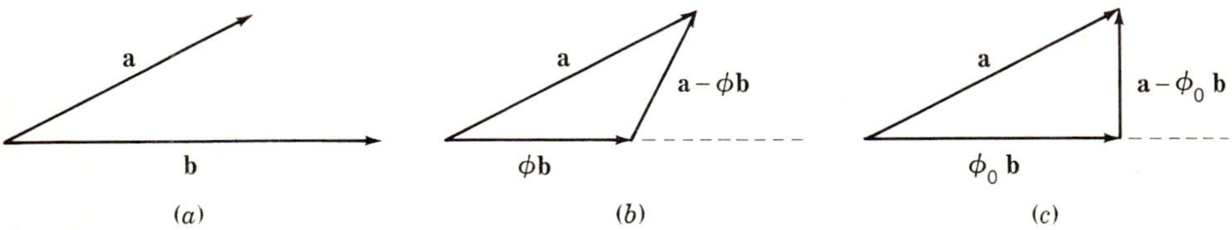

FIG. 11.9-1 Choice of the value ϕ_0 of the scale factor ϕ for minimum vector difference $\mathbf{a} - \phi\mathbf{b}$. (a) *The vectors* **a** *and* **b**. (b) *The difference vector* $\mathbf{a} - \phi\mathbf{b}$. (c) *The minimum distance.*

nal. The reader might enjoy inventing a "quiet" radar employing a transmitted waveform of low power but large total energy. Such considerations arise in a wide variety of signal-processing applications, from radio astronomy to ultrasonic medical diagnostics.

11.10 THE SUPERPOSITION INTEGRAL

Suppose that we want to produce a smoothed version of some signal $x(t)$. In particular, let $y(t)$ be the mean value of $x(t)$ over the immediately preceding time interval of unit duration:

$$y(t) = \int_{t-1}^{t} x(\xi)\, d\xi \tag{1}$$

Equation (1) has the alternate form,

$$y(t) = \int_{0}^{1} x(t-\tau)\, d\tau \tag{2}$$

The integral in Eq. (2) may be approximated by the summation

$$y(t) \approx \left[\sum_{k=0}^{n} x(t-kT)\right] T \tag{3}$$

in which $T = 1/(n+1)$. The geometric interpretation is given in Fig. 11.10-1 for $n = 3$. The approximation improves with increasing n. Indeed, the value of the integral in Eq. (2) may be defined as the limit approached by the summation (3) as n becomes large.

Now, with Eqs. (1) to (3) serving as background, we can find the integral representing the limiting form of Eq. (2) of Sec. 11.8 as nT becomes

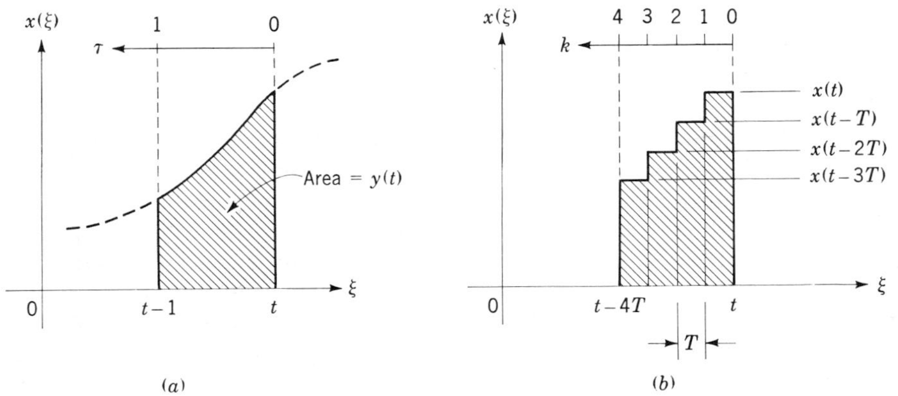

FIG. 11.10-1 Approximation of integral. (a) *The area.* (b) *The approximate area.*

large and T becomes small. Let the weights h_k be treated as samples of some function $h(t)$,

$$h_k = h(kT) \tag{4}$$

whence Eq. (2) of Sec. 11.8 is immediately recognizable as the series-summation approximation to the integral

$$y(t) = \int_0^\infty h(\tau)x(t-\tau)\,d\tau \tag{5}$$

which has the alternate form,

$$y(t) = \int_{-\infty}^t x(\xi)h(t-\xi)\,d\xi \tag{6}$$

Equations (1) to (3) are analogous respectively to Eqs. (6) and (5); and Eq. (2) of Sec. 11.8. The former were introduced first because the limiting process is easier to visualize when the integrand is a single function, instead of a product of two functions.

Equation (5) or its alternate form (6) is called the *superposition integral,* because it superposes the effects of past input excitation to account for the present output. For any linear system, the superposition integral expresses the output directly in terms of an explicit integral operation upon the input. In particular, the present value of the output is a weighted integral over the history of the input, with weighting function $h(t)$.

The superposition integral is the input-output relationship for a limiting form of the system in Fig. 11.8-1 with nT large and T small. Hence $h(t)$ is the response to an input pulse of unit area and infinitesimal duration, called a *unit impulse*, and $h(t)$ is therefore the *impulse response* of the linear system. The reader should try to understand how this discussion is related to Theorem 7.6-1, and to the narrow-pulse response, of RC networks (Sec. 7.7), by reexamining the above sections.

If the impulse response of any linear system is known by analysis or approximate measurement, then the response to any other input signal can be computed directly from the superposition integral.

11.11 THE CONVOLUTION OPERATION

The superposition integral represents an application of a mathematical operation called *convolution*. The convolution of two functions $a(\cdot)$ and $b(\cdot)$ yields a third function $c(\cdot)$. The operation is indicated by the following notation:

$$c = a \otimes b \text{ means } c(t) = \int_{-\infty}^\infty a(\xi)b(t-\xi)\,d\xi \tag{1}$$

The limits of integration in Eqs. (5) and (6) of Sec. 11.10 differ from those in Eq. (1) above because the impulse response $h(t)$ of a realizable linear system is

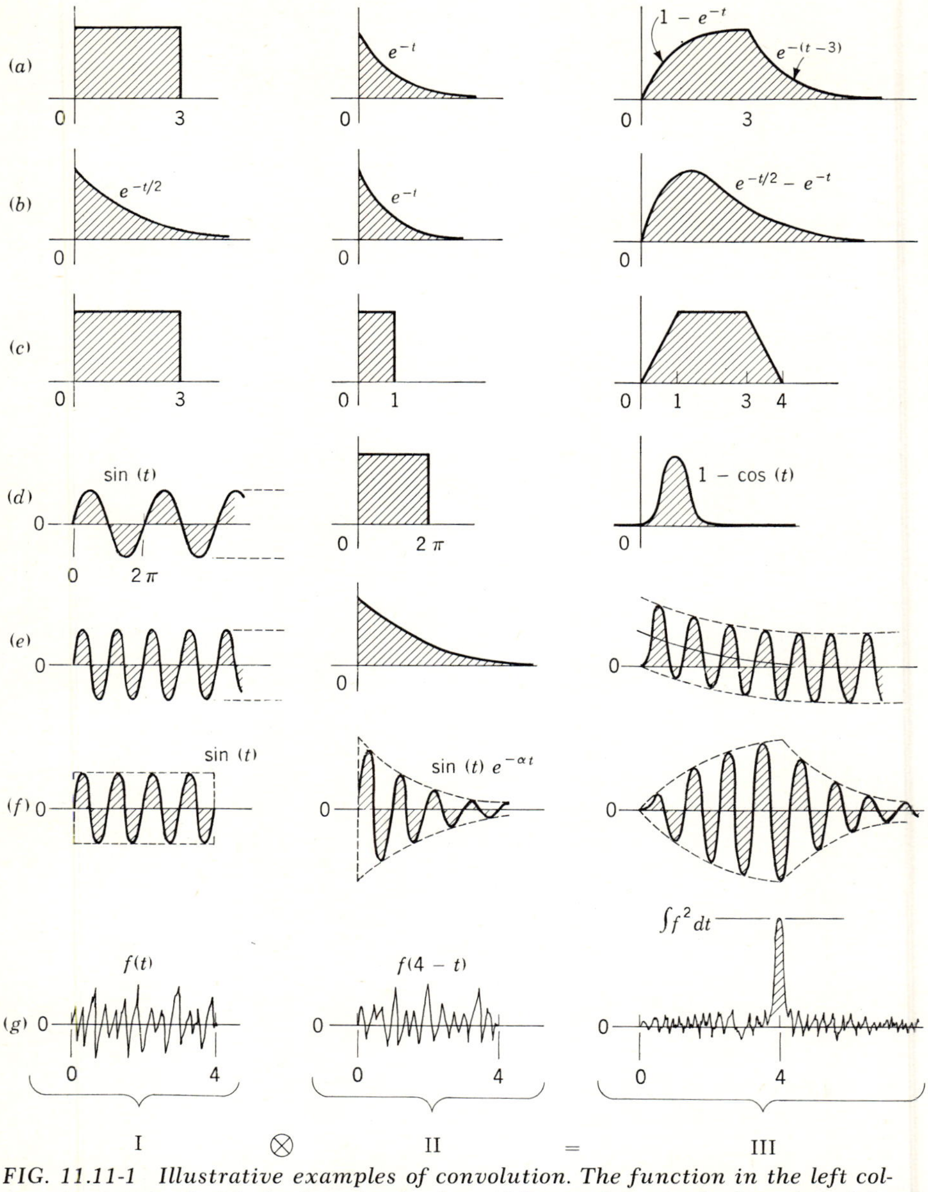

FIG. 11.11-1 Illustrative examples of convolution. The function in the left column is convolved with the function in the middle, resulting in the function of the right column.

assumed to vanish for negative values of its argument (the system does not respond before it is excited), whereas the functions $a(t)$ and $b(t)$ in Eq. (1) have no such restriction.

Selected examples of waveform convolution are plotted in Fig.

11.11-1. A more detailed graphical interpretation of the example in Fig. 11.11-1*b* is plotted in Fig. 11.11-2. One of the two waveforms to be convolved is first reversed left for right about its origin and then shifted past the other function. The area under the product curve, plotted as a function of the amount of shift, gives the third waveform, that is, the convolution of the first two.

Figure 11.11-1*g* shows the convolution of a noiselike waveform with its reversed replica. The result is a noiselike waveform containing a spike whose height is the energy (the integral of the square) of the original waveform. This example illustrates the continuous counterpart of the discrete-sample matched filter described earlier.

The reader is urged to verify, at least qualitatively, each of the results in Fig. 11.11-1 in order to develop intuition for the process of convolution.

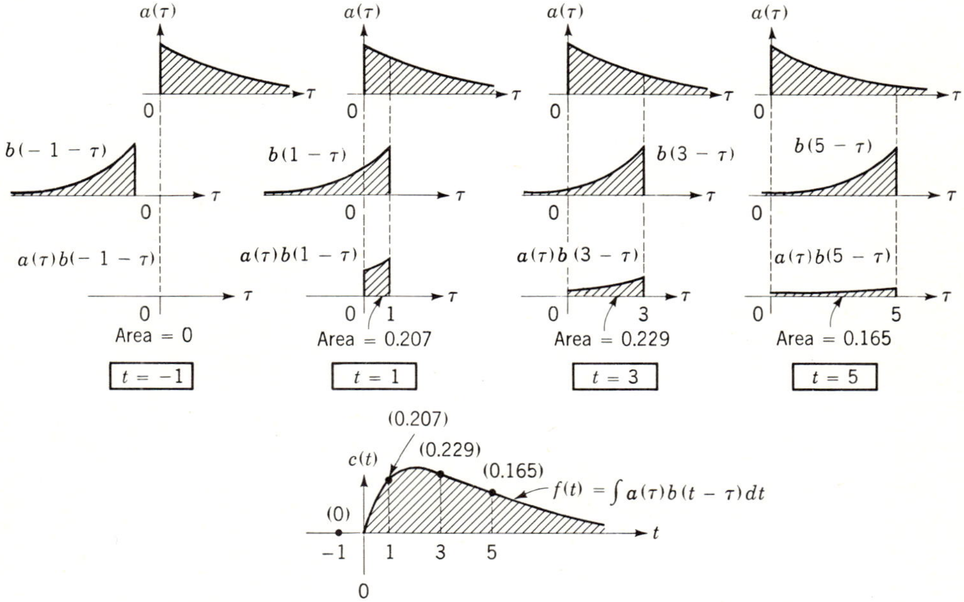

FIG. 11.11-2 Graphical interpretation of convolution.

11.12 LINEAR-SYSTEM EQUIVALENCES REPRESENTING THE BASIC PROPERTIES OF CONVOLUTION

The output waveform of a linear system is the convolution of the input waveform and the impulse response. Therefore, as indicated in Fig. 11.12-1, the impulse response of a cascade combination of two linear systems is the convolution of their individual impulse responses.

Figure 11.12-2 shows four equivalences that can be proved directly from the definition of convolution. In each case functions *a, b, c, d, e* are the impulse responses of subsystems *A, B, C, D, E;* and ϕ is a scale factor.

Convolution is commutative, as shown in Fig. 11.12-2*a*. Inter-

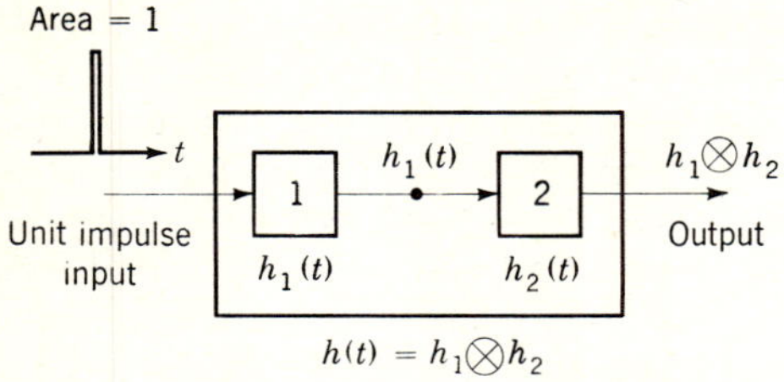

FIG. 11.12-1 Convolution of the impulse responses h_1 and h_2 of cascaded subsystems 1 and 2 to obtain the impulse response h of the resultant system.

changing the order of cascaded subsystems does not affect the overall impulse response.

Convolution obeys scalar associativity, as indicated in Fig. 11.12-2*b*. Here the diagrammatic interpretation is simply that we may compute either the impulse response *c* or the impulse response *d* as a first step toward computation of the overall impulse response.

Convolution is associative, as represented in Fig. 11.12-2*c*. In short, $d \otimes c = a \otimes e$, where $d = a \otimes b$ and $e = b \otimes c$. This result really includes scalar associativity as a special case, obtainable by letting the impulse response

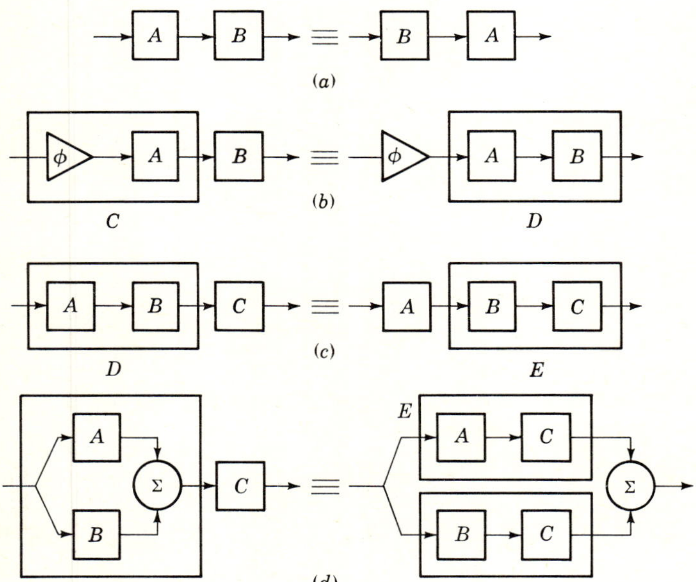

FIG. 11.12-2 Linear system equivalences. In each case, the system on the left is equivalent to the system on the right. (a) $a \otimes b = b \otimes a$. (b) $(\phi a) \otimes b = \phi(a \otimes b)$. (c) $(a \otimes b) \otimes c = a \otimes (b \otimes c)$. (d) $(a + b) \otimes c = (a \otimes c) + (b \otimes c)$.

of subsystem A be an impulse of area ϕ. (Similarly, one or both of the boxes in Fig. 11.12-2a may be a scaler.)

Convolution is distributive over addition, as shown in Fig. 11.12-2d. Thus, $d \otimes c = e + f$, where $d = a + b$, $e = a \otimes c$, and $f = b \otimes c$.

In summary, convolution behaves like a special kind of multiplication. It obeys the same laws of ordinary algebra as multiplication and, together with addition, forms the same kind of mathematical operational structure. The variables in this algebra are, of course, *names for waveforms* rather than names for numbers.

11.13 CONVOLUTIONS INVOLVING DELAY, DIFFERENTIATION, AND INTEGRATION

We will now make use of convolution for the operational definition of certain basic functions representing the impulse responses of certain basic linear systems. Continuity and differentiability are tacitly assumed wherever needed to make an operation meaningful.

Let $d_1(t)$ be defined as that function whose convolution with an arbitrary function $f(t)$ produces a delayed replica $f(t-T)$ of the function. Similarly, let $d_{-1}(t)$ be associated with negative delay, and let d_k be defined by a convenient recursive relationship. That is,

$$d_1(t) \otimes f(t) = f(t-T) \tag{1}$$

$$d_{-1}(t) \otimes f(t) = f(t+T) \tag{2}$$

$$d_{j+k}(t) = d_j(t) \otimes d_k(t) \tag{3}$$

In like manner, we can define the functions $u_k(t)$ by means of the convolution relationships,

$$u_1(t) \otimes f(t) = \frac{df(t)}{dt} \tag{4}$$

$$u_{-1}(t) \otimes f(t) = \int_{-\infty}^{t} f(\xi)\, d\xi \tag{5}$$

$$u_{j+k}(t) = u_j(t) \otimes u_k(t) \tag{6}$$

It follows that

$$d_0(t) \otimes f(t) = f(t) \tag{7}$$

$$u_0(t) \otimes f(t) = f(t) \tag{8}$$

for an arbitrary $f(t)$, and that

$$d_k(t) = u_0(t-kT) \tag{9}$$

Observe further that

$$d_k(t) \otimes f(t) = f(t-kT) \tag{10}$$

$$u_k(t) \otimes f(t) = \frac{d^k f(t)}{dt^k} \tag{11}$$

$$u_{-k}(t) \otimes \frac{d^k f(t)}{dt^k} = f(t) \tag{12}$$

Function u_0 is evidently the impulse response of a unit scaler and is therefore itself a unit impulse. Functions u_1 and u_{-1} are evidently the impulse responses of an ideal differentiator and an ideal integrator, respectively.

As pointed out previously in the development of the superposition integral, the unit impulse may be visualized as a pulse of unit area and short duration. In the present context let a unit pulse be defined as

$$p(t) = \begin{cases} \dfrac{1}{T} & \text{for } 0 \leq t < T \\ \\ 0 & \text{elsewhere} \end{cases} \tag{13}$$

so that

$$p(t) \otimes f(t) = \frac{1}{T} \int_0^T f(t - T)\, dT = \frac{1}{T} \int_{t-T}^{t} f(\xi)\, d\xi \tag{14}$$

and it follows that

$$p(t) \otimes f(t) \rightarrow u_0(t) \otimes f(t) \qquad \text{as } T \rightarrow 0 \tag{15}$$

The character of $u_1(t)$, the impulse response of an ideal differentiator, may be examined by approximating the derivative of a function as the difference of that function and its delayed replica, divided by the delay

$$u_1 \otimes f \approx \frac{1}{T}(d_0 \otimes f - d_1 \otimes f) \tag{16}$$

The desired impulse response is therefore approximated by an impulse of area $1/T$ at the origin and a delayed impulse of area $-1/T$ located at $t = T$.

$$u_1 \approx \frac{1}{T}(d_0 - d_1) \tag{17}$$

In the limit of small T, this pair of huge opposite impulses is the familiar doublet. Similarly, the next higher-order u function, representing the impulse response of a cascade of two differentiators, is called a triplet and is approximated by the convolution of the right side of Eq. (17) with itself:

$$u_2 \approx \frac{1}{T^2}(d_0 - 2d_1 + d_2) \tag{18}$$

The u functions of negative index are the impulse responses of one or more cascaded integrators. In particular,

$$u_{-1} \otimes f = \int_{-\infty}^{\infty} f(\xi) u_{-1}(t - \xi)\, d\xi \tag{19}$$

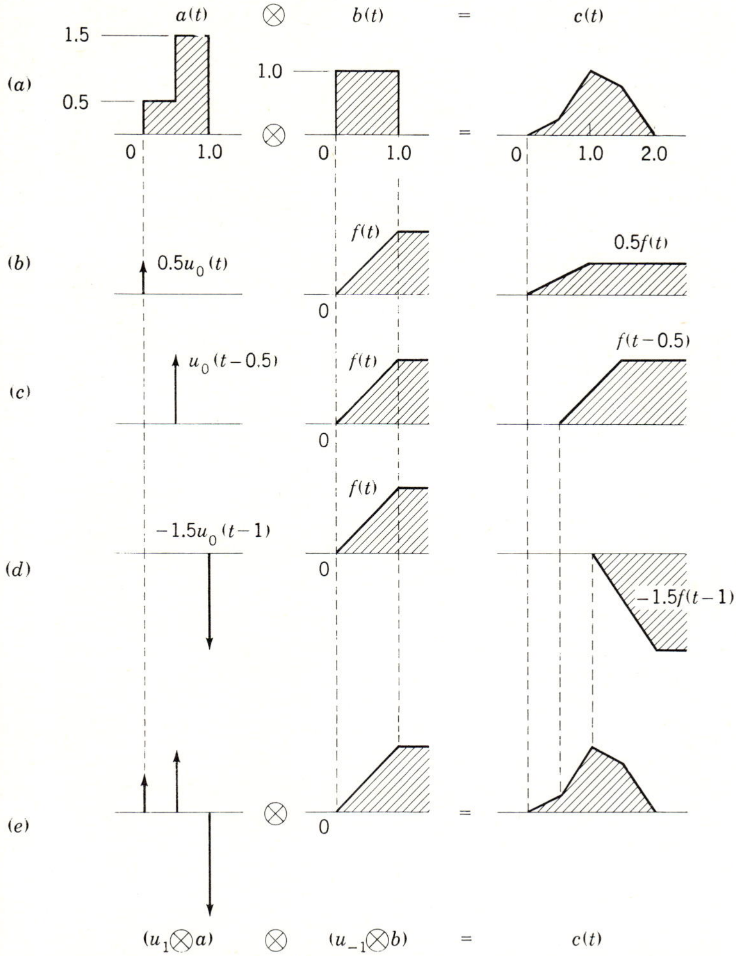

FIG. 11.13-1 Convolution of the derivative of a(t) with the integral of b(t) results in convolution of a(t) with b(t). The horizontal axis is time t, in all cases.

and it follows from Eq. (5) that

$$u_{-1}(t) = \begin{cases} 1 & t > 0 \\ 0 & t \leq 0 \end{cases} \tag{20}$$

Function u_{-1} is the impulse response of an ideal integrator and is the familiar *unit step*. The next lower-index u function is, of course, the unit ramp,

$$u_{-2}(t) = tu_{-1}(t) \tag{21}$$

and for lower index,

$$u_{-k}(t) = \frac{t^k}{k!}\, u_{-1}(t) \tag{22}$$

Hence, the u_k are the singularity functions of Sec. 8.3.

As a practical matter, a convolution calculation is often facilitated if one of the two functions is first differentiated and the other integrated before the convolution is carried out. The legitimacy of such preparatory process can be justified from the observation that

$$(u_1 \otimes a) \otimes (u_{-1} \otimes b) = (u_1 \otimes u_{-1}) \otimes (a \otimes b) \tag{23}$$

and therefore

$$\frac{da(t)}{dt} \otimes \int_{-\infty}^{t} b(\xi)\, d\xi = a \otimes b \tag{24}$$

Figure 11.13-1 offers an example of convolution of $a(t)$ and $b(t)$ shown in Fig. 11.13-1*a*, after prior differentiation of $a(t)$ and integration of $b(t)$. The derivative of $a(t)$ is the set of three impulses shown in Fig. 11.13-1*e*. These impulses appear separately in Fig. 11.13-1*b* through *d*, and the convolution of each impulse with $f(t)$, the integral of $b(t)$, is a suitably scaled and delayed replica of $f(t)$. Finally, $c(t)$ is simply the super-position of these three components, as shown in Fig. 11.13-1*b* through *e*.

As an alternative to the procedure illustrated in Fig. 11.13-1, integration of $b(t)$ could have been omitted, whence the result of the convolution would have been $dc(t)/dt$, awaiting final integration to yield $c(t)$. Had $a(t)$ been a piecewise-linear curve, or a piecewise-linear approximation to some function, then two differentiations would have been required to reduce $a(t)$ to a train of impulses, and two compensating integrations would have been needed on $b(t)$ or on the result of the convolution.

11.14 COMPARISON OF INTEGRATION AND DELAY AS SYSTEM MEMORY ELEMENTS

For linear systems whose memory consists of ideal delay the important system models and equivalences are presented in Figs. 11.3-1, 11.4-1, 11.5-1, 11.5-2; and Eq. (5) of Sec. 11.3 through Eq. (17) of Sec. 11.5. In these figures and equations $f^{(1)}$ and $(1/\lambda)f$ both denote an ideal delay $f(t - T)$ of the original waveform $f(t)$.

We may now make the very important observation that *the entire analysis holds in exactly the same form when the ideal delay operation is replaced by any other specific linear operation.*

In particular, if the ideal delay blocks are replaced by ideal integrators in Figs. 11.3-1 through 11.5-2, and if $f^{(1)}$ and $(1/\lambda)f$ both denote integration whenever they appear in Eq. (5) of Sec. 11.3 through Eq. (17) of Sec. 11.5, then the entire analysis including the partial fraction expansion, remains intact and carries through for systems whose memory consists of ideal integrators. The basic input-output difference equation (4) of Sec. 11.4 becomes an ordinary integral equation and may be differentiated twice to obtain the ordinary differential equation,

$$\frac{d^2y}{dt^2} - b_1\frac{dy}{dt} - b_2y = a_0\frac{d^2x}{dt^2} + a_1\frac{dx}{dt} + a_2x \tag{1}$$

Partial fraction expansion, with lambda now playing the role of the time-differentiation operator, places the system in the form of Fig. 11.5-2, wherein each delay element is to be replaced by an ideal integrator.

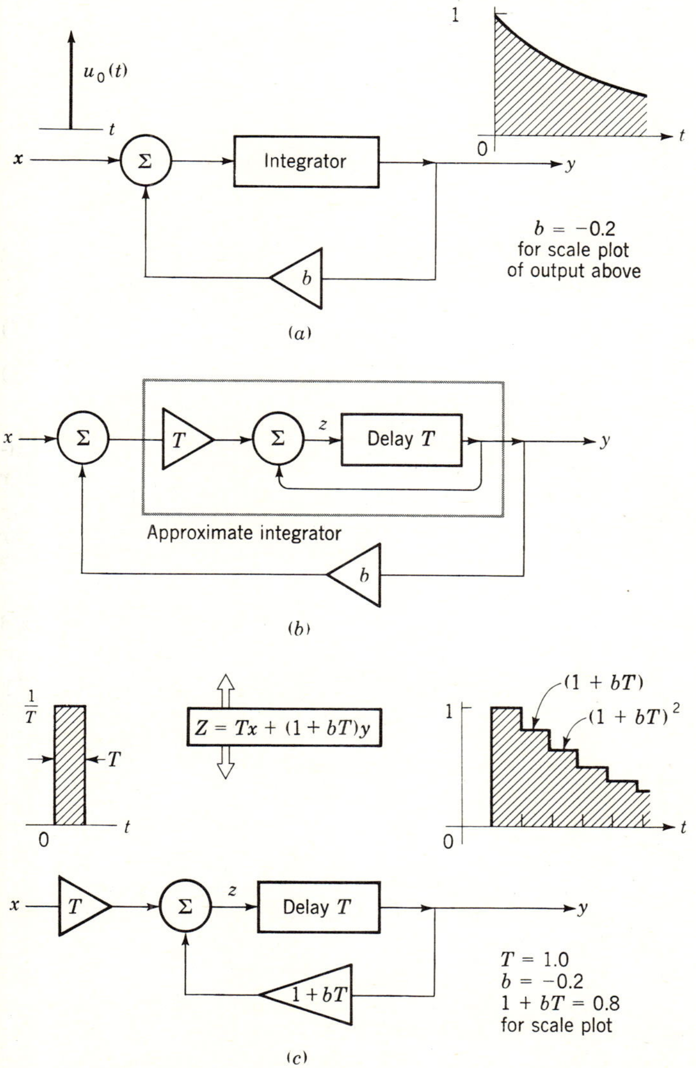

FIG. 11.14-1 Comparison of the basic integrator and delay feedback systems.

The basic parallel building block is now the first-order integrator system shown in Fig. 11.14-1*a*. Its input-output differential equation is

$$\frac{dy}{dt} - by = x \tag{2}$$

and for a unit-impulse input

$$x = u_0(t) \tag{3}$$

the output becomes the impulse response

$$y = h(t) = e^{bt}u_{-1}(t) \tag{4}$$

To verify Eq. (4) first rewrite Eq. (2) in the form

$$(u_1 - bu_0) \otimes y = x \tag{5}$$

and substitute Eqs. (3) and (4),

$$(u_1 - bu_0) \otimes e^{bt}u_{-1} = u_0 \tag{6}$$

Expansion and straightforward manipulation lead to the equality:

$$u_1 \otimes e^{bt}u_{-1} - bu_0 \otimes e^{bt}u_{-1} = u_0 \tag{7}$$

$$be^{bt}u_{-1} + u_0 - be^{bt}u_{-1} = u_0 \tag{8}$$

The impulse response is plotted in Fig. 11.14-1*a*.

The overall impulse response of a higher-order linear system containing integrators is obtainable by superposition of the parallel first-order responses, just as in Fig. 11.7-1, and the response to an arbitrary input waveform may then be calculated directly from the superposition integral. The reader may find it instructive to consider each of the illustrative examples in Sec. 11.7 with the delays replaced by integrators.

As a final point of contact and comparison between delays and integrators, the system in Fig. 11.14-1*b* offers an interesting link. First approximate the integrator in Fig. 11.14-1*a* by the box in Fig. 11.14-1*b*, and then simplify to the form shown in Fig. 11.14-1*c*. Observe that a feedback scaler b around an integrator is approximated by a feedback scaler $1 + bT$ around a small delay T, together with a cascaded scale factor T.

11.15 THE SYSTEM FUNCTION

The impulse response $h(t)$ is a complete input-output characterization of a linear system, because the response to any input can be computed when $h(t)$ is known. A different characterization results from the search for an input signal that produces an output signal having *exactly the same waveform*, apart from a scale factor. The output waveform may be visualized as the result of a linear transformation upon the input waveform. Any waveform whose transformation is a scaled replica of itself is called an *eigenfunction* of the transformation, and the scale factor is called an *eigenvalue*.†

†Eigenfunctions and eigenvalues are discussed in greater detail in Volume 2.

Construction of an eigenfunction proceeds as follows. Let H stand for the unknown eigenvalue, i.e., for a number

$$y(t) = Hx(t) \qquad \text{for all } t \tag{1}$$

and substitute this constraint into the superposition-integral equation

$$Hx(t) = \int_0^\infty h(\tau)x(t-\tau)\,d\tau \tag{2}$$

If Eq. (2) is to hold for all t, then a factor $x(t)$ must be recoverable from within the integral. In short, we must have

$$x(t-\tau) = f(\tau)x(t) \tag{3}$$

where $f(\tau)$ is as yet undetermined. Cancellation of $x(t)$ gives

$$H = \int_0^\infty h(\tau)f(\tau)\,d\tau$$

To linearize the difference equation, let

$$p(t) = \log x(t) \tag{5}$$

$$q(\tau) = \log f(\tau) \tag{6}$$

whence Eq. (3) may be replaced by

$$p(t-\tau) = q(\tau) + p(t) \tag{7}$$

Now assume that $p(t)$ is sufficiently well behaved for representation as a power series,

$$p(t) = a_0 + a_1 t + a_2 t^2 + a_3 t^3 + \cdots \tag{8}$$

whereupon substitution into Eq. (7) yields

$$a_0 + a_1(t-\tau) + a_2(t-\tau)^2 + \cdots = q(\tau) + a_0 + a_1 t + a_2 t^2 + \cdots \tag{9}$$

By equating like terms on each side of Eq. (9) we find that

$$p(t) = a_0 + a_1 t \tag{10}$$

$$q(\tau) = -a_1\tau \tag{11}$$

Hence, from Eqs. (5) and (6)

$$x(t) = e^{a_0 + a_1 t} = Ce^{st} \tag{12}$$

$$f(\tau) = e^{-a_1\tau} = e^{-s\tau} \tag{13}$$

The eigenfunction, therefore, is an exponential function of time, having an arbitrary (perhaps complex) exponential coefficient s and an arbitrary (perhaps complex) amplitude C. The eigenvalue H is a function of s and upon substitution of Eq. (13) into Eq. (4) we find that $H(s)$ has the form

$$H(s) = \int_0^\infty h(\tau)e^{-s\tau}\,d\tau \tag{14}$$

For a tapped delay line system like that in Fig. 11.8-1*a*, the superposition integral would be replaced by the summation [Eq. (2) of Sec. 11.8]; and Eq. (14) above would become

$$H(s) = \left(\sum_{k=0}^{n} h_k e^{-k\tau s}\right)\tau \tag{15}$$

In either case, the result is that a linear system excited by the input signal,

$$x(t) = Ce^{st} \qquad \text{for all } t \tag{16}$$

produces the output signal,

$$y(t) = H(s)Ce^{st} \qquad \text{for all } t \tag{17}$$

where s and C are arbitrary (perhaps complex) constants.

Observe that function $H(s)$ is the familiar *system function* of Chap. 8. In the domain of s for which integral (14) exists, that integral defines a function $H(s)$. For a linear system containing integrators, $h(t)$ is a sum of exponentials, and $H(s)$ is a rational function of s. The rational function introduced in Sec. 11.5 is, in fact, exactly the same rational function as $H(s)$.

PROBLEMS

11.1-1 For the system of Fig. 11.1-1*b* suppose that the input waveform is an ideal unit ramp, that is, $x(t) = 0$ for negative t and $x(t) = t$ otherwise. Assume that $T = 1$. Sketch the waveforms of $y(t)$ and $z(t)$.

11.1-2 For the system of Fig. 11.1-1*c* in which the function block "saw" is defined by Fig. 11.1-2, suppose that the input waveform is an ideal unit ramp, that is, $x(t) = 0$ for negative t and $x(t) = t$ otherwise. Assume that $T = 1$. Sketch the waveforms of $y(t)$, $y_1(t)$, $y_2(t)$, $z(t)$.

11.1-3 For the system in Fig. 11.1-1*b* suppose that the input signal is a suddenly applied sinusoid. In particular, let $x(t) = 0$ for negative t and $x(t) = \sin t$ otherwise. Assume that delay T is equal to the period of the sinusoid. Sketch the waveforms of $y(t)$ and $z(t)$. Comment on the character of "information transmission" under these particular conditions.

11.2-1 Suppose that in the system of Fig. 11.2-2*b* the input signal $x(t)$ is a pulse of unspecified shape but of known area P. Prove that the output $y(t)$ is a pulse signal having the same area P. *Hint:* First show that for a constant input the output is equal to the input. Then show that the time-average values of the input and output are identical. Then consider an input consisting of a periodic train of widely spaced pulses.

11.4-1 In the system of Fig. 11.4-1 let $a_{01} = a_{10} = a_{21} = a_{02} = 1$, $a_{12} = -1$, and set all others to the value zero. Calculate the resulting values of the parameters in the equivalent model (Fig. 11.5-1) having cascaded delay and parallel feedback. Can you find a periodic input signal (of nonzero amplitude) such that the resulting output signal is zero?

11.5-1 For the system of Fig. 11.5-2 set $k_0 = 0$, $k_1 = k_2 = 1$, $c_1 = 1/2$, $c_2 = -1/2$.

(*a*) Produce an equivalent system in the form of Fig. 11.5-1.

(*b*) By means of step-by-step calculations at successive times T apart, determine and sketch the pulse response of each of the two equivalent forms.

11.6-1 Reflection of radio waves from the ionosphere and reflection of diagnostic ultrasonic acoustic waves from a tissue layer within the ailing body of a medical patient are examples of wave-propagation systems for which an ideal delay model is often a useful approximation. With reference to part I of the figure, suppose that the interface reflection coefficient r can be predicted from the physical properties of the two media. The two variables involved in acoustic wave propagation are velocity of the medium and differential pressure in the medium. We will take velocity as the reference variable, so that the incident, transmitted, and reflected pulses are "velocity pulses." There is a constraint, $t = 1 + r$, between the reflection and transmission coefficients at a contact boundary because of continuity of velocity of motion of the medium, at the boundary as well as elsewhere. Moreover, because of the nature of the physical laws governing the wave motion, we have $r_{21} = -r_{12}$ in part II of the figure.

Assume that a pulse of any waveform is undistorted after reflection from or transmission through any boundary, but scaled in size according to r or t. Assume that the system is linear so that superposition of component waves is applicable. Let $r_{21} = 0.1$.

(*a*) Using ideal delay elements, make an idealized system model that describes the transmission and reflection of waves from a tissue layer.

(*b*) Putting the model in some appropriate equivalent form, discuss the distorting effect of the layer upon a transmitted pulse whose duration is several times longer than twice the transit time through the layer.

(*c*) At what frequencies is the layer most transparent? least transparent?

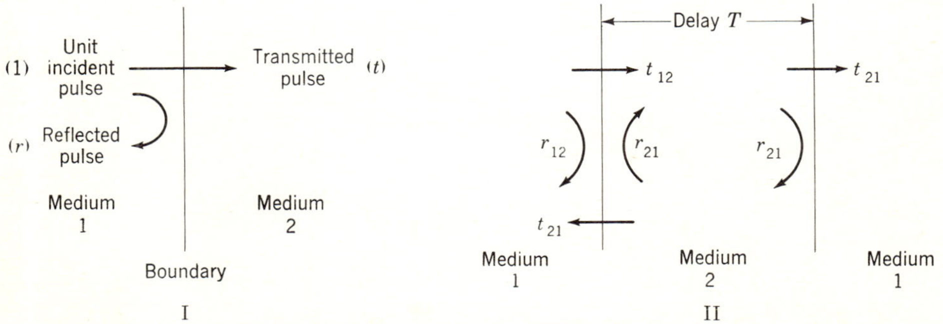

PROB. FIG. 11.6-1 Reflection and transmission of acoustic waves by a layer. I, Interface reflection and transmission. II, Multiple reflections within a layer.

11.7-1 An ideal delay system is characterized by the input-output difference equation,

$$y(t) = x(t) + 0.5y(t-1) - y(t-2)$$

(*a*) Make a state model of this system, utilizing ideal delay elements.

(*b*) Transform the model to the parallel delay form, and find the step response.

11.7-2 Consider a system consisting of n ideal delay elements, all of delay T, connected in a simple closed loop, with all elements oriented in the same direction around the loop. Signal-transit time around the loop is therefore nT. Now insert a summing element, so that an input signal can be injected at some point in the loop, and provide an output from some point in the loop. How is the character of the input-output difference equation influenced by the choice of the signal injection and output points?

11.7-3 For the system in Fig. 11.7-1*a* suppose that the output point is relocated between the two delay elements instead of at the right end of the second delay element. What is the effect upon the equivalent parallel model and upon the pulse response?

11.7-4 A particular ideal delay system model has the associated rational function,

$$f(\lambda) = \frac{1}{\lambda^3 - \lambda^2 + \lambda - 1}$$

Find a representation as a first-order system cascaded with a second-order system, and briefly describe the pulse response of each cascaded part and also that of the complete system.

11.8-1 The system model of a "pulse-forming" electric transmission-line

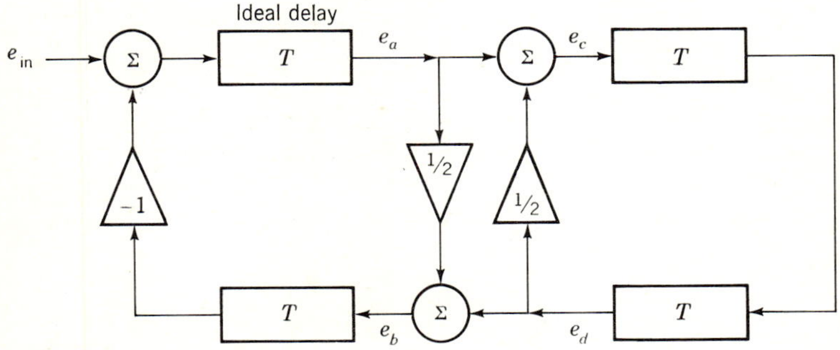

PROB. FIG. 11.8-1

device is shown in the figure. The output voltage of the device is the voltage difference

$$e_{out} = (e_a + e_b) - (e_c + e_d)$$

where e_a and e_b are the amplitudes of right-going and left-going waves at one point on the transmission line and e_c and e_d are similar components at another point

The applied input voltage is held at some fixed value until the system reaches equilibrium, after which the input is suddenly changed to the value zero and held at that value, resulting in the generation of a square pulse of output voltage. Analyze the system.

11.8-2 The pulse response of a linear system containing ideal delay memory elements is known to be a finite sequence of pulses, as in Fig. 11.8-1. There are eight pulses in the set and

$$h_k = (0.9)^k \qquad \text{for } k = 0, 1, \ldots, 7$$

(*a*) Synthesize a nonfeedback system model, utilizing scalers of value 0.9 and ideal delays T.

(*b*) Synthesize a system model, utilizing no more than three delays and three scalers.

In (*b*) neither the delays nor the scalers need be the same. In both (*a*) and (*b*) utilize as many summers as necessary.

11.9-1 Design a "matched filter" for a signal that consists of a square wave of finite duration. Utilize a number of delay elements approximately equal to twice the number of square-wave periods contained in the signal. After you design the system model, discuss its steady-state response to sinusoids of various frequencies. In particular, at what frequencies does it exhibit null response?

11.11-1 Verify each of the convolutions exhibited in Fig. 11.11-1.

11.11-2 Show that the convolution of two pulse waveforms produces a pulse whose area is the product of their two areas.

11.11-3 The input $x_1(t)$ to a linear system is a square pulse of unit duration. For each of the following outputs $x_2(t)$, find the implied impulse response $h(t)$, and synthesize a system model.

(*a*) $x_2(t)$ is a tent-shaped triangular pulse of duration 2, starting at $t = 0$.

(*b*) $x_2(t)$ is a square pulse of duration 1.3, starting at $t = 0$.

(*c*) $x_2(t)$ is a square pulse of unit duration, centered at the t origin.

(*d*) $x_2(t) = \sin \pi t$ for t between 0 and 1, and $x_2(t)$ vanishes elsewhere.

(*e*) $x_2(t)$ is a ramp; that is, it equals t for positive t, and vanishes for negative t.

Appendix A
Complex Numbers and Exponentials

The purpose of this appendix is to review some basic properties of complex numbers and exponentials. It is particularily appropriate to investigate the properties of certain exponential signals that the reader will frequently encounter. These signals are complex exponentials, *written as* Xe^{st}, *where* X *and* s *are complex numbers, and* t *is a real variable. Signals of this form are eigenfunctions*† *of linear time-invariant systems. Moreover, the complex exponential forms the basis for several powerful techniques of signal representation.*

†That is, if the input is of the form Xe^{st}, the output will be of the same form, with typically different values for X and s.

Complex numbers are useful in general algebraic theory because the real numbers do not provide a representation for the square root of a negative real number. A convenient way to represent complex numbers is as an ordered pair (x,y) of real numbers. When a complex number z is represented as (x,y), x is called the *real part* of z, $x = \text{Re}\,[z]$; while *y is called the imaginary part* of z, $y = \text{Im}\,[z]$. It is important to remember that *both the real and imaginary part of a complex number are real quantities.* Common operations, such as addition and multiplication, are defined for complex numbers so that if the imaginary parts of the complex numbers involved are zero, the results for reals are obtained. Thus,

$$(x_1,0) + (x_2,0) = (x_1 + x_2, 0)$$

$$(x_1,0)(x_2,0) = (x_1x_2,0)$$

For arbitrary complex numbers, it is necessary to define the various operations. Equality of complex numbers is defined as

$$(x_1,y_1) = (x_2,y_2) \qquad \text{if and only if } x_1 = x_2 \text{ and } y_1 = y_2.$$

Thus, an equality involving complex numbers represents two equalities involving real numbers, one for the real parts and another for the imaginary parts.

Multiplication for complex numbers is defined as

$$(x_1,y_1)(x_2,y_2) \triangleq (x_1x_2 - y_1y_2,\ x_1y_2 + x_2y_1)$$

from which we see that

$$(0,y_1)(0,y_1) = (-y_1^2,0)$$

Addition of complex numbers is defined as

$$(x_1,y_1) + (x_2,y_2) \triangleq (x_1 + x_2,\ y_1 + y_2)$$

Moreover, the same associative, commutative, and distributive laws for addition and multiplication that hold for real numbers hold for complex numbers.

The representation of a complex number by an ordered pair of reals is somewhat abstract and formal. An alternative way, more suitable for manipulation, is

$$(x,y) = x + jy$$

where j is a special quantity† defined to have the property $j^2 = -1$. Using this representation, we can manipulate complex numbers as reals, always remembering that the imaginary part is distinguished by the appearance of the quantity j. For example,

$$(x_1 + jy_1) + (x_2 + jy_2) = x_1 + x_2 + j(y_1 + y_2)$$

which is consistent with the definition of addition. For multiplication, we need to observe that regardless of what powers of j appear in the result, a reduction can be performed so that j appears at most to the first power. This is because

†In the mathematical literature, i is used instead of j.

$j^3 = -j$, $j^4 = 1$, etc. Thus for multiplication we have

$$(x_1 + jy_1)(x_2 + jy_2) = x_1x_2 + jy_1x_2 + j^2y_1y_2 + jx_1y_2$$
$$= x_1x_2 - y_1y_2 + j(x_1y_2 + x_2y_1)$$

again consistent with our earlier definition. We will postpone a discussion of division until we have developed another and still more convenient representation of complex numbers.

Since we can represent a complex number by an ordered pair of real numbers, we can establish a one-to-one correspondence between complex numbers and points in a two-dimensional coordinate system. One axis of the system will correspond to the real part, or first component of the complex number, while the other axis will correspond to the imaginary part or second component. Thus the complex number $(-2,3) = -2 + j3$ is represented as shown in Fig. A-1.

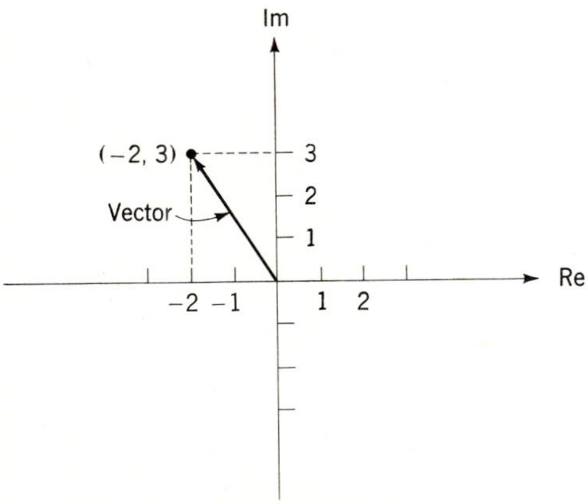

FIG. A-1 The number $-2 + j3$.

If we think of a complex number as being represented by a *vector* from the origin to the appropriate point, we can develop yet another representation. Knowing the length of this vector and the angle it makes with some axis allows us to specify the complex number. The length of the vector defining a complex number z is called the magnitude of the complex number, denoted $|z|$. If $z = x + jy$, then

$$|z| = \sqrt{x^2 + y^2}$$

Although the magnitude of any complex number is unique, the angle defining it is not. If we agree to measure the angle ϕ (in radians) in a counterclockwise direction from the positive real axis, we see that changes of 2π rad leave the defined number unchanged. If, however, we agree to restrict the angle $-\pi < \phi \leq \pi$, the resulting representation is unique. Thus, the complex number $z = (1, -1) = 1 - j$ has magnitude $|z| = \sqrt{2}$ and angle $\phi = -\pi/4$. If the magnitude $|z|$ and angle ϕ of a number z are specified, we have

$$z = (|z| \cos \phi, |z| \sin \phi)$$

Operations among complex numbers can be viewed as operations among the appropriate vectors. Although we will not develop this point of view, the reader is encouraged to do so for himself.

With this review completed we can now turn to a discussion of complex exponentials. We want to develop a relation between the expression e^z, where $z = x + jy$.

This is conveniently done, for our purposes, with the definition†

$$e^z \triangleq e^x(\cos y + j \sin y)$$

This relation permits the use of the law of exponents, namely, $e^{z_1}e^{z_2} = e^{z_1+z_2}$, a fact that should be verified by the reader. If we let $x = 0$, we have the Euler relation,

$$e^{jy} = \cos y + j \sin y$$

Thus, e^{jy} *is a complex number of unit magnitude and angle* y. This allows us to write any complex number z as a scaled version of $e^{j\phi}$ for suitable ϕ. The scale factor will be the magnitude of z, and ϕ will be its angle. Thus, the complex number $(1,-1)$ can be written as

$$\sqrt{2}\, e^{-j\pi/4}$$

This is called the *polar form* and the general expression is

$$(x,y) = \sqrt{x^2 + y^2}\, e^{j\phi}$$

where ϕ is the angle of the vector representing (x,y), i.e., $\phi = \arctan (y/x)$.

The polar form is convenient for multiplication and division of complex numbers. Let $z_1 = |z_1|e^{j\phi_1}$ and $z_2 = |z_2|e^{j\phi_2}$; then

$$z_1z_2 = |z_1|\, |z_2|e^{j(\phi_1+\phi_2)}$$

and

$$\frac{z_1}{z_2} = \frac{|z_1|}{|z_2|}\, e^{j(\phi_1-\phi_2)}$$

The reader is encouraged to use this division relationship to find an equivalent one using ordered pairs.

The last operation to be discussed is that of taking the complex conjugate of a number. If $z = x + jy$, we define the *complex conjugate* of z, written z^*, as $z^* \triangleq x - jy$. We see immediately that

$$z + z^* = 2 \text{ Re } [z]$$

and

$$z - z^* = 2j \text{ Im } [z]$$

Use of the polar form gives us, if $z = |z|\, e^{j\phi}$,

$$z^* = |z|e^{-j\phi}$$

†Although we treat this equality as a definition, it is actually a result of the complex power-series expansion for e^z.

and

$$zz^* = |z|\, e^{j\phi}\, |z|\, e^{-j\phi} = |z|^2$$

Some identities involving the complex conjugate are listed below and should be verified by the reader.

$$(z_1z_2)^* = (z_1)^*(z_2)^*$$

$$(z_1 + z_2)^* = (z_1)^* + (z_2)^*$$

$$\left(\frac{z_1}{z_2}\right)^* = \frac{z_1{}^*}{z_2{}^*}$$

$$(e^s)^* = e^{s^*}$$

We are now equipped to investigate the complex exponential Xe^{st}, where X and s are complex numbers, and for the purpose of representing time functions, t will be a real variable. X will be called the *complex amplitude*, and s the *complex frequency* of the exponential. Our purpose will be to examine the variety of real-time functions that can be produced, using the complex exponential and various operations on it.

If X and s are both real, we see that $x(t) = Xe^{st}$ is simply a real exponential of one of the three types shown in Fig. A-2. If both X and s have

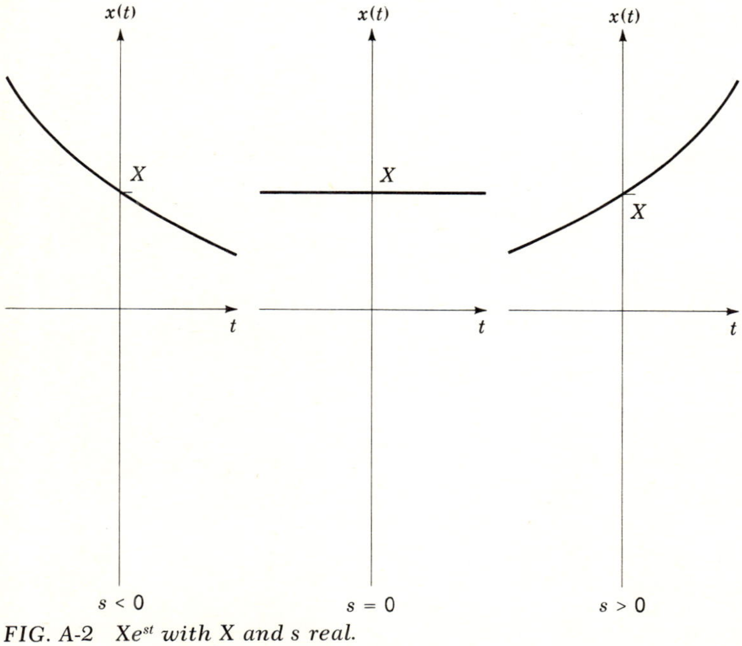

FIG. A-2 Xe^{st} with X and s real.

nonzero imaginary parts, Xe^{st} is not generally real. Since we are interested in real-time functions, a suitable operation must be performed to generate a real signal. A convenient operation is to add Xe^{st} to its complex conjugate, giving a result that we know is always real. In this case we have

$$y(t) = Xe^{st} + X^*e^{s^*t} = 2\ \text{Re}\ [Xe^{st}]$$

It is standard convention to let $s \triangleq \sigma + j\omega$ where σ and ω are real.

Writing $X = |X|e^{j\phi}$ we have $Xe^{st} = |X|e^{j\phi}e^{\sigma t}e^{j\omega t}$. Hence,

$$\begin{aligned} y(t) = 2\ \text{Re}\ [Xe^{st}] &= |X|e^{j\phi}e^{\sigma t}e^{j\omega t} + |X|e^{-j\phi}e^{\sigma t}e^{-j\omega t} \\ &= |X|e^{\sigma t}(e^{j(\phi+\omega t)} + e^{-j(\phi+\omega t)}) \\ &= |X|e^{\sigma t}(2\ \text{Re}\ [e^{j(\phi+\omega t)}]) \\ &= |X|e^{\sigma t}\ [2\cos(\omega t + \phi)] \\ &= 2|X|e^{\sigma t}\cos(\omega t + \phi) \end{aligned}$$

Now we can investigate the form of this signal for various values of $|X|$, ϕ, ω, and σ. If $\omega = 0$, we have a signal of the form of Fig. A-2 except that the value for $t = 0$ is $y(0) = 2|X| \cos \phi$. If $\omega \neq 0$ and $\sigma = 0$, we have

$$y(t) = 2|X| \cos(\omega t + \phi)$$

which is a cosinusoid of phase angle ϕ and angular frequency ω (radians per second). If $\sigma \neq 0$, this oscillatory signal is multiplied by the term $e^{\sigma t}$, resulting in signals $y(t)$ of the form shown in Fig. A-3, where for convenience we have set $\phi = 0$.

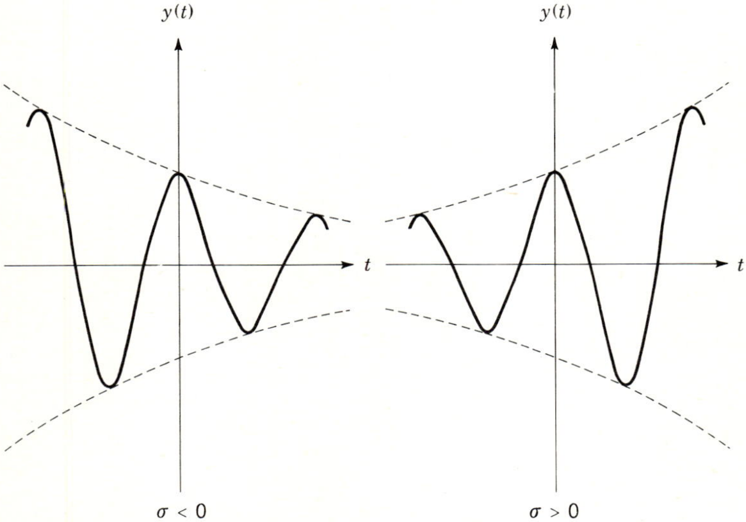

FIG. A-3 Other types of Xe^{st} waveforms.

We have shown that by choosing the parameters in various ways we can represent a variety of real-time functions $y(t) = Xe^{st} + X^*e^{s^*t}$. It should be pointed out that there are many signals which cannot be represented by Xe^{st}. Such signals, however, can be represented by sums of exponentials—a topic which is beyond the scope of this book.

Index